The Chemical Kinetics
of Enzyme Action

The Chemical Kinetics of Enzyme Action

KEITH J. LAIDLER

AND

PETER S. BUNTING

DEPARTMENT OF CHEMISTRY
UNIVERSITY OF OTTAWA

SECOND EDITION

CLARENDON PRESS · OXFORD

1973

Oxford University Press, Ely House, London W.1

GLASGOW NEW YORK TORONTO MELBOURNE WELLINGTON
CAPE TOWN IBADAN NAIROBI DAR ES SALAAM LUSAKA ADDIS ABABA
DELHI BOMBAY CALCUTTA MADRAS KARACHI LAHORE DACCA
KUALA LUMPUR SINGAPORE HONG KONG TOKYO

ISBN 0 19 855353 6

© OXFORD UNIVERSITY PRESS 1958, 1973

FIRST EDITION 1958

SECOND EDITION 1973

PRINTED IN GREAT BRITAIN
BY J. W. ARROWSMITH LTD., BRISTOL, ENGLAND

Preface

T HE field of enzyme kinetics has developed greatly since the first edition of this book appeared in 1958. As a result, the book has been completely rewritten. Several completely new topics—such as isotope effects and sigmoid kinetics— have been introduced. The present edition places somewhat greater emphasis on general principles than did the first; there are many reviews of individual enzyme systems, and it was felt that it would be more useful to develop the underlying theory.

We are grateful to many colleagues for help with the preparation of the present edition. Dr. W. E. Hornby of the University of St. Andrews has read much of the manuscript and has made many valuable suggestions. Several of our colleagues at the University of Ottawa—especially Mr. N. H. Hijazi and Mr. I. Hinberg— have also been very helpful in the same way.

Ottawa, 1972
K.J.L.
P.S.B.

Contents

LIST OF MAIN SYMBOLS EMPLOYED ix

1. The chemical characteristics of enzymes 1
2. General kinetic principles 35
3. The steady state in enzyme kinetics 68
4. Two-substrate reactions 114
5. The influence of hydrogen ion concentration 142
6. The time course of enzyme reactions 163
7. Molecular kinetics 196
8. Kinetic isotope effects 233
9. Some deductions about mechanisms 254
10. Some reaction mechanisms 292
11. Sigmoid kinetics and allostery 352
12. Supported enzyme systems 382
13. The denaturation of proteins 413

AUTHOR INDEX 461
SUBJECT INDEX 469

List of Main Symbols Employed

THE conventional Michaelis constant is employed in the present edition, instead of its reciprocal as in the first edition; the symbol K_A has been employed for the Michaelis constant with respect to the substrate A. Similarly, for an inhibitor Q the *dissociation* constant of the enzyme-inhibitor complex EQ is represented by the symbol K_Q. The reciprocal quantities employed in the first edition have the great advantage of simplifying the form of the kinetic equations, but the conventional symbols seem to be too deeply entrenched for any change to be practicable.

A, B; [A], [B]	substrate; concentration of substrate ⎫ small letters, a, a_0 etc. are
	⎬ used in equations involv-
$[A]_0$, $[B]_0$	substrate concentration at zero time ⎭ ing time-dependence
$[A]'$	concentration of substrate inside a support
AQ	substrate-modifier addition complex;
[AQ]	concentration of complex
β	voidage of a column
C	reaction capacity
ε_r	dielectric constant
D	diffusion constant
D	denatured protein
e	base of natural logarithms
e	charge on the electron
E; [E]; {E}	uncombined enzyme; its concentration; total amount of enzyme
E	activation energy
E_0	activation energy at 0 K
$[E]_0$	total concentration of all forms of enzyme
$[E]_m$	concentration of enzyme in a support
EA; [EA]	enzyme-substrate complex; its concentration
EA'; [EA']	second intermediate (e.g. acyl enzyme); its concentration
ΔE	change in internal energy
EQ; [EQ]	enzyme-modifier complex; its concentration
EAQ; [EAQ]	enzyme-substrate-modifier complex; its concentration
EA'Q; [EA'Q]	complex formed from EA' and modifier; its concentration
F	function employed for supported enzymes
F, G	constants in transient-phase equations
$\Delta G^{\ominus}$	standard Gibbs-free-energy change
$\Delta G^{\ne}$	free energy of activation
h	Planck's constant
H^+; $[H^+]$	hydrogen ion; concentration of hydrogen ions
$\Delta H^{\ominus}$	standard change in enthalpy
$\Delta H^{\ne}$	enthalpy of activation
k, k_1, k_{-1}, k_2 etc.	rate constant; subscripts refer to numbered reactions
k	Boltzmann constant
k_0	second-order rate constant at low substrate concentrations
k_c	limiting rate constant at high substrate concentrations
$\tilde{k}_c$	pH-dependent k_c

$K_1, K_2, \ldots$ equilibrium constant for reaction 1, 2, etc.

K_a, K_b acid dissociation constants

$K_A, K_B, \ldots$ Michaelis constant for substrates A, B, ...

$\tilde{K}_A, \tilde{K}_B$ pH-dependent Michaelis constants

K_Q dissociation constant for enzyme-modifier complex (for $EQ \rightleftharpoons E + Q$)

K_Q^i, K_Q^s constants in general inhibition equation

K_Q^A dissociation constant for a substrate-modifier complex ($AQ \rightleftharpoons A + Q$)

K_{AQ} dissociation constant for enzyme-substrate-inhibitor complex ($EAQ \rightleftharpoons EA + Q$)

$K_{A'Q}$ dissociation constant for EA'Q ($EA'Q \rightleftharpoons EA' + Q$)

m, n concentration of a reaction intermediate (e.g. EA, EA')

pK_a, pK_b negative logarithm of acid dissociation constant for free enzyme

pK_a', pK_b' negative logarithm of dissociation constant of enzyme-substrate complex

pK_a'', pK_b'' negative logarithm of acid dissociation constant of second intermediate (e.g. acyl enzyme)

pK_A negative logarithm of Michaelis constant

$p\tilde{K}_A$ negative logarithm of pH-dependent Michaelis constant

l thickness of solid support

$N; [N]$ native protein; concentration of native protein

P partition coefficient

$P; [P]$ protein; its concentration

$P, Q, R, P',$ etc.   constants in transient-phase equations

ψ electrical potential

q partition function

$Q; [Q]$ modifier (activator or inhibitor); its concentration

Q rate of flow

r distance of approach of charge centres in bimolecular ionic reactions

R gas constant

s cross-sectional area

$\Delta S^{\ominus}$ standard entropy increase

$\Delta S^{\neq}$ entropy of activation

$\Delta S^{\neq}_{es}$ electrostatic contribution to entropy of activation

$\Delta S^{\neq}_{nes}$ nonelectrostatic contribution to entropy of activation

t time

t^0 induction period

T temperature (Kelvins)

τ relaxation time

θ fraction of solid consisting of pores

u ionic strength

$U; [U]$ urea; its concentration

v, v_1, v_{-1} velocity

v_0 initial velocity

$\Delta V^{\neq}$ volume of activation

V_A, V_B limiting velocity, as A, B is varied

x	distance
x_e	concentration of X at equilibrium
X, Y; [X], [Y]	product of reaction; its concentration (small letters used for time dependence)
$z, z_A, z_B \ldots$	ionic charge, in units of electronic charge (actual charges are $z_A e$, $z_B e$, etc.)

The reference abbreviations employed in this book are those of the World List of Scientific Periodicals.

1. The Chemical Characteristics of Enzymes

THE enzymes are important and essential components of biological systems, their function being to catalyse the chemical reactions that are essential to life. Without the efficient aid of the enzymes these chemical processes would occur at greatly diminished rates, or not at all.

This book is concerned mainly with the kinetics of the reactions catalysed by enzymes, and with the mechanisms by which the reactions occur. The general properties of enzymes have been discussed in a number of books and review articles[1]. In the present chapter will be given only a very brief account of those characteristics of enzymes about which some knowledge is necessary for an understanding of the kinetic aspects of the subject. The main topics that will be considered are the chemical nature of enzymes, some qualitative features of enzyme-catalysed reactions, and the nature of the active centres of the enzyme molecules.

The general nature of enzymes

A considerable number of enzymes have been prepared in highly purified form, and their chemical and physical properties studied in some detail. A very important advance was made in 1926 by Sumner[2], who was the first to crystallize an enzyme, urease, and to show that it is a pure protein. With the advances in techniques that have been made since then it has proved possible to prepare a number of enzymes in a high degree of purity. Sumner's procedure with urease was to extract jack bean meal with acetone, followed by crystallization from acetate buffer in the cold. In performing this extraction he utilized some of the basic properties of enzymes. The present section will outline some of these properties, with an indication as to how they have been utilized in the separation and purification of enzymes[3].

Most enzymes are soluble in water, in dilute salt solutions, and in dilute solutions of alcohol in water. They are, however, insoluble in water containing high proportions of alcohol or other organic solvents, and are salted out of solution by neutral salts, such as ammonium sulphate, in higher concentrations. They are precipitated out of solution by heavy metal ions such as lead and mercury, and also by changing the pH of a solution drastically, as by adding trichloracetic acid. These properties are often used to obtain crude preparations of enzymes as a first step in the purification process.

[1] See, for example, M. Dixon and E. C. Webb, *Enzymes*, Academic Press, New York, 1964; H. R. Mahler and E. H. Cordes, *Biological Chemistry*, Harper International, New York and Tokyo, 1966; R. E. Dickerson and I. Geis, *The Structure and Action of Proteins*, Harper and Row, New York, 1969.
[2] J. B. Sumner, *J. biol. Chem.*, **69**, 435 (1926); **70**, 97 (1926).
[3] For further details see J. Leggett Bailey, *Techniques in Protein Chemistry*, American Elsevier Publishing Co., New York, 1967.

More recently-developed methods for purification include absorption. Proteins are absorbed to different degrees on materials such as silica and starch, and gels of these materials have been used for the separation of enzymes[1].

The essential structural unit in the proteins is a long chain arising from the condensation of a number of α-amino acids. These residues are held together by peptide ($-CO-NH-$) linkages, and the structure may be represented as

$$\ldots NH-\underset{\underset{R}{|}}{CH}-CO-NH-\underset{\underset{R}{|}}{CH}-CO-NH-\underset{\underset{R}{|}}{CH}-CO-NH-\ldots$$

There is considerable variation between enzymes in the number of amino acids that are linked together in these polypeptide chains, the molecular weights ranging from a few thousands to several millions. This variation in size, and also in shape, has been used for the separation of enzymes and other proteins; in the ultracentrifuge, for example, molecules of different weights undergo sedimentation at different speeds. Size difference is also utilized in the molecular sieving (gel filtration) of proteins, a technique which provides good separation[2].

Other characteristic properties of the enzymes and of other proteins arise from their electrical nature. The amino acids themselves normally exist as zwitterions

$$H_3N^+-\underset{\underset{H}{|}}{\overset{\overset{R}{|}}{C}}-COO^-$$

at physiological pH values, but the charges due to these $-NH_2$ and $-COOH$ groups are lost on peptide bond formation; the groups at the ends of the chains are, however, free, and exist in the charged or uncharged forms according to the pH; typical pK values for these terminal groups are:

$$-COOH \rightleftharpoons -COO^- + H^+ \qquad pK \approx 4$$

$$-NH_3^+ \rightleftharpoons -NH_2 + H^+ \qquad pK \approx 9.$$

In addition, many of the R-side groups on the amino acids contain ionizing groups such as $-COOH$, $-NH_2$, and $-OH$. Table 1.1 lists the amino acids commonly found in proteins (including enzymes) and the following are the

---

[1] It is of interest that Sumner (*J. biol. Chem.*, **70**, 97 (1926)) was convinced that absorption could never be utilized for such separations.

[2] H. Determann, *Adv. Chromatogr.*, **8**, 3 (1969).

TABLE 1.1
Amino acids commonly found in enzymes

Name	Structure
Glycine (gly)	H_2N-CH_2-COOH
Alanine (ala)	CH_3 | $H_2N-CH-COOH$
Valine (val)	$CH_3 \quad CH_3$ \ / CH | $H_2N-CH-COOH$
Leucine (leu)	$CH_3 \quad CH_3$ \ / CH | CH_2 | $H_2N-CH-COOH$
Isoleucine (ile or ilu)	CH_3 \ $CH_2 \quad CH_3$ \ / CH | $H_2N-CH-COOH$
Serine (ser)	CH_2OH | $H_2N-CH-COOH$
Threonine (thr)	CH_3 | $CHOH$ | $H_2N-CH-COOH$
Aspartic acid (asp)	$COOH$ | CH_2 | $H_2N-CH-COOH$
Asparagine (asn)	$CONH_2$ | CH_2 | $H_2N-CH-COOH$
Glutamic acid (glu)	$COOH$ | CH_2 | CH_2 | $H_2N-CH-COOH$
Glutamine (gln)	$CONH_2$ | CH_2 | CH_2 | $H_2N-CH-COOH$
Lysine (lys)	$(CH_2)_4-NH_2$ | $H_2N-CH-COOH$

TABLE 1.1—continued

Name	Structure

Histidine (his)

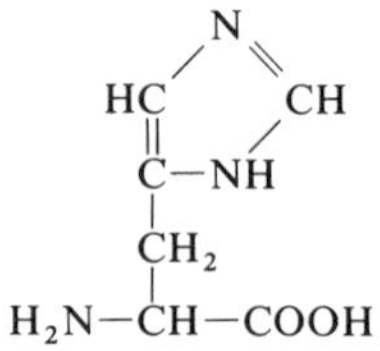

$$\begin{array}{c} N \\ \diagdown \\ HC \qquad CH \\ \diagdown \quad \diagup \\ C-NH \\ | \\ CH_2 \\ | \\ H_2N-CH-COOH \end{array}$$

Arginine (arg)

$$\begin{array}{c} NH \\ \| \\ (CH_2)_2-NH-C \\ | \qquad\qquad \diagdown \\ H_2N-CH-COOH \quad NH_2 \end{array}$$

Phenylalanine (phe)

$$\begin{array}{c} \text{(benzene ring)} \\ | \\ CH_2 \\ | \\ H_2N-CH-COOH \end{array}$$

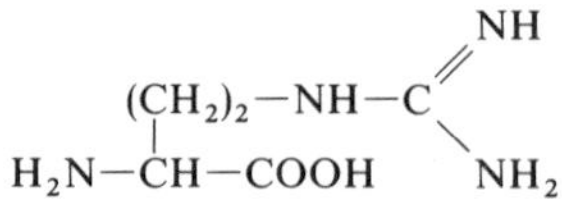

Tyrosine (tyr)

$$\begin{array}{c} OH \\ | \\ \text{(benzene ring)} \\ | \\ CH_2 \\ | \\ H_2N-CH-COOH \end{array}$$

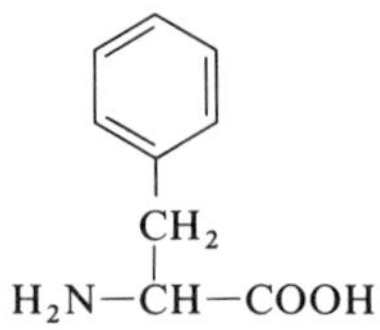

Tryptophan (try or trp)

$$\begin{array}{c} \text{(indole ring)} \\ N-H \\ | \\ CH_2 \\ | \\ H_2N-CH-COOH \end{array}$$

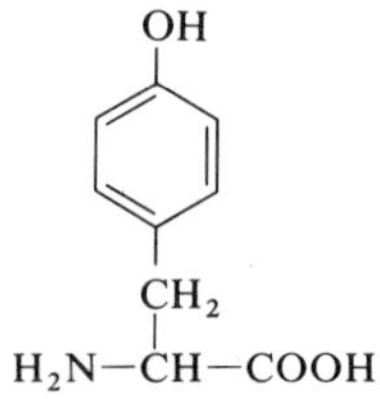

Cystein (cys)

$$\begin{array}{c} CH_2-SH \\ | \\ H_2N-CH-COOH \end{array}$$

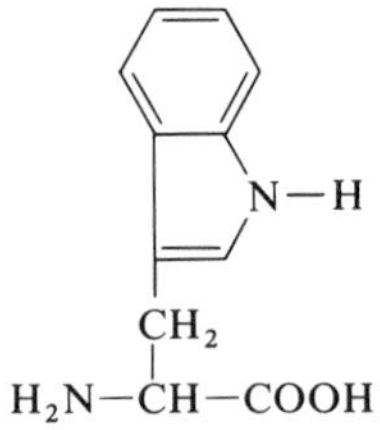

Cystine (cys-cys)

$$\begin{array}{c} CH_2-S-S-CH_2 \\ | \qquad\qquad\qquad\quad | \\ H_2N-CH-COOH \quad H_2N-CH-COOH \end{array}$$

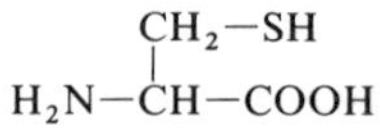

Methionine (met)

$$\begin{array}{c} (CH_2)_2-SCH_3 \\ | \\ H_2N-CH-COOH \end{array}$$

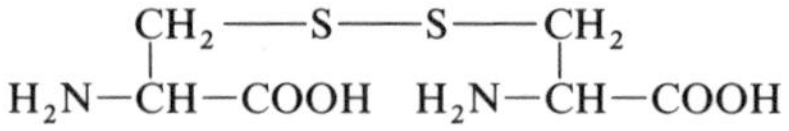

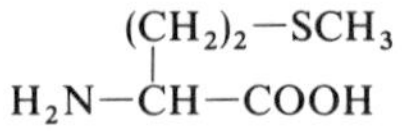

Proline (pro)

$$\begin{array}{c} \text{(pyrrolidine ring)}-COOH \\ N \\ H \end{array}$$

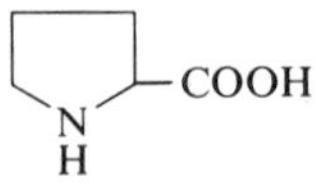

approximate pK values:

$$-\text{COOH} \rightleftharpoons -\text{COO}^- + \text{H}^+ \text{ (glu, asp)} \qquad \text{p}K \approx 4$$

$$-\text{NH}_3{}^+ \rightleftharpoons -\text{NH}_2 + \text{H}^+ \text{ (lys)} \qquad \text{p}K \approx 9$$

$$-\text{OH} \rightleftharpoons -\text{O}^- + \text{H}^+ \text{ (thr, ser, tyr)} \qquad \text{p}K \approx 10$$

$$-\text{SH} \rightleftharpoons \text{S}^- + \text{H}^+ \text{ (cys)} \qquad \text{p}K \approx 8$$

$$
\begin{array}{c}
\text{CH}\!-\!\text{N}^+\text{H} \\
\| \qquad \| \\
-\text{C} \diagdown \quad \diagup \text{CH} \\
\text{NH}
\end{array}
\rightleftharpoons
\begin{array}{c}
\text{CH}\!-\!\text{N} \\
\| \qquad \| \\
-\text{C} \diagdown \quad \diagup \text{CH} + \text{H}^+ \text{ (his)} \\
\text{NH}
\end{array}
\qquad \text{p}K \approx 7
$$

If sufficiently acid conditions prevail, the carboxyl, hydroxyl and sulphydryl groups will be in their neutral forms, while the amino and imidazole groups will be in their protonated (positively charged) forms; the molecule will therefore have a net positive charge and migrate towards the cathode if a potential is applied. Conversely, at low acidity there will be an overall negative charge, and the molecule will migrate towards the anode. At some intermediate pH value the number of positive groups will be equal to the number of negative ones. The molecule will then have no net charge, and there will be no movement in an electric field; this pH at which there is no migration is known as the *isoelectric point*, and is a characteristic property of a protein.

This property of protein molecules has been exploited in isolating and identifying proteins, peptides and amino acids. Different species are readily separable by electrophoresis, owing to their different mobilities in an electric field. Two recent applications of this are electrofocussing[1], which involves the molecular sieve method combined with a pH gradient, and polyacrylamide electrophoresis[2], which combines molecular sieving with electrophoresis, without the pH gradient. These provide a sensitive method for separating proteins. Another technique involves the use of ion-exchange resins[3], which contain either positively-charged (such as $-\text{NH}_3^+$) or negatively-charged (such as $-\text{SO}_3^-$) groups; the amino acids or proteins attach themselves to these resins with different firmness depending on their overall charge, and this permits a separation.

There is much evidence, to be discussed later, that these ionizing groups on proteins have much to do with the catalytic activity of proteins. The remarkable variations of catalytic activity are to be correlated with changes of ionization of certain of these groups (see Chapter 5). Another point of importance is that the charged groups of proteins tend to bind foreign ions (such as those of metals), and that the binding of such ions is sometimes closely related to enzyme action.

[1] D. H. Leaback and A. C. Rutter, *Biochem. biophys. Res. Commun.*, **32**, 447 (1968).
[2] B. J. Davis, *Ann. N.Y. Acad. Sci.*, **121**, 404 (1964); *Techniques in Biochemistry and Molecular Biology* (T. S. Work and E. Work, eds.), John Wiley and Sons, New York, 1969.
[3] P. B. Hamilton, *Adv. Chromatogr.*, **2**, 3 (1966).

The structure of enzyme molecules

A detailed understanding of the catalytic function of an enzyme requires that as much information as possible be available about the structure of the enzyme molecule. There are several important aspects to this problem. In the first place, about 20 different amino acids are present in protein molecules, and it is necessary to know what amino acids are present, and in what number, in a given molecule. In addition, the sequence in which the amino acids occur along the polypeptide chain is important, since this determines the overall conformation† of the protein and this in turn has a great influence on the catalytic action. Finally, it is important to have direct information about the conformation of the molecule.

Amino-acid sequence

It is a comparatively simple matter, by the use of standard analytical techniques, to determine the proportion of the various amino acids in a protein molecule; if the molecular weight of the protein is known the number of the different amino acids in the molecule can then be determined. Automatic amino-acid analysers[1] are now available for solving this problem.

The determination of the sequence of amino acids in a protein chain is of much greater difficulty, in view of the very large number of units in even the smallest protein molecules. One of the smallest is insulin, which contains 51 amino acids and has a molecular weight of a little under 6000‡. The establishment of the amino-acid sequence in a protein molecule was first achieved for insulin in 1955 by Sanger. Since this pioneering work the sequences have been worked out for several other proteins, including the enzymes trypsin[2] (M.W. 23 800), chymotrypsin (M.W. 24 800), ribonuclease[3] (M.W. 12 600) and lysozyme[4] (M.W. 14 000). The methods employed are briefly as follows. Chains held together by disulphide (—S—S—) bonds are separated by oxidation or reduction of the bonds. These chains are then split into smaller peptide fragments, either by partial acid hydrolysis, by specific chemical cleavage[5] or by using one of the several proteolytic enzymes such

---

† The term *conformation* is used to refer to the general shape of a molecule, and is to be distinguished from configuration, which refers to the relative arrangements of the atoms or groups in a molecule. The conformations of a given molecule, also referred to as rotational isomers, are converted into one another merely by rotation about single bonds, and without the breaking and making of major bonds: in some instances, however, as with proteins, it is permissible to make and break hydrogen bonds and other weak bonds in passing from one conformation to another. For a review of the significance of conformations in organic chemistry, see D. H. R. Barton and R. C. Cookson, *Quart. Revs.*, **10**, 44 (1956).

‡ Insulin usually exists in solutions as a dimer, of molecular weight about 12 000.

[1] See D. H. Spackman, W. H. Stein, and S. Moore, *Analyt. Chem.*, **30**, 1190 (1958).

[2] K. A. Walsh, D. L. Kauffman, D. S. V. S. Kumar, and H. Neurath, *Proc. natn. Acad. Sci. U.S.A.*, **51**, 301 (1964).

[3] R. R. Redfield and C. B. Anfinsen, *J. biol. Chem.*, **221**, 385 (1956); C. H. W. Hirs, S. Moore, and W. H. Stein, *ibid.*, **221**, 151 (1956).

[4] J. Jolles, J. Jauregui-Adell, and P. Jolles, *Biochim. biophys. Acta*, **71**, 488 (1963); R. E. Canfield, *J. biol. Chem.*, **238**, 2684, 2691, 2698 (1963).

[5] T. F. Spande, B. Witkop, Y. Degani, and A. Patchornik, *Adv. Protein Chem.*, **24**, 98 (1970).

as trypsin and chymotrypsin. Some easily identifiable group can then be caused to react with one part of the molecule; the chain can then be broken down and a study made to determine the amino acid to which the group has become attached. Ultimately, by ways such as this, and at the cost of much painstaking work, the entire sequence of amino acids in the original molecule can be established[1].

As an example, the amino-acid sequence determined for chymotrypsin is shown in Fig. 1.1. It is to be seen that there are three chains of amino acids, referred to as the A, B and C chains. These are held together by —S—S—

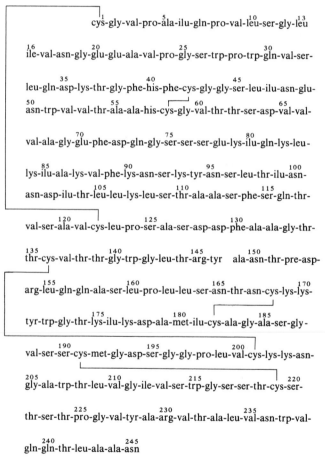

FIG. 1.1. The amino acid sequence in α-chymotrypsin (cf. Table 1.1 for list of amino acids). The numbering of amino acids is the sequence in the precursor chymotrypsinogen.

[1] For automated techniques see P. Edman and C. Begg, *Eur. J. Biochem.*, **1**, 80 (1967); G. E. Van Lear and F. W. McLafferty, *A. Rev. Biochem.*, **38**, 289 (1969).

linkages which are found in the amino acid cysteine (cf. Table 1.1). These —S—S— linkages play a very important role in protein structure, since they bring into close spatial proximity groups that might otherwise be far apart. This has a profound effect on the whole conformation, and on the biological activity.

At one time it was suspected that the amino acids in proteins are arranged in some regular and repeating pattern. Now that the sequence has been determined in a number of cases, however, it has become apparent that there is no evidence for such a conclusion. It has also been found that the sequence is the same for all samples of enzyme obtained from a given source.

Protein conformations

The properties of a protein cannot be understood entirely on the basis of the sequence of amino acids; they also depend very critically on the three-dimensional structure of the molecule. This three-dimensional structure determines the way in which the various side-groups on the molecule are brought into close proximity with one another, and this has an important effect on chemical and physical properties. In particular, if the protein is an enzyme, the relative positions of certain groups have a very important effect on the catalytic action.

Certain physical methods have been used for many years for gaining a very general idea of the overall shapes of protein molecules. These include the measurements of the viscosities of solutions, and of their light-scattering. These properties are very different for the molecules that are long and thin from what they are for molecules that have a more or less spherical form. By the use of such methods, proteins have been grouped into two main classes, the *fibrous* proteins, in which the molecules are fairly extended, and the *globular* proteins, which are very roughly spherical in shape.

Very much more detailed information about protein structure is provided by the technique of X-ray crystallography[1]. The reason that X-rays are the most suitable for work of this kind is that the detail with which an object can be observed, i.e. the resolving power of an instrument, depends in a fundamental way on the wavelength of the radiation employed; as a rough rule, no two objects can be seen separately if they are closer together than one-half the wavelength of the radiation used. The lengths of chemical bonds are between 0·1 and 0·2 nanometers (1 to 2×10^{-8} cm), so that one must use wavelengths not much longer than this; consequently, X-rays must be used. The employment of X-rays, however, introduces new difficulties, because no satisfactory way has yet been found to make lenses and mirrors that will focus X-rays. The way the technique has to be employed is to record and study the diffraction pattern that is produced when X-rays strike the material under

[1] For a discussion of basic principles of X-ray crystallography see G. Kartha, *Acct. chem. Res.*, **1**, 374 (1968). Many results for proteins are included in *Cold Spring Harb. Symp. quant. Biol.*, **36** (1972).

study. The analysis of an X-ray diffraction pattern is a matter of very considerable difficulty for a molecule as large as a protein, because of the very large number of interatomic distances that are involved. In 1953 Perutz[1] made a very important contribution to the analysis of X-ray patterns by his method of *isomorphous replacement*; this method depends upon the preparation and study of protein crystals into which heavy atoms, such as atoms of uranium, have been introduced without otherwise altering the crystal structure. Since that time a considerable number of protein structures have been worked out[2].

It is the electrons within the molecules that scatter the X-rays, so that the image calculated from the diffraction pattern reveals the distribution of electrons within the molecule. The usual procedure is to calculate, using high-speed computers, the electron density at a regular array of points, and to make the image visible by drawing contour lines through points of equal electron density. These contour maps can be drawn on clear plastic sheets, and a three-dimensional image can then be obtained by stacking the maps one above the other. The amount of detail that can be seen depends upon the resolving power of the effective microscope, and if this is sufficiently good the atoms appear as individual peaks in the image map. At lower resolutions, groups of unresolved atoms appear which can frequently be recognized.

There are a number of difficulties associated with work of this kind. The establishment of a protein structure by X-ray methods is an extremely time-consuming process, usually requiring several years before a conclusive result can be obtained. The enzyme must, of course, be available in pure crystalline form, and obtaining suitable crystals frequently presents considerable difficulty. The need to make an isomorphous replacement intensifies this problem, since the heavy-metal derivatives must be isomorphous with the native enzyme crystals. In addition, it is desirable that the heavy atom interacts with only one site in the enzyme, or at the most two. Obviously, the introduction of the heavy atom should not produce too much change in the conformation of the enzyme. A further difficulty is that one cannot be certain that the crystalline structure of the enzyme molecule is the same as its structure in solution. This problem can sometimes be overcome by the use of techniques available for the study of the three-dimensional conformations of molecules in solution; one of these is optical rotatory dispersion, which is referred to later (p. 13). Another approach that has been made is comparison of catalytic rates with the enzyme in solution compared with the wet crystal; for chymotrypsin[3] these rates have been found to be the same.

[1] D. W. Green, V. M. Ingram, and M. F. Perutz, *Proc. Roy. Soc.*, **A225**, 287 (1954); for a review see C. C. F. Blake, *Adv. Protein Chem.*, **23**, 59 (1968).

[2] For reviews see J. Kraut, *A. Rev. Biochem.*, **34**, 247 (1965); D. R. Davies, *A. Rev. Biochem.*, **36**, 321 (1967); L. Stryer, *A. Rev. Biochem.*, **37**, 25 (1968).

[3] G. L. Rossi and S. A. Bernhard, *J. molec. Biol.*, **49**, 85 (1970).

The enzymes for which reliable X-ray structures have been determined include lysozyme[1], ribonuclease[2], chymotrypsin[3], papain[4], carboxypeptidase A[5], cytochrome c[6], subtilisin[7], elastase[8], carbonic anhydrase[9], lactate dehydrogenase[10] and staphylococcus nuclease[11]. A schematic view of the chymotrypsin structure, as determined by Blow and his coworkers, is given in Fig. 1.2. It is to be seen that there are three polypeptide chains, A, B and C, held together by $-S-S-$ linkages. At the left of the diagram, at the $-COO^-$ end of the C chain, there is a helical or spiral structure; the nature of this will be discussed later (p. 15). The region of the molecule that is catalytically active is also indicated; the evidence that this is indeed the active centre is also considered later (p. 25).

Other techniques[12] that provide information about protein conformations will be discussed only briefly:

(1) *Ultraviolet spectroscopy*[13]: The absorption by proteins in the ultraviolet, at about 280 nm, results largely from the absorption of the radiation by tyrosine, tryptophan and phenylalanine residues. These three amino acids have different absorption maxima and absorbancies at 280 nm (ε_{280}), and can therefore be distinguished from one another. If changes are made in the pH, dielectric constant, refractive index, and ionic strength of the medium the ultraviolet absorbance provides information about the environment of these amino acids, and hence about the overall structure of the protein. For example, exposure of the protein to solvents of higher refractive index than water usually produces a small red shift in the absorption maximum for those

[1] C. C. F. Blake, D. F. Koenig, G. A. Mair, A. C. T. North, D. C. Phillips, and V. R. Sarna, *Nature*, **206**, 757 (1965); C. C. F. Blake, G. A. Mair, A. C. T. North, D. C. Phillips, and V. R. Sarna, *Proc. R. Soc.*, **B167**, 365 (1967); for a popular account see D. C. Phillips, *Scient. Am.*, Nov. 1966, p. 76.

[2] G. Kartha, J. Bello, and D. Harker, *Nature*, **213**, 862 (1967).

[3] B. W. Matthews, P. B. Sigler, R. Henderson, and D. M. Blow, *Nature*, **214**, 652 (1967); P. B. Sigler, D. M. Blow, B. W. Matthews, and R. Henderson, *J. molec. Biol.*, **35**, 143 (1968); D. M. Blow, J. J. Birktoft, and B. S. Hartley, *Nature*, **221**, 337 (1969); R. Henderson, *J. molec. Biol.*, **54**, 341 (1970); for work on the precursor chymotrypsinogen see J. Kraut, D. F. High, and L. C. Sieber, *Proc. natn. Acad. Sci. U.S.A.*, **51**, 839 (1964); S. T. Freer, J. Kraut, J. D. Roberts, H. T. Wright, and Ng. H. Xuong, *Biochemistry*, **9**, 1997 (1970).

[4] J. Drenth, J. N. Jansonius, R. Koekoek, H. M. Swen, and B. G. Wolthers, *Nature*, **218**, 929 (1968); J. Drenth, J. N. Jansonius, and B. G. Wolthers, *J. molec. Biol.*, **24**, 449 (1967).

[5] G. N. Reeke, J. A. Hartsuck, M. L. Ludwig, F. A. Quiocho, T. A. Steitz, and W. N. Lipscomb, *Proc. natn. Acad. Sci.*, **58**, 2220 (1967); W. N. Lipscomb, J. A. Hartsuck, G. N. Reeke, F. A. Quiocho, P. H. Bethge, M. L. Ludwig, T. A. Steitz, H. Muirhead, and J. C. Coppola, *Brookhaven Symp. Biol.*, **21**, 24 (1968).

[6] R. E. Dickerson, M. L. Kopka, J. E. Weinzierl, J. C. Varnum, D. Eisenberg, and E. Margoliash, *J. biol. Chem.*, **242**, 3015 (1967).

[7] C. S. Wright, R. A. Alden, and J. Kraut, *Nature*, **221**, 235 (1969).

[8] D. M. Shotton and H. C. Watson, *Nature*, **225**, 811 (1970).

[9] K. Fridborg, K. K. Kannan, A. Liljas, J. Lundin, B. Strandberg, R. Strandberg, B. Tilander, and G. Wiren, *J. molec. Biol.*, **25**, 505 (1967).

[10] M. J. Adams, G. C. Ford, R. Koekoek, P. J. Lentz, A. McPherson, M. G. Rossman, I. E. Smiley, R. W. Schevitz, and A. J. Wonacott, *Nature*, **227**, 1098 (197).

[11] A. Arnone, C. J. Bier, F. A. Cotton, V. W. Bay, E. E. Hazen, D. C. Richardson, J. S. Richardson, and A. Yonath, *J. biol. Chem.*, **246**, 2302 (1971).

[12] For a general review see S. N. Timascheff, in *The Enzymes* (ed. P. D. Boyer), Academic Press, New York, 1970, Vol. 2, p. 371.

[13] For a review, see D. B. Wetlaufier, *Adv. Protein Chem.*, **17**, 303 (1962).

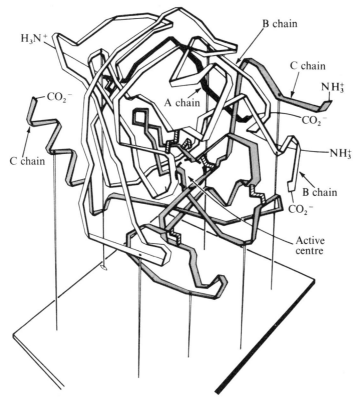

FIG. 1.2. A schematic drawing showing the general conformation of the polypeptide chains in chymotrypsin. There are three polypeptide chains, A, B, and C, held together by S—S linkages, by hydrogen bonds, and by hydrophobic bonds. The dotted circle shows the active centre, i.e. the portion of the molecule that is mainly concerned with the catalytic action.

residues which are exposed to the influence of the solvent[1]; those groups that are not exposed (buried groups) do not undergo significant changes in absorption. Unfortunately, conclusions of this kind are not always reliable. Markland[2], for example, divided the tyrosine residues of the enzyme subtilisin into exposed, buried and partially-buried residues using this technique; subsequent X-ray work[3], however, has revealed that all of the tyrosine residues are exposed.

(2) *Infrared spectroscopy*[4]: The frequencies of vibration of the N—H, C=O and C—N bonds in proteins can be measured by infrared spectroscopy

[1] M. Laskowsky, *Fedn. Proc.*, **25**, 20 (1966); E. J. Williams, T. T. Herskovitz, and M. Laskowsky, *J. biol. Chem.*, **240**, 3474 (1965).

[2] F. S. Markland, *J. biol. Chem.*, **244**, 694 (1969).

[3] C. S. Wright, R. A. Alden, and J. Kraut, *Nature*, **221**, 235 (1969).

[4] For a review, see J. A. Schellman and C. Schellman, in *The Proteins* (ed. H. Neurath), Academic Press, New York, 1964.

and compared with standard compounds or with the values for proteins known to be unfolded. In this way information can be obtained about the environment of the group; for example, it can be deduced whether the group is in a helical or a pleated-sheet structure (see p. 16). This technique has not yet found much application in the study of the nature of the active centres of enzymes.

(3) *Nuclear magnetic resonance spectroscopy*: This technique has recently been applied to studies of active centres. For example, the protons in the carbon atoms C^2 and C^4 in histidine can be easily distinguished from other protons by their chemical shifts. Changes induced by variation of pH have also been followed by studying the changes in chemical shifts; one can obtain a titration curve similar to that given by kinetic studies of the change in activity with pH (see Chapter 5).

As an example of this type of study may be mentioned work done with ribonuclease[1]. An n.m.r. scan of this enzyme revealed the presence of four peaks which were distinguishable as histidine residues. The problem is how to relate the peaks to the four histidine residues known from sequence studies to be present in the molecule; these are his–12, his–19, his–105 and his–119. The reader is referred to the original paper for details; in brief the procedure was as follows. The enzyme was alkylated to form 1-carboxy-methyl-histidine–119; the n.m.r. peaks 2 and 3 were found to have shifted, while 1 and 4 remained unchanged. The same thing happened when 3-CBM-histidine–12 was made. These results showed that peaks 2 and 3 corresponded to histidines–12 and –119, but did not prove which was which. In order for this to be done, the ribonuclease was digested with the enzyme subtilisin, which cleaves the bond between residues 20 and 21 and produces ribonuclease S; this consists of a small S-peptide and a larger S-protein, held loosely together. It was found that replacement of H_2O by D_2O shifted peak 4 in the case of ribonuclease S, but not with the original ribonuclease. It is known from X-ray studies that his–48 is more exposed in the S-enzyme, so that one can conclude that peak 4 corresponds to his–48. Deuteration of the S-peptide, which contains his–12 led to a disappearance of peak 2; deuterium gives no n.m.r. spectrum so that peak 2 must be due to his–12. By elimination peak 3 must correspond to his–119, and peak 1 to his–105. This work confirmed the evidence from chemical modification that histidines–12 and –119 are necessary for catalysis (see also p. 295).

Somewhat similar studies have been made with lysozyme[2].

Electron magnetic spin resonance (spin labelling) has also been used in conformational studies[3].

[1] D. H. Meadows, O. Jardetzky, R. Epand, H. Ruterjans, and H. A. Scheraga, *Proc. natn. Acad. Sci. U.S.A.*, **60**, 766 (1968); D. H. Meadows and O. Jardetzky, *Proc. natn. Acad. Sci.*, **61**, 406 (1968).

[2] J. S. Cohen, *Nature*, **223**, 43 (1969).

[3] T. J. Stone, T. Buckman, P. L. Nordio, and H. M. McConnell, *Proc. natn. Acad. Sci. U.S.A.*, **54**, 1010 (1965); H. M. McConnell, S. Ogawa, and A. Horwitz, *Nature*, **220**, 787 (1968).

(4) *Optical rotatory dispersion*[1] : The dependence of the optical rotation of a substance on the wavelength of the light employed is known as its optical rotatory dispersion (ORD). For proteins it is convenient to work with a mean residue rotation $[R]_\lambda$, which is related to the specific rotation $[\alpha]$ as follows:

$$[R]_\lambda = \frac{3}{n^2 + 2}\left(\frac{M_0}{100}\right)[\alpha] \tag{1}$$

where n is the refractive index of the medium and M_0 is the average weight (about 125) of an amino-acid residue. The optical rotatory dispersion of proteins frequently exhibits behaviour described as *complex*, the variation of the mean residue rotation with the wavelength λ being of the following form:

$$[R]_\lambda = \frac{a_0\lambda_0^2}{\lambda^2 - \lambda_0^2} + \frac{b_0\lambda_0^4}{(\lambda^2 - \lambda_0^2)^2}. \tag{2}$$

The quantities a_0, b_0 and λ_0 are known as Moffitt parameters[2]; their values can be obtained from suitable linear plots.

The interpretation of the ORD results is complicated and by no means unambiguous. Proteins and peptides consisting entirely of helical structures have sometimes been found to have b_0 values of -630, while those with no helical structure have b_0 values of zero; unfortunately this correlation does not always hold. Also, some indication of the amount of pleated-sheet structure (see p. 16) has been obtained from the a_0 and b_0 values; structures consisting entirely of pleated sheets frequently have $b_0 = 0$ and $a_0 = 400$–700. Such correlations are complicated by the fact that a_0 (but not b_0) is solvent dependent.

In spite of these difficulties, ORD studies have been very helpful. For example, the fact that they suggest similar amounts of helical structure for crystalline proteins and dissolved proteins suggests that X-ray structures of crystals may also apply to the protein in solution.

(5) *Hydrogen exchange*[3] : When a protein is dissolved in heavy water, D_2O, there is rapid exchange of certain of the hydrogen atoms with the deuterium atoms of the solvent. The hydrogen atoms that undergo exchange, referred to as *reactive* hydrogen atoms, include those present in the $-OH$, $-NH_2$ and $-SH$ groups. This exchange is found to be much slower for the hydrogen atoms that are buried inside the protein, a result that provides another method of obtaining information about conformations in solution. This is potentially a very powerful tool for conformation study, since the

[1] For reviews see C. Tanford, *The Physical Chemistry of Macromolecules*, John Wiley, New York, 1961; J. A. Schellman and C. Schellman, in *The Proteins* (ed. H. Neurath), Academic Press, New York, 1964.
[2] W. Moffitt, *Proc. natn. Acad. Sci. U.S.A.*, **42**, 735 (1956); *J. chem. Phys.*, **25**, 467, (1956); see also I. P. Carver, E. Shechter, and E. R. Blout, *J. Am. chem. Soc.*, **2550**, 3562 (1966).
[3] For a review, see G. Di Sabato and M. Ottesen, *Methods in Enzymology*, **11**, 734 (1967).

effects can be studied without the addition of other materials such as solvents or salts. The exchange rates can be followed by infrared and nuclear magnetic resonance techniques, and by density gradient centrifugation. Small changes in pH will bring about changes in exchange rates, leading to effects that would be too small to detect by optical rotatory dispersion methods, so that the technique is very sensitive.

Principles of protein structure

Before any detailed protein structures had been worked out by X-ray methods, a number of important conclusions had been drawn about the structures and many general principles had evolved. Many of these suggestions continue to be of great value, and will now be briefly reviewed.

In the first case, a distinction was made between different kinds of bonds present in a protein. The ordinary covalent bonds are referred to as *primary* bonds. In addition there are *secondary* bonds, examples of which are intra-helix bonds, and the bonds that hold pleated sheets together (see p. 17). There are also bonds that hold together various regions of the amino-acid chains and which are known as *tertiary* bonds. Included in the tertiary bonds are *interhelix* bonds.

One suggestion that has proved particularly useful was made many years ago by Rideal and Langmuir, according to whom a protein molecule is an oil drop with a polar coat. Non-polar groups such as alkyl groups tend to stick together in an aqueous environment, a phenomenon referred to as *hydrophobic* bonding (see p. 20). As seen in Table 1.1, some of the side groups that form the polypeptide chains are non-polar groups (in alanine, phenyl-alanine, etc.), while some (e.g. tyrosine, cystine, histidine) contain polar groups such as $-OH$, $-CO_2H$ and $-NH_2$, which tend to form hydrogen bonds with water. The essence of the suggestion of Rideal and Langmuir is that the polypeptide chains in proteins will tend to become folded in such a way that the non-polar groups will come into contact with each other as much as possible in the interior of the molecule, and that the polar groups will be as far as possible on the exterior where they can form hydrogen bonds with the surrounding water molecules. It is significant in this connection that proteins which tend to assume a globular form, such as chymotrypsin, usually have a larger proportion of non-polar groups than those that are fibrous. The tendency of proteins having a larger proportion of non-polar groups to assume globular structures is undoubtedly connected with the fact that numerous hydrophobic bonds can be formed in the interior of the molecule.

The fibrous proteins, on the other hand, show a considerable amount of helical structure. A very important suggestion relating to helical structures was made by Pauling and Corey[1], who pointed out that certain helical struc-

[1] L. Pauling, R. B. Corey, and H. R. Branson, *Proc. natn. Acad. Sci. U.S.A.*, **37**, 205 (1951); L. Pauling and R. B. Corey, *ibid.*, **37**, 235, 241, 251, 256, 261, 272, 282 (1951); **38**, 86 (1952); *Nature*, **168**, 550 (1951); **169**, 494, 920 (1952); **171**, 59 (1953).

tures formed by polypeptide chains allow a considerable amount of hydrogen bonding between the carboxyl group on one part of the chain and the amino group on the other:

$$\text{\textbackslash}\atop{C=O \ldots H-N}{}\,.$$

They considered in detail the known bond lengths and angles in the flexible chains and concluded that two different helices, corresponding to about 3·7 and 5·1 amino-acid groups per turn of the spiral, provided the maximum number of hydrogen bonds. The first of these, the so-called α helix, has in fact been shown to be quite common in protein structures, particularly in the fibrous proteins. In this arrangement, each N—H group is hydrogen-bonded to the third C=O group beyond it along the helix. Such a helix may be regarded as a spiral staircase in which the amino-acid residues form the steps; the height of each step in the α helix is 1·5 Å. A protein helix can be right-

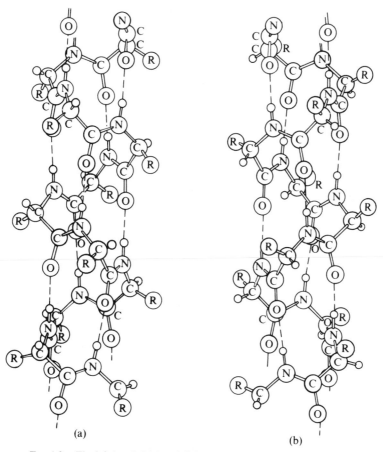

(a)   (b)

FIG. 1.3.   The left-handed (a) and right-handed (b) forms of the α-helix.

handed or left-handed; Fig. 1.3 shows the two forms of the α helix. The right-handed helix appears to occur more commonly in protein structures, but in insulin, for example, there is a region where there is a left-handed helix. A section of right-handed α helix is to be seen to the left of the diagram (Fig. 1.2) of the structure of α-chymotrypsin.

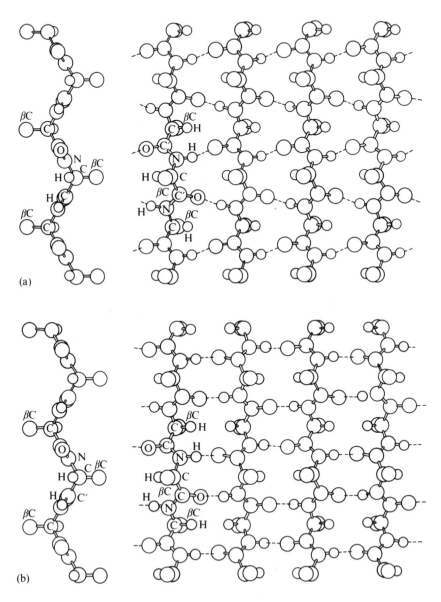

FIG. 1.4.   The parallel (a) and antiparallel (b) pleated sheets.

It was originally thought that the α helix comprised a large part of a protein structure. A considerable amount of α helix is in fact found in the fibrous proteins such as α-keralin, and in haemoglobin and myoglobin. Some of the globular proteins, however, have very little helical content.

Another structure suggested by Pauling and Corey, and one commonly found in proteins, is the beta-pleated sheet. This is illustrated in Fig. 1–4. It arises as a result of inter-chain hydrogen bonding, in contrast to the intra-chain bonding of the α helix. The two chains may have the same sense (the parallel pleated sheet) or the opposite sense (the antiparallel pleated sheet); both are illustrated in Fig. 1.4. The latter is more commonly found, since the angle between the two atoms involved in the inter-chain hydrogen bonding is very nearly 180°, which makes for stable bonding. In the parallel pleated sheet, on the other hand, the angle is less than 180° and the hydrogen bonding is weaker; this structure is therefore less common, but it has been observed in the enzyme carboxypeptidase A[1].

A useful analysis of the possible structures found in proteins has been made by Ramakrishnan and Ramachandran[2], in terms of two angles ϕ and ψ shown in Fig. 1.5. The six atoms in the peptide skeleton all lie in the same

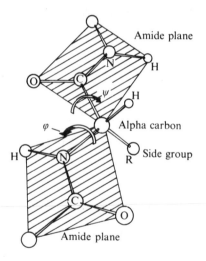

FIG. 1.5.   Relative positions of two amide planes, as defined by the angles ϕ and ψ.

plane, owing to resonance, and two adjacent peptide bonds make angles ϕ and ψ with the bonds for the α carbon atom in the middle. When steric and other factors are taken into account, only certain values of ϕ and ψ are permissible, as shown in Fig. 1-6. Within these permissible areas fall the α helix and the β-pleated sheet.

[1] W. N. Lipscomb, *Acct. chem. Res.*, 3, 81 (1970).
[2] C. Ramakrishnan and G. N. Ramachandran, *Biophys. J.*, 5, 909 (1965).

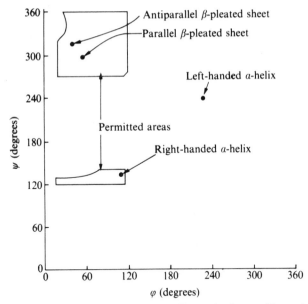

FIG. 1.6.   Permitted values of ϕ and ψ, according to the theory of Ramachandran.

Role of individual amino acids

Specific roles are sometimes played by amino acids in maintaining the secondary structures of proteins. The classic example is proline,

$$H_2C \text{———} CH_2$$

This is rarely if ever present in an α helix, but is often found as the residue immediately following a length of helix. This is because proline, on account of its peculiar structure, forces the chain to bend.

Another special case is glycine, which has no side chain and is therefore very small and compact. As a result, the backbone of the protein chain can bend without steric interference, and in addition there is easy formation of hydrogen bonds. A correlation has in fact been observed[1] between the presence of glycine and the presence of β pleated sheets in various cytochrome c molecules from different species. Threonine and serine are also found[2] to be contributors to hydrogen bonding in the pleated sheets.

[1] E. L. Smith and E. Margoliash, *Fedn. Proc.*, **23**, 1243 (1964).
[2] R. E. Dickerson and I. Geis, *The Structure and Action of Proteins*, Harper and Row, New York, 1969.

Secondary and tertiary bonding

The final topic to be considered in this section is the question of the particular bonds that are concerned in maintaining the secondary and tertiary structures of proteins[1]:

(a) *Sulphur-sulphur bonds*: Two molecules of cysteine may combine under oxidizing conditions to give cystine, which contains a disulphide bond:

$$2H_2NCHCO_2H \xrightarrow{-2H} \begin{array}{cc} CH_2-S-S-CH_2 \\ | \quad\quad | \\ H_2NCH \quad H_2NCH \\ | \quad\quad | \\ CO_2H \quad CO_2H \end{array}$$

Such disulphide bonds have been seen to be involved in the chymotrypsin structure (Fig. 1.2, p. 11) and they play an important role in many enzymes. They are often essential to enzymic activity; if they are ruptured an inactive or a partially active enzyme usually results. This has been demonstrated with ribonuclease by Anfinsen and coworkers[2], who broke the disulphide bond and then caused the enzyme to unfold in 8 M urea solution; the enzyme lost its activity. There is, however, an indication that for some enzymes a particular conformation may be maintained without S–S bridges. Chymotrypsin and subtilisin have very similar catalytic sites and show similar kinetic behaviour[3]; the former has 4 S–S bridges in each molecule (see Fig. 1.2), while the latter contains no cystine or cysteine residues.

(b) *Coulombic forces*: It has been seen that many of the amino-acid residues bear charges under certain pH conditions. The attraction that will occur between a neighbouring positive and a negative group depends to an important extent on the position of the groups on the enzyme. If the groups are at the surface of the molecule, in contact with the aqueous surroundings, the force of attraction is substantially reduced for two related reasons: the high dielectric constant of the water reduces the force, and in addition there is strong hydration by water molecules at charged groups, which tends to keep them apart. The X-ray studies of several proteins including some enzymes, indicate a diffuseness in electron density in the region of groups such as lysine, arginine, glutamic acid and aspartic acid, when they are at the surface of the molecule; this indicates a reduction in the strength of the coulombic bond. In the interior of the molecules, on the other hand, such interaction with the surrounding medium will be much less important, and the effective dielectric constant will be much lower. X-ray crystallographic studies have indicated that in the interior of proteins, groups of opposite charge are held together by electrostatic attraction, forming what are known as *ion pairs* or *salt bridges*. A good example of this is provided by the X-ray results on

[1] For a review, see F. M. Richards. *A. Rev. Biochem.*, **32**, 269 (1963).
[2] C. J. Epstein, R. F. Goldberger, and C. B. Anfinsen, *Cold Spring Harb. Symp. quant. Biol.*, **28**, 439 (1963).
[3] A. N. Glazer, *J. biol. Chem.*, **241**, 635 (1966); **242**, 433 (1967).

chymotrypsin[1] which show that an ion pair is formed between the $-COO^-$ group of asp–194 and the $-NH_3^+$ group of ile–16. This has an important significance in connection with the active centre; thus at higher pH values the $-NH_3^+$ group becomes deprotonated and there is a conformational change resulting from the disappearance of the ion pair. Chymotrypsin exists in the digestive system in the form of a precursor, chymotrypsinogen (see p. 307), in which the ile–16 group is not exposed; the exposure of this group during the activation of chymotrypsinogen thus has important mechanistic consequences.

(c) *Hydrogen bonding*: It has already been noted that structures like the α helix and the β pleated sheet owe their existence to hydrogen bonding between C=O and N—H groups. In addition, the X-ray evidence indicates that hydrogen bonds play a very important role in maintaining protein conformations. It appears that these hydrogen-bonded structures tend to be located in the interior of the molecule, rather than at the surface. This is understandable in view of the fact that the hydrogen bond is largely electrostatic in character, so that its strength will be reduced in an aqueous environment; aside from this, hydrogen bonding between groups on the protein molecule and water molecules will compete with the intermolecular hydrogen bonding. This conclusion has been supported by studies of model compounds in different solvent environments; Klotz and Franzen[2] found that *N*-methylacetamide, $CH_3CONHCH_3$, did not tend to form hydrogen bonds in aqueous solution, but did in non-polar solvents.

(d) *Hydrophobic bonds*[3]: The attractive forces between isolated non-polar groups, such as hydrocarbon chains, are relatively weak, being dispersion or van der Waals forces; the energies of attraction between two methyl groups amount only to about $0 \cdot 5$ kcal mol^{-1}. When, however, non-polar groups are present in an aqueous environment there is a much stronger tendency for them to come together, since when they do so many more hydrogen bonds can be formed between neighbouring water molecules; the term *hydrophobic bond* is commonly used to describe this effect. This phenomenon explains why hydrocarbons are insoluble in water; their molecules are forced together by the strong attractive forces between water molecules. This results in a strong apparent attraction between non-polar groups. There are many non-polar groups on the side chains of amino acids, and in aqueous solution there is therefore a tendency for these to come together.

When isolated non-polar groups are present in an aqueous environment the surrounding water molecules are believed to have a fairly ordered structure, as in a clathrate compound. When these non-polar structures come

[1] P. B. Sigler, D. M. Blow, B. W. Matthews, and R. Henderson, *J. molec. Biol.*, **35**, 143 (1968); D. M. Blow, J. J. Birktoft, and B. S. Hartley, *Nature*, **221**, 337 (1969).
[2] I. M. Klotz and J. S. Franzen, *J. Am. chem. Soc.*, **84**, 3461 (1962); however, see I. M. Klotz and S. B. Farnham, *Biochemistry*, **7**, 3879 (1968).
[3] For reviews, see W. J. Kauzmann, *Adv. Protein Chem.*, **14**, 1 (1959); I. M. Klotz, *Brookhaven Symp. Biol.*, **25** (1960).

together, with the formation of a hydrophobic bond, these cage-like water structures are broken down, the water molecules gaining more freedom of movement. As a result there is an increase of entropy when hydrophobic bonds are formed; such entropy increases have been observed experimentally[1]. Favourable enthalpy changes are also probably involved[2].

It will be clear from the above discussion that the whole problem of protein conformations is an exceedingly complex matter, the final structure being the result of a number of competing tendencies. The most important of these are:

(1) The tendency of non-polar groups to form hydrophobic bonds and to remain in the interior of the protein molecule;

(2) The tendency of polar groups to remain at the exterior of the molecule, where they can form hydrogen bonds with water molecules;

(3) The tendency of $C=O$ and $N-H$ groups to form hydrogen bonds, with the formation of a helical or pleated-sheet structure.

The actual structure of a given protein is determined in a very subtle way by the demands of these different effects, and it obviously depends to a considerable extent on the nature and positions of the various amino-acid side groups. There still remains a question as to whether a protein molecule assumes the structure that is thermodynamically most stable, or whether there is kinetic control of the structure as the protein is being synthesized. The former view is supported by the work of Anfinsen and coworkers[3].

Since the conformation of a protein is due to the resultant of a number of forces many of which are sensitive to small changes of temperature, ionic strength, dielectric constant, pH, etc., it is not surprising that the structure is very sensitive to the environment, and that loss of enzymic activity can easily result from a change in one of these factors.

Prosthetic groups

Certain enzymes, like pepsin, chymotrypsin and urease, are found on detailed analysis to be pure proteins. Others, however, have been found to consist of a non-protein part in addition to the protein. In such cases the protein part of the enzyme is known as the apoenzyme, and the non-protein part as the prosthetic group. The enzyme as a whole, consisting of the apoenzyme and the prosthetic group, is known as the holoenzyme. Sometimes the prosthetic group is easily separated from the apoenzyme, in which case it is sometimes known as a coenzyme. An example is the enzyme lactate dehydrogenase which is easily split into its apoenzyme and a substance known as nicotinamide adenine dinucleotide (NAD), as diphosphopyridine nucleo-

[1] I. M. Klotz, *Brookhaven Symp. Biol.*, **25** (1960).

[2] See W. J. Kauzmann, *Adv. Protein Chem.*, **14**, 1 (1959).

[3] C. J. Epstein, R. F. Goldfinger, and C. B. Anfinsen, *Cold Spring Harb. Symp. quant. Biol.*, **28**, 439 (1963); for a review, see R. E. Canfield and C. B. Anfinsen, in *The Proteins*, Vol. 1, p. 311 (ed. H. Neurath), Academic Press, New York, 1963; see also R. D. Singhal and M. Z. Atassi, *Biochemistry*, **9**, 4252 (1970).

tide (DPN) or as Coenzyme I; for enzymic activity the presence of this coenzyme is essential. In the case of other enzymes it is found that ions are attached to the protein and that removal of these causes loss of enzymic activity; such ions are usually known as activators rather than as coenzymes. In the above cases an easy way of separating the protein from the prosthetic group is by dialysis; the prosthetic group passes readily through the membrane while the apoenzyme remains behind.

In other enzymes the apoenzyme is firmly bound to the prosthetic group and cannot be separated from it by dialysis or by other simple means. This is the case with the enzyme catalase, which consists of a protein very firmly bound to haematin, its prosthetic group. In some enzymes the apoenzyme is very firmly bound to metal atoms that cannot easily be removed. It appears that the distinction between the coenzymes and the prosthetic groups that are firmly bound is simply one of degree, the function of the enzyme apparently not being related to the strength of binding to the apoenzyme.

Enzymes as catalysts

The essential characteristic of a catalyst[1] is that it influences the rate of chemical reaction but is not itself used up during the process and can in ideal cases be recovered at the end of the reaction. Indeed, a simple and informative definition of a catalyst is that it is a substance which is both a reactant and a product of the reaction. This implies that the catalysts do not act by virtue of some external effect which they exert, without themselves entering into reaction: on the contrary it is well established that in all types of catalysis the catalyst forms some kind of complex with the substrate (i.e. the reacting substance) and that this complex finally breaks down into the products of reaction and the catalyst. It is frequently found that the rates of catalysed reactions are directly proportional to the concentration of the catalyst; in the case of reactions that do not occur at an appreciable rate in the absence of catalyst this means that a plot of rate against catalyst concentration is simply a straight line passing through the origin.

It is a characteristic property of catalysts that they catalyse reactions to the same degree in forward and reverse directions. The following simplified argument (which is not to be taken as a rigorous proof) gives an indication of why this is so. Consider the chemical equilibrium: $A + B \rightleftharpoons X + Y$. The rate of this reaction may under certain conditions be given by:

$$v_1 = k_1[A][B], \tag{3}$$

where k_1 is the rate constant and [A] and [B] the concentration of the two reactants. The rate from right to left is similarly:

$$v_{-1} = k_{-1}[X][Y]. \tag{4}$$

---

[1] For a general account of catalysis, with special reference to enzyme systems, see W. P. Jencks, *Catalysis in Chemistry and Enzymology*, McGraw-Hill, New York, 1969.

At equilibrium these rates must be equal, so that

$$k_1[A][B] = k_{-1}[X][Y],$$ (5)

whence

$$\frac{[X][Y]}{[A][B]} = \frac{k_1}{k_{-1}} = K,$$ (6)

where K, equal to k_1/k_{-1}, is the equilibrium constant for the reaction. Since the catalyst does not enter into the overall reaction it cannot have any effect on the extent to which the reaction occurs, i.e. it cannot affect K. If on the additon of a catalyst the rate v_1 is increased by a certain factor, k_1 must be increased by the same factor, and it therefore follows that k_{-1} and v_{-1} are also increased by the same factor.

In all of the above respects the enzymes exhibit the typical properties of catalysts. In some ways, however, they show unusual characteristics which are not found with other types of catalysts. Among these are those properties, like sensitivity to heat and extremes of pH, in which they show the typical behaviour of proteins. Another way in which they are unusual is that on a molecular basis they are remarkably effective in comparison with other types of catalysts. A very crude but sometimes useful way of giving an indication of the efficiency of a catalyst is in terms of its *turnover number*, which is defined as the number of molecules that are caused to react in one minute by one molecule of catalyst. Under certain conditions the enzyme catalase has a turnover number of about 5 000 000 and this is higher by many powers of ten than that shown by other catalysts for the decomposition of hydrogen peroxide. This is rather an extreme case, but in every instance it appears that the natural enzyme has a higher turnover number than has any other catalyst.

Specificity

Another respect in which enzymes have unusual catalytic properties is in their specificity. Various different types of effects have been observed, of which the following are particularly important:

(1) *Absolute specificity*: When an enzyme will bring about reaction in only a single substrate it is said to exhibit absolute specificity. For example, succinic dehydrogenase is absolutely specific with respect to the oxidation of succinic acid. Absolute specificity with respect to coenzyme is frequently encountered.

(2) *Group specificity*: A lower degree of specificity is shown by certain enzymes that act upon a number of substrates but which have definite requirements with respect to the types of atomic groupings that must be present in these molecules. The enzyme pepsin, for example, will hydrolyse certain peptide linkages but requires among other things that an aromatic

group be present in a certain position with respect to the peptide linkage. Most of the proteolytic enzymes show such group specificity.

(3) *Reaction specificity* : The lowest degree of specificity is shown by enzymes that catalyse a certain type of reaction irrespective of what groups are present in the neighbourhood of the critical linkages in the substrate. For example, the enzymes known as lipases will catalyse the hydrolysis of any organic ester (including the lipids). There are usually differences in rates for the different substrates; some lipases, for example, act more rapidly on the esters of short-chain acids than on those of long-chain acids, while others do just the reverse. Group and absolute specificity may indeed be regarded as special cases of reaction specificity, certain substrates reacting with zero velocity.

(4) *Stereochemical specificity* : It is frequently found that an enzyme will catalyse the reaction of only one stereochemical form of the substrate; for example, the proteolytic enzymes usually only act on peptides that are made up of amino acids in the L-forms. Similarly, lactate dehydrogenase catalyses the oxidation of L-lactic acid, but not that of the D-form.

Much work still remains to be done on the question of the molecular explanation of specificity, and full understanding will only come when more knowledge has developed about the exact structures of enzyme molecules. Some of the ideas that have been put forward are discussed in Chapter 9, and here it may simply be noted that the exact fitting together of enzyme and substrate molecules may be essential for reaction. The replacement of one group by another may affect this fit in various ways; a group may simply, by virtue of its shape and size, get in the way and prevent the intimate contact of the vital parts of the enzyme and substrate molecules; or it may bring about a change in the general conformation of the substrate or enzyme molecule.

The active centres of enzymes

During the course of the various investigations that have been carried out on enzyme systems it has become apparent that the catalytic properties depend primarily upon the existence of certain small areas on the enzyme molecules. These small areas, believed to account for the main effects, are referred to as *active centres*, or active sites. It is at these active centres that the enzymes form addition compounds with the substrates, and at which chemical reaction occurs. These active centres are usually structures of some complexity, containing a number of different groups arranged according to a particular pattern. In cases when detailed investigations have been made, as with chymotrypsin (see Fig. 1.2, p. 11), it is found that the different parts of the active centre belong to widely different regions of the peptide chain— sometimes to different chains—these parts being brought into suitable proximity by the specific folding of the chains.

Although the active centre is the catalytically most important part of the enzyme molecule, it must be recognized that other regions are also extremely important. Indeed the whole molecule may be said to play a role in catalysis, since minor changes in the peptide chains may lead to a conformational change which will destroy the integrity of the active centre. Aside from this, there may also exist on the enzyme surface certain *allosteric sites*, which are sites for activation or inhibition of the enzymic activity by molecules other than the substrate; these sites are frequently situated at a considerable distance from the active centre. Allosteric action is considered in some detail in Chapter 11.

It is naturally of some importance to investigate the nature of active centres, and the number that exist on each enzyme molecule, since this information strongly supports the purely kinetic studies in leading to conclusions about mechanisms. There is considerable overlap between methods which provide information about the nature and number of active centres. Some of the main methods will now be discussed.

Reaction with diisopropylfluorophosphate

The use of this method may be illustrated by an investigation by Jansen, Nutting, Jang, and Balls[1] on the inhibition of chymotrypsin by diisopropyl-fluorophosphate (DFP),

$$\text{i-C}_3\text{H}_7\text{O} \diagdown \diagup \text{O}$$
$$\text{P}$$
$$\text{i-C}_3\text{H}_7\text{O} \diagup \diagdown \text{F}.$$

From the form of the curves of activity against inhibitor concentration they deduced that complete inhibition would involve the attachment of one mole of inhibitor to 25 000 g of enzyme, which is close to the molecular weight. The product was isolated and crystallized; it was completely inactive enzymically and each mole contained one gramme atom of phosphorus, a result that was confirmed using radioactive phosphorus[1]. The enzyme–inhibitor compound was also found[2] to contain two equivalents of the isopropyl group, but no fluorine. These results lead to the conclusion that the inhibition reaction is of the type:

$$\text{i-C}_3\text{H}_7\text{O} \diagdown \diagup \text{O} \qquad\qquad \text{i-C}_3\text{H}_7\text{O} \diagdown \diagup \text{O}$$
$$\text{P} \qquad\rightarrow\qquad \text{P} \quad + \text{HF}.$$
$$\text{i-C}_3\text{H}_7\text{O} \diagup \diagdown \text{F} + \text{H} - \text{X} \qquad \text{i-C}_3\text{H}_7\text{O} \diagup \diagdown \text{X}$$

[1] E. F. Jansen, M. D. F. Nutting, R. Jang, and A. K. Balls, *J. biol. Chem.*, **179**, 189 (1949); for reviews, see A. K. Balls and E. F. Nutting, *Adv. Enzymol.*, **13**, 321 (1952); B. J. Jandorf, H. O. Michel, N. K. Schaffer, R. Egan, and W. H. Summerson, *Discuss. Faraday Soc.*, **20**, 134 (1955).
[2] E. F. Jansen, M. D. F. Nutting, and A. K. Balls, *J. biol. Chem.*, **179**, 201 (1949).
[3] E. F. Jansen, M. D. F. Nutting, R. Jang, and A. K. Balls, *J. biol. Chem.*, **185**, 209 (1950).

It is a significant fact that reaction with DFP only occurs when the active centre exists intact in the molecule; thus there is no reaction with the precursor chymotrypsinogen[1], with denatured chymotrypsin[2], or with the constituent amino acids[2].

The DFP studies have also provided information about the chemical nature of the active centres. When the DFP–chymotrypsin compound is hydrolysed enzymically or by acid the phosphorus is found to be bound to the amino acid serine[3].

These results with DFP have been confirmed by Hartley, Kilby and their coworkers, using other types of inhibitors. Hartley and Kilby[4], for example, studied diethyl *p*-nitrophenyl phosphate (known as E 600, or Paraoxon) as an inhibitor; when this combines with chymotrypsin, nitrophenol is liberated according to the reaction

$$ChTr{-}H + (EtO)_2PO_2C_6H_4NO_2 \rightarrow ChTr{-}PO(OEt)_2 + NO_2C_6H_4OH,$$

and it was found that the maximum number of moles of nitrophenol liberated was equal to the number of moles of enzyme present. A similar conclusion that there is a single active centre on the molecule was obtained from work with other *p*-nitrophenyl esters[5].

Equilibrium dialysis and ultracentrifugation

In some cases the complex formed between enzyme and activator or inhibitor undergoes dissociation too readily to permit an analysis to be made of its composition, and the nature of the attachment must then be studied using less direct methods. The enzyme may be allowed to come into contact with inhibitor or activator at a known total concentration, and the enzyme then separated from the solution by ultracentrifugation or equilibrium dialysis[6]. From the composition of the resulting solution it is possible to calculate how the inhibitor or activator is bound to the enzyme. This may be repeated over a range of concentrations, and by suitable extrapolation procedures one can then determine the extent of binding that corresponds to saturation of the enzyme. An example of the use of this method is the work of Doherty and Vaslow[7], using α-chymotrypsin. In this investigation they found that complete inhibition was achieved when 1 mole of inhibitor combined with 22 000 g of enzyme. Since the molecular weight is close to this value it

[1] E. F. Jansen, M. D. F. Nutting, R. Jang, and A. K. Balls, *J. biol. Chem.*, **179**, 189 (1949); E. F. Jansen, M. D. F. Nutting, and A. K. Balls, *J. biol. Chem.*, **179**, 201 (1949).

[2] E. F. Jansen, M. D. F. Nutting, R. Jang, and A. K. Balls, *J. biol. Chem.*, **185**, 209 (1950).

[3] N. K. Schaffer, S. C. May, and W. H. Summerson, *J. biol. Chem.*, **202**, 67 (1953); N. K. Schaffer, S. Harshman, and R. R. Engle, *J. biol. Chem.*, **214**, 799 (1955).

[4] B. S. Hartley and B. A. Kilby, *Biochem. J.*, **50**, 672 (1952).

[5] B. S. Hartley and B. A. Kilby, *Biochem. J.*, **56**, 288 (1954); B. A. Kilby and G. Youatt, *Biochem. J.*, **57**, 303 (1954).

[6] Cf. I. M. Klotz, F. M. Walker, and R. B. Pivan, *J. Am. chem. Soc.*, **68**, 1486 (1946); J. E. Hayes and S. F. Velick, *J. biol. Chem.*, **207**, 225 (1954).

[7] D. G. Doherty and F. Vaslow, *J. Am. chem. Soc.*, **74**, 931 (1952); see also M. W. Loewus and D. R. Briggs, *J. biol. Chem.*, **199**, 857 (1952).

follows that there is one active centre to each molecule of this enzyme. The method can be equally well used with activators; the procedure is then to determine how much activator is bound when maximum activation is achieved. Similar techniques can be applied to two-substrate systems and to substrate–coenzyme systems[1].

Active site titration

Bender and coworkers[2] devised a method for determining the *operational normality*, i.e. the number of active centres per litre, and applied it to chymotrypsin. They allowed the enzyme to interact with N-transcinnamoylimidazole as substrate; this forms an acyl–enzyme intermediate with the enzyme, as do other substrates (for further details see Chapter 10), but the rate of deacylation is low. By measuring absorption at suitable wavelengths it was possible to distinguish between the substrate and the enzyme–substrate intermediate; inactive protein did not interfere, since the intermediate had a specific absorption peak which was not observed with inactive enzyme. By correlating the intensity of the absorption peak with the substrate concentration it was possible to determine the amount of enzyme that combined with the substrate. There was found to be a one-to-one ratio, indicating one active site per enzyme molecule. Similar methods have been devised for trypsin[3], papain[4], and other proteolytic enzymes[5].

Substrate labelling of active centres

The methods discussed so far have been mainly concerned with determining the number of active centres, although some of them have also provided information about their nature. The procedures which follow are chemical methods designed to provide information more particularly about the *nature* of the active centres[6]. Reference has already been made to the X-ray methods which give the overall structure, and which can also be used to identify the active centre by structural determinations of enzyme–substrate complexes; this has been done in particular with lysozyme (see p. 292) and chymotrypsin (see p. 306).

An important chemical method for obtaining information about the nature of the active centre involves labelling them by use of the substrate. As an example may be considered the work of Horecker and coworkers[7] on aldolase, an enzyme which catalyses the conversion of fructose-1,6-diphosphate into dihydroxyacetone phosphate and glyceraldehyde-3-

[1] C. S. Vestling, *Methods of Biochemical Analysis*, **10**, 137 (1962).
[2] G. R. Schonbaum, B. Zerner, and M. L. Bender, *J. biol. Chem.*, **236**, 2930 (1961).
[3] M. R. Jackson and M. L. Bender, *Biochem. biophys. Res. Commun.*, **39**, 1157 (1970).
[4] K. Brocklehurst and G. Little, *FEBS Letters*, **9**, 113 (1970).
[5] M. L. Bender *et al.*, *J. Am. chem. Soc.*, **88**, 5890 (1966).
[6] For a review, see S. J. Singer, *Adv. Protein Chem.*, **22**, 1 (1967); B. L. Vallee and J. F. Riordan, *A. Rev. Biochem.*, **38**, 733 (1969).
[7] B. L. Horecker, S. Pontremoli, C. Ricci, and T. Chang, *Proc. natn. Acad. Sci.*, **47**, 1949 (1961); J. C. Speck, P. T. Rowley, and B. L. Horecker, *J. Am. chem. Soc.*, **85**, 1012 (1963).

phosphate, and the reverse process. Dihydroxyacetone phosphate labelled with C^{14} was caused to become attached to the enzyme, and it was found that the label, C^{14}, was incorporated into the enzyme. The enzyme–substrate complex was not stable enough to be isolated, but it was stabilized by reduction with sodium borohydride. Analysis of the now stable intermediate revealed that the labelling was at a lysine group, so that it appears that the site of attachment is at this group.

Milstein and Sanger[1] performed similar experiments on phosphoglucomutase, using glucose-1-phosphate labelled with P^{32}, as the substrate. In this case the site of attachment of the phosphorus was a serine residue, which is concluded to be part of the active centre. Work has also been done with alkaline phosphatase[2].

Labelling by substrate analogues and inhibitors

The use of DFP has already been described as providing a way of determining the number and nature of active centres; it was seen to lead to the conclusion that serine is an important constituent of the active centre of chymotrypsin.

Another interesting example of this method of identifying groups present at active centres is provided by the work of Shaw and coworkers[3] on chymotrypsin and trypsin. The former enzyme hydrolyses the substrate N-tosyl-L-phenylalanine ethyl ester, while the latter hydrolyses the corresponding lysine derivative. Replacement of the ester grouping of each of these substrates by the chloromethylketone group converted them into potent specific inhibitors of the enzyme; there was no cross inhibition. In each case the inhibitor was found to attach itself to a histidine residue in the active centre. It was also found that the inhibitors did not become attached to the histidine residue if denatured enzyme is used. These results clearly implicate histidine as part of the active centre of both chymotrypsin and trypsin.

Non-specific labelling of active centres

By far the greatest number of chemical techniques for studying active centres involve the use of labels not necessarily specific for a particular active site. The most useful reagents are those which react with a particular group, such as histidine or serine. These reagents, however, usually combine with all of the available groups of this kind, and not merely those at the active centre. Information regarding this is provided by measurements of the catalytic activity of the enzyme as the labelling is carried out. However, a complication is that it is not always correct to assume that, because a particular reagent combines with the enzyme and inactivates it, the group or

[1] C. Milstein and F. Sanger, *Biochem. J.*, **79**, 456 (1961).
[2] J. H. Schwartz, *Proc. natn. Acad. Sci. U.S.A.*, **49**, 872 (1963); D. Levine, T. W. Reid, and I. B. Wilson, *Biochemistry*, **8**, 2374 (1969).
[3] G. Schoellman and E. Shaw, *Biochemistry*, **2**, 252 (1963); E. Shaw, M. Mares-Guia, and W. Cohen, *ibid.*, **4**, 2219 (1965).

groups so modified are essential to the enzymic activity; there is also the possibility that introduction of the reagent into the enzyme will bring about a change of conformation and a resulting inactivation. An additional problem is that a particular reagent may not react specifically with one type of amino-acid residue, but may combine with several having similar chemical reactivities; for example, lysine, histidine, tyrosine and cysteine may all react with fluorodinitrobenzene. A further difficulty is that it is not always clear how inactivation takes place. For example, a given reagent may reduce the enzymic activity to 10 per cent of the normal activity. This may be the result of complete inactivation of 90 per cent of the enzyme molecules, the remaining 10 per cent being completely active; or it may be that all of the molecules have been 90 per cent inactivated; or a combination of both effects may be involved.

There are various ways by which these difficulties can be overcome. The complication that the reagent may combine with groups other than those at the active centre has been dealt with by first masking the active site with a competitive inhibitor and then allowing the reagent to combine with all of the available groups on the enzyme[1]. The excess reagent is then removed, and then the competitive inhibitor; finally the reagent labelled with, for example C^{14}, is introduced. Any label incorporated into the enzyme is then most likely to enter the active centre.

The possibility that a reagent may bring about inactivation by inducing a conformational change is often rather difficult to eliminate. It is possible to employ some of the techniques referred to earlier (pp. 10–14) in order to detect conformational changes; however, conformational changes may be too small to be detected by such techniques, but important enough to induce inactivation.

The difficulties that there may not be specific interaction of the reagent with one type of amino-acid residue, and that the nature of inactivation is uncertain, are rather closely connected. Work with a variety of reagents is often helpful; for example, whereas fluorodinitrobenzene reacts with a number of residues, O-methylisourea generally attacks only ε-amino groups of lysine[2]. This method is, however, not always feasible. Ray and Koshland[3] devised a method for distinguishing between two modified residues as far as their extent of inactivation is concerned. This method involves an all-or-none assay of enzymic activity, and comparison with the so-called efficiency assay. The former involves the determination of the number of active centres having a certain efficiency, while the latter determines the overall catalytic efficiency of a given solution. By means of an extrapolation procedure they distinguish between completely inactivated enzymes and partially inactivated

[1] J. A. Cohen and M. G. P. J. Warringa, *Biochim. biophys. Acta*, **11**, 52 (1953).
[2] B. Kassell and R. B. Chow, *Biochemistry*, **5**, 3449 (1966); see, however, G. S. Shields, R. B. Hill, and E. L. Smith, *J. biol. Chem.*, **234**, 1747 (1959).
[3] W. J. Ray and D. E. Koshland, *J. Amer. chem. Soc.*, **85**, 1977 (1963).

enzymes. For example, they studied the photo-oxidation[1] of phosphogluco-mutase and found that loss of methionine caused complete inactivation of the enzyme, whereas loss of histidine caused partial inactivation.

The procedures considered above apply to both one-substrate and two-substrate systems; with the latter, however, there are some special problems, which will now be considered briefly. These stem from the presence of two or more active sites of a different nature. For example, a given non-specific reagent might cause a two-substrate enzyme to lose all of its activity, but it could not be said, without further experimentation, whether this is due to a modification of a group in the substrate binding site or in the coenzyme binding site or possibly in some other site.

Such problems can be overcome in the following ways. The modification to the enzyme is first carried out in the usual way. It is then carried out, in separate experiments, in the presence of excess of the following: (a) substrate, (b) coenzyme, (c) competitive inhibitor for substrate, and (d) coenzyme analogue (often a competitive inhibitor for the coenzyme). If loss of enzyme activity due to modification is significantly reduced in cases (a) and (c), it is reasonable to conclude that the modified group is at or near the substrate-binding site; if the reduction is in cases (b) and (d) the group modified is probably near the coenzyme binding site.

It is possible that groups in both sites, or elsewhere in the enzyme, are modified. In the former case, excess of both substrate and coenzyme should reduce the loss of activity due to the modification process. In the latter case, no decrease in activity loss would be expected.

By the use of such methods as these it has been possible, in some instances, to relate the group modified to a particular site. Examples where this has been done are included in Table 1.3.

Kinetic methods

The preceding discussion of methods for the determination of the nature and number of active centres on enzyme molecules has been confined to non-kinetic methods. In addition, considerable information has been provided by kinetic studies. In particular, studies of the pH-dependence of the rates of enzyme reactions, under various conditions of temperature, substrate concentration and solvent composition, have revealed pK values of the groups that are involved in the catalytic action; such values provide very valuable clues as to the nature of the participating groups. Investigations of this kind are described in some detail in Chapter 5. Numbers of active centres have also been revealed by transient-phase studies, as will be discussed in Chapter 6.

Summary of results

A summary of some of the conclusions that have been drawn about the

[1]For a discussion of photo-oxidation see W. J. Ray, *Meth. Enzym.*, **11**, 490 (1967).

groups that are involved with the catalytic action at the active centre is included in Tables 1.2 and 1.3.

TABLE 1.2

The number of the active centres on enzyme molecules

Enzyme	Molecular weight	Number of active centres per molecule	Type of investigation
α-Chymotrypsin	24 800	1	Reaction with DFP [1]; reaction with other inhibitors [2]; equilibrium dialysis [3]
Trypsin	23 800	1	Reaction with DFP [4]; reaction with other inhibitors [5]
Urease	483 000	3–4	Inhibition by silver ions [6]
Alkaline phosphatase (*E. coli*)	86 000	2	P-32 labelled orthophosphate [7]
Fumarase	194 000	4	Reaction with labelled inhibitor [8]; substrate and inhibitor binding [9]; affinity labelling with substrate analogues [10]; N-terminal amino acid analysis [11]; molecular weight (sedimentation equilibrium) and tryptic peptide mapping [11]
Phosphoglycerate mutase (yeast)	112 000	4	Chemical modification [12]; end-group amino acid analysis [13]
Glycogen phosphorylase-*a*	495 000	4	Sedimentation equilibrium molecular weight determinations [14]
Glycogen phosphorylase-*b*	242 000	2	ditto
Liver alcohol dehydrogenase	80 000	2	Affinity labelling (iodoacetate) and amino acid analysis [15]; coenzyme binding studies [16]; fluorescence quenching [17]
L-Lactate dehydrogenase	140 000	4	Coenzyme binding [18]; thiol modification [19]; substrate analogues [20]

[1] E. F. Jansen, M. D. F. Nutting, R. Jang, and A. K. Balls, *J. biol. Chem.*, **179**, 189 (1949); see also p. 25.

[2] B. S. Hartley and B. A. Kilby, *Biochem. J.*, **50**, 672 (1952); **56**, 288 (1954); B. A. Kilby and G. Youatt, *ibid.*, **57**, 303 (1954).

[3] D. G. Doherty and F. Vaslow, *J. Am. chem. Soc.*, **74**, 931 (1952); M. W. Loewus and D. R. Briggs, *J. biol. Chem.*, **199**, 857 (1952).

[4] E. F. Jansen and A. K. Balls, *J. biol. Chem.*, **194**, 721 (1952).

[5] B. A. Kilby and G. Youatt, *Biochem. J.*, **57**, 303 (1954).

[6] J. F. Ambrose, G. B. Kistiakowsky, and A. G. Kridl, *J. Am. chem. Soc.*, **73**, 1232 (1951).

[7] C. Lazdunski, C. Petitclerc, D. Chappelet, and M. Lazdunski, *Biochem. biophys. Res. Commun.*, **37**, 744 (1969).

[8] R. A. Laursen, J. B. Baumann, K. B. Linsley, and W. C. Shen, *Archs. Biochem. Biophys.*, **130**, 688 (1969).

[9] J. Teipel and R. L. Hill, *J. biol. Chem.*, **243**, 5679 (1968).

[10] R. A. Bradshaw, G. W. Robinson, G. M. Hass, and R. L. Hill, *J. biol. Chem.*, **244**, 1755 (1969).

[11] L. Kanarek, E. Marler, R. A. Bradshaw, R. E. Fellows, and R. L. Hill, *J. biol. Chem.*, **239**, 4207 (1964).

[12] R. Sasaki, E. Sugimolo, and H. Chiba, *Biochim. biophys. Acta*, **227**, 584 (1971).

[13] R. Sasaki, E. Sugimolo, and H. Chiba, *Agric. and biol. Chem.*, **34**, 135 (1970).

[14] E. G. Krebs and E. H. Fischer, *Biochim. biophys. Acta*, **20**, 150 (1956); G. T. Cori and C. F. Cori, *J. biol. Chem.*, **158**, 321 (1945); P. J. Keller and G. T. Cori, *ibid.*, **214**, 127 (1955); H. B. Madsen and C. F. Cori, *ibid.*, **223**, 1055 (1956); L. L. and J. Kastenschmidt and E. Helmreich, *Biochemistry*, **7**, 3590 (1968); S. Shaltiel, J. L. Hedrich, and E. H. Fischer, *Biochemistry*, **5**, 2108 (1966); V. L. Seers, E. H. Fischer, and D. C. Teller, *ibid.*, **6**, 3315 (1967).

[15] H. Jornvall and J. I. Harris, *Eur. J. Biochem.*, **13**, 565 (1970); H. Jornvall, *ibid.*, **16**, 25, 41 (1970).

[16] H. Theorell and R. Bonnischen, *Acta chem. scand.*, **5**, 1105 (1951).

[17] H. Theorell and K. Tatemoto, *Archs. Biochem. Biophys.*, **142**, 69 (1971).

[18] S. F. Velick, *J. biol. Chem.*, **233**, 1455 (1958); A. D. Winer, G. W. Schwert, and D. B. S. Millar, *ibid.*, **234**, 1149 (1959).

[19] Y. Takenaka and G. W. Schwert, *ibid.*, **223**, 157 (1956).

[20] J. van Eys, F. E. Stolzenbach, L. Sherwood, and N. O. Kaplan, *Biochim. biophys. Acta*, **27**, 63 (1958); G. P. Heiderer, D. Jeckel, and P. Weiland, *Biochem. Z.*, **238**, 187 (1956).

TABLE 1.3

Groups identified as present at active centres of enzyme molecules

Enzyme	Groups at active centre	Evidence
α-Chymotrypsin	Serine–195	Reaction with DFP [1]
	Histidine–57	Oxidative inactivation [2]; pH studies [3]; chemical inhibition [4]
	Isoleucine–16	Acetylation [5]
	Serine–195; histidine–57; isoleucine–16, aspartic acid–194 and aspartic acid–102	X-ray studies [6]
Trypsin	Serine–183	Alkylation [7]; DFP [8]
	Histidine–46	pH studies [9]; chemical inhibition [10]
	Lysine not needed	Acetylation [11]
Subtilisin	Serine	DFP [12] Modification to –SH [13]
	Histidine	pH studies [14]
	Group of pK_a 8·7	Titration [15]
Ribonuclease	Histidine	Alkylation [16]; pH studies [17]
	Lysine	Arylation (FDNB) [18]
	Aspartic acid	[19]
Alkaline phosphatase (*E. coli*)	Serine	Inorganic phosphate (P^{32}) [20]
	Zn^{++}	Chelating agents [21]
	–SH (for intestinal only)	Various chemical methods [22]
	Histidine	pH studies [23]; active-site inhibitors [24]

TABLE 1.3—continued

Enzyme	Groups at active centre	Evidence
Papain	Cystine Histidine Carboxyl group	Chemical modification [25] Inhibition studies [26] pH studies [27]; inhibitor studies [28]
Lactate dehydrogenase	Cysteine, histidine (probably at substrate binding site) Tyrosine (possibly at coenzyme site)	Chemical modification and amino-acid analysis [29]; X-ray studies [30] Chemical modification using iodine [31]
Fumarase	Methionine, histidine	Chemical modification using iodoacetate and 4-bromocrotonate [32]
Glucose-6-phosphate dehydrogenase (yeast)	Lysine Histidine 2 tyrosines	Reversible inactivation by pyridoxal-5'-phosphate [33] Photo-oxidation [34] Chemical modification using tetranitromethane and acetyl amidazole [35]
Phosphoglucomutase (yeast)	An amino group (possibly lysine)	Chemical modification using trinitrobenzene sulphonate [36]
L-malate dehydrogenase	Tyrosine (at least one)	Chemical modification using tetranitromethane and acetylamidazole [37]

[1] N. K. Schaffer, S. C. May, and W. H. Summerson, *J. biol. Chem.*, **202**, 67 (1953); N. K. Schaffer, S. Harshman, and R. Engle, *ibid.*, **214**, 799 (1955).
[2] L. Weil, S. James, and A. R. Buchert, *Archs. Biochem. Biophys.*, **46**, 266 (1953).
[3] K. J. Laidler and M. L. Barnard, *Trans. Faraday Soc.*, **51**, 497 (1956); H. Kaplan and K. J. Laidler, *Can. J. Chem.*, **45**, 547 (1967).
[4] G. Schoellman and E. Shaw, *Biochemistry*, **2**, 252 (1963).
[5] H. L. Oppenheimer, B. Labouesse, and G. P. Hess, *Biochem. biophys. Res. Commun.*, **14**, 318 (1964); H. L. Oppenheimer, B. Labouesse, K. Carleson, and G. P. Hess, *Fedn Proc.*, **23**, 315 (1964); M. L. Bender and F. J. Kédzy, *J. Am. chem. Soc.*, **86**, 3705 (1964); *idem.*, *A. Rev. Biochem.*, **34**, 49 (1965).
[6] B. W. Matthews, P. B. Sigler, R. Henderson, and D. M. Blow, *Nature*, **214**, 652 (1967); D. M. Blow, J. J. Bicktoft, and B. S. Hartley, *Nature*, **221**, 337 (1969).
[7] W. B. Lawson, M. D. Leafer, A. Tewes, and G. J. S. Rao, *Hoppe-Seyler's Z. physiol. Chem.*, **349**, 251 (1968).
[8] G. H. Dixon, D. L. Kauffman, and H. Neurath, *J. biol. Chem.*, **233**, 1373 (1958).
[9] H. Gutfreund, *Trans. Faraday Soc.*, **51**, 441 (1955); B. R. Hammond and H. Gutfreund, *Biochem. J.*, **61**, 181 (1955); S. A. Bernhard, *Biochem. J.*, **59**, 506 (1955).
[10] E. Shaw and S. Springham, *Biochem. biophys. Res. Commun.*, **27**, 391 (1967); E. Shaw, M. Mares-Guia, and W. Cohen, *Biochemistry*, **4**, 2219 (1965).
[11] J. Labouesse and M. Gervais, *Eur. J. Biochem.*, **2**, 215 (1967).
[12] F. Sanger and D. C. Shaw, *Nature*, **187**, 872 (1960).
[13] L. Polgar and M. L. Bender, *Biochemistry*, **6**, 610 (1967); K. E. Neet and D. E. Koshland, *Proc. natn. Acad. Sci. U.S.A.*, **56**, 1606 (1966).
[14] A. N. Glazer, *J. biol. Chem.*, **242**, 433 (1967); *Proc. natn. Acad. Sci. U.S.A.*, **59**, 996 (1968).
[15] S. H. Bernhard and J. Keizer, *Biochemistry*, **5**, 4127 (1965); S. H. Bernhard and Z. Tashjian, *J. Am. chem. Soc.*, **87**, 1806 (1965).
[16] W. D. Stein and E. A. Barnard, *J. molec. Biol.*, **1**, 350 (1959); A. M. Crestfield, W. H. Stein, and S. Moore, *J. biol. Chem.*, **238**, 2413 (1963).

[17] D. Findlay, A. P. Mathias, and B. R. Rabin, *Biochem. J.*, **85**, 139 (1962); E. N. Ramsden and K. J. Laidler, *Can. J. Chem.*, **44**, 2597 (1966).

[18] J. F. Riordan, W. E. C. Wacker, and B. L. Vallee, *Biochemistry*, **4**, 1758 (1965); C. B. Anfinsen, *Brookhaven Symp. Biol.*, **15**, 184 (1962).

[19] W. H. Stein, *Brookhaven Symp. Biol.*, **13**, 104 (1960); C. B. Anfinsen, *J. biol. Chem.*, **221**, 405 (1956).

[20] J. H. Schwartz and F. Lipmann, *Proc. natn. Acad. Sci.*, **47**, 1996 (1961); J. H. Schwartz, *ibid.*, **49**, 871 (1963).

[21] D. J. Plocke and B. L. Vallee, *Biochemistry*, **1**, 1039 (1962); R. T. Simpson and B. L. Vallee, *Ann. N.Y. Acad. Sci.*, **166**, 670 (1969); *Biochemistry*, **7**, 4343 (1968).

[22] W. H. Fishman and N. K. Ghosh, *Biochem. J.*, 1163 (1967); N. K. Ghosh, *Ann. N.Y. Acad. Sci.*, **166**, 620 (1969).

[23] C. Lazdunski and M. Lazdunski, *Biochim. biophys. Acta*, **113**, 551 (1966).

[24] H. Czopak and G. Folsch, *Acta chem. scand.*, **24**, 1025 (1970).

[25] S. S. Hussain and G. Lowe, *Biochem. J.*, **108**, 861 (1968).

[26] S. S. Hussain and G. Lowe, *ibid.*, p. 855.

[27] J. R. Whitaker and M. L. Bender, *J. Am. chem. Soc.*, **87**, 2728 (1965); L. A. A. E. Sluyterman, *Biochim. biophys. Acta*, **151**, 178 (1968).

[28] M. L. Bender and L. J. Brubacher, *J. Am. chem. Soc.*, **88**, 5880 (1966); K. Wallenfels and B. Eisele, *Eur. J. Biochem.*, **3**, 267 (1968).

[29] C. Wenckhaus, J. Berghauser, and G. Pfleiderer, *Hoppe-Seyler's Z. physiol. Chem.*, **350**, 473 (1969); T. P. Fondy, J. Everse, G. A. Driscoll, F. Castillo, F. E. Stolzenbach, and N. O. Kaplan, *J. biol. Chem.*, **240**, 4219 (1965); J. J. Holbrook, G. Pfleiderer, K. Mella, M. Volz, M. Leskowac, and R. Jeckel, *Eur. J. Biochem.*, **1**, 476 (1967).

[30] M. J. Adams and coworkers, *Nature*, **227**, 1098 (1970).

[31] G. DiSabato, *Biochemistry*, **4**, 2288 (1965); W.-C. Shen and P. M. Wasserman, *Biochim. biophys. Acta*, **221**, 405 (1970).

[32] R. A. Bradshaw, G. W. Robinson, G. M. Hass, and R. L. Hill, *J. biol. Chem.*, **244**, 1755 (1969).

[33] W. Domschke and G. E. Domagk, *Hoppe-Seyler's physiol. Chem.*, **350**, 1111 (1969).

[34] W. Domschke, H. J. Engel, and G. F. Domagk, *ibid.*, 1117 (1969).

[35] W. Domschke, C. Von Hinueben, and G. F. Domagk, *Biochim. biophys. Acta*, **207**, 485 (1970).

[36] R. Sasaki, E. Sugimolo, and H. Chiba, *Biochim. biophys. Acta*, **227**, 584 (1971).

[37] L. Siegel and J. S. Ellison, *Biochemistry*, **10**, 2856 (1971).

2. General Kinetic Principles

KNOWLEDGE of the kinetics and mechanisms of enzyme reactions owes much to the large amount of kinetic work that has been done on non-enzymic systems. The present chapter gives an account of a few aspects of general chemical kinetics that are particularly relevant to enzyme kinetics. No attempt is made to give a complete review of the field, for which the various modern texts should be consulted[1].

Analysis of kinetic data

A conventional kinetic study consists essentially of mixing the reactants and following the concentration of a reactant or product as a function of time; this is the *static* method. Various methods are available for analyzing the results of such a study so as to arrive at the kinetic law that is followed by the reaction, and the values of the kinetic constants. The method that is most useful for a reaction where there may be some kinetic complexity, and therefore for an enzyme-catalyzed reaction, is the *differential* method. The essential feature of this method is that tangents are drawn to the concentration–time curves, and that the rates obtained from the slopes of these lines are introduced into the kinetic equations in their differential forms.

The theory of the method is as follows. The instantaneous rate of a reaction of the nth order involving only one reacting substance is proportional to the nth power of its concentration,

$$-\frac{da}{dt} = ka_0^n. \tag{1}$$

Taking logarithms of both sides gives

$$\ln\left(-\frac{da}{dt}\right) = \ln k + n \ln a_0. \tag{2}$$

A plot of $\ln(-da/dt)$ against $\ln a_0$ will therefore give a straight line if the reaction is of simple order; the slope will be of the order n.

There are two different ways in which this procedure can be applied. One method is to carry out a single run—that is, to allow the reaction to proceed and determine a at various times. Tangents can then be drawn at different concentrations, as shown schematically in Fig. 2.1, and the slopes, da/dt, determined. A plot of $\log_{10}(-da/dt)$ is then made against $\log_{10} a$ as shown in Fig. 2.2 for a particular reaction. The value of the slope determined in this

[1] See, for example, C. N. Hinshelwood, *Kinetics of Chemical Change*, Clarendon Press, Oxford (1941); K. J. Laidler, *Chemical Kinetics*, McGraw-Hill Book Co., New York, 2nd edition (1965); A. A. Frost and R. G. Pearson, *Kinetics and Mechanism*, John Wiley and Sons, New York, 2nd edition (1961); K. J. Laidler, *Theories of Chemical Reaction Rates*, McGraw-Hill Book Co., New York, 1969.

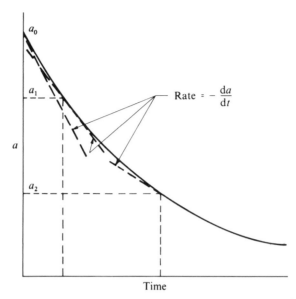

FIG. 2.1.   Schematic plots of concentration against time, illustrating the use of the differential method. Tangents are shown drawn at the initial concentration a_0 and at the concentrations a_1 and a_2.

way is known as the *order with respect to time*[1], since it has been determined in a run in which time is the variable. When this procedure is employed there may be interference from the products of reaction.

In the second method, slopes are only measured at the very beginning of the reaction, and the reaction is run at various initial concentrations. This

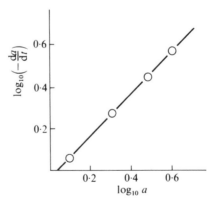

FIG. 2.2.   A plot of $\log_{10}(-da/dt)$ against $\log_{10} a$ for the conversion of N-chloracetanilide into p-chloracetanilide.

[1] M. Letort, Thesis, University of Paris (1937); *J. Chim. phys.*, **34**, 206 (1937); *Bull. Soc. chim. Fr.*, **9**, 1 (1942).

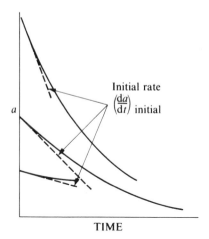

Initial rate
$\left(\dfrac{\mathrm{d}a}{\mathrm{d}t}\right)_{\text{initial}}$

a

TIME

FIG. 2.3. Schematic plots of concentration against time, for three initial concentrations. The differential method can be applied to the initial slopes.

type of procedure is represented schematically in Fig. 2.3. The order, determined by plotting $\log_{10}(-\mathrm{d}a/\mathrm{d}t)_{\text{initial}}$ against $\log_{10} a_0$, is now known as the *order with respect to concentration*, since it is now the concentration that is varied. This order has been called by Letort the *true order*, since it is concerned only with the reacting substances, and not with the products of reaction, which have no effect on the initial rates.

The differential method, which was introduced by van't Hoff[1], is a very valuable method, and is particularly useful for enzyme reactions, for which the products frequently interfere and render other methods unreliable. It is often very convenient in studying enzyme systems to cause the reaction to occur sufficiently slowly (by reducing the enzyme concentration, for example) so that a number of measurements can be made during the very early stages of reaction. In this way it is possible for the initial rates to be determined very accurately. A disadvantage of the differential method is that it is not always easy to obtain accurate values of the slopes of the rate curves.

There are many reactions that do not admit to the assignment of an order: this is frequently true of enzyme reactions. The differential method provides a convenient means of dealing with this situation; rates (slopes) are measured accurately in the initial stages of the reaction, and runs are carried out at a series of initial concentrations. In this way a plot of velocity against concentration can be prepared. The manner in which the constants are determined from results of this type, and their theoretical significance, is discussed in later chapters.

[1] J. H. van't Hoff, *Etudes de dynamique chimique*, F. Muller and Company, Amsterdam (1884), p. 87.

Flow methods[1]

The conventional method for following the course of a chemical reaction is the so-called *static* method; the reactants are introduced into a reaction vessel, and concentration changes in that vessel are followed as a function of time. Difficulties arise, however, if the reaction proceeds very rapidly, since the time required for mixing may be comparable with the half life† of the reaction. Alternative methods are then necessary. If the rates are not too great, flow methods are satisfactory; for exceedingly rapid reactions, however, relaxation methods must be employed (see following section).

For details of flow methods, and of the analysis of the results of flow experiments, the reader is referred to the textbooks of kinetics. Here it will simply be pointed out that two experimental techniques, the continuous-flow and the stopped-flow techniques, can be used. In the *continuous-flow* technique the solutions (e.g. of enzyme and substrate) are injected into a specially-designed mixing chamber in which mixing is rapid. They then pass along a tube in which at various positions a measurement can be made of the concentration of a reactant or product; this is usually done by spectro-photometry. From the rate of flow and other data one knows the time that it takes for the solution to travel from the mixing chamber to the observation point. By altering the rate of flow and the position of the observation point this time can be varied, and allows the kinetic equation to be established.

In the *stopped-flow* technique the solutions are again injected into a mixing chamber, after which they enter a reaction cell. After this cell has been filled with freshly mixed solution the flow is stopped and the concentration of a reactant or product is measured as a function of time, usually by means of a spectrophotometer and recorder. Strictly speaking this is not really a flow method since the kinetic study is made of the static system; the flow is simply utilized to bring about rapid mixing.

Relaxation methods

Because of the limitations to speeds of mixing, the continuous-flow and stopped-flow methods cannot be used for reactions having half lives of less than about 10^{-3} s. There are many examples of reactions that are much faster, the classical example being the neutralization of hydrogen and hydroxide ions, $H^+ + OH^- \rightarrow H_2O$. Many processes that are important in enzyme mechanisms, such as conformational changes, are also too rapid to be measurable by flow methods.

It is obviously necessary for the studies on these very rapid reactions to be made by techniques that do not require the initial mixing of the reactants.

† The half life of a reaction is the time taken for it to go half way to completion.

[1] For a review, with special reference to enzyme systems, see B. Chance and F. J. W. Roughton, in *Investigations of Rates and Mechanisms of Reactions* (ed. F. L. Friess, E. S. Lewis and A. Weiss-berger), Interscience, New York, 1961 and 1963.

The relaxation methods, devised by Eigen[1], avoid this difficulty. In these methods the reaction is allowed to go to a state of equilibrium, and is then rapidly disturbed in some way; its approach to a new equilibrium is then followed using high-speed techniques. In order for the methods to be used, the perturbation from equilibrium must be accomplished in a time that is much less than the relaxation time. A number of ways of doing this have been employed; these include the temperature-jump and pressure-jump methods, in which temperature and pressure changes are brought about very rapidly.

The theory of the method may be illustrated with reference to the very simple reaction scheme

$$A \underset{k_{-1}}{\overset{k_1}{\rightleftharpoons}} X,$$

the reactions in the two directions being of the first order. Suppose that such a reaction is allowed to come to equilibrium, and that its temperature is suddenly altered so that it is no longer at equilibrium. Let a_0 be the total concentration of A and X, and x that of X; that of A is $a_0 - x$. The kinetic equation is thus

$$\frac{dx}{dt} = k_1(a_0 - x) - k_{-1}x. \tag{3}$$

At equilibrium

$$k_1(a_0 - x_e) = k_{-1}x_e \tag{4}$$

where x_e is the concentration of X at equilibrium. The deviation Δx from equilibrium is $x - x_e$, and its time derivative is

$$\frac{d\Delta x}{dt} = \frac{dx}{dt} = k_1(a_0 - x) - k_{-1}x \tag{5}$$

$$= k_1 a_0 - (k_1 + k_{-1})x \tag{6}$$

$$= k_1 a_0 - (k_1 + k_{-1})(x_e + \Delta x). \tag{7}$$

From (4) and (7) it follows that

$$\frac{d\Delta x}{dt} = -(k_1 + k_{-1})\Delta x. \tag{8}$$

Integration of (8), subject to the boundary condition that $\Delta x = (\Delta x)_0$ when $t = 0$, leads to

$$\ln \frac{(\Delta x)_0}{\Delta x} = (k_1 + k_{-1})t. \tag{9}$$

---

[1] For reviews, see M. Eigen, *Discuss. Faraday Soc.*, **24**, 25 (1957); M. Eigen and K. Kustin, *ICSU Rev.*, **5**, 97 (1963); for recent reviews on the application of this technique to enzyme systems see G. G. Hammes, *Adv. Protein Chem.*, **23**, 1 (1968); *Acc. chem. Res.*, **1**, 321 (1968).

The relaxation time τ is defined as the time corresponding to

$$\frac{(\Delta x)_0}{\Delta x} = e. \tag{10}$$

or to

$$\ln \frac{(\Delta x)_0}{\Delta x} = 1. \tag{11}$$

From (9) and (11)

$$\tau = \frac{1}{k_1 + k_{-1}} \tag{12}$$

The relaxation time therefore allows $k_1 + k_{-1}$ to be calculated. Since k_1/k_{-1} can be determined (it is the equilibrium constant) the individual rate constants can be calculated.

Different relationships between relaxation times and rate constants are obtained for different reaction schemes. If a number of different reactions are involved in the one mechanism the system will exhibit several reaction times, which can be determined on the basis of mathematical analysis of the results.

Mechanisms of chemical reactions

The main purpose of a kinetic investigation is usually to determine something about the way in which the reaction occurs. So far we have considered the analysis of kinetic results with a view to discovering how the rate varies with the concentration of reactants (and possibly of other substances such as products) and what the various proportionality constants are. From this information alone certain conclusions about mechanisms can be drawn. Additional information may then be obtained by finding how the constants vary with such factors as temperature, pressure, and pH; some of these matters are considered later.

One question of very considerable importance in connection with reaction mechanisms is whether the reactant molecules are converted into the final products in a single step, or whether a number of independent steps are involved. In the former case the reaction is frequently spoken of as an *elementary* one, while in the latter case the reaction is said to be *complex*. Evidently a complex reaction can be broken down into a number of elementary steps, and such an analysis is an important part of the kinetic study of a complex reaction.

The determination of whether a reaction is simple or complex is sometimes a matter of difficulty. If the rate varies with the reactant concentrations according to a very simple law, such as that of a first-order or a second-order reaction, the obvious conclusion to draw is that it occurs in a single step, the reaction simply involving the participation of one, or two, reactant molecules. This may indeed be the case, but there are a number of factors which make it unwise to draw this simple conclusion without a considerable amount of additional evidence. In the first place, even a complex reaction may obey a simple kinetic law over a certain range of conditions; it is therefore necessary to ensure that the kinetics are simple over a wide range of conditions before concluding that a reaction is elementary. Another circumstance which may lead to difficulty is that it sometimes happens that a complex kinetic scheme will give rise to a very simple overall kinetic law.

The history of kinetics is full of cases of reactions that obey simple laws and were originally thought to be elementary but which turned out, on more detailed study, to be complex. One way in which a reaction can be shown to be complex is by the detection, by various methods, of reaction intermediates. If these are unstable substances, such as free radicals or enzyme–substrate complexes, special techniques, such as mass spectrometry or ultraviolet spectroscopy, may have to be used in their detection. Otherwise, if they are stable and present in appreciable concentrations, ordinary methods of chemical analysis may be adequate.

When the kinetics are simple, and all tests fail to reveal the presence of reaction intermediates, it is reasonable to draw the tentative conclusion that the reaction is an elementary one. The question then to be settled is how many molecules enter into reaction. With certain exceptions it is correct to assume that the order of the reaction indicates the number of molecules, i.e. that the order and the molecularity of an elementary reaction are the same. If, for example, an elementary reaction is of the first order with respect to a reactant A, and of the first order with respect to another substance B, the conclusion is that one molecule of A and one of B enter into the reaction.

Under certain circumstances, however, this procedure may lead to incorrect conclusions. In some cases, for example, one reactant may be present in large excess, so that its concentration does not change as the reaction proceeds; moreover its concentration may be the same in different runs. If this is so the kinetic studies will not reveal any dependence of the rate on the concentration of this substance, which would therefore not be considered to be a reactant. This is frequently the case in solution reactions, where the solvent may act as a reactant; in hydrolyses in water, for example, a water molecule must undergo reaction with the substrate. Unless special procedures are used, the kinetics will not reveal the participation of the solvent. If the solvent appears in the stoichiometric equation, however, it must participate in the reaction.

Another circumstance in which the kinetics may not reveal that a substance enters into reaction is when a catalyst is involved. Since a catalyst by definition is not used up during the reaction its concentration remains constant during a run; its concentration therefore need not enter explicitly into the rate equation. The participation of the catalyst in the reaction may, however, be revealed by measuring the rate at a variety of catalyst concentrations;

generally a linear dependence will be found. This situation arises not only with obvious catalysts such as enzymes, but also with substances that may be present fortuitously in the reaction system; an enzyme may, for example, contain an impurity that has a marked effect on the reaction, but the concentration of which remains constant during a run. The participation of such substances frequently eludes detection, and it is particularly necessary to investigate such factors as this in enzyme systems.

The Arrhenius law

Extremely valuable information about reaction mechanisms is provided by studies of the effect of temperature on kinetic constants. The law that applies almost universally to this temperature effect is the Arrhenius law,

$$k = A \, e^{-E/RT} \tag{13}$$

where A and E are constants, R is the gas constant and T is the temperature in kelvins. The significance of E, which is usually expressed in kilocalories per mole or in joules (J) per mole, is that it is the height of the energy barrier which the system must surmount in order to pass from reactants to products. This is illustrated in Fig. 2.4. The system existing at the top of the barrier is known as an *activated complex* or *transition state*.

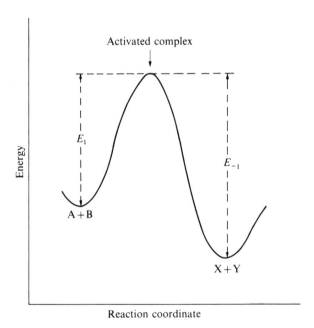

Fig. 2.4.   Schematic representation of the energy changes in a chemical reaction.

Equation (13) can be expressed as

$$\ln k = \ln A - \frac{E}{RT} \tag{14}$$

or as

$$\log_{10} k = \log_{10} A - \frac{E}{2 \cdot 303 RT}. \tag{15}$$

The magnitude of the activation energy E is therefore obtained from the slope of the line when $\log_{10} k$ is plotted against $1/T$. The value of A can be calculated from the intercept.

Potential-energy surfaces

Considerable insight into the way in which elementary processes occur is provided by the construction of potential-energy surfaces. Consider a general reaction

$$A + B\text{–}C \to A\text{–}B + C$$

in which A, B and C may be atoms or groups of atoms. In the overall process there is a breaking of a bond and the making of a bond. Processes of this type are known as *abstraction* or *single-displacement* reactions; A abstracts B from C, and there is a displacement of C by A. It turns out that a great many reactions, including some enzyme reactions, occur in this way. For energetic reasons it is most likely for A to approach B–C along the line of centres, forming a linear A ... B ... C complex, and such linear complexes will be considered in the following discussion.

In the initial state of the reaction the A–B distance is large, while in the final state B–C is large. All intermediate states of the linear complex are represented by specifying the distances A–B and B–C. We can then plot the potential energy of the system against these A–B and B–C distances. The result, illustrated in Fig. 2.5 for a typical type of reaction, is known as a potential-energy surface.

On the left-hand face of the diagram appears a dissociation-energy curve for the B–C molecule, the group A being sufficiently far away that it does not influence the shape of this curve. On the right-hand face, where C is sufficiently far away, is the dissociation energy curve for the molecule A–B. The initial and final states correspond to points towards the bottoms of these two curves, as represented in the diagram. Potential-energy surfaces are frequently represented as contour diagrams; an example is to be found in Fig. 8.3, p. 238, where it is applied to a discussion of kinetic-isotope effects.

The path by which the system passes from the initial to the final state is determined by the shape of the potential-energy surface. As seen in Fig. 2.5,

---

[1] For a more detailed treatment see K. J. Laidler, *Theories of Chemical Reaction Rates*, McGraw-Hill Book Co., New York, 1969.

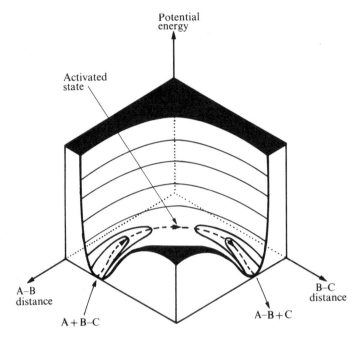

FIG. 2.5.   A potential-energy surface for the system A + B–C → A–B + C. The potential energy is plotted against the A–B and B–C distances.

there are two valleys, meeting at a *col* or *saddle point* in the interior of the diagram. The system at the col is referred to as the *activated complex* or *transition state*. The path of least resistance, which the reaction will follow, corresponds to motion up the left-hand valley, over the col, and into the other valley.

The rate of the reaction is determined primarily by the height of the barrier at the col, with reference to the initial energy. If the barrier is high very many collisions between A and B–C will have insufficient energy to surmount the barrier; a low barrier will be surmounted by a larger proportion of colliding systems. The col represents a point of no return; once a system has reached the col it almost inevitably passes down the second valley and gives rise to products.

A complete understanding of the mechanism of an elementary process involves knowing the potential-energy surface for the system, and calculating the rate of passage of systems over the barrier. Unfortunately, it is still quite impossible to calculate reliable potential-energy surfaces except for exceedingly simple reactions, such as H + H_2; it is therefore impossible to estimate activation energies. In spite of this, however, it is extremely useful to think of reactions in terms of potential-energy surfaces, which provide much valuable insight into mechanisms.

Activated-complex theory

It has been seen that rate constants for chemical reactions are conveniently interpreted in terms of the Arrhenius equation

$$k = A\,e^{-E/RT}. \tag{16}$$

The quantity E, the activation energy, is interpreted in terms of potential-energy surfaces, being the height of the barrier at the col. There remains the problem of how to interpret, and if possible calculate, the value of the frequency factor A.

The most satisfactory treatment of the frequency factor has been given by Eyring[1], and is now known either as *transition-state theory* or as *activated-complex theory*; the latter expression will be employed in this book. Details cannot be given here[2], but a general account will be presented of the main concepts that underlie the theory. The basic idea is that rates are controlled exclusively by the situation at the activated state; once the system has reached this state it is bound to pass into reactants. The properties of the activated complexes therefore play the paramount role in the theory.

There are two formulations of activated-complex theory; both are of great value in interpreting reaction rates, including those of enzyme-catalyzed reactions. The first is a statistical-mechanical formulation, the second a thermodynamical one. The former is particularly useful for interpreting kinetic-isotope effects (see Chapter 8); the second provides a convenient way of treating the general aspects of solution reactions and of interpreting them in terms of environmental effects (see especially Chapter 7).

In the statistical-mechanical formulation the rate expression for a reaction $A + B$ is

$$k = \frac{\mathbf{k}T}{h}\frac{Q_{\neq}}{Q_A Q_B}e^{-E_0/RT}. \tag{17}$$

Here $\mathbf{k}$ is the Boltzmann constant, h Planck's constant and T the temperature in kelvins. The quantity E_0 is the difference between the zero-point levels of the initial and activated states. The Q's are partition functions for A, B and the activated complex, which is denoted by the symbol $\neq$.

The partition functions are the products of factors for the various degrees of freedom in the molecule. Thus for the three degrees of translational freedom possessed by a gas molecule the partition function is

$$q_t = \frac{(2\pi m\mathbf{k}T)^{\frac{3}{2}}}{h^3} \tag{18}$$

[1] H. Eyring, *J. chem. Phys.*, 3, 107 (1935); W. F. K. Wynne-Jones and H. Eyring, *J. chem. Phys.*, 3, 492 (1935).
[2] For detailed treatments, see S. Glasstone, K. J. Laidler, and H. Eyring, *The Theory of Rate Processes*, McGraw-Hill Book Co., New York (1941); K. J. Laidler, *Theories of Chemical Reaction Rates*, McGraw-Hill Book Co., New York, 1969; for an elementary discussion see K. J. Laidler, *Chemical Kinetics*, McGraw-Hill Book Co., New York, 2nd ed. (1965), chap. 3.

where m is the mass of the molecule. For a linear molecule of moment of inertia I there is a contribution for two degrees of rotational freedom of

$$q_{r(z)} = \frac{8\pi^2 I \mathbf{k} T}{h^2}. \tag{19}$$

A more complicated expression applies to a non-linear molecule. For each degree of vibrational freedom there is a contribution

$$q_v = \frac{1}{1 - e^{-h\nu/\mathbf{k}T}} \tag{20}$$

where ν is the frequency of the vibration. The number of vibrational degrees of freedom is determined from the fact that the total number is $3 \times$ (no. of atoms in the molecule), and that there are three degrees of translational freedom and either two or three degrees of rotational freedom (according to whether the molecule is linear or non-linear). For molecules in solution the translations and rotations must be replaced by more restricted types of motion; a satisfactory formulation is a matter of some difficulty.

The thermodynamical formulation of the rate constant leads to the expression

$$k = \frac{\mathbf{k}T}{h} e^{\Delta S^{\neq}/R} e^{-\Delta H^{\neq}/RT} \tag{21}$$

where $\Delta S^{\neq}$ is the *entropy of activation*, i.e. the entropy increase when the activated complex is formed, and $\Delta H^{\neq}$ is the *enthalpy of activation*. The enthalpy of activation is not quite the same as the experimental activation energy E, as determined from an Arrhenius plot; in terms of E the rate equation is, for a reaction in solution,

$$k = e\frac{\mathbf{k}T}{h} e^{\Delta S^{\neq}/R} e^{-E/RT}. \tag{22}$$

The frequency factor A is thus

$$A = e\frac{\mathbf{k}T}{h} e^{\Delta S^{\neq}/R} \tag{23}$$

so that the entropy of activation $\Delta S^{\neq}$ can be calculated from the experimental value of A as obtained from an Arrhenius plot.

Complex reactions

Many reactions are found to occur not by a simple rearrangement of the atoms on a single collision, but in two or more well-defined steps. In such cases an important part of the kinetic study is the elucidation of the nature of these steps and of how the overall reaction is made up of these individual steps.

There are various indications of this type of complexity. An obvious piece of evidence is when the kinetic law is more complex than would be consistent with the occurrence of the reaction in a single step. For example, if an enzyme reaction involving a single substrate occurred by a simple bimolecular reaction between enzyme and substrate the kinetics would be first order with respect to substrate. Usually, however, the rates are not directly proportional to the substrate concentration, and it can therefore be concluded that the reaction occurs in more than one stage.

Another indication of complexity is provided when intermediates can be detected, by chemical or other means, during the course of reaction. When this is so a kinetic scheme must be developed which will account for the existence of these intermediates. Sometimes these intermediates are relatively stable substances, while in other cases they are labile substances such as free radicals. Enzyme–substrate complexes are of the latter class in that they cannot be isolated and preserved, and can only be detected by special methods such as spectroscopic ones. Free radicals can sometimes be observed by spectroscopic methods, and evidence for their existence may be obtained by causing them to undergo certain specific reactions which less active substances cannot bring about.

When the nature of the reaction intermediates has been determined by methods such as those outlined above, the next step is to devise a reaction scheme that will involve these intermediates and will account for the kinetic features of the reaction. If such a scheme fits the data satisfactorily it can be assumed tentatively that the mechanism is the correct one. It should be emphasized, however, that additional kinetic work frequently leads to the overthrow of schemes that had previously been supposed to be firmly established.

In order to obtain the overall kinetic equation for a complex reaction scheme it is necessary to write down the differential rate equations for the reactants, products, and intermediates, to solve these for the overall rate, and to integrate the resulting equation to obtain an expression for the concentrations of reactants and products at various times. The methods for doing this have been described in detail elsewhere[1] and will not be discussed here. Table 2.1 gives solutions for some useful cases.

In some cases there are no mathematical methods available for the solution of the differential equations. When this is so numerical solutions may be obtained using a computer. It is sometimes possible to avoid this procedure by using an approximate method, known as the *steady-state* method, to obtain the required equation.

The steady-state treatment is based on the hypothesis that the concentrations of certain reaction intermediates, such as free radicals and enzyme–

[1] See, for example, A. A. Frost and R. G. Pearson, *Kinetics and Mechanism*, John Wiley and Sons, New York, 2nd ed. (1953), chap. 8; Z. G. Szabó, in *Comprehensive Chemical Kinetics* (eds. C. H. Bamford and C. H. F. Tipper), Elsevier Publishing Co., Amsterdam (1969), vol. 2.

TABLE 2.1

Kinetic laws for some complex reaction systems

$A \xrightarrow{k_1} X + \dots$ $A \xrightarrow{k_2} Y + \dots$ $A \xrightarrow{k_3} Z + \dots$	$[X] = [X]_0 + \dfrac{k_1[A]_0}{k_1 + k_2 + k_3}(1 - e^{-(k_1 + k_2 + k_3)t})$
$A \xrightarrow{k_1} X + \dots$ $B \xrightarrow{k_2} X + \dots$	$[X] = [A]_0 + [B]_0 - [A]_0\, e^{-k_1 t} - [B]_0\, e^{-k_2 t}$
$A \xrightarrow{k_1} B \xrightarrow{k_2} C$	$[A] = [A]_0\, e^{-k_1 t}$ $[B] = [B]_0 + \dfrac{k_1[A]_0}{k_2 - k_1}(e^{-k_1 t} - e^{-k_2 t})$ $[C] = [C]_0 + [A]_0\left[1 - \dfrac{1}{k_2 - k_1}(k_2\, e^{-k_1 t} - k_1\, e^{-k_2 t})\right]$

The subscript 0 indicates the concentration to be that at $t = 0$

substrate complexes, do not change rapidly during the course of reaction[1]. If [J] is the concentration of such an intermediate, for example, the approximation is made that

$$\frac{d[J]}{dt} = 0. \qquad (24)$$

Such relationships lead to a considerable simplification in the differential equations, and allow a solution to be obtained readily.

Suppose, for example, that the reaction scheme under consideration is

(1) $$A + B \underset{k_{-1}}{\overset{k_1}{\rightleftharpoons}} J,$$

(2) $$J \xrightarrow{k_2} X.$$

The differential rate equations that apply to this set of reactions are:

$$-\frac{d[A]}{dt} = -\frac{d[B]}{dt} = k_1[A] \cdot [B] - k_{-1}[J], \qquad (25)$$

$$\frac{d[J]}{dt} = k_1[A] \cdot [B] - k_{-1}[J] - k_2[J], \qquad (26)$$

$$\frac{d[X]}{dt} = k_2[J]. \qquad (27)$$

In order to treat this problem exactly it would be necessary to eliminate [J] and to solve the resulting differential equation to get [X] as a function of t.

[1] For a brief discussion of the exact conditions under which this is true, with special reference to enzyme systems, see p. 73.

Unfortunately, however, in spite of the simplicity of the kinetic scheme, this cannot be done by known mathematical procedures.

The steady-state treatment involves using eqn (24), so that, from (26),

$$k_1[A].[B] - k_{-1}[J] - k_2[J] = 0. \tag{28}$$

The concentration of [J] is therefore given by

$$[J] = \frac{k_1[A].[B]}{k_{-1} + k_2} \tag{29}$$

and insertion of this in eqn (27) gives

$$\frac{d[X]}{dt} = \frac{k_1 k_2 [A].[B]}{k_{-1} + k_2}. \tag{30}$$

If the amounts of A and B are initially a_0 and b_0, and the amount of X at time t is x this equation may be written as

$$\frac{dx}{dt} = \frac{k_1 k_2 (a_0 - x)(b_0 - x)}{k_{-1} + k_2} \tag{31}$$

and this is readily soluble by simple methods. Several examples of the application of the steady-state method to enzyme systems will be given later.

Reactions in solution

Since enzyme reactions are special types of reactions in solution, some consideration will be given in the present section to the factors that determine the magnitudes of the rates of reactions in solution. This topic can be divided into two parts, one of which is concerned with the activation energy, the other with the entropy of activation $\Delta S^{\neq}$. As mentioned earlier, very little can be said about the magnitudes of activation energies, and this is especially true of reactions in solution. The best that one can do is to correlate energies of activation for certain homologous types of reactions, a matter that is considered later in connection with the effects of substituents on rates.

With regard to entropies of activation (or frequency factors) rather more can be said. No exact treatment is possible at the present owing to the inherent complexity of the problem of reactions in solution. A number of important generalizations are, however, well recognized, and it is now possible to make fairly reliable estimates of the entropy of activation of a solution reaction, and hence of the frequency factor, in terms of the various factors involved in the reaction. Conversely, it is possible, from the known values of entropies of activation, to draw some inferences about the nature of a reaction (such as its ionic character) in cases where this is not already known. Such inferences are particularly valuable with enzyme reactions, where one very frequently does not know the nature of the active centre of the enzyme and where information on this point is very valuable.

Reactions in solution are of a variety of types. There are certain reactions, involving non-polar molecules, in which it seems evident that the solvent plays a relatively subsidiary role. Many such reactions, in fact, can be made to occur in the gas phase as well as in solution, and when they do so they occur with very much the same rates as in solution. Evidently the interactions between reactant molecules and solvent molecules are not of great importance for such reactions, and the solvent can be regarded as merely filling up space between the reactant molecules. Theory[1], and experiments with mechanical models[2], both indicate that in such cases the number of effective collisions between reactant molecules is hardly affected by the presence of solvent. The entropy of activation is therefore close to zero, and the frequency factor may be described as a 'normal' one; its magnitude for a bimolecular reaction is of the order of 10^9 to 10^{11} $M^{-1} s^{-1}$.

More often, however, and especially in enzyme systems, the solvent does not act as an inert space-filler, but is involved in a significant way in the reaction itself. This is particularly the case when the reactants are ions or polar molecules and when the solvent also has some polarity. In such cases, changes in polarity occurring during the course of reaction will cause the solvent molecules to reorient themselves, and hence will bring about 'abnormalities' in the entropies of activation. The way in which this occurs will first be discussed for reactions between ions and dipoles, and between dipoles and dipoles.

Reactions between ions

The reason why reactions between ions are easier to treat theoretically than are the other reactions is that the thermodynamics of ions in solution have been reasonably well worked out, so that estimates of entropy changes during the formation of the activated complex can fairly easily be made. The Debye–Hückel theory of the activity coefficients of electrolytic solutions has been particularly valuable in this connection. The relationship between the rate constant of a reaction in solution and that of a gas-phase reaction must first be developed. This type of theory was first put forward by Brønsted[3] and by Bjerrum[4], it will here be presented in more modern terms.

The treatment is based on the assumption, which can be shown to be well justified, that the activated complexes are in equilibrium with the reactant molecules, and that the rate of the reaction is proportional to the concentra-

[1] P. Debye and H. Mecke, *Z. Phys.*, **31**, 797 (1931); **32**, 593 (1932); cf. W. H. Keesom and J. de Smedty, *Proc. Amsterdam Acad.*, **25**, 118 (1927); F. Zernickel and J. A. Prins, *Z. Phys.*, **41**, 184 (1927); E. Rabinowitch, *Trans. Faraday Soc.*, **33**, 1225 (1937); E. Rabinowitch and W. C. Wood, *Trans. Faraday Soc.*, **32**, 1381 (1936).
[2] E. Rabinowitch and W. C. Wood, *loc. cit.*
[3] J. N. Brønsted, *Z. phys. Chem.*, **102**, 169 (1922).
[4] N. Bjerrum, *Z. phys. Chem.*, **108**, 82 (1924).

tion of activated complexes. A reaction between A and B may be represented as

$$A + B \rightleftharpoons X^{\neq} \rightarrow \text{products},$$

where $X^{\neq}$ is the activated complex. The equilibrium between A, B, and $X^{\neq}$ is expressed by

$$K^{\neq} = \frac{[X^{\neq}]\alpha_{\neq}}{[A].[B]\alpha_A\alpha_B}, \tag{32}$$

where $K^{\neq}$ is the equilibrium constant and the α's are the activity coefficients; these would be unity for the case of the perfect gas, but otherwise, and particularly in solution, they must be included. The rate, proportional to $[X^{\neq}]$, is given by

$$v = k[X^{\neq}] \tag{33}$$

$$= k_g[A].[B]\frac{\alpha_A\alpha_B}{a_{\neq}} \tag{34}$$

where k_g, equal to $kK^{\neq}$, is the proportionality constant. Since the rate constant k is related to the rate by $v = k[A].[B]$ it follows that

$$k = k_g\frac{\alpha_A\alpha_B}{\alpha_{\neq}}. \tag{35}$$

In the case of the ideal gas the α's are all unity so that k_g is simply the rate constant when the reaction occurs in the absence of solvent and under ideal conditions. Equation (35) therefore relates the rate constant under any conditions (e.g. in solution) to that in the ideal gaseous state. In order to determine the specific effects of the solvent it is therefore necessary to have some theory about the magnitudes of the activity coefficients; for ionic reactions this problem can be treated quite satisfactorily.

The problem is conveniently treated in two parts, since the properties of an ion are affected by its environment in two ways. In the first place, the ionic atmosphere surrounding each ion affects its activity coefficient and hence affects the rate. Secondly, quite apart from the atmosphere effect, the solvent itself, by virtue of its dielectric constant, affects the activity coefficients and therefore the rate. It will thus be convenient to treat the problem by first considering how the rate constant k of a reaction in solution is related to that, k_0, of the reaction occurring in solution but under the conditions of infinite dilution, where there is no ionic atmosphere. The rate constant k_0 at infinite dilution will then be related to the rate constant k_g for the reaction in the gas phase.

This may be done by splitting the activity coefficient α into the product of two factors β and f,

$$\alpha = \beta f. \tag{36}$$

Of these activity coefficients, f relates the actual solution to the infinitely dilute solution, and β the infinitely dilute solution to the gas phase. Introduction of these activity coefficients into the rate equation (35) gives rise to

$$k = k_g \frac{\beta_A \beta_B}{\beta_{\neq}} \frac{f_A f_B}{f_{\neq}}. \tag{37}$$

In the infinitely dilute solution the f's are unity; it follows that the rate constant in the infinitely dilute solution is given by

$$k_0 = k_g \frac{\beta_A \beta_B}{\beta_{\neq}}. \tag{38}$$

The constants k and k_0 are therefore related by

$$k = k_0 \frac{f_A f_B}{f_{\neq}} \tag{39}$$

so that a theory of the f's will give an explicit relationship between k and k_0.

The Debye–Hückel theory provides such a theory of the f's. According to it the activity coefficient f of an ion of valence z is given by[†]

$$\log_{10} f = -z^2 Q u^{\frac{1}{2}} \tag{40}$$

where u is the ionic strength defined by

$$u = \tfrac{1}{2} \sum c z^2 \tag{41}$$

(c is the concentration of each ion, the summation being taken over all of the ions). Q is given by

$$Q = \frac{N^2 e^3 (2\pi)^{\frac{1}{2}}}{2.303(\varepsilon_r R T)^{\frac{3}{2}}(1000)^{\frac{1}{2}}} \tag{42}$$

where N is Avogadro's number, e is the charge on the electron, ε_r is the dielectric constant, R is the gas constant, and T is the temperature. Introduction of eqn (40) into eqn (39) for each of the ionic species gives rise to

$$\log_{10} k = \log_{10} k_0 + \log_{10} f_A + \log_{10} f_B - \log_{10} f_{\neq}, \tag{43}$$

$$= \log_{10} k_0 - Q(z_A^2 + z_B^2 - z_{\neq}^2) u^{\frac{1}{2}}. \tag{44}$$

However, $z_{\neq}$ is necessarily equal to $z_A + z_B$ so that

$$\log_{10} k = \log_{10} k_0 - Q\{z_A^2 + z_B^2 - (z_A + z_B)^2\} u^{\frac{1}{2}} \tag{45}$$

$$= \log_{10} k_0 + 2Q z_A z_B u^{\frac{1}{2}}. \tag{46}$$

[†] z is the integral charge on the ion; it is positive in the case of a positive ion and negative for a negative ion (e.g. for Na^+, $z = 1$; for SO_4^{--}, $z = -2$; for Al^{+++}, $z = 3$).

The value of Q is approximately 0·51 for aqueous solutions at 25°C; eqn (46) may therefore be written as

$$\log_{10} k = \log_{10} k_0 + 1{\cdot}02 z_A z_B u^{\frac{1}{2}}. \tag{47}$$

This equation has been subjected to a considerable amount of experimental test, and has been found to be very satisfactory. It can conveniently be applied to the data by plotting $\log k$ against $u^{\frac{1}{2}}$, when a straight line should be obtained, of slope $1{\cdot}02 z_A z_B$. This has been found to be so for a number of different types of reactions, with $z_A z_B$ varying from 4 to -2.

The problem of relating the rate at infinite dilution to that in the gas phase, using eqn (38), has been treated in two different ways, which differ according to the model assumed for the activated complex. Scatchard[1] treated the problem by considering the reactant molecules A and B to be spherical ions, and supposing that in the activated complex these two spheres are touching one another, forming a double sphere. Application of the Debye–Hückel theory to the equilibrium between A + B and $X^{\neq}$, for a dielectric constant of ε_r and then for $\varepsilon_r = 1$ (ideal gas), gives the relationship:

$$\ln \frac{\beta_A \beta_B}{\beta_{\neq}} = \frac{e^2 z_A z_B}{k T r_{\neq}} \left(1 - \frac{1}{\varepsilon_r} \right), \tag{48}$$

where $r_{\neq}$ is the distance between the centres of the spheres in the activated complex. Equation (38) can be written as

$$\ln k_0 = \ln k_g + \ln \frac{\beta_A \beta_B}{\beta_{\neq}}, \tag{49}$$

whence

$$\ln k_0 = \ln k_g + \frac{e^2 z_A z_B}{k T r_{\neq}} \left(1 - \frac{1}{\varepsilon_r} \right). \tag{50}$$

This treatment would predict a linear relationship between the logarithm of the rate constant and the reciprocal of the dielectric constant. Such behaviour has been observed in several instances, and the slopes of the lines have been shown to be consistent with a reasonable value for $r_{\neq}$, the distance between the centres.

According to the second type of treatment applied to this problem[2] the activated complex is assumed to be single sphere rather than a double one, the two spherical ions becoming fused together during the reaction. The activity coefficient β_i for a spherical ion of radius r_i is given by

$$\ln \beta_i = \frac{z_i^2 e^2}{2 k T r_i} \left(\frac{1}{\varepsilon_r} - 1 \right) \tag{51}$$

[1] G. Scatchard, *Chem. Revs.*, **10**, 229 (1932); *J. chem. Phys.*, **7**, 657 (1939).
[2] K. J. Laidler and H. Eyring, *Ann. N.Y. Acad. Sci.*, **39**, 303 (1940).

and substitution of the appropriate expressions for $\ln \beta_A$, $\ln \beta_B$, and $\ln \beta_{X*}$ into eqn (49) gives

$$\ln k_0 = \ln k_g + \frac{e^2}{2kT}\left[\frac{(z_A + z_B)^2}{r_{\neq}} - \frac{z_A^2}{r_A} - \frac{z_B^2}{r_B}\right]\left(1 - \frac{1}{\varepsilon_r}\right). \tag{52}$$

Here r_A and r_B are the radii of the spherical reactant molecules, and $r_{\neq}$ is the radius of the activated complex. This equation reduces to Scatchard's equation (50) if one puts $r_{\neq} = r_A = r_B$.

The procedure for testing equations such as (50) and (52) with respect to the experimental data consists of plotting $\ln k_0$ or $\log_{10} k_0$ against $1/\varepsilon_r$ and seeing if a straight line is obtained. It may then be ascertained whether by taking reasonable values of r (in eqn (50)) and of r_A, r_B, and $r_{\neq}$ (in eqn (52)) it is possible to account for the slope of the line. When applied to some extensive data for the reaction between bromacetate and thiosulphate ions both equations were found to give reasonable agreement with the results. The truth probably lies somewhere between the two models for the complex.

A matter of great importance with reactions of this type is the electrostatic contribution to the entropy of activation. When two ions come together and form a complex there will be some re-orientation of the neighbouring water molecules, and this will involve some entropy change. The theory already outlined for ionic reactions enables an expression to be obtained for the entropy of activation, but some error arises from the fact that the solvent is treated as a continuous dielectric instead of an assembly of highly polar water molecules.

For simplicity it will be assumed that the data have been extrapolated down to zero ionic strength, and that eqn (50) is the one that applies. The rate constant of a reaction may be written as

$$k_0 = \frac{kT}{h}\, e^{\Delta S_0^{\neq}/R}\, e^{-\Delta H_0^{\neq}/RT} \tag{53}$$

$$= \frac{kT}{h}\, e^{-\Delta G_0^{\neq}/RT} \tag{54}$$

where $\Delta G_0^{\neq}(= \Delta H_0^{\neq} - T\,\Delta S_0^{\neq})$ is the Gibbs free energy of activation for the reaction at zero ionic strength. Comparison of eqn (50) and (54), making use of the fact that $R = Nk$, shows that the free energy of activation is given by

$$\Delta G_0^{\neq} = RT\left(\ln\frac{kT}{h} - \ln k_g\right) - \frac{N^2 z_A z_B}{r_{\neq}}\left(1 - \frac{1}{\varepsilon_r}\right). \tag{55}$$

The last term in this expression is the contribution to the free energy of activation arising from the electrostatic interactions,

$$\Delta G_{es}^{\neq} = -\frac{N^2 z_A z_B}{r_{\neq}}\left(1 - \frac{1}{\varepsilon}\right). \tag{56}$$

The entropy of activation is obtained from the free energy of activation by making use of the thermodynamical relationship

$$\Delta S^{\neq} = -\left(\frac{\partial \Delta G^{\neq}}{\partial T}\right)_P, \tag{57}$$

so that the electrostatic contribution is given by

$$\Delta S_{es}^{\neq} = -\left(\frac{\partial \Delta G_{es}^{\neq}}{\partial T}\right)_P. \tag{58}$$

Application of this relationship to eqn (55) gives

$$\Delta S_{es}^{\neq} = \frac{Ne^2 z_A z_B}{r_{\neq} \varepsilon_r^2}\left(\frac{\partial \varepsilon_r}{\partial T}\right)_P. \tag{59}$$

In an aqueous solution ε_r is about 80 and $\partial \ln \varepsilon_r / \partial T$ is about -0.0046, and taking $r_{\neq}$ as 2×10^{-8} cm it is found that

$$\Delta S_{es}^{\neq} \approx -10 z_A z_B. \tag{60}$$

For a reaction between ions having unit charges of the same sign the entropy of activation should be lower by about 10 units than the value if electrostatic effects were unimportant; if the charges are of opposite sign it should be higher by about 10 entropy units. As a general rule the entropy of activation should decrease by 10 units for each unit of $z_A z_B$. Moreover, since the frequency factor A is proportional to $e^{\Delta S^{\neq}/R}$, it follows that its value should decrease by a factor of about $e^{10/R}$, i.e. about a hundredfold, for each unit of $z_A z_B$.

TABLE 2.2
Entropy factors for ionic reactions

Reaction	$z_A z_B$	Frequency factor
$Co(NH_3)_5Br^{++} + OH^-$	-2	5×10^{17}
$R_4N^+ + Br^-$	-1	3×10^{16}
$CH_2BrCO_2H + S_2O_3^{--}$	0	1×10^{14}
$CH_2ClCO_2^- + OH^-$	1	6×10^{10}
$CH_2BrCO_2^- + S_2O_3^{--}$	2	1×10^9

This rough rule is in fact found to be obeyed approximately, as is shown by the figures given in Table 2.2. Deviations may, of course, arise from many causes, of which may be mentioned in particular: (1) the inapplicability of the double-sphere model for the complex, (2) the inadequacy of the assumption that the solvent is a continuous dielectric, (3) value of $r_{\neq}$ differing markedly from 2×10^{-8} cm, and (4) salt effects (if the data are not extrapolated to zero ionic strength).

Applications of the above type of treatment to enzyme reactions will be discussed later (p. 216).

Reactions between an ion and a neutral molecule

The theoretical treatment of the influence of environment on the kinetics of reactions between an ion and a neutral molecule cannot be carried out as satisfactorily as with two ions. If the neutral molecule is a completely non-polar substance it may in some cases be satisfactory to employ the equations developed in the previous section, with z_B set equal to zero. If the neutral molecule has a dipole moment, on the other hand, it will be necessary to take explicit account of its interaction with the ion and the surrounding solvent.

The simplest case, in which there is no dipole moment in either of the reactants or the complex, will be considered first. If in eqn (47) z_B is set equal to zero there is no ionic strength effect. With regard to the effect of changing the dielectric constant, eqn (50) is no longer applicable to the case of $z_B = 0$. Equation (52), on the other hand, reduces to

$$\ln k_0 = \ln k_g + \frac{e^2 z_A^2}{2kT}\left(\frac{1}{r_{\neq}} - \frac{1}{r_A}\right)\left(1 - \frac{1}{\varepsilon_r}\right) \tag{61}$$

and a plot of $\ln k_0$ against $1/\varepsilon_r$ should therefore give a straight line. Such straight line plots have been found for a number of reactions, and the slopes are consistent with reasonable values of r_A and $r_{\neq}$.

This simplified theory gives no ionic strength effect for reactions of this type; the dielectric constant effect is, moreover, small since r_A and $r_{\neq}$ are usually not very different. If dipole interactions are involved, however, there may be additional effects comparable with, or even more important than, those predicted by the simplified treatment. Such effects have been discussed along the following lines.

Equation (47) has been seen to be based on the Debye–Hückel expression (40) for the activity coefficient of an ion. This expression can be improved by the addition of a term which is linear in the ionic strength, and which was introduced by Hückel[1]:

$$\log f_i = -z_i^2 Q u^{\frac{1}{2}} + b_i u. \tag{62}$$

This expression may, for a reaction between an ion and a neutral molecule, be applied to the ion and the activated complex, which has the same charge as the ion;

$$\log f_A = -z_A^2 Q u^{\frac{1}{2}} + b_A u, \tag{63}$$

$$\log f_{\neq} = -z_A^2 Q u^{\frac{1}{2}} + b_{\neq} u. \tag{64}$$

[1] E. Hückel, *Z. Phys.*, **26**, 93 (1925).

For a neutral molecule the activity coefficient is given by the approximate expression of Debye and McAulay[1],

$$\log f_{\mathrm{B}} = b_{\mathrm{B}} u, \tag{65}$$

which may be regarded as arising from eqn (62) when $z_i = 0$. Introduction of these expressions into eqn (43) gives

$$\log k = \log k_0 + (b_{\mathrm{A}} + b_{\mathrm{B}} - b_{\neq})u. \tag{66}$$

The square-root term has disappeared, since it occurs in both equations (63) and (64), and the theory therefore predicts a linear dependence of $\log k$ on the ionic strength. Such relationships have been observed, but unfortunately there exists no adequate theory of the magnitudes of the b terms.

The problem of the influence of dielectric constant on the rates of reactions of this type has been treated by Laidler and Landskroener[2]. On the basis of a theory due to Kirkwood[3] it can be shown that the activity coefficient of any species is given approximately by the equation:

$$\ln \beta_i = \frac{z_i^2 e^2}{2kTr_i}\left(\frac{1}{\varepsilon_r} - 1\right) + \frac{3G_i e^2}{8kTr_i^3}\left(\frac{2}{\varepsilon_r} - 1\right) \tag{67}$$

where r_i is the radius of the spherical molecule, z_i its charge, and G_i is a quantity that can be calculated from the distribution of charges in the molecule; its magnitude is proportional to the square of the dipole moment of the molecule. This equation reduces to eqn (52) when the molecule is non-polar ($G_i = 0$). It applies either to an ion or to a neutral molecule ($z_i = 0$).

The application of this equation to a reaction between an ion A and a neutral molecule B ($z_{\mathrm{B}} = 0$) proceeds as follows. The equations for the activity coefficients of A, B, and the complex are:

$$\ln \beta_{\mathrm{A}} = \frac{z_{\mathrm{A}}^2 e^2}{2kTr_{\mathrm{A}}}\left(\frac{1}{\varepsilon_r} - 1\right) + \frac{3G_{\mathrm{A}} e^2}{8kTr_{\mathrm{A}}^3}\left(\frac{2}{\varepsilon_r} - 1\right), \tag{68}$$

$$\ln \beta_{\mathrm{B}} = \frac{3G_{\mathrm{B}} e^2}{8kTr_{\mathrm{B}}^3}\left(\frac{2}{\varepsilon_r} - 1\right), \tag{69}$$

$$\ln \beta = \frac{z_{\mathrm{A}}^2 e^2}{2kTr_{\neq}}\left(\frac{1}{\varepsilon_r} - 1\right) + \frac{3G_{\neq} e^2}{8kTr_{\neq}^3}\left(\frac{2}{\varepsilon_r} - 1\right). \tag{70}$$

Introduction of these into the rate equation gives:

$$\ln k_0 = \ln k_{\mathrm{g}} + \frac{z_{\mathrm{A}}^2 e^2}{2kT}\left(\frac{1}{r_{\mathrm{A}}} - \frac{1}{r_{\neq}}\right)\left(\frac{1}{\varepsilon_r} - 1\right) + \frac{3e^2}{8kT} \times$$
$$\times \left(\frac{G_{\mathrm{A}}}{r_{\mathrm{A}}^3} + \frac{G_{\mathrm{B}}}{r_{\mathrm{B}}^3} - \frac{G_{\neq}}{r_{\neq}^3}\right)\left(\frac{2}{\varepsilon_r} - 1\right). \tag{71}$$

[1] P. Debye and J. McAulay, *Z. Phys.*, **26**, 22 (1925).
[2] K. J. Laidler and P. A. Landskroener, *Trans. Faraday Soc.*, **52**, 200 (1956); cf. also K. J. Laidler, *Suomen Kemistilehti*, A33, 44 (1960); K. Hiromi, *Bull. chem. Soc. Japan*, 33, 1251, 1264 (1960).
[3] J. G. Kirkwood, *J. chem. Phys.*, **2**, 351 (1934).

A plot of $\ln k_0$ against $1/\varepsilon_r$ should therefore be a straight line, its slope depending upon the magnitudes of the G's and of the r's. This equation has been applied successfully to the hydrolysis of esters, amides, and anilides, and the results shown to be consistent with reasonable values of those quantities.

The situation regarding the magnitudes of frequency factors and of entropies of activation for reactions of this type is not as straightforward as for ion–ion reactions. Application of the same kind of treatment as discussed earlier (eqns (54) to (59)) leads to the result that the entropies of activation for these reactions should be fairly close to zero, so that the frequency factors should correspond to the predictions of the kinetic theory of collisions. This is indeed found to be the case for a number of reactions of this class, including the hydrolyses of alkyl chlorides and similar compounds. The acid and base catalysed hydrolyses of esters, amides, and anilides are, however, anomalous, in that quite negative values of the entropy of activation (-10 to -30 e.u.) are generally obtained, so that the frequency factors are abnormally low. The simple equations developed for these reactions do not provide an explanation for this behaviour. It may be suggested, however, that the effect is related to the presence of the carbonyl group in these compounds, which almost certainly ionizes when the complex is formed to give a structure such as

$$\begin{array}{c} O^- \\ | \\ -C- \\ | \\ OH \end{array}$$

The $-O^-$ ion so produced may bind water molecules strongly, and so bring about a considerable loss in entropy when the complex is formed.

Reactions between two neutral molecules

Reactions between two neutral molecules that are sufficiently polar for the electrostatic interactions between them and the solvent to be of predominant importance may be treated by a modification of the methods outlined above. The ionic strength effect, for example, may be predicted by applying the Debye–McAulay eqn (65) to the two reactants and to the activated complex,

$$\log f_A = b_A u, \tag{72}$$

$$\log f_B = b_B u, \tag{73}$$

$$\log f_{\neq} = b_{\neq} u. \tag{74}$$

Introduction of these equations into eqn (43) gives rise to

$$\log k = \log k_0 + (b_A + b_B - b_{\neq})u, \tag{75}$$

which is identical with eqn (66). A linear dependence of the logarithm of the rate constant on the ionic strength is therefore expected.

The effect of the dielectric constant on the rate constants at zero ionic strength may be treated by applying eqn (67) with $z_i = 0$, to the two reactants and the activated complex, and introducing the equations into eqn (49). The resulting relationship is

$$\ln k_0 = \ln k_g + \frac{e^2}{8kT}\left(\frac{G_A}{r_A^3} + \frac{G_B}{r_B^3} - \frac{G_{\neq}}{r_{\neq}^3}\right)\left(\frac{2}{\varepsilon_r} - 1\right). \tag{76}$$

This equation is, of course, the special case of eqn (71) with $z_A = 0$. It predicts a linear dependence of the logarithm of the rate on the reciprocal of the dielectric constant, and such relationships have been observed in a number of instances.

The modifications to the entropy of activation resulting from electrostatic interactions between neutral molecules are generally fairly small, but important effects can arise if a reactant or the complex has a high dipole moment. A good example of this is to be found in the reaction between a tertiary amine and an alkyl iodide, with the formation of a quaterary ammonium salt. The product is a salt and the activated complex is highly polar; consequently the complex tends to bind solvent molecules much more strongly than do the reactant molecules. This effect results in a considerable negative entropy of activation for reactions of this kind; values of -20 to -30 cal $\deg^{-1}\mathrm{mol}^{-1}$ are common.

Catalysis[1]

All that has been said in the preceding sections on the kinetics of solution reactions applies to catalysed reactions in solution. These reactions do, however, exhibit some special features which it is convenient to consider separately.

Intermediates in catalysed reactions

All catalysed reactions, including enzyme reactions, appear to involve the formation of some kind of intermediate, formed by reaction between the catalyst and the substrate. In some cases, as in many surface and enzyme reactions, this complex is a simple addition complex, and the overall reactions can then sometimes be written as

(1) $$C + A \rightleftharpoons CA$$

(2) $$CA \rightarrow X + C$$

[1] For further details see, for example, P. G. Ashmore, *Catalysis and Inhibition of Chemical Reactions*, Butterworths, New York, 1963; W. P. Jencks, *Catalysis in Chemistry and Enzymology*, McGraw-Hill, New York, 1969.

Here C is the catalyst, A the substrate, CA the addition complex and X the product, which is formed in reaction (2) by elimination of the catalyst. In other cases, as with catalysis by acids and bases and by certain enzymes, the complex is not an addition compound of catalyst and substrate but is formed along with some other substance W. The reaction scheme may then be something like

(1) $$C + A \rightleftharpoons J + W$$

(2) $$J \rightarrow X + C$$

Two important possibilities exist with regard to the stability of the inter-mediate complex J, and the kinetic laws obtained depend in an important manner upon what is assumed in this connection. In the first place, the complex may be one that is reconverted into the catalyst and the substrate at a rate which is significantly greater than the rate with which it undergoes reaction (2) and gives the final products. The overall rate of reaction can then be calculated by obtaining the concentration of J from the equilibrium (1) alone, and multiplying this by the rate constant for reaction (2). Since this situation corresponds to Arrhenius's concept of a chemical reaction as involving equilibrium between the reactants and the activated complex, the complexes in this type of catalysis are frequently known as *Arrhenius complexes*.

The second possibility is that the intermediate complex reacts to give final products at a rate which is not small compared with the reverse rate of reaction (1). It is then not permissible to calculate the concentration of J by considering only the equilibrium (1). Since the rate of (2) cannot be neglected the concentration of J is, however, usually fairly constant over the course of the reaction, and the steady-state treatment may therefore be applied. A complex of this type is frequently known as a *van't Hoff complex*.

Acid–base catalysis

A considerable number of chemical reactions are catalysed by acids and bases. Since an understanding of the mechanism of such action is very important from the standpoint of enzyme catalysis a brief account will here be presented.

When catalysis is brought about by acids and bases the most important effects are usually those directly produced by the hydrogen and hydroxide ions in the solution. The rate of the reaction varies markedly with the pH of the solution and, if there is catalysis by both hydrogen and hydroxide ions, will pass through a minimum as the pH is varied. By an analysis of the varia-tion of rate with pH it is possible to determine the separate effects of the hydrogen and hydroxide ions.

In addition to the specific effects of hydrogen and hydroxide ions it is frequently possible to detect the effects of other acidic and basic species present in solution. According to the definitions of acids and bases proposed by Brønsted and by Lowry, an acid is any substance that will donate a proton to a suitable receptor, and a base is any substance that will accept a proton. Substances like undissociated acids and ammonium ions are acids since they will undergo the following reactions:

$$HA + H_2O \rightleftharpoons H_3O^+ + A^-$$

$$NH_4^+ + H_2O \rightleftharpoons NH_3 + H_3O^+$$

Similarly ammonia and the anions of acids are bases since they are involved in reactions such as the following:

$$NH_3 + H_2O \rightleftharpoons NH_4^+ + OH^-$$

$$A^- + H_2O \rightleftharpoons HA + OH^-.$$

As may be seen from the above equations water may act as either an acid or a base according to circumstances.

When catalysis is brought about by species other than hydrogen and hydroxide ions it is customary to speak of *general* catalysis. As an example may be mentioned the iodination of acetone, which is catalysed, among many other substances, by monochloracetic acid. A study of this reaction revealed that the catalysis was brought about not only by the hydrogen ions produced in the dissociation of some of the monochloracetic acid, but also by the undissociated acid molecules. The rate constant for the reaction was in fact found to vary according to the equation

$$k = k_{H^+}[H_3O^+] + k_{CH_2ClCO_2H}[CH_2ClCO_2H], \tag{77}$$

where k_{H^+} and $k_{CH_2ClCO_2H}$ are the catalytic coefficients for the two catalysing species. Values of the catalytic coefficients of various substances have been determined for a number of different reactions.

When results of this type are available it is natural to seek for correlations between catalytic coefficients and other properties of the catalysing species. It is to be expected that there would be some relationship between the catalytic power of an acid and its acid strength, and similarly between that of a base and its basic strength. Various attempts to formulate such correlations have been made, the most important being that of Brønsted, who proposed the relationships

$$k_a = G_a K_a^\alpha \tag{78}$$

and

$$k_b = G_b K_b^\beta. \tag{79}$$

Here k_a is the catalytic coefficient of an acid and K_a its acid strength, while G_a and α are constants; eqn (79) is the equivalent relationship for a base. The exponents α and β are always less than unity. The constants G_a and G_b

are the same for a given chemical reaction, but vary from reaction to reaction. The Brønsted relationships require some modification for acids and bases containing more than one ionizing group.

Neutral salts usually have a very marked effect on the rates of reactions catalysed by acids and bases. Two types of effect, known as *primary* and *secondary*, are to be distinguished. The primary effect arises from the fact that the rate of a bimolecular interaction in solution between A and B is given by an equation of the form (see eqn (39))

$$v = k_0[A] . [B] \frac{f_A f_B}{f_{\neq}} \tag{80}$$

where the f's are the activity coefficients. Since the ionic strength affects the f's, in accordance with the laws discussed previously, the velocity will be influenced by the addition of salt. When the ionic nature of the reaction is known the explicit equations relating rate to ionic strength can be worked out using the procedure outlined above.

The secondary salt effect is concerned not with the direct influence of a salt on the rate of reaction, but with its influence on the concentrations of reactants; it is therefore operative even when no reaction is occurring. The equilibrium constant for the dissociation of a weak acid HA can be written, for example, as

$$K_a = \frac{[H^+] . [A^-]}{[HA]} \frac{f_{H^+} f_{A^-}}{f_{HA}}, \tag{81}$$

where K_a is a true constant at a given temperature and in a given solvent. Since salts affect these f's, they also influence the concentrations H^+, A^-, and HA, and in this way indirectly influence the rate of a reaction involving any one of these species. The actual variations of the concentrations can be predicted theoretically by application of the Debye–Hückel and other equations to the activity coefficients.

Enzyme kinetics

The study of the kinetics of enzyme reactions has been actively pursued during the past few decades, and the subject has now reached a very interesting phase. The investigation of the mechanism of an enzyme reaction has, for the most part, been carried out from a somewhat different point of view than has the study of an ordinary chemical reaction. With most chemical reactions we have knowledge of the structures of the reacting molecules, and are not concerned with obtaining information about their structures from the kinetic investigations. With enzyme reactions, on the other hand, there has hitherto been little knowledge of the detailed structure of the enzyme (and sometimes of the substrate also), and we have had to obtain what information we could from the kinetic study. For example, studies of the pH dependence of the rate have given the values of the pK's of those groups in the enzyme that are concerned in the catalytic action. From a study of the way in which these pKs vary with solvent composition, it is possible in some cases to make a firm identification of the groups. Similar information is provided by the effects of various inhibitors on the rates.

Until very recently such kinetic results have been the main source of information about the nature of the active centres of enzymes. X-ray investigations of protein structures have now, however, reached the stage at which we are learning the exact structures of some important enzymes, and it will therefore be possible to correlate the structural and kinetic information. Already there have been X-ray studies in which not only has the structure of the enzyme been determined, but also that of the enzyme-substrate complex. No doubt during the next few years there will be many more structural results on enzymes, enzyme-substrate complexes and enzyme-inhibitor complexes, and it will be possible to relate them to the kinetics. The structural studies will never, however, entirely supersede the kinetic ones. Kinetic investigations will always be needed to provide information about the detailed course of enzyme reactions and about the nature of the very short-lived activated complexes; purely structural studies can never provide such information.

Measurement of the rates of enzyme reactions

The present section gives a very brief treatment of the methods most commonly used in studying the rates of enzyme reactions. The special advantages and drawbacks of various techniques will be discussed in relation to the special problems likely to be encountered in the measurement of enzyme reaction rates.

General precautions

There are three factors which are likely to be critical in carrying out an enzyme kinetic study, namely temperature, pH and ionic strength. These factors require even more stringent control than in the case of ordinary chemical reactions, since enzymes are particularly sensitive to their environment, and readily undergo subtle changes in structure which lead to an alteration in activity and kinetic properties. Enzymes undergo conformation changes very easily, so that stricter control over temperature is necessary than with ordinary reactions. An example of such a conformational change taking place at room temperature is to be found with ribonuclease[1].

Changes in pH can also produce drastic changes in activity, both directly (by altering the ionization of groups involved in catalysis) and indirectly (through a change in conformation). Ionic strength is also known to have

[1] J. F. Brandts and L. Hunt, *J. Am. chem. Soc.*, **89**, 4826 (1967); J. Hermans and H. A. Scheraga, *ibid.*, **83**, 3283 (1961).

marked effects on the activities of many enzymes, such as chymotrypsin[1], trypsin and ribonuclease[2]. This is particularly important when buffers of constant concentration, but different pH values, are required. Changes in the ionic composition of the buffer, and hence the ionic strength of the solution, must be compensated for with other ionic species such as salts.

The more important methods used for following the progress of an enzymic reaction will now be briefly discussed.

(1) SPECTROPHOTOMETRY: When there is a significant difference between the wavelength at which product and reactant of a reaction absorb radiation, the progress of the reaction may conveniently be followed as a function of time by using a spectrophotometer. Such a method is continuous, manual sampling being avoided; by use of a recorder a continuous visual record of the progress is obtained. There are two special points associated with this technique that merit attention when it is applied to enzyme systems: (a) measurement of the rate of an enzyme reaction in the ultraviolet part of the spectrum is complicated by the fact that proteins absorb strongly in the region around 280 nm, so that absorption of ultraviolet light from 260 to 300 nm can often not be used to follow the rate of an enzyme reaction. Sometimes, if the binding of substrate to enzyme causes light to be absorbed at a wavelength different from that associated with either enzyme or substrate, this may be used to advantage, as in the case of chymotrypsin (see p. 312). It is then possible to follow the rate of change of the enzyme–substrate complex; (b) in order to avoid this difficulty in the ultraviolet region of the spectrum, several derivatives of substrates absorbing in the visible region have been designed and synthesized. This, however, produces a constraint on the system, since such substrates may not behave in the way that the natural substrate does.

It has sometimes been possible to use an indicator to follow the course of a reaction, the indicator interacting with the enzyme to produce a characteristic absorption band, which can be employed in the usual way to follow the reaction rate. Examples of this are to be found for ribonuclease (p. 305) and trypsin (p. 333).

There are various kinetic techniques involving fluorescence[3]. Many compounds emit light of a characteristic wavelength (the fluorescent light) when light of another wavelength (usually in the ultraviolet region) is incident on them. Use of a special three-windowed quartz cell enables the emitted light to be recorded without interference from the incident light. Here again, the fluorescent emission may be followed as a function of time. Such methods employ special compounds, but they are generally far more

[1] G. Royer, R. Wildnauer, C. C. Cuppett, and W. J. Canady, *J. biol. Chem.*, **246**, 1129 (1971); and references therein; P. Valenzuela and M. L. Bender, *Proc. natn. Acad. Sci.*, **63**, 1214 (1969).

[2] J. A. Winstead and F. Wold, *J. biol. Chem.*, **240**, 3694 (1965).

[3] S. Udenfriend, *Fluorescence Assay in Biology and Medicine*, Academic Press, N.Y. and London (1962).

sensitive than ordinary spectrophotometric methods, since the light emitted is generally more intense. This technique has had useful employment with enzyme reactions involving coenzymes, which are amenable to fluorescence studies.

A double-beam or split-beam spectrophotometer is commonly used to follow the progress of an enzyme reaction. The use of a reference cell cuts out interfering signals from components of the system other than the product (or reactant) compound being monitored, leading to less variability and great accuracy of measurement.

(2) METHODS EMPLOYING ELECTRODES: Electrodes of various types are used in a variety of techniques for the measurement of enzyme reaction rates. Among such techniques are the pH-stat method, conductivity measurements, and the oxygen electrode; these will be briefly discussed.

(a) *pH measurements*: Many enzyme reactions produce changes in pH during their progress. A typical example is the hydrolysis of an ester substrate by the proteolytic enzymes, the products being an alcohol and an acid. Prior to the introduction of instruments to record continuous changes in pH, use was made of indicator dyes[1], the light absorption of which was sensitive to pH changes. The advantage of this method is that there is a greater time resolution for the following of rapid reactions via pH changes. This is necessary since only small changes in pH can be tolerated by the enzyme without a significant change in rate. The disadvantage is that the indicators may interfere with the enzyme reaction, for example by inhibition. Such problems may be avoided by the use of a pH-stat, illustrated in Fig. 2.6. The glass-calomel electrode system monitors the pH, and addition of either acid or base automatically compensates for any changes in pH which would otherwise be caused by the enzyme reaction. Apart from eliminating the need for an indicator, such a system also requires no buffer. Gutfreund[2] has described a system which allows a rapid reaction (with a half life of a few seconds), to be studied in this way.

(b) Apart from changes in pH, various reactions produce changes in the ionic composition of the medium on a microscopic scale. Recently, advantage has been taken of such changes to study enzyme reactions using conductimetry measurements, such as with urease and cholinesterase[3]. While this method has not yet been fully worked out, results differing for different buffers, it has considerable potential, the time response being excellent.

(c) Another electrode method involves the measurement of changes in the concentration of oxygen in a given reaction system. Clark[4] designed such an

---

[1] See e.g. R. P. Davis, *Methods of Biochemical Analysis*, **11**, 307 (1963) and references therein.

[2] H. Gutfreund, *An Introduction to the Study of Enzymes*, Blackwell Scientific Publications, Oxford (1965), p. 142ff.

[3] M. Hanss and A. Rey, *Biochim. biophys. Acta*, **227**, 618, 630 (1971); see also A. J. Lawrence, *Eur. J. Biochem.*, **18**, 221 (1971).

[4] L. C. Clarke, *Trans. Am. Soc. Art. Int.*, **2**, 41 (1956).

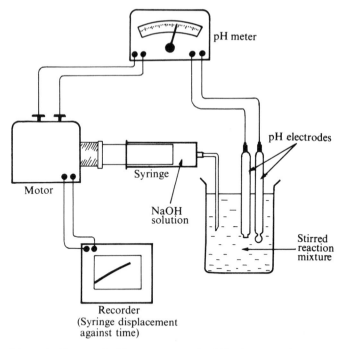

FIG. 2.6.   Schematic diagram of pH-stat apparatus for following a reaction in which acid is produced.

electrode based on a polarographic system, and it has been used in the study of a variety of reactions such as the reduction of oxygen by mitochondrial systems and by glucose oxidase. The technique has considerable advantages over alternative methods such as manometry, in that smaller quantities may usually be used, the time involved is less, and the procedure much simpler. Electrodes for the ammonium ion have also been developed.

(3) MAGNETIC MEASUREMENTS: Two techniques which have received increasing attention in recent years in biochemical systems are electron spin resonance (e.s.r.) and nuclear magnetic resonance (n.m.r.), particularly proton magnetic resonance. These have been discussed in considerable detail in a recent symposium[1], and the latter technique has been recently reviewed with particular emphasis on the information obtainable about enzyme structure and mechanism[2]. Certain nuclei, and unpaired electrons, when subjected to magnetic fields of considerable strength, produce a signal at specific values of the frequency, which is varied. These signals are generally very sensitive to the exact nature of the environment of the resonating

[1] *Magnetic Resonance in Biological Systems*, Volume 9 in the Int. Symp. Series of the Wenner-Gren Centre, Pergamon Press (1966).
[2] G. C. K. Roberts and O. Jardetzky, *Adv. Protein Chem.*, **24**, 447 (1970).

nucleus or (unpaired) electron. It has been possible to investigate enzyme-catalysed phosphoryl transfer reactions using electron paramagnetic resonance (e.p.r.)[1]. The same technique has been used to study free-radical intermediates in certain oxidation processes[2], and to investigate the kinetics of haemoglobin reactions[3], the haem iron producing e.p.r. signals. Nuclear magnetic resonance has been used to study the reaction of aspartate trans-carbamylase with various substrates[4] and numerous studies have been carried out on ribonuclease, as discussed in Chapter 10. In addition, resonant probes have frequently been used to investigate the nature of the active sites and binding sites of various enzymes[5]. In many cases, the information obtained could not have been obtained in any other way.

(4) CALORIMETRY[6]: Calorimetry has occasionally been used for enzyme kinetic studies. The main disadvantage is that complex systems have heat changes from several different processes, and these are not always separable.

Many thermodynamic (equilibrium) studies have, however, been carried out with this method. The advantage is that small changes can be measured even in the presence of large concentrations of reactants. In other methods, such as spectrophotometry, sensitivity is lost if small changes are monitored at high absorbances.

(5) OPTICAL ROTATION: The change in optical rotation as a function of time has occasionally been used to follow the progress of a reaction where substrate and product have different optical activities.

It has been seen earlier in this chapter (p. 38) that when reactions are fast they cannot be studied by conventional static methods, but that flow and relaxation techniques must be used. Such techniques have frequently been utilized in enzyme studies, and a number of examples are to be found in later chapters. Methods used for monitoring reactions include chemical and physical quenching[7], electron spin resonance[8] and spectrophotometry.

---

[1] M. Cohn and J. Reyben, *Acc. chem. Res.*, **4**, 214 (1971).

[2] T. Nakamura and Y. Ogura, p. 205 of the quoted Symposium.

[3] R. G. Shulman, *Science*, **165**, 251 (1969); H. Watasi, A. Hayashi, H. Morimoto, and M. Kotani, in *Recent Development of Magnetic Resonance in Biological Systems*, p. 128 (Hirokowa, Tokyo, 1968).

[4] P. G. Schmidt, G. R. Stark, and J. D. Baldescheieler, *J. biol. Chem.*, **244**, 1860 (1969).

[5] L. H. Piette and G. P. Rabold, p. 351 of Symposium.

[6] For detailed discussions, see J. M. Sturtevant, in *Techniques in Organic Chemistry*, part 1, 3rd edn. (1959), Interscience, N.Y.; H. Gutfreund, *Introduction to the Study of Enzymes*, pp. 152–9.

[7] See for instance C. G. Miller and M. L. Bender, *J. Am. chem. Soc.*, **90**, 6850 (1968) and T. E. Barman and H. Gutfreund, *Proc. natn. Acad. Sci.*, **53**, 1243 (1965) for pH quenching; and P. Douzou, R. Sireix, and F. Travers, *Proc. natn. Acad. Sci.*, **66**, 787 (1970) and R. C. Bray, *Biochem. J.*, **81**, 189 (1961) for low-temperature quenching.

[8] B. Venkataraman and G. Fraenkel, *J. Am. chem. Soc.*, **77**, 2707 (1955); I. Yamazaki, H. S. Mason, and L. Piette, *Biochem. biophys. Res. Commun.*, **1**, 336 (1959).

3. The Steady State in Enzyme Kinetics

DURING the latter part of the nineteenth century a number of investigations were made of the kinetics of enzyme reactions, and it soon became apparent that the behaviour observed was frequently different from that found with the simple homogeneous reactions studied up to that time. In particular it was found that very often the rate of reaction was not simply proportional to the concentrations of the reacting substances.

Many of the earlier observations became clarified on the publication, in 1902, of the results of a very important investigation made by Adrian Brown[1]. Previous studies by this author[2] on the fermentation of yeast had shown that a given amount of yeast brings about the breakdown of a constant amount of sucrose in unit time in solutions of varying concentration; in other words, the reaction is of zero order with respect to concentration. In the 1902 paper Brown confirmed that this result is also obtained when an invertase solution is allowed to act upon sucrose. He found, however, that in a given run the amount transformed is not proportional to time, but that there is a falling-off from linearity; the apparent order with respect to time is therefore greater than zero. Brown found that the fraction x of sugar converted into product varied with time according to the equation

$$2kt = \log \frac{1 + x}{1 - x},\tag{1}$$

an empirical relationship that had previously been shown by Victor Henri[3] to apply to the same reaction.

The apparent discrepancy between the facts that the rates in different runs are independent of concentration and that the rate in a given run falls off as the reaction proceeds was traced by Brown to be due to inhibition by the products of reaction; as these gradually accumulate they cause a slowing-down of the reaction. The important point that remained to be explained was therefore that the initial rates are independent of the initial sucrose concentration.

In order to explain this fact Brown proposed the hypothesis that there is a definite complex formed between enzyme and substrate. In his own words:

There is reason to believe that during inversion of cane sugar by invertase the sugar combines with enzyme previous to inversion. C. O'Sullivan and Tompson (*loc. cit.*)[4] have shown that the activity of invertase in the presence of cane sugar survives a temperature which completely destroys it if cane sugar is not present, and regard this as

[1] A. J. Brown, *J. chem. Soc.*, **81**, 373 (1902).
[2] A. J. Brown, *J. chem. Soc.*, **61**, 380 (1892).
[3] V. Henri, *C.R. Acad. Sci.*, **133**, 891 (1901).
[4] C. O'Sullivan and F. W. Tompson, *J. chem. Soc.*, **57**, 865 (1890).

indicating the existence of a combination of the enzyme and sugar molecules. Wurtz (*Compt. rend.*, 1880, **91**, 787) has also shown that papain appears to form an insoluble compound with fibrin previous to hydrolysis. Moreover the more recent conception of E. Fischer with regard to enzyme configuration and action also implies some form of combination of enzyme and reacting substance.

This concept of the enzyme–substrate complex has been of enormous importance in enzyme kinetics, and has been amply supported by more recent evidence.

In order to explain how the idea of the enzyme–substrate complex provides an explanation for the zero-order kinetics exhibited by the sucrose inversion, Brown argued as follows:

Let it be assumed, therefore, that one molecule of an enzyme combines with one molecule of a reactive substance, and that the compound molecule exists for a brief interval of time during the further actions which end in disruption and change. Let it be assumed also that the interval of time during which the compound molecule of enzyme and reacting substance exists is 1/100th of a time unit. Then it follows that a molecule of the enzyme may assist in effecting 100 completed molecular changes in unit time, but that this is the limit to its power of change.

Again, let it be assumed that the number of molecular collisions between the active and reacting molecules which lead to their combination bears some proportion to the number of possible completed molecular changes in unit time. Let the number of collisions be 20, then there may be 20 complete molecular changes; if 40, there may be 40 changes. In fact the action of the mass law is observed for, other conditions being equal, the average number of molecular collisions must depend on the number of molecular, or mass, of the matter present.

But now assume that the mass of reacting substance is increased, so that the number of molecular collisions in unit time exceeds 100; let it be 150, 1,000, or any other number larger than 100. Then, although the number of molecular collisions may exceed 100 by a number following the law of mass action, 100 molecular changes cannot be exceeded, for the compound enzyme and sugar molecule is only capable of effecting 100 complete changes in unit time.

It follows, therefore, that if, in a series of changes like the imaginary ones described, a constant quantity of enzyme is in the presence of varying quantities of a reacting substance, and in all cases the quantity of reacting substances present ensures a greater number of molecular collisions in unit time than the possible number of molecular changes. Then a *constant weight* of substance may be changed in unit time in all the actions.

This explanation resembles quite closely the explanation that would be given at the present day, but the emphasis is now not on the number of collisions between enzyme and substrate, but rather on the equilibrium constant between enzyme and substrate. A more modern version of Brown's theory is as follows. Enzyme reactions proceed in two well-defined steps; the first is the formation of a complex between enzyme E and substrate A:

(1) $$E + A \rightleftharpoons EA$$

and the second is the decomposition of the complex into the products of the reaction, the enzyme being regenerated in this step:

(2) $EA \rightarrow E + X$

At high substrate concentrations the enzyme, the supply of which is usually limited compared with that of substrate, becomes saturated with substrate, so that practically all of the enzyme is in the form of complex. Increase in the substrate concentration can therefore cause no further increase in the concentration of complex so that the rate, proportional to the concentration of complex, is independent of the substrate concentration.

At very low substrate concentrations, on the other hand, most of the enzyme molecules are in the free state, only a very small fraction being combined with substrate. Under these conditions the amount of complex formed is proportional to the amount of substrate, so that the rate of formation of products, i.e. the rate of reaction (2), is proportional to the concentration of substrate. Brown confirmed experimentally that at sufficiently low concentrations of sucrose the inversion follows the first-order law.

Brown's theory, and the deductions from it that the kinetics should be first order with respect to concentration at low substrate concentrations and zero order at high substrate concentrations, have been fully confirmed for a large variety of reactions involving a single substrate. A general formulation of the theory, in terms of a rate law that reduces to the simple laws at low and high substrate concentrations, was later given by Michaelis and Menten, and further developed by Briggs and Haldane (see pp. 72–73).

Evidence for enzyme–substrate intermediates

Since the original proposal of Brown that a definite complex is formed between an enzyme and a substrate molecule, a considerable amount of experimental evidence has been obtained in support of this idea. Extensive evidence comes from the kinetic laws found for a large number of enzyme systems and under a variety of experimental conditions. Later in this chapter, and in subsequent chapters, the kinetic equations for several different types of reactions will be derived on the basis of the hypothesis that compounds are formed between enzymes and substrates, and between enzymes and such substances as inhibitors and activators. It will be seen that these equations provide a satisfactory interpretation of the experimental results. Attempts to formulate kinetic equations on the basis of any other type of hypothesis have not been successful.

Apart from such kinetic evidence, based on variations of overall rates with concentrations of substrates and inhibitors, there is also now a good deal of other evidence for enzyme–substrate addition complexes and for other types of reaction intermediates. The earliest of these observations appears to have been the demonstration, in 1880, of an insoluble complex formed between papain and its substrate fibrin[1]. With the haemoprotein enzymes catalase and peroxidase, Chance[2] obtained direct spectroscopic evidence for the occurrence of intermediates during the course of reaction. In these and other reactions spectral shifts have been observed as the process takes place, and these shifts have been satisfactorily correlated with the kinetics.

Other techniques that have been employed for the demonstration of the existence of intermediates in enzyme reactions are fluorescence[3], light scattering[4], ultracentrifugal separation[5] and chromatography[6]. Some of these techniques provide information as to whether or not the intermediates are simple addition compounds.

Rate equations

The remainder of this chapter is concerned with the development of rate equations for various types of enzyme systems, on the basis of the hypothesis of enzyme–substrate complexes. The treatment in the present chapter is confined to single-substrate systems and to obtaining the rate equations in their differential forms, i.e. to seeing how the rate of change of substrate concentration, $-d[A]/dt$, varies with the concentrations of substrate and of other substances present, such as activators and inhibitors. A separate chapter (Chapter 5) will be devoted to a detailed treatment of pH effects.

The equations developed in the present chapter are directly applicable to experimental studies in which rates are measured in the fairly early stages of reaction†, since the conditions are known in a precise manner. The rate equations in their integrated forms, which account for the variations of substrate and product concentrations with the time, are considered in Chapter 6.

The situations dealt with in the present chapter are as follows:

(1) Uncomplicated one-substrate reactions in which there is a single intermediate, the enzyme–substrate complex (p. 72).
(2) One-substrate reactions in which there are two intermediates (p. 77).
(3) Reversible one-substrate reactions (p. 81).
(4) Inhibition by substrate (p. 84).
(5) Inhibition (p. 89).
(6) Activation (p. 110).
(7) Reactions of two competing substrates (p. 110).

† Not, however, in the pre-steady state period; this is discussed in Chapter 6.

[1] A. Wurtz, *Comp rend.*, **91**, 787 (1880).
[2] B. Chance, *J. biol. Chem.*, **151**, 553 (1943); *Acta chem. Scand.*, **1**, 236 (1947); cf. also K. G. Stern, *J. biol. Chem.*, **114**, 473 (1936); D. Keilin and T. Mann, *Proc. R. Soc.*, **B122**, 119 (1936).
[3] S. F. Velick, *J. biol. Chem.*, **233**, 1455 (1958).
[4] J. J. Blum and M. F. Morales, *Archs Biochem. Biophys.*, **43**, 208 (1953).
[5] Y. Takenaka and G. W. Schwartz, *J. biol. Chem.*, **223**, 157 (1956); S. F. Velick, J. E. Hayes, and J. Hasting, *J. biol. Chem.*, **203**, 527 (1953); S. F. Velick, *J. biol. Chem.*, **203**, 563 (1953).
[6] R. S. Anthony and L. B. Spector, *J. biol. Chem.*, **245**, 6739 (1970); Y. Tsuda, R. A. Stephani, and A. Meister, *Biochem.*, **10**, 3186 (1971).

Reactions between two or more substrates are considered separately in Chapter 4.

One-Substrate reactions having a single intermediate

The reaction scheme that may apply to some one-substrate systems, un-complicated by the reverse reactions or the effects of modifiers (activators or inhibitors) is the one proposed by Brown:

(1) $$E + A \underset{k_{-1}}{\overset{k_1}{\rightleftharpoons}} EA$$

(2) $$EA \overset{k_2}{\rightarrow} E + X.$$

Michaelis and Menten[1] formulated a rate equation based on this mechanism, but made the assumption that the second reaction does not disturb the first equilibrium (i.e. that $k_{-1} \gg k_2$). A more general formulation was given by Briggs and Haldane[2] on the basis of the steady-state hypothesis (p. 48) that the net rate of change of [EA], the concentration of complexes, is zero.

The rate of formation of complexes is made up of three terms. Complexes are formed by reaction (1) going from left to right, the rate of which is $k_1[E].[A]$. They are removed by reaction (2), of rate $k_2[EA]$, and by the reverse of reaction (1), of rate $k_{-1}[EA]$. The net rate of formation of complexes, set equal to zero, is thus

$$k_1[E].[A] - k_{-1}[EA] - k_2[EA] = 0. \tag{2}$$

In most experiments in enzyme kinetics the concentration of substrate is much greater than that of enzyme; indeed, if this condition is not satisfied steady-state conditions may never be attained during the course of reaction. Only a small proportion of the substrate is involved in attachment to the enzyme, but an appreciable fraction of the enzyme may be tied to substrate. The total enzyme concentration, $[E]_0$, is equal to

$$[E]_0 = [E] + [EA]. \tag{3}$$

Equation (2) may therefore be written as

$$k_1([E]_0 - [EA])[A] - k_{-1}[EA] - k_2[EA] = 0 \tag{4}$$

whence

$$[EA] = \frac{k_1[E]_0.[A]}{k_{-1} + k_2 + k_1[A]}. \tag{5}$$

The rate of reaction is therefore

$$v = k_2[EA] \tag{6}$$

$$= \frac{k_1 k_2 [E]_0.[A]}{k_{-1} + k_2 + k_1[A]} \tag{7}$$

[1] L. Michaelis and M. L. Menten, *Biochem. Z.*, **49**, 333 (1913).
[2] G. E. Briggs and J. B. S. Haldane, *Biochem. J.*, **19**, 338 (1925).

$$= \frac{k_2[\text{E}]_0 \cdot [\text{A}]}{(k_{-1} + k_2)/k_1 + [\text{A}]} \tag{8}$$

$$= \frac{k_2[\text{E}]_0 \cdot [\text{A}]}{K_\text{A} + [\text{A}]}. \tag{9}$$

In this last equation K_A has been written for $(k_{-1} + k_2)/k_1$. The constant K_A is generally known as the Michaelis constant; it has the same units as a concentration, i.e. almost invariably moles per litre.

In the Michaelis–Menten derivation the assumption was made that k_{-1} was much greater than k_2, so that the constant K_A had the significance of the equilibrium constant k_{-1}/k_1. Alternatively, Van Slyke and Cullen[1] formulated the rate equation on the assumption that reaction (1) is essentially irreversible, i.e. that $k_{-1} \ll k_2$. Their treatment was therefore another special case of the Briggs–Haldane treatment, with K_A reducing to k_2/k_1. Studies of the variation of rate with substrate concentration obviously cannot provide information as to the relative magnitudes of k_{-1} and k_2; this can be done by studies of the kinetics in the very early stages of reaction (see Chapter 6).

Justification of the steady-state hypothesis

There have been a number of discussions of the validity of the steady-state hypothesis in enzyme kinetics[2]. Unfortunately the matter is mathematically rather complicated, since no exact solution of the differential rate equations can be given even for the simplest one-intermediate mechanism. Only very brief comments will be made here; for further details the reader is referred to the original publications.

The differential equations to be solved for the single intermediate mechanism (see p. 72) are

$$\frac{d[\text{EA}]}{dt} = k_1[\text{E}] \cdot [\text{A}] - (k_{-1} + k_2)[\text{EA}] \tag{10}$$

and

$$\frac{d[\text{X}]}{dt} = k_2[\text{EA}]. \tag{11}$$

Since $[\text{E}]_0 = [\text{E}] + [\text{EA}]$, eqn (10) can be written as

$$\frac{d[\text{EA}]}{dt} = k_1[\text{E}]_0 \cdot [\text{A}] - (k_1[\text{A}] + k_{-1} + k_2)[\text{EA}] \tag{12}$$

or

$$\frac{d[\text{EA}]}{dt} + (k_1[\text{A}] + k_{-1} + k_2)[\text{EA}] = k_1[\text{E}]_0 \cdot [\text{A}]. \tag{13}$$

[1] D. D. Van Slyke and G. E. Cullen, *J. biol. Chem.*, **19**, 141 (1914).
[2] K. J. Laidler, *Can. J. Chem.*, **33**, 1614 (1955); F. A. Hommes, *Archs Biochem. Biophys.*, **96**, 28 (1962); C. Walter and M. F. Morales, *J. biol. Chem.*, **239**, 1277 (1964); C. Walter, *J. theor. Biol.*, **11**, 181 (1966); J. T. Wong, *J. Am. chem. Soc.*, **87**, 1788 (1965).

If the first term on the left-hand side is much less than the second

$$(k_1[A] + k_{-1} + k_2)[EA] = k_1[E]_0 . [A] \tag{14}$$

which is the steady-state equation (eqn (5)). It is to be emphasized that strictly speaking the steady-state approximation is not $d[EA]/dt = 0$, but is

$$\frac{d[EA]}{dt} \ll (k_1[A] + k_{-1} + k_2)[EA]. \tag{15}$$

It can also be expressed as

$$\frac{d[EA]}{dt} \ll k_1[E]_0 . [A]. \tag{16}$$

It can be seen at once that the inequalities (15) and (16) can always be satisfied by making the concentration [A] sufficiently large. When this is done the magnitude of [EA] attains a limiting value, since [EA] cannot become greater than $[E]_0$. The magnitude of $d[EA]/dt$ therefore reaches a limiting value as [A] is made large. The expressions on the right-hand sides of (15) and (16), however, continue to rise as [A] is increased, so that eventually the inequalities will be satisfied. Quantitative calculations indicate that the steady-state hypothesis is satisfactory provided that $[A]/[E]_0$ is 10^3 or greater.

The converse situation in which $[E]_0/[A]$ is very large does *not* lead to the establishment of a steady state. The whole problem of the kinetics in the pre-steady-state region, and the establishment of the steady-state, is considered further in Chapter 6.

Analysis of experimental results

It is readily seen that eqn (9) predicts the extreme types of behaviour observed and explained by Brown and in many subsequent investigations. At sufficiently low concentrations of substrate the concentration [A] in the denominator is much smaller than K_A and can therefore be neglected; the rate is then

$$v^0 = \frac{k_2[E]_0}{K_A}[A] \tag{17}$$

and the kinetics are first-order in [A]. At sufficiently high concentrations, on the other hand, K_A can be neglected in comparison with [A]. The rate is now

$$V = k_2[E_0] \tag{18}$$

and the kinetics are zero-order in [A]. In view of eqn (18) the rate equation (9) is conveniently written as

$$v = \frac{V[A]}{K_A + [A]}. \tag{19}$$

Equations (9) and (19) are the equations of a rectangular hyperbola. A schematic plot of v against [A] is shown in Fig. 3.1. At low concentrations the rate rises, at first linearly, but at high concentrations it levels off at a limiting rate, equal to $V(=k_2[E]_0)$. If $[E]_0$ is known (which means that the enzyme must be of known purity and that its molecular weight must be known), k_2 can be calculated from the limiting rate at high substrate concentrations. It is seen from eqn (19) that when $[A] = K_A$, $v = V/2$; this relationship is shown on the diagram and provides a means of determining K_A from plots of v against [A].

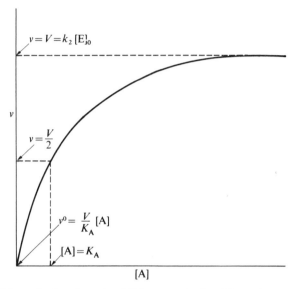

FIG. 3.1.   Schematic plot of v against [A] for a reaction involving a single intermediate.

Linear plots are, of course, much more useful than non-linear ones in testing equations and determining parameters from them. Various linear plots have been suggested. A commonly-employed procedure is one proposed by Lineweaver and Burk[1]. Equation (19) may be written in reciprocal form as

$$\frac{1}{v} = \frac{K_A}{V} \cdot \frac{1}{[A]} + \frac{1}{V} \tag{20}$$

and a plot of $1/v$ against $1/[A]$ will give a straight line of slope K_A/V and intercept $1/V$ on the $1/v$ axis (Fig. 3.2). K_A and V can therefore be calculated from the slopes and intercepts of such plots.

[1] H. Lineweaver and D. Burk, *J. Am. chem. Soc.*, **56**, 658 (1934).

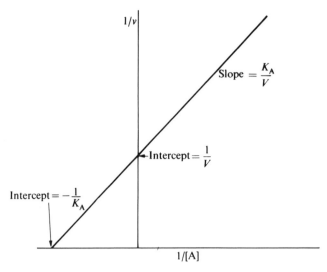

FIG. 3.2.   Schematic Lineweaver–Burk plot of $1/v$ against $1/[A]$.

A second method of obtaining a linear plot is due to Eadie[1]. Equation (19) can be written as

$$vK_A + v[A] = V[A] \tag{21}$$

whence

$$\frac{v}{[A]} = \frac{V}{K_A} - \frac{v}{K_A}. \tag{22}$$

If the equation is obeyed a straight line will therefore be obtained if $v/[A]$ is plotted against v. The slopes and intercepts are shown in Fig. 3.3. Alternatively, one can plot $[A]/v$ against $[A]$.

The question of which is the best type of plot is frequently discussed. If rate measurements are made at fairly regular intervals of $[A]$, the Lineweaver–Burk method has the slight disadvantage of leading to a plot in which most of the points are towards one end of the line; a more even distribution is provided by the Eadie plot. However, the method of plotting should not be of great importance since the evaluation of kinetic parameters should be done by statistical methods. Various discussions of this problem have been given[2], and Cleland[3] has provided computer programmes. It is important to

[1] G. S. Eadie, *J. biol. Chem.*, **146**, 85 (1942); cf. B. H. J. Hofstee, *Science*, **116**, 329 (1952); *Nature*, **184**, 1296 (1959).
[2] G. N. Wilkinson, *Biochem. J.*, **80**, 325 (1961); G. Johansen and G. Lumry, *C.R. Trav. Lab.*, *Carlsberg*, **32**, 185 (1961).
[3] W. W. Cleland, *Adv. Enzymol.*, **29**, 1 (1967); see C. I. Bliss and A. T. James, *Biometrics*, **22**, 573 (1966); K. R. Hanson, R. Ling, and E. Havir, *Biochem. biophys. Res. Commun.*, **29**, 194 (1967).

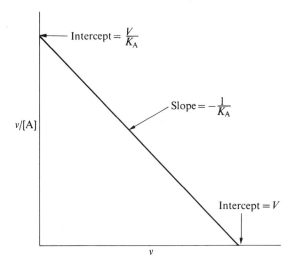

FIG. 3.3. Schematic Eadie plot of $v/[A]$ against v.

note that a simple least-squares analysis of the linear plots is not adequate; appropriate weighting of the points is necessary.

A considerable number of enzyme systems obey the hyperbolic law represented by eqn (19), but this is not to be taken as evidence that the single-intermediate mechanism applies. It will be seen later in this chapter that when two or more intermediates are involved the rate equation is of the same general form as (19), although the significance of the parameters V and K_A is different. Furthermore, even reactions between two or more substrates frequently exhibit hyperbolic behaviour with respect to one substrate when the concentrations of other substrates are held constant. Mechanisms that are consistent with this behaviour are considered in Chapter 4.

Deviations from the behaviour represented by eqn (19) are frequently observed. Sometimes, for example, the rate does not reach a limiting value at high substrate concentrations, but passes through a maximum. Interpretations of this and other more complex behaviour are considered later.

Reactions involving two intermediates

During recent years it has become increasingly clear that although the hyperbolic eqn (19) is applicable to a wide variety of enzyme-catalysed reactions, the actual mechanism is frequently more complicated than the single-intermediate mechanism. In particular, for a considerable number of reactions there is evidence that a second intermediate is formed after the addition complex. For example, the hydrolysis of esters by proteolytic enzymes and esterases has frequently been found to involve first the formation of an addition complex and then of an intermediate in which a hydroxyl group on the enzyme has become acylated; thus if the enzyme is represented as E-OH,

$$E\text{-OH} + RCO_2R' \rightleftharpoons \underset{\text{addition complex}}{E\text{-OH}.RCO_2R'} \longrightarrow$$
$$R'OH$$

$$\underset{\text{acyl enzyme}}{E\text{-OCOR}} \xrightarrow{+H_2O} EOH + RCO_2H.$$

It is to be seen that the alcohol is split off during the formation of the acyl enzyme from the addition complex, while the acid is formed by hydrolysis of the acyl enzyme. Similar mechanisms have been found to be involved in a number of other enzyme systems, such as the hydrolysis of amides and peptides by proteolytic enzymes, and the hydrolysis of phosphates by phosphatases.

The evidence for the formation of the second intermediate is of various kinds, and will be discussed later. The present section is concerned with the derivation of the steady-state equation that applies to this type of mechanism. It will be seen that the resulting equation is of exactly the same form as that given by the single-intermediate mechanism, so that steady-state studies cannot provide information about the number of intermediates.

The two-intermediate mechanism may be represented as follows:

$$E + A \underset{k_{-1}}{\overset{k_1}{\rightleftharpoons}} EA \xrightarrow{k_2} EA' \xrightarrow{k_3} E + Y$$
$$X$$

where EA' is the second intermediate (e.g. the acyl enzyme), and X and Y are two products of reaction. The total concentration of enzyme is

$$[E]_0 = [E] + [EA] + [EA']. \tag{23}$$

The steady-state equation for [EA] is

$$k_1[E].[A] = (k_{-1} + k_2)[EA] \tag{24}$$

and that for [EA'] is

$$k_2[EA] = k_3[EA']. \tag{25}$$

Equations (24) and (25) allow [E] and [EA'] to be expressed in terms of [EA], and the resulting equations may be inserted into (23); the result is

$$[E]_0 = [EA]\left\{\frac{k_{-1} + k_2}{k_1[A]} + 1 + \frac{k_2}{k_3}\right\} \tag{26}$$

whence

$$[EA] = \frac{[E]_0.[A]}{\{(k_{-1} + k_2)/k_1\} + \{(k_2 + k_3)/k_3\}[A]}. \tag{27}$$

The rate of formation of X is $k_2[EA]$, and the rate of formation of Y is $k_3[EA']$. These are equal by eqn (25); this is necessarily the case after the steady state has been established. The overall rate of product formation is thus

$$v = \frac{k_2[E]_0 \cdot [A]}{\{(k_{-1} + k_2)/k_1\} + \{(k_2 + k_3)/k_3\}[A]}. \tag{28}$$

This may be put into the conventional form by multiplying numerator and denominator by $k_3/(k_2 + k_3)$:

$$v = \frac{\{k_2 k_3/(k_2 + k_3)\}[E]_0 \cdot [A]}{\{(k_{-1} + k_2)/k_1\} \cdot \{k_3/(k_2 + k_3)\} + [A]}. \tag{29}$$

Comparison of this equation with eqn (19) shows that

$$V = \frac{k_2 k_3}{k_2 + k_3}[E]_0 \quad \text{and} \quad K_A = \frac{k_{-1} + k_2}{k_1} \cdot \frac{k_3}{k_2 + k_3}. \tag{30}$$

Two special cases of this are of interest:
(1) Suppose that $k_3 \gg k_2$; equation (30) then reduces to

$$V = k_2[E]_0 \quad \text{and} \quad K_A = \frac{k_{-1} + k_2}{k_1}. \tag{31}$$

This is the result obtained on the basis of the single intermediate; the second intermediate is now unimportant because of the large rate constant for its disappearance.
(2) Suppose that $k_2 \gg k_3$; equations (30) now reduce to

$$V = k_3[E]_0 \quad \text{and} \quad K_A = \frac{k_{-1} + k_2}{k_1} \cdot \frac{k_3}{k_2}. \tag{32}$$

The rate constant at high $[A]$, generally known as k_c or k_{cat}, is now k_3, the rate constant for the slower process; it is to be noted that with both $k_3 \gg k_2$ and $k_2 \gg k_3$, the rate constant at high $[A]$ is the rate constant for the slower process. The Michaelis constant in this case is more complicated. It is less than $(k_{-1} + k_2)/k_1$, since by hypothesis k_3/k_2 is a small fraction. The quantity $(k_{-1} + k_2)/k_1$ is sometimes written as K_s, the symbol K_A being reserved for the experimental quantity. However K_s is sometimes used also for the equilibrium constant k_{-1}/k_1. In the present book, to avoid confusion, the expressions will always be written out in terms of the rate constants; K_A will always refer to the experimental quantity, irrespective of the mechanism.

Schematic method of deriving rate equations

As kinetic mechanisms for enzyme-catalysed reactions become more complicated, there is a corresponding increase in the amount of labour required to obtain the final rate expression in terms of the concentrations and the individual rate constants. It is therefore useful to have a simplified but reliable method of arriving at the expression. A very useful procedure has been devised by King and Altman[1], and is now widely used in work of this kind. These workers have proved the validity of their method by a determinant procedure. The method will be explained and illustrated with respect to the two-intermediate mechanism just considered.

The procedure is first to write the mechanism in cyclic form

There are two intermediates, EA and EA′, and two steady-state equations for them. The concentrations of the three species E, EA and EA′ are proportional to the sums of terms that are obtained by writing down reaction steps which individually or in sequence lead to the species in question. Thus for E the terms are

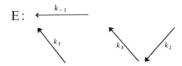

and the concentration of E is therefore proportional to $k_{-1}k_3 + k_2k_3$. The number of rate constants in each term must be one less than the number of species in the cycle; in this example there are three species and each term is therefore the product of two rate constants. Similarly for EA and EA′

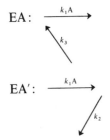

Thus

$$[E] \propto (k_{-1}k_3 + k_2k_3) \tag{33}$$

$$[EA] \propto k_1k_3[A] \tag{34}$$

$$[EA'] \propto k_1k_2[A]. \tag{35}$$

[1] E. L. King and C. Altman, *J. phys. Chem.*, **60**, 1375 (1956); H. J. Fromm, *Biochem. biophys. Res. Commun.*, **40**, 692 (1970); N. Seshagiri, *J. theor. Biol.*, **34**, 469 (1972).

Since $[E]_0 = [E] + [EA] + [EA']$ it follows that, for example,

$$\frac{[EA]}{[E]_0} = \frac{k_1 k_3 [A]}{k_{-1} k_3 + k_2 k_3 + k_1 k_3 [A] + k_1 k_2 [A]}. \tag{36}$$

The final expression for the rate is therefore

$$v = k_2 [EA] \tag{37}$$

$$= \frac{k_1 k_2 k_3 [E]_0 \cdot [A]}{k_3 (k_{-1} + k_2) + (k_2 + k_3) k_1 [A]} \tag{38}$$

which is equivalent to eqn (29). Further examples of the use of this method will be given later (e.g., p. 122).

Reversible one-substrate reactions

The steady-state treatment of a reversible reaction, in which a single substrate is involved on both sides of the equation, was first given by Haldane[1] on the basis of the mechanism

$$E + A \underset{k_{-1}}{\overset{k_1}{\rightleftharpoons}} EA \underset{k_{-2}}{\overset{k_2}{\rightleftharpoons}} EX \underset{k_{-3}}{\overset{k_3}{\rightleftharpoons}} E + X.$$

The steady-state equations for $[EA]$ and $[EX]$ are respectively

$$\frac{d[EA]}{dt} = k_1 [E] \cdot [A] + k_{-2} [EX] - (k_{-1} + k_2)[EA] = 0 \tag{39}$$

$$\frac{d[EX]}{dt} = k_{-3} [E] \cdot [X] + k_2 [EA] - (k_3 + k_{-2})[EX] = 0. \tag{40}$$

Since the total concentration of enzyme is

$$[E]_0 = [E] + [EA] + [EX] \tag{41}$$

these equations become

$$k_1 ([E]_0 - [EA] - [EX])[A] + k_{-2}[EX] - (k_{-1} + k_2)[EA] = 0 \tag{42}$$

$$k_{-3}([E]_0 - [EA] - [EX])[X] + k_2 [EA] - (k_3 + k_{-2})[EX] = 0. \tag{43}$$

The solution of these equations for $[EA]$ and $[EX]$ is

$$\frac{[EA]}{[E]_0} = \frac{k_1 (k_{-2} + k_3)[A] + k_{-2} k_{-3} [X]}{k_{-1} k_{-2} + k_{-1} k_3 + k_2 k_3 + k_1 (k_2 + k_{-2} + k_3)[A] + k_{-3}(k_{-1} + k_2 + k_{-2})[X]} \tag{44}$$

$$\frac{[EX]}{[E]_0} = \frac{k_1 k_2 [A] + k_{-3}(k_{-1} + k_2)[X]}{k_{-1} k_{-2} + k_{-1} k_3 + k_2 k_3 + k_1 (k_2 + k_{-2} + k_3)[A] + k_{-3}(k_{-1} + k_2 + k_{-2})[X]}. \tag{45}$$

[1] J. B. S. Haldane, *Enzymes*, Longmans, Green and Co., London (1930), p. 81, see also F. Vaslow, *Cr. Trav. Lab. Carlsberg, Ser. chim.*, **30**, 45 (1956).

The net rate of reaction from left to right is therefore

$$v = k_2[EA] - k_{-2}[EX] \tag{46}$$

$$= \frac{(k_1 k_2 k_3[A] - k_{-1} k_{-2} k_{-3}[X])[E]_0}{k_{-1} k_{-2} + k_{-1} k_3 + k_2 k_3 + k_1(k_2 + k_{-2} + k_3)[A] +}{+ k_{-3}(k_{-1} + k_2 + k_{-2})[X]} \tag{47}$$

This may be written as

$$v = \frac{V_A K_X[A] - V_X K_A[X]}{K_A K_X + K_X[A] + K_A[X]} \tag{48}$$

where

$$V_A = \frac{k_2 k_3}{k_2 + k_{-2} + k_3}[E]_0 \tag{49}$$

$$V_X = \frac{k_{-1} k_{-2}}{k_{-1} + k_2 + k_{-2}}[E]_0, \tag{50}$$

$$K_A = \frac{k_{-1} k_{-2} + k_{-1} k_3 + k_2 k_3}{k_1(k_2 + k_{-2} + k_3)} \tag{51}$$

$$K_X = \frac{k_{-1} k_{-2} + k_{-1} k_3 + k_2 k_3}{k_{-3}(k_{-1} + k_2 + k_{-2})}. \tag{52}$$

From eqn (48) it may be seen that in an experiment with pure A (i.e. $[X] = 0$) the initial rate is given by

$$v_1 = \frac{V_A[A]}{K_A + [A]}. \tag{53}$$

Similarly, the initial rate starting with pure X is given by

$$v_{-1} = \frac{V_X[X]}{K_X + [X]}. \tag{54}$$

These rate equations are seen to be of the same form as the simple hyperbolic equations for an irreversible process, but the constants have a different significance. It may be noted, however, that if the reaction is actually irreversible the rate constant k_3 will be much greater than both k_2 and k_{-2} (thus preventing any accumulation of EX); under these conditions V_A is found to be equal to $k_2[E]_0$, and K_A to $(k_{-1} + k_2)/k_1$, so that the equation has reduced to the simple case.

A study of a reversible reaction of this type will permit a determination of the quantities V_A, V_X, and K_A, and K_X, but will not allow a separate evaluation of the individual rate constants; these can only be obtained by the use of special methods. One interesting result that emerges from the use of eqn (48) concerns the relationship between the constants and the equilibrium constant K_0 for the overall reaction. At equilibrium $v = 0$, and so from eqn (48),

$$V_A K_X [A] = V_X K_A [X] \qquad (55)$$

whence

$$K_0 = \frac{[X]}{[A]} = \frac{V_A K_X}{V_X K_A} = \frac{k_1 k_2 k_3}{k_{-3} k_{-2} k_{-1}}. \qquad (56)$$

The ratio $V_A K_X / V_X K_A$ must therefore be a constant at a given temperature, and independent of the amount of catalyst. Equation (56) is sometimes known as the Haldane equation.

The equation has been verified for several enzyme systems. For the β-glucosidase-catalysed hydrolysis of β-methyl glucoside Euler and Josephson[1] calculated from the values of V_A and V_X, and from the equilibrium constant K_0 determined by Bourquelot[2], that the quantity $V_A / V_X K_0$ is equal to 4·0. According to eqn (56) this should be equal to K_A / K_X. Josephson's results[3] for these quantities give 0·71 for the glucoside and 0·1775 for glucose, and the ratio is exactly 4·0 as expected.

Another reaction for which the relationship has been verified is the fumarase system, studied from this point of view by Bock and Alberty[4]. The results obtained at 25°C and pH 7·4 (0·05 M sodium phosphate buffer) are as follows, the values relating to the reaction written as fumarate $\rightleftharpoons$ L-malate:

$$K_A \text{ (fumarate)} = 1\cdot37 \times 10^{-3} \text{ M}$$

$$K_X \text{ (malate)} = 4\cdot81 \times 10^{-3} \text{ M}$$

$$\frac{V_A}{V_X} = 1\cdot19. \qquad (57)$$

From these results it follows that

$$\frac{V_A K_X}{V_X K_A} = 4\cdot2, \qquad (58)$$

and this is in satisfactory agreement with the directly determined value of 4·45 for K_0 under the same conditions. The ratio of V_A / V_X is found[5] to vary markedly with pH over the range from 6·0 to 9·0, but the equilibrium constant does not vary in this range. It follows from eqn (56) that K_X / K_A must also vary markedly; its variation has actually been shown[6] to be consistent with

[1] H. von Euler and K. Josephson, *Z. physiol. Chem.*, **136**, 30 (1924).
[2] E. Bourquelot, *J. Pharm. et Chim.*, **10**, 361, 393 (1914).
[3] K. Josephson, *Z. physiol. Chem.*, **147**, 1, 155 (1925).
[4] R. M. Bock and R. A. Alberty, *J. Am. chem. Soc.*, **75**, 1921 (1953); see also R. A. Alberty, *J. Am. chem. Soc.*, **75**, 1928 (1953), where some extensions of the procedure are treated.
[5] R. M. Scott and R. Powell, *J. Am. chem. Soc.*, **79**, 1104 (1948).
[6] R. A. Alberty, V. Massey, C. Frieden, and A. R. Fuhlbrigge, *J. Am. chem. Soc.*, **76**, 2485 (1954).

the Haldane equation. On the other hand, the variation of V_A/V_X with temperature over the range from 8° to 45°C nearly parallels the variation of K_0, so that the ratio K_X/K_A must remain nearly constant with changing temperature.

A special case of the above mechanism is when there is a common Michaelis complex for the forward and reverse reactions:

$$E + A \underset{k_{-1}}{\overset{k_1}{\rightleftharpoons}} EA \underset{k'_1}{\overset{k'_{-1}}{\rightleftharpoons}} E + X.$$

The equations for this case can be obtained directly, or from eqns (48)–(52) by making k_2 and k_{-2} much larger than the other rate constants. The resulting expression in either case is

$$v = \frac{k_1 k_2[A] - k_{-1} k_{-2}[X]}{k_{-1} + k_2 + k_1[A] + k_{-2}[X]}. \tag{59}$$

Inhibition by substrate

In a number of single-substrate reactions it has been found that the hyperbolic law is not obeyed, but that the rate passes through a maximum as the substrate concentration is increased, and then falls[1]. An example of this type of behaviour is shown in Fig. 3.4. In some cases the rate approaches zero at very high substrate concentrations; however, with some reactions, such as the one shown in the figure, the rate approaches a value of greater than zero.

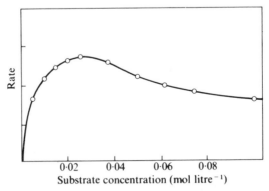

FIG. 3.4.   A plot of rate against substrate concentration for the hydrolysis of carbobenzoxy-glycyl-L-tryptophan, catalysed by carboxypeptidase. The rate passes through a maximum and reaches a limiting finite value {data of Lumry, Smith, and Glantz, *J. Am. chem. Soc.*, **73**, 4330 (1951)}.

[1] See, for example, K. J. Laidler and J. P. Hoare, *J. Am. chem. Soc.*, **71**, 2699 (1949), for urease; I. B. Wilson and F. Bergmann, *J. biol. Chem.*, **186**, 683 (1950), for acetylcholinesterase; R. A. Alberty, V. Massey, C. Frieden, and A. R. Fuhlbrigge, *J. Am. chem. Soc.*, **76**, 2485 (1954), for fumarase; R. Lumry, E. L. Smith, and R. R. Glantz, *J. Am. chem. Soc.*, **73**, 4330 (1951), and J. R. Whitaker, F. Menger, and M. L. Bender, *Biochemistry*, **5**, 386 (1966), for carboxypeptidase.

The simplest mechanism that explains this behaviour was first suggested independently by Haldane[1] and by Murray[2]. It involves the formation, subsequent to the ordinary Michaelis complex, of a second complex which contains two molecules of substrate attached to one of enzyme:

$$
\begin{array}{c}
\text{A} \\
+ \\
\text{E} + \text{A} \underset{k_{-1}}{\overset{k_1}{\rightleftharpoons}} \text{EA} \overset{k_2}{\rightarrow} \text{E} + \text{X} \\
k_{-a} \Big\Updownarrow k_a \\
\text{EA}_2 \overset{k_2'}{\rightarrow} \text{EA} + \text{X}.
\end{array}
$$

At very high substrate concentrations EA will all be converted into EA_2, and if this does not react further the rate approaches zero; if it reacts less rapidly than EA the rate approaches a limiting value that is lower than the maximum value, as in Fig. 3.4.

The steady-state analysis is as follows. The total concentration of enzyme is

$$
[E]_0 = [E] + [EA] + [EA_2] \tag{60}
$$

and the steady-state equations for EA and EA_2 are

$$
k_1[E].[A] - (k_{-1} + k_2 + k_a[A])[EA] + k_{-a}[EA_2] = 0 \tag{61}
$$

and

$$
k_a[EA].[A] - (k_{-a} + k_2')[EA_2] = 0. \tag{62}
$$

Equations (61) and (62) allow E and EA_2 to be expressed in terms of EA and A, and insertion of the expressions into (60) leads to

$$
[E]_0 = [EA]\left\{\frac{k_{-1} + k_2 + k_a[A]}{k_1[A]} - \frac{k_a k_{-a}}{k_1(k_{-a} + k_2')} + 1 + \frac{k_a}{k_{-a} + k_2'}[A]\right\}. \tag{63}
$$

If $k_2' = 0$ the rate is $k_2[EA]$, and reduces to

$$
v = \frac{k_2[E]_0.[A]}{\{(k_{-1} + k_2 + k_a[A])/k_1\} + [A] + (k_a[A]^2/k_{-a})}. \tag{64}
$$

This may be written as

$$
v = \frac{k_2[E]_0.[A]}{K_A' + [A] + K_A''[A]^2} \tag{65}
$$

where K_A' is a modified Michaelis constant and K_A'' is the equilibrium constant for the formation of EA_2 from $EA + A$.

Equation (65) is of the right form to explain the cases where the rate approaches zero at high substrate concentrations; the term in $[A]^2$ predominates in the denominator and the rate is inversely proportional to $[A]$. The

---

[1] J. B. S. Haldane, *Enzymes*, Longmans, Green and Co., London (1930); reprinted by the M.I.T. Press, Cambridge, Mass., 1965.
[2] D. R. P. Murray, *Biochem. J.*, **24**, 1890 (1930).

equation predicts that the rate passes through a maximum at a substrate concentration given by

$$[A]_{max} = \left\{\frac{K'_A}{K''_A}\right\}^{\frac{1}{2}}. \tag{66}$$

If k'_2 is not zero the overall rate of product formation is

$$v = k_2[EA] + k'_2[EA_2] \tag{67}$$

$$= \left\{k_2 + k'_2 \frac{k_a[A]}{k_{-a} + k'_2}\right\}[EA]. \tag{68}$$

This reduces to an equation of the form

$$= \frac{k'(k_2 + k'_2 K'_a[A])[E]_0 \cdot [A]}{K'_A + [A] + K''_A[A]^2} \tag{69}$$

where $K'_a = k_a/(k_{-a} + k'_2)$. At sufficiently high substrate concentrations the rate is

$$v_{lim} = \frac{k'_2 K'_a}{K''_A}[E]_0. \tag{70}$$

Equation (69) provides a satisfactory explanation of the results shown in Fig. 3.4.

Three different kinds of behaviour are special cases of eqn (70), and are illustrated in Fig. 3.5. Curve (*b*) shows the behaviour when the rate of breakdown of EA_2 is small compared with that of EA. Curve (*c*) corresponds to a high rate of reaction of EA_2; this case could be referred to as 'activation by substrate', but there appear to be few clear-cut examples.[1] Curve (*a*) shows an intermediate situation where hyperbolic kinetics are obeyed.

Other mechanisms also provide an explanation for inhibition by substrate. Laidler and Hoare[2] suggested an extension of the Haldane–Murray mechanism in which explicit consideration was given to the existence of two sites on the enzyme, each one of which can accept a substrate molecule. There can therefore be two different enzyme–substrate complexes, but only the one in which the substrate was on one of the sites was considered to be capable of breaking down into the products of reaction. This mechanism led to an equation of similar form to (65), and accounted for the three types of behaviour shown in Fig. 3.5.

An entirely different mechanism to explain substrate inhibition was suggested by Krupka and Laidler[3] in connection with their investigations of

---

[1] See, for example, J. R. Whitaker, F. Menger and M. L. Bender, *Biochemistry*, **5**, 386 (1966) for carboxypeptidase; H. Nakata and S. Ishü, *J. Biochem.* (*Japan*), **72**, 281 (1972), for trypsin.
[2] K. J. Laidler and J. P. Hoare, *J. Am. chem. Soc.*, **71**, 2699 (1949); see also R. J. Foster and C. Niemann, *Proc. natn. Acad. Sci.*, **39**, 371 (1953).
[3] R. M. Krupka and K. J. Laidler, *J. Am. chem. Soc.*, **83**, 1448 (1961).

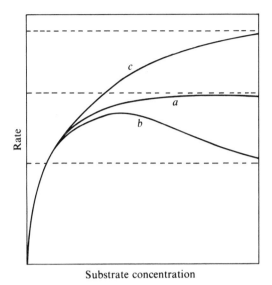

FIG. 3.5. Schematic curves showing (a) the simple hyperbolic behaviour, (b) inhibition by the substrate, and (c) activation by the substrate.

acetylcholinesterase. There is good evidence that the reactions catalysed by this enzyme proceed by a mechanism involving first an addition complex and then an acyl intermediate (see pp. 77–78). Furthermore, there are various lines of evidence that indicate that the portion of the active centre that is concerned with the binding of the substrate in the addition complex is not involved in the attachment of the acyl group to the enzyme. This suggests that this binding centre, free in the acyl enzyme, can bind a substrate molecule, and that it is this that causes substrate inhibition rather than the formation of a complex containing two substrate molecules. The latter possibility is somewhat difficult to envisage in view of specificity requirements for substrate attachment, since it seems to require that the enzyme is provided with two similar neighbouring sites for attachment.

The Krupka–Laidler mechanism, which avoids this difficulty, can be represented as

$$
\begin{array}{c}
\text{A} \\
+ \\
\text{E} + \text{A} \underset{k_{-1}}{\overset{k_1}{\rightleftharpoons}} \text{EA} \xrightarrow{k_2} \text{EA}' \xrightarrow{k_3} \text{E} + \text{X} \\
\text{X} \quad k_a \big\updownarrow k_{-a} \\
\text{EA}'\text{A}
\end{array}
$$

where EA$'$A represents the substrate attached to the acyl enzyme; this species is here assumed to be unreactive. The steady-state equations for

EA, EA′ and EA′A are

$$k_1[E].[A] - (k_{-1} + k_2)[EA] = 0 \qquad (71)$$

$$k_2[EA] - k_3[EA'] - k_a[EA'].[A] + k_{-a}[EA'A] = 0 \qquad (72)$$

$$k_a[EA'].[A] - k_{-a}[EA'A] = 0. \qquad (73)$$

Addition of (72) and (73) gives†

$$k_2[EA] - k_3[EA'] = 0. \qquad (74)$$

Solution of these equations for E, EA′ and EA′A in terms of EA, and insertion in the equation

$$[E]_0 = [E] + [EA] + [EA'] + [EA'A] \qquad (75)$$

leads to

$$[E]_0 = [EA]\left\{\frac{k_{-1} + k_2}{k_1[A]} + 1 + \frac{k_2}{k_3} + \frac{k_a k_2[A]}{k_{-a}k_3}\right\}. \qquad (76)$$

The rate is $k_2[EA]$, or

$$v = \frac{k_2[E]_0.[A]}{\{(k_{-1} + k_2)/k_1\} + \{(k_2 + k_3)/k_3\}[A] + (k_a k_2/k_{-a}k_3)[A]^2}. \qquad (77)$$

This equation is of the same form as eqn (65). It can be extended to include the possibility that EAA′ also undergoes reaction, as was done above. In the case of reactions catalysed by acetylcholinesterase Krupka and Laidler[1] were able to show, by means of experiments involving inhibitors, that the results were inconsistent with the mechanism involving the formation of EA_2. They were, however, consistent with the EA′A mechanism. This EA′A mechanism does not appear to have been tested for other systems, but it seems reasonable to suppose that it will apply to those cases in which an acyl or second intermediate is involved, particularly if the binding group is freed when the second intermediate is formed.

Analysis of experimental results

Systems in which there is inhibition by substrate are usually analysed by making Lineweaver–Burk plots, in which $1/v$ is plotted against $1/[A]$. Equation (65) can be written as

$$1/v = \frac{K'_A}{V_A}.\frac{1}{[A]} + \frac{1}{V_A} + \frac{K''_A}{V_A}[A] \qquad (78)$$

---

† This illustrates a useful general principle; when an intermediate, like EA′A, is at the end of a branch in a mechanism (i.e. it does not react further) the terms for its formation and removal can be omitted from the steady-state equations for other intermediates (e.g. EA′).

[1] *Loc. cit.*

where $V_A = k_2[E]_0$. A plot of $1/v$ against $1/[A]$ is shown schematically in Fig. 3.6; it is linear at sufficiently large values of $1/[A]$, but shows a deviation from linearity at low $1/[A]$ values when the final term in eqn (78) becomes important. The deviation is in the direction of making $1/v$ abnormally large at these high $[A]$ values. From the slope of the linear part of the curve one can determine K'_A/V_A, and from the intercept of the extrapolated straight line $1/V_A$; V_A and K'_A can therefore be determined separately. An Eadie plot ($v/[A]$ against v) can be employed in a similar manner to determine V_A and K'_A from the results in the low concentration regions.

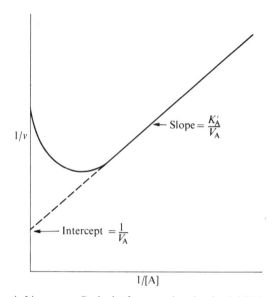

$1/v$

← Slope $= \dfrac{K'_A}{V_A}$

— Intercept $= \dfrac{1}{V_A}$

$1/[A]$

FIG. 3.6.   Schematic Lineweaver–Burk plot for a reaction showing inhibition by the substrate.

Various plots can be used to determine K''_A; these depend on making use of data in the high concentration region. For example, a plot of $1/v$ against $[A]$ will be linear at high $[A]$ values as shown in Fig. 3.7, and the slope gives K''_A/V_A, from which K''_A can be calculated. A plot of $v/[A]$ against v can be employed in a similar way.

Again, the most reliable procedure for determining kinetic parameters is not to measure slopes of plots, but to carry out a proper statistical analysis, with appropriate weighting of values (see p. 76).

Inhibition

It is commonly found that foreign substances influence the rates of enzyme-catalysed reactions. When there is a reduction of rate there is said to be *inhibition*, and the substance bringing about the effect is called an *inhibitor*. The rate is sometimes increased by an added substance; there is

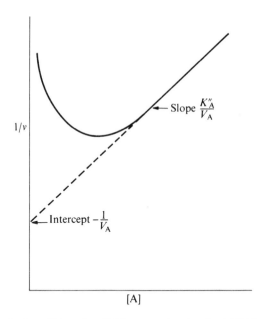

Fig. 3.7. Schematic plot of $1/v$ against [A] for a reaction showing inhibition by the substrate.

then said to be *activation*. Substances that are either activators or inhibitors are commonly referred to as *moderators*; the symbol Q will be used to signify a moderator.

In dealing with inhibition it is important to distinguish sharply between the effects that are observed experimentally, and the mechanisms that are proposed to explain them. It is always possible to put forward a number of plausible mechanisms to explain a given set of inhibition results. It has been common in the past to discuss inhibition effects entirely in terms of rather simple mechanisms, which we shall refer to as *classical* inhibition mechanisms; however, these mechanisms are undoubtedly simpler than those usually involved in enzyme systems. In the present treatment we discuss first the phenomenology only, without reference to mechanisms, and then consider possible mechanisms that explain the results.

When inhibition occurs in enzyme systems three different kinds of behaviour are to be distinguished, according to the way in which the degree of inhibition is affected by the substrate concentration. The degree of inhibition, i, is related to the velocity of the inhibited reaction, v, and that of the uninhibited reaction, v_0. It is defined as the reduction in velocity, $v_0 - v$, brought about by a given amount of inhibitor divided by the velocity of the uninhibited reaction:

$$i = \frac{v_0 - v}{v_0}. \tag{79}$$

The three situations that are to be distinguished are:

(1) The degree of inhibition may be unaffected by the concentration of substrate; in this case one speaks of *pure non-competitive* inhibition.

(2) The degree of inhibition may depend upon the amount of substrate present. Usually the degree of inhibition is reduced as the substrate concentration is increased; in this case one says that the inhibition has some *competitive* character.

(3) The degree of inhibition may be increased as the substrate concentration is increased. The adjectives *anticompetitive, coupling* and *uncompetitive* have been used to refer to this type of behaviour. The term *anticompetitive* will be employed in this book.

The use of these terms competitive, non-competitive and anticompetitive may have been somewhat unfortunate, since they imply conclusions about mechanisms that are not necessarily justified. The term competitive for example, was used because it was thought that when there was a reduction of degree of inhibition with substrate concentration the inhibitor and substrate molecules were competing with one another for a site on the enzyme surface; when the degree of inhibition was independent of substrate concentration it was thought that there was no such competition, the substrate and inhibitor molecules being able to become simultaneously attached to the enzyme. Such conclusions are only justified if the mechanism is of the simple Michaelis type, with a single intermediate. It will be seen later that if a second intermediate is involved, or there is more than one substrate, there is no such simple correlation between the behaviour observed and the question of competition for enzyme sites; a given mechanism can, in fact, lead to different types of behaviour depending upon the relative magnitudes of rate constants (see especially pp. 102–108). In spite of this it is convenient to employ the terms competitive, non-competitive and anticompetitive as descriptions of the behaviour observed, with no implication as to mechanisms.

The rate equations relating velocity to substrate and inhibitor concentrations are often rather complex, but there are three particularly simple equations that merit consideration, especially since there are many systems to which they apply satisfactorily. The first of these can be written as

$$v = \frac{V[A]}{(K_A + [A])(1 + [Q]/K_Q)} \tag{80}$$

where $[Q]$ is the concentration of inhibitor and V, K_A and K_Q are constants. The constant K_A is the Michaelis constant for the reaction in the absence of inhibitor, in which case the rate is

$$v_0 = \frac{V[A]}{K_A + [A]}. \tag{81}$$

Similarly V is the rate of the uninhibited reaction at high substrate concentrations. The constant K_Q gives some indication of the strength of the inhibitor Q; a powerful inhibitor will have a low value of K_Q and vice versa. If eqn (80) applies the degree of inhibition is given by

$$i = \frac{v_0 - v}{v_0} = 1 - \frac{v}{v_0} \tag{82}$$

$$= 1 - \frac{1}{1 + [Q]/K_Q} \tag{83}$$

$$= \frac{[Q]}{K_Q + [Q]} \tag{84}$$

and is therefore independent of [A]. The behaviour is therefore *non-competitive*. We shall employ the term *pure non-competitive* to describe the behaviour of systems obeying eqn (80).

The second simple type of behaviour is represented by the equation

$$v = \frac{V[A]}{K_A\left(1 + \dfrac{[Q]}{K_Q}\right) + [A]}. \tag{85}$$

Again, V and K_A are kinetic parameters for the reaction occurring in the absence of inhibitor, when eqn (81) applies. The degree of inhibition is

$$i = \frac{K_A\dfrac{[Q]}{K_Q}}{K_A\left(1 + \dfrac{[Q]}{K_Q}\right) + [A]}. \tag{86}$$

The degree of inhibition is now reduced by addition of substrate, and the behaviour is *competitive*. Systems which follow an equation of the form of eqn (85) will be said to show *pure competitive* behaviour.

The third simple type of behaviour, which relates to what we shall call *pure anticompetitive* inhibition, is described by the equation

$$v = \frac{V[A]}{K_A + \left(1 + \dfrac{[Q]}{K_Q}\right)[A]}. \tag{87}$$

This corresponds to little inhibition at low substrate concentrations and to maximum inhibition at high substrate concentrations. The degree of inhibition is

$$i = \frac{[A]\dfrac{[Q]}{K_Q}}{K_A + \left(1 + \dfrac{[Q]}{K_Q}\right)[A]} \tag{88}$$

and rises from zero to $[Q]/(K_Q + [Q])$ as the substrate concentration is raised from zero to a high value.

These three pure types of inhibition can readily be detected in experimental systems by means of Lineweaver–Burk plots of $1/v$ against $1/[A]$. This is illustrated in Fig. 3.8, which summarizes the rate equations, including the

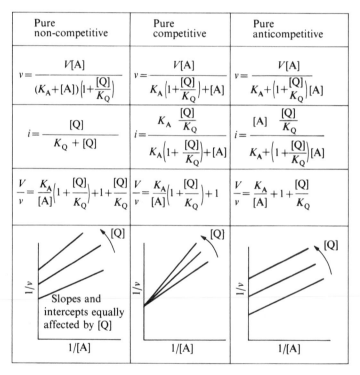

Pure non-competitive	Pure competitive	Pure anticompetitive
$v = \dfrac{V[A]}{(K_A+[A])\left(1+\dfrac{[Q]}{K_Q}\right)}$	$v = \dfrac{V[A]}{K_A\left(1+\dfrac{[Q]}{K_Q}\right)+[A]}$	$v = \dfrac{V[A]}{K_A+\left(1+\dfrac{[Q]}{K_Q}\right)[A]}$
$i = \dfrac{[Q]}{K_Q+[Q]}$	$i = \dfrac{K_A\dfrac{[Q]}{K_Q}}{K_A\left(1+\dfrac{[Q]}{K_Q}\right)+[A]}$	$i = \dfrac{[A]\dfrac{[Q]}{K_Q}}{K_A+\left(1+\dfrac{[Q]}{K_Q}\right)[A]}$
$\dfrac{V}{v} = \dfrac{K_A}{[A]}\left(1+\dfrac{[Q]}{K_Q}\right)+1+\dfrac{[Q]}{K_Q}$	$\dfrac{V}{v} = \dfrac{K_A}{[A]}\left(1+\dfrac{[Q]}{K_Q}\right)+1$	$\dfrac{V}{v} = \dfrac{K_A}{[A]}+1+\dfrac{[Q]}{K_Q}$

FIG. 3.8. Summary of equations and plots for the three pure forms of inhibition. In the pure non-competitive case the lines intersect on the $1/[A]$ axis.

reciprocal forms, and shows the types of Lineweaver–Burk plots that are obtained in these three cases.

It is also of interest to compare the way in which, according to the three equations (80), (85) and (87), the effective $k_c(Q)(= V/[E]_0)$ and $K_A(Q)$ values for the inhibited reactions vary with the inhibitor concentrations. The

TABLE 3.1

Effective $k_c(Q)$ and $K_A[A]$ values for systems showing the pure types of inhibition

	$k_c(Q)$	$K_A(Q)$
Pure non-competitive inhibition		
$v = \dfrac{k_c[E]_0 \cdot [A]}{(K_A + [A])\left(1 + \dfrac{[Q]}{K_Q}\right)}$	$\dfrac{k_c}{1 + \dfrac{[Q]}{K_Q}}$	K_A
Pure competitive inhibition		
$v = \dfrac{k_c[E]_0 \cdot [A]}{K_A\left(1 + \dfrac{[Q]}{K_Q}\right) + [A]}$	k_c	$K_A\left(1 + \dfrac{[Q]}{K_Q}\right)$
Pure anticompetitive inhibition		
$v = \dfrac{k_c[E]_0 \cdot [A]}{K_A + \left(1 + \dfrac{[Q]}{K_Q}\right)[A]}$	$\dfrac{k_c}{1 + \dfrac{[Q]}{K_Q}}$	$\dfrac{K_A}{1 + \dfrac{[Q]}{K_Q}}$

expressions for $k_c[Q]$ and $K_A(Q)$ are shown in Table 3.1. It is to be noted that for pure non-competitive inhibition $K_A(Q)$ is independent of Q but $k_c(Q)$ depends on Q; for pure competitive inhibition the dependencies are reversed, while for pure anticompetitive inhibition both $k_c(Q)$ and $K_A(Q)$ are functions of Q.

So far we have considered only the three pure types of inhibition. Intermediate behaviour can, of course, also arise. Many situations are covered by the general equation

$$v = \frac{V[A]}{K_A\left(1 + \dfrac{[Q]}{K_Q^s}\right) + \left(1 + \dfrac{[Q]}{K_Q^i}\right)[A]} \tag{89}$$

which in reciprocal form becomes

$$\frac{V}{v} = \frac{K_A}{[A]}\left(1 + \frac{[Q]}{K_Q^s}\right) + 1 + \frac{[Q]}{K_Q^i}. \tag{90}$$

A Lineweaver–Burk plot for this equation is shown schematically in Fig. 3.9. The slope of this plot, for a given value of [Q], depends upon the magnitude

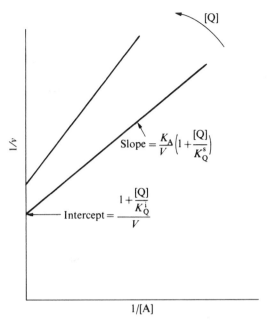

FIG. 3.9. Schematic plot of $1/v$ against $1/[A]$ for the general inhibition equation (89).

of K_Q^s, and the intercept on the magnitude of K_Q^i. This more general equation reduces to the pure forms under the following conditions:

(1) $K_Q^s = K_Q^i$; pure non-competitive

(2) $K_Q^i = \infty$; pure competitive

(3) $K_Q^s = \infty$; pure anticompetitive.

When neither of these three conditions is satisfied the behaviour can be described as *mixed*. The degree of inhibition is, when eqn (89) applies, given by

$$i = \frac{K_A\left(\dfrac{[Q]}{K_Q^s}\right) + \left(\dfrac{[Q]}{K_Q^i}\right)[A]}{K_A\left(1 + \dfrac{[Q]}{K_Q^s}\right) + \left(1 + \dfrac{[Q]}{K_Q^i}\right)[A]}. \tag{91}$$

The degree of inhibition at very low $[A]$ values is given by

$$i([A] \to 0) = \frac{[Q]}{K_Q^s + [Q]}. \tag{92}$$

That at very high $[A]$ values is given by

$$i([A] \to \infty) = \frac{[Q]}{K_Q^i + [Q]}. \tag{93}$$

It is clear that two cases are to be distinguished:

(1) If $K_Q^i > K_Q^s$ the degree of inhibition is reduced as [A] is increased; the inhibition therefore has some competitive character, and it will therefore be described in the present book as *mixed non-competitive-competitive inhibition.*

(2) If $K_Q^s > K_Q^i$ the degree of inhibition is increased as [A] is increased; the inhibition therefore has some anticompetitive character. It will here be described as *mixed non-competitive-anticompetitive inhibition*[†].

Figure 3.10 illustrates this classification in a schematic fashion.

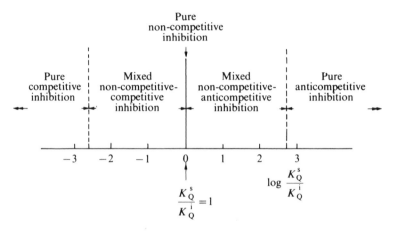

FIG. 3.10. Classification of the different types of inhibition. The boundaries are of course arbitrary, depending upon the precision with which the K_Q values can be determined.

Analysis of experimental results

By far the most useful way of analysing experimental data is to make Lineweaver–Burk plots of $1/v$ against $1/[A]$, as discussed with reference to Fig. 3.9. Such plots may be referred to as primary plots. The magnitudes of the inhibition constants K_Q^s and K_Q^i can then be determined by secondary plots of slopes and intercepts against [Q], as shown schematically in Fig. 3.11. The slopes of such secondary plots are K_A/VK_Q^s for slope and $1/K_Q^iV$ for intercept. Since K_A and V can be determined from studies in the absence of inhibitor, the K_Q^s and K_Q^i values are readily calculated.

[†] W. W. Cleland (*Biochim. biophys. Acta*, **67**, 173 (1963), who has made important contributions to the theory of inhibition, especially for multisubstrate systems (see Chapter 4), has used the term *linear non-competitive* to describe behaviour represented by eqn (89) in which K_Q^s and K_Q^i have values which are experimentally distinguishable from infinity (i.e., in which the slopes and intercepts of the Lineweaver–Burk plots both show a significant linear variation with [Q]). The term *linear* is useful in indicating that there is a linear variation with [Q] (more complicated dependencies are also observed), but the simple term *non-competitive* does not seem sufficiently descriptive in such cases, giving no indication of whether the degree of inhibition is increased or decreased as [A] is increased. For example, it seems misleading to describe as non-competitive a system in which $K_Q^i \gg K_Q^s$, the degree of inhibition being strongly reduced by increasing [A]; such inhibition in fact has more competitive than non-competitive character.

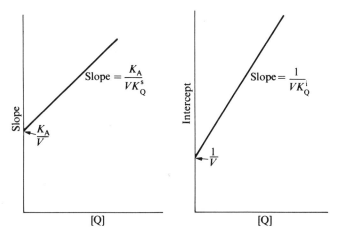

FIG. 3.11. Secondary plots of the slopes and intercepts of a primary-inhibition plot (Fig. 3.9) against [Q].

Sometimes the secondary plots are not linear. This can arise from various causes; examples, and mechanisms which explain this more complex behaviour, will be given later. The type of plots to be used varies with the kind of behaviour observed.

Several other graphical methods have also been suggested and employed, but on the whole they have not proved as useful as the primary and secondary plots described above. One such procedure[1] involves plotting v_0/v, or $1/v$, against [Q]. Such plots are often useful in determining reliable K_Q values after the type of inhibition has been clearly established by the Lineweaver–Burk plots.

Classical inhibition mechanisms

Until recently it has been usual to interpret inhibition in terms of schemes involving only a single intermediate, the Michaelis complex. There are some mechanisms which appear to be of this kind (for example, ribonuclease, considered in Chapter 10). We therefore give a brief account of these schemes, which will be referred to as classical inhibition mechanisms.

The essential difference between the classical schemes for simple competitive and simple non-competitive inhibition is that in the former there cannot exist a complex EAQ in which both substrate and inhibitor molecules are simultaneously attached to the enzyme. In the non-competitive mechanism, however, such a complex can be formed. It can be imagined that in competitive inhibition there is a single site that can bind either a substrate or an inhibitor molecule but not both; in non-competitive inhibition one envisages two sites, one for the substrate and one for the inhibitor, and it must be assumed

[1] M. Dixon, *Biochem. J.*, **55**, 170 (1953).

that the bound inhibitor molecule interferes with the subsequent reaction of the substrate.

The classical scheme of reactions applicable to competitive inhibition may be written as follows:

$$
\begin{array}{c}
A \\
+ \\
Q + E \underset{k_{-Q}}{\overset{k_Q}{\rightleftharpoons}} EQ \\
k_1 \big\updownarrow k_{-1} \\
EA \\
\big\downarrow k_2 \\
E + X.
\end{array}
$$

The steady-state equations for EA and EQ are†

$$k_1[E] \cdot [A] - (k_{-1} + k_2)[EA] = 0 \tag{94}$$

$$k_Q[E] \cdot [Q] - k_{-Q}[EQ] = 0. \tag{95}$$

Insertion of the resulting expressions for E and EQ, in terms of EA, into the expression for the total enzyme concentration leads to

$$[E]_0 = [E] + [EA] + [EQ] \tag{96}$$

$$= [EA]\left\{\frac{k_{-1} + k_2}{k_1[A]} + 1 + \frac{k_Q[Q](k_{-1} + k_2)}{k_{-Q}k_1[A]}\right\}. \tag{97}$$

The rate, equal to $k_2[EA]$, is

$$v = \frac{k_2[E]_0 \cdot [A]}{\{(k_{-1} + k_2)/k_1\} + [A] + \{(k_Q(k_{-1} + k_2))/k_{-Q}k_1\}[Q]}. \tag{98}$$

This may be written as

$$v = \frac{V[A]}{K_A\left(1 + \dfrac{[Q]}{K_Q}\right) + [A]} \tag{99}$$

where $K_A\{= (k_{-1} + k_2)/k_1\}$ is the Michaelis constant for the reaction occurring in the absence of inhibitor, and $K_Q(= k_{-Q}/k_Q)$ is the equilibrium constant for the *dissociation* of the complex EQ into E + Q.

---

† Note that in the steady-state equation for E the reactions involving EQ can be neglected; see footnote on p. 88.

The classical mechanism for simple non-competitive inhibition may be written as

$$
\begin{array}{c}
\text{A} \\
+ \\
\text{Q} + \text{E} \underset{}{\overset{K_Q}{\rightleftharpoons}} \text{EQ} \\
k_1 \big\updownarrow k_{-1} \\
\text{Q} + \text{EA} \overset{K_{AQ}}{\rightleftharpoons} \text{EAQ} \\
\big\downarrow k_2 \\
\text{E} + \text{X}
\end{array}
$$

K_Q and K_{AQ} are the equilibrium constants for the dissociations of EQ and EAQ.

The steady-state equations for EA, EQ and EAQ are

$$k_1[\text{E}].[\text{A}] - (k_{-1} + k_2)[\text{EA}] = 0 \tag{100}$$

$$[\text{E}].[\text{Q}] - K_Q[\text{EQ}] = 0 \tag{101}$$

$$[\text{EA}].[\text{Q}] - K_{AQ}[\text{EAQ}] = 0. \tag{102}$$

The total enzyme concentration is

$$[\text{E}]_0 = [\text{E}] + [\text{EQ}] + [\text{EA}] + [\text{EAQ}] \tag{103}$$

$$= [\text{EA}]\left\{ \frac{k_{-1} + k_2}{k_1[\text{A}]}\left(1 + \frac{[\text{Q}]}{K_Q}\right) + 1 + \frac{[\text{Q}]}{K_{AQ}}\right\} \tag{104}$$

$$= \frac{[\text{EA}]}{[\text{A}]}\left\{ K_A\left(1 + \frac{[\text{Q}]}{K_Q}\right) + [\text{A}]\left(1 + \frac{[\text{Q}]}{K_{AQ}}\right)\right\} \tag{105}$$

where $K_A = (k_{-1} + k_2)/k_1$. The rate, $k_2[\text{EA}]$, is

$$v = \frac{k_2[\text{E}]_0.[\text{A}]}{K_A\left(1 + \dfrac{[\text{Q}]}{K_Q}\right) + \left(1 + \dfrac{[\text{Q}]}{K_{AQ}}\right)[\text{A}]}. \tag{106}$$

If $K_{AQ} = K_Q$, which means that EA binds the inhibitor to the same extent as E, this equation reduces to

$$v = \frac{V[\text{A}]}{(K_A + [\text{A}])\left(1 + \dfrac{[\text{Q}]}{K_Q}\right)} \tag{107}$$

where $V = k_2[\text{E}]_0.[\text{A}]$. This is the equation for pure non-competitive inhibition.

If K_Q and K_{AQ} are different from one another the inhibition has some competitive or anticompetitive character, as discussed earlier (p. 96). The constant K_Q is, in fact, equivalent to the K_Q^s considered earlier, and K_{AQ} is K_Q^i. The pure competitive and pure anticompetitive forms are, of course, obtained if $K_{AQ} \gg K_Q$ and $K_Q \gg K_{AQ}$ respectively.

Several further points are worthy of note with respect to these classical inhibition mechanisms, since some of them involve principles which apply to the more complex systems. In the first place, it is to be seen that in the general scheme given on p. 99 there are no arrows connecting EQ and EAQ. Inclusion of these reactions complicates the steady-state treatment very considerably, and leads to rate expressions that cannot satisfactorily be tested with reference to the experimental results. One case, however, is particularly simple; this is when EQ + A is converted into EAQ with the same rate constant as for the conversion of E + A into EA. If this is so the formation of EQ is kinetically irrelevant; the behaviour is the same as in the case in which EQ is not formed at all. If EAQ is completely unreactive the behaviour is pure anticompetitive. If, on the other hand, EAQ gives rise to products with a rate constant k_2, there is no inhibition; the formation of EQ and EAQ has no effect on the rates.

Secondly, the possibility that Q combines with the *substrate* has not been considered. If there is no combination with either E or EA the scheme is

$$Q + A \underset{}{\overset{K_Q^A}{\rightleftharpoons}} AQ$$
$$+$$
$$E$$
$$k_1 \Big\updownarrow k_{-1}$$
$$EA$$
$$k_2 \Big\downarrow$$
$$E + X$$

K_Q^A is the dissociation constant of AQ,

$$K_Q^A = \frac{[A] \cdot [Q]}{[AQ]} \tag{108}$$

so that the total substrate concentration is

$$[A]_0 = [A] + [AQ] = [A]\left(1 + \frac{[Q]}{K_Q^A}\right). \tag{109}$$

It will be assumed that excess of Q is present, so that $Q \simeq Q_0$. The steady-state equation for EA is

$$k_1[E] \cdot [A] - (k_{-1} + k_2)[EA] = 0 \tag{110}$$

and the total enzyme concentration is

$$[E]_0 = [E] + [EA] \tag{111}$$

$$= \left\{ \frac{k_{-1} + k_2}{k_1} \cdot \frac{1}{[A]_0} \left(1 + \frac{[Q]}{K_Q^A} \right) + 1 \right\} [EA]. \tag{112}$$

The rate, $k_2[EA]$, is therefore

$$v = \frac{k_2[E]_0 \cdot [A]_0}{\{(k_{-1} + k_2)/k_1\}(1 + [Q]/K_Q^A) + [A]_0}. \tag{113}$$

This is of exactly the same form as the equation for competitive inhibition. In interpreting inhibition results it is therefore important to establish whether there is any combination between the inhibitor and the substrate.

Thirdly, the possibility of irreversible inhibition effects must be borne in mind. Suppose, for example, that Q combined irreversibly with E but not at all with EA. The result is simply a diminution of the concentration of effective enzyme, to an extent depending on how much Q is present. Addition of A would have no restoring effect, in contrast to the reversible case. The limiting rate V_A would be lowered, but K_A would be unaffected. The behaviour would therefore appear to be non-competitive.

Table 3.2 summarizes the conclusions of the present section, and includes some additional cases.

TABLE 3.2
Summary of inhibition results obtained
for single intermediate systems

Inhibitor combines with	Type of inhibition law	
	Reversible combination	Irreversible combination
E only	Pure competitive	Pure non-competitive
A only	Pure competitive	Non-linear
EA only	Pure anticompetitive	Pure non-competitive
E and EA	Mixed	Pure non-competitive
A and EA	Mixed	Non-linear

Non-linear means that the slopes and intercepts of the plots of $1/v$ against $1/[A]$ do not vary linearly with [Q].

Reactions involving a second intermediate

In view of the prevalence of reaction mechanisms in which, besides the addition complex, a second intermediate plays an important kinetic role, it is natural to explore inhibition mechanisms in which the existence of this intermediate is taken into account. This was first done by Krupka and Laidler[1] in terms of the mechanism:

$$
\begin{array}{c}
\text{A} \\
+ \\
\text{Q + E} \xrightleftharpoons{K_Q} \text{EQ} \\
k_1 \updownarrow k_{-1} \\
\text{Q + EA} \xrightleftharpoons{K_{AQ}} \text{EAQ} \\
k_2 \searrow \text{X} \\
\text{Q + EA}' \xrightleftharpoons{K_{A'Q}} \text{EA}'\text{Q} \\
k_3 \downarrow \\
\text{E + Y.}
\end{array}
$$

Here the inhibitor is allowed to combine with the free enzyme, with EA, and with the second intermediate EA'. A special case of this, and one that was found to apply to acetylcholinesterase[2], is when the inhibitor combines with E and EA' but is unable to become attached to the addition complex $\text{EA}(K_{AQ} = 0)$:

$$
\begin{array}{c}
\text{A} \\
+ \\
\text{Q + E} \xrightleftharpoons{K_{AQ}} \text{EQ} \\
k_1 \updownarrow k_{-1} \\
\text{EA} \\
k_2 \searrow \text{X} \\
\text{Q + EA}' \xrightleftharpoons{K_{A'Q}} \text{EA}'\text{Q} \\
k_3 \downarrow \\
\text{E + Y.}
\end{array}
$$

If this is the situation it is easy to see, without carrying out the steady-state analysis, that either competitive or non-competitive behaviour can arise according to whether reaction 2 or 3 is slow and rate-determining:

(1) If $k_2 \gg k_3$ the kinetics will be controlled by the breakdown of EA', and since EA'Q can be formed the kinetics will be non-competitive (pure non-competitive if $K_{A'Q} = K_{AQ}$).

---

[1] R. M. Krupka and K. J. Laidler, *J. Am. chem. Soc.*, **83**, 1454 (1961); cf. also H. Kaplan and K. J. Laidler, *Can. J. Chem.*, **45**, 539, 559 (1967).
[2] R. M. Krupka and K. J. Laidler, *loc. cit.*

(2) If $k_3 \gg k_2$ the conversion of EA into EA′ is slow and rate-determining, and since EAQ is not formed the behaviour will be competitive.

The steady-state analysis of the general scheme is as follows. The steady-state equations for EA and EA′ are

$$k_1[A] - (k_{-1} + k_2)[EA] = 0 \tag{114}$$

$$k_2[EA] - k_3[EA'] = 0. \tag{115}$$

The equilibrium equations for EQ, EAQ and EA′Q are

$$K_Q[EQ] = [E].[Q] \tag{116}$$

$$K_{AQ}[EAQ] = [EA].[Q] \tag{117}$$

$$K_{A'Q}[EA'Q] = [EA'].[Q]. \tag{118}$$

The total enzyme concentration is given by

$$[E]_0 = [E] + [EQ] + [EA] + [EAQ] + [EA'] + [EA'Q]$$

$$= [E]\left(1 + \frac{[Q]}{K_Q}\right) + [EA]\left(1 + \frac{[Q]}{K_{AQ}}\right) + [EA']\left(1 + \frac{[Q]}{K_{A'Q}}\right). \tag{119}$$

It is obvious that the final expression for the rate will be the same as that for the reaction in the absence of inhibitor (eqn (28)) except that the term in the denominator arising from E will be multiplied by $(1 + [Q]/K_Q)$, that from EA by $(1 + [Q]/K_{AQ})$ and that from EA′ by $(1 + [Q]/K_{A'Q})$; the result is

$$v = \frac{k_2[E]_0.[A]}{\dfrac{k_{-1}+k_2}{k_1}\left(1 + \dfrac{[Q]}{K_Q}\right) + \left(1 + \dfrac{[Q]}{K_{AQ}}\right)[A] + \dfrac{k_2}{k_3}\left(1 + \dfrac{[Q]}{K_{A'Q}}\right)[A]} \tag{120}$$

$$= \frac{\{k_2k_3/(k_2+k_3)\}[E]_0.[A]}{\left(\dfrac{k_{-1}+k_2}{k_1}\right)\left(\dfrac{k_3}{k_2+k_3}\right)\left(1+\dfrac{[Q]}{K_Q}\right) + \dfrac{k_3}{k_2+k_3}\left(1+\dfrac{[Q]}{K_{AQ}}\right)[A] + \dfrac{k_2}{k_2+k_3}\left(1+\dfrac{[Q]}{K_{A'Q}}\right)[A]} \tag{121}$$

$$= \frac{V[A]}{K_A\left(1+\dfrac{[Q]}{K_Q}\right) + \dfrac{k_3}{k_2+k_3}\left(1+\dfrac{[Q]}{K_{AQ}}\right)[A] + \dfrac{k_2}{k_2+k_3}\left(1+\dfrac{[Q]}{K_{A'Q}}\right)[A]} \tag{122}$$

where K_A, equal to $(k_{-1} + k_2)k_3/k_1(k_2 + k_3)$, is the Michaelis constant for the system in the absence of inhibitor. $V\{= k_2k_3[E]_0/(k_2 + k_3)\}$ is the limiting maximum velocity in the absence of inhibitor. It can be seen from eqn (120) that the effective Michaelis constants and limiting velocity in the presence of inhibitor are

$$K_A(Q) = \frac{\left(\dfrac{k_{-1} + k_2}{k_1}\right)\left(1 + \dfrac{[Q]}{K_Q}\right)}{\left(1 + \dfrac{[Q]}{K_{AQ}}\right) + \dfrac{k_2}{k_3}\left(1 + \dfrac{[Q]}{K_{A'Q}}\right)} \tag{123}$$

and

$$V(Q) = \frac{k_2[E]_0}{\left(1 + \dfrac{[Q]}{K_{AQ}}\right) + \dfrac{k_2}{k_3}\left(1 + \dfrac{[Q]}{K_{A'Q}}\right)} \tag{124}$$

Some special cases of these equations are of interest:

(1) If the inhibitor is bound equally strongly to E, EA and EA' ($K_Q = K_{AQ} = K_{A'Q}$) the rate equation (122) becomes

$$v = \frac{V[A]}{(K_A + [A])\left(1 + \dfrac{[Q]}{K_Q}\right)}. \tag{125}$$

This equation corresponds to pure non-competitive inhibition under all conditions.

(2) Suppose that the inhibitor is bound equally strongly to E and EA', but not at all to EA($K_Q = K_{A'Q}$; $K_{AQ} = \infty$). Equation (122) now reduces to

$$v = \frac{V[A]}{K_A\left(1 + \dfrac{[Q]}{K_Q}\right) + \left(\dfrac{k_3}{k_2 + k_3}\right)[A] + \left(\dfrac{k_2}{k_2 + k_3}\right)\left(1 + \dfrac{[Q]}{K_Q}\right)}. \tag{126}$$

Two subcases of this are:

(a) Reaction 2 is rate-controlling; i.e., $k_3 \gg k_2$. Equation (126) now reduces to

$$v = \frac{V[A]}{K_A\left(1 + \dfrac{[Q]}{K_Q}\right) + [A]} \tag{127}$$

which is the equation for pure competitive inhibition (see eqn (86)).

(b) Reaction 3 is rate-controlling; i.e., $k_2 \gg k_3$. Equation (128) now becomes

$$v = \frac{V[A]}{(K_A + [A])\left(1 + \dfrac{[Q]}{K_Q}\right)} \tag{128}$$

which is the equation for pure non-competitive inhibition (see eqn (80)).

The important point that emerges from the preceding treatment is that a given inhibitor for a particular enzyme may be either competitive, non-competitive or mixed, depending upon the substrate employed. This has, in fact been demonstrated by Krupka and Laidler[1] for the inhibition of the enzyme acetylcholinesterase by *cis*-2-dimethylaminocyclohexanol. When methylaminoethyl acetate was used as substrate, reaction 2 was rate controlling and the inhibition was competitive. When, on the other hand, acetylcholine was used, $k_2 \gg k_3$, and the inhibition was non-competitive. The evidence is that the inhibitor binds equally strongly to the free enzyme and the acetyl enzyme, and not at all to the Michaelis complex, as in the scheme shown above.

TABLE 3.3
Summary of inhibition relationships for the two-intermediate mechanism (unreactive inhibitor complexes)

Conditions	Rate equation with	
	$k_3 \gg k_2$	$k_2 \gg k_3$
$K_Q = K_{AQ} = K_{A'Q}$	$v = \dfrac{V[A]}{(K_A + [A])\left(1 + \dfrac{[Q]}{K_Q}\right)}$ Pure non-competitive	
$K_Q = K_{A'Q}; K_{AQ} = \infty$	$v = \dfrac{V[A]}{K_A\left(1 + \dfrac{[Q]}{K_Q}\right) + [A]}$ Pure competitive	$v = \dfrac{V[A]}{(K_A + [A])\left(1 + \dfrac{[Q]}{K_Q}\right)}$ Pure non-competitive
$K_Q = K_{AQ}; K_{A'Q} = \infty$	$v = \dfrac{V[A]}{(K_A + [A])\left(1 + \dfrac{[Q]}{K_Q}\right)}$ Pure non-competitive	$v = \dfrac{V[A]}{K_A\left(1 + \dfrac{[Q]}{K_Q}\right) + [A]}$ Pure competitive
$K_{AQ} = K_{A'Q} = \infty$	$v = \dfrac{V[A]}{K_A\left(1 + \dfrac{[Q]}{K_Q}\right) + [A]}$ Pure competitive	

Table 3.3 summarizes the rate equations to which eqn (122) reduces, under several conditions including those considered above, and indicates the type of inhibition observed in each case.

[1] R. M. Krupka and K. J. Laidler, *J. Am. chem. Soc.*, **83**, 1445 (1961).

A more general treatment

The preceding treatment has involved the assumption that the species EQ, EAQ and EA'Q do not undergo any further reaction. However, there is evidence for some acetylcholinesterase systems[1] that the acetyl intermediate EA'Q is formed but breaks down at the same rate as EA'. It is therefore necessary to develop equations for the scheme:

The following steady-state analysis of this problem was developed by Kaplan and Laidler[2].

An exact treatment of this type of mechanism, in which there are reaction cycles, leads to very complicated rate equations which are not useful for the analysis of experimental results. A satisfactory approximate solution for cases of this kind is to treat the horizontal reactions as rapid and therefore in equilibrium, and to apply the steady-state treatment to the vertical reactions. This procedure is justified by the fact that the horizontal reactions are simply binding and dissociation processes, whereas the vertical reactions involve chemical change.

The equilibrium equations corresponding to the three horizontal reactions are

$$K_Q[EQ] = [E].[Q] \tag{129}$$

$$K_{AQ}[EAQ] = [EA].[Q] \tag{130}$$

$$K_{A'Q}[EA'Q] = [EA'].[Q]. \tag{131}$$

The sum of the steady-state equations for EA and EAQ is

$$k_1[E].[A] + ak_1[EQ].[A] = (k_{-1} + k_2)[EA] + (bk_2 + a'k_{-1})[EAQ]. \tag{132}$$

[1] R. M. Krupka and K. J. Laidler, *J. Am. chem. Soc.*, **83**, 1454 (1961).
[2] H. Kaplan and K. J. Laidler, *Can. J. Chem.*, **45**, 539 (1967); see also R. M. Krupka and K. J. Laidler, *J. Am. chem. Soc.*, **83**, 1445, 1457 (1961).

From (129), (130) and (132),

$$k_1[\text{E}] \cdot [\text{A}]\left(1 + \frac{a[\text{Q}]}{K_\text{Q}}\right) = (k_{-1} + k_2)\left(1 + \frac{\bar{a}[\text{Q}]}{K_\text{AQ}}\right)[\text{EA}] \qquad (133)$$

where

$$\bar{a} = \frac{bk_2 + a'k_{-1}}{k_{-1} + k_2}. \qquad (134)$$

Similarly, the sum of the steady-state equations for EA' and EA'Q is

$$k_2[\text{EA}]\left(1 + \frac{b[\text{Q}]}{K_\text{AQ}}\right) = k_3[\text{EA}']\left(1 + \frac{c[\text{Q}]}{K_\text{A'Q}}\right). \qquad (135)$$

The total enzyme concentration is

$$[\text{E}]_0 = [\text{E}] + [\text{EQ}] + [\text{EA}] + [\text{EAQ}] + [\text{EA}'] + [\text{EA}'\text{Q}] \qquad (136)$$

and the rate is

$$v = k_2[\text{EA}] + bk_2[\text{EAQ}]. \qquad (137)$$

These equations lead finally to the rate equation

$$v = \frac{V(\text{Q})[\text{A}]}{K_\text{A}(\text{Q}) + [\text{A}]} \qquad (138)$$

where

$$V(\text{Q}) = \frac{k_2[\text{E}]_0}{\left(1 + \dfrac{[\text{Q}]}{K_\text{AQ}}\right)\bigg/\left(1 + \dfrac{b[\text{Q}]}{K_\text{AQ}}\right) + k_2\left(1 + \dfrac{[\text{Q}]}{K_\text{A'Q}}\right)\bigg/ k_3\left(1 + \dfrac{c[\text{Q}]}{K_\text{A'Q}}\right)} \qquad (139)$$

and

$$\frac{1}{K_\text{A}(\text{Q})} = \frac{k_1}{k_{-1} + k_2}\left\{\frac{\left(1 + \dfrac{a[\text{Q}]}{K_\text{Q}}\right)\left(1 + \dfrac{[\text{Q}]}{K_\text{AQ}}\right)}{\left(1 + \dfrac{[\text{Q}]}{K_\text{Q}}\right)\left(1 + \dfrac{\bar{a}[\text{Q}]}{K_\text{AQ}}\right)} + \right. \qquad (140)$$

$$\left. + \frac{k_2\left(1 + \dfrac{a[\text{Q}]}{K_\text{Q}}\right)\left(1 + \dfrac{b[\text{Q}]}{K_\text{AQ}}\right)\left(1 + \dfrac{[\text{Q}]}{K_\text{A'Q}}\right)}{k_3\left(1 + \dfrac{\bar{a}[\text{Q}]}{K_\text{AQ}}\right)\left(1 + \dfrac{[\text{Q}]}{K_\text{AQ}}\right)\left(1 + \dfrac{c[\text{Q}]}{K_\text{A'Q}}\right)}\right\}.$$

A considerable number of special cases are now possible; some of the more interesting ones are summarized in Tables 3.4 and 3.5. It is to be noted again that the same type of mechanism can lead to different types of inhibition laws for $k_2 \ll k_3$ and for $k_3 \ll k_2$. The last entry in Table 3.4 deserves special mention. It is to be seen that for $k_3 \gg k_2$, $a = a' = 0$, $b = 1$ and $K_\text{AQ} = K_\text{Q}$ the behaviour is pure competitive; however the apparent

TABLE 3.4

Summary of inhibition relationships for systems obeying the general two-intermediate mechanism, with $k_3 \gg k_2$

Condition†	$k_c (= V(Q)/[E]_0)$	$K_A(Q)$	Type of inhibition
None	$\dfrac{\left(1 + \dfrac{b[Q]}{K_{AQ}}\right)k_2}{1 + \dfrac{[Q]}{K_{AQ}}}$	$\dfrac{(k_{-1} + k_2)\left(1 + \dfrac{[Q]}{K_Q}\right)\left(1 + \dfrac{\bar{a}[Q]}{K_{AQ}}\right)}{k_1\left(1 + \dfrac{a[Q]}{K_Q}\right)\left(1 + \dfrac{[Q]}{K_{AQ}}\right)}$	Non-linear
$a = a' = b = 0;\ K_{AQ} = K_Q$	$\dfrac{k_2}{1 + \dfrac{[Q]}{K_{AQ}}}$	$\dfrac{k_{-1} + k_2}{k_1}$	Pure non-competitive
$a = a' = 0;\ K_{AQ} = \infty$	k_2	$\dfrac{k_{-1} + k_2}{k_1}\left(1 + \dfrac{[Q]}{K_Q}\right)$	Pure competitive
$a = a' = 1,\ b = 1$	k_2	$\dfrac{k_{-1} + k_2}{k_1}$	No inhibition
$a = a' = 1,\ b = 0$	$\dfrac{k_2}{1 + \dfrac{[Q]}{K_{AQ}}}$	$\dfrac{(k_{-1} + k_2)\left(1 + \dfrac{k_{-1}}{k_{-1} + k_2}\cdot\dfrac{[Q]}{K_{AQ}}\right)}{k_1\left(1 + \dfrac{[Q]}{K_{AQ}}\right)}$	Non-linear (pure non-competitive if $k_{-1} \gg k_2$)
$a = a' = 0,\ b = 1$	k_2	$\dfrac{(k_{-1} + k_2)\left(1 + \dfrac{[Q]}{K_Q}\right)\left(1 + \dfrac{k_2}{k_{-1} + k_2}\cdot\dfrac{[Q]}{K_{AQ}}\right)}{k_1\left(1 + \dfrac{[Q]}{K_{AQ}}\right)}$	Non-linear (pure competitive if $k_2 \gg k_{-1}$ or $K_{AQ} = K_Q$)

† When $k_3 \gg k_2$ the values of c and $K_{A'Q}$ are irrelevant.

TABLE 3.5
Summary of inhibition relationships for systems obeying the general two-intermediate mechanism, with $k_2 \gg k_3$

Condition†	$k_c = V(Q)/[E]_0$	$K_A(Q)$	Type of inhibition
None	$\dfrac{k_3\left(1+\dfrac{c[Q]}{K_{A'Q}}\right)}{1+\dfrac{[Q]}{K_{A'Q}}}$	$\dfrac{k_3(k_{-1}+k_2)\left(1+\dfrac{a[Q]}{K_Q}\right)\left(1+\dfrac{[Q]}{K_{AQ}}\right)\left(1+\dfrac{c[Q]}{K_{A'Q}}\right)}{k_1k_2\left(1+\dfrac{\bar{a}[Q]}{K_{AQ}}\right)\left(1+\dfrac{b[Q]}{K_{AQ}}\right)\left(1+\dfrac{[Q]}{K_{A'Q}}\right)}$	Non-linear
$b = c = 0; K_{A'Q} = K_A[Q]$	$\dfrac{k_3}{1+\dfrac{[Q]}{K_{A'Q}}}$	$\dfrac{k_3(k_{-1}+k_2)\left(1+\dfrac{a[Q]}{K_Q}\right)}{k_1k_2\left(1+\dfrac{\bar{a}[Q]}{K_{AQ}}\right)}$	Non-linear (pure non-competitive if $\bar{a}=a$)
$b = 0; K_{A'Q} = \infty$ or $b = 0, c = 1$	k_3	$\dfrac{k_3(k_{-1}+k_2)\left(1+\dfrac{a[Q]}{K_Q}\right)\left(1+\dfrac{[Q]}{K_{AQ}}\right)}{k_1k_2\left(1+\dfrac{\bar{a}[Q]}{K_{AQ}}\right)}$	Non-linear (pure competitive if $\bar{a}=a$)
$b = 1, c = 1$	k_3	$\dfrac{k_3(k_{-1}+k_2)\left(1+\dfrac{a[Q]}{K_Q}\right)}{k_1k_2\left(1+\dfrac{\bar{a}[Q]}{K_Q}\right)}$	Non-linear (no inhibition if $\bar{a}=a$)
$b = 1, c = 0$	$\dfrac{k_3}{1+\dfrac{[Q]}{K_{A'Q}}}$	$\dfrac{k_3(k_{-1}+k_2)\left(1+\dfrac{\bar{a}[Q]}{K_{AQ}}\right)\left(1+\dfrac{a[Q]}{K_Q}\right)}{k_1k_2\left(1+\dfrac{\bar{a}[Q]}{K_{AQ}}\right)\left(1+\dfrac{[Q]}{K_{A'Q}}\right)}$	Non-linear (pure anticompetitive if $\bar{a}=a$)

† When $k_2 \gg k_3$ the values of a and K_Q are irrelevant.

inhibition constant is $K_{AQ}(k_{-1} + k_2)/k_2$ and therefore does not correspond to a simple dissociation constant for a complex. The fact that this can arise should always be borne in mind in interpreting experimental results.

Another point of importance relates to the situation when $k_2 \gg k_3$. It is to be seen from Table 3.5 that if $\bar{a} = a$ (which can arise if $a = b = a'$), $K_{AQ} = K_Q$, a and K_Q do not enter into the equations for this case, so that no information about these quantities can be obtained. There is, for example, no inhibition in this case if Q is only bound to the free enzyme. The rate equation for $k_2 \gg k_3$ with $\bar{a} = a$ is

$$v = \frac{\left\{ k_3 \left(1 + \dfrac{c[Q]}{K_{A'Q}} \right) \right\} \Big/ \left(1 + \dfrac{[Q]}{K_{A'Q}} \right) [E]_0 \cdot [A]}{\left\{ (k_{-1} + k_2)k_3 \left(1 + \dfrac{[Q]}{K_{AQ}} \right)\left(1 + \dfrac{c[Q]}{K_{A'Q}} \right) \right\} \times \left\{ k_1 k_2 \left(1 + \dfrac{b[Q]}{K_{AQ}} \right)\left(1 + \dfrac{[Q]}{K_{A'Q}} \right) \right\} + [A]} \tag{141}$$

and the slope of a plot of $1/v$ against $1/[A]$ is

$$\text{Slope} = \left\{ k_1 k_2 \left(1 + \frac{[Q]}{K_{AQ}} \right) \right\} \Big/ \left\{ (k_{-1} + k_2)[E]_0 \left(1 + \frac{b[Q]}{K_{AQ}} \right) \right\}. \tag{142}$$

In the classical schemes (and in the general case with $k_3 \gg k_2$) this slope involves K_Q. When reaction 3 is the slow step, however, the slope depends upon K_{AQ} and upon b. A slope independent of Q suggests that $b = 1$ or $K_{AQ} = \infty$; that is, either EAQ is not formed or, if it is formed, the addition of Q to EA has no effect on the subsequent reaction. A slope linearly dependent on Q implies that EAQ is formed and that the subsequent reaction is blocked by the addition of Q to EA.

Table 3.6 summarizes some systems which have been classified by analysis of the available inhibition data.

Activation

Activation, in which the moderator brings about an increase in rate, requires only very brief comments, since the general eqns (138), (139) and (140) apply to it as well as to inhibition. If the species EQ, EAQ and EA'Q react more rapidly than the corresponding species to which Q is not attached, there can result an increase in rate as Q is added. One possibility, the equations for which could be worked out in a similar way as for inhibition, is that there is complete blockage of one of the paths $E + A \rightarrow EA$, $EA \rightarrow EA' + X$ or $EA' \rightarrow E + Y$, and that Q provides an alternative route.

TABLE 3.6

Classification of enzyme systems in terms of the type of inhibition

Enzyme	Substrate	Inhibitor	Type of inhibition	Ref.
Cholinesterase	Methylaminoethyl acetate	*cis*-2-Dimethyl-amino-cyclo-hexanol	$k_3 \gg k_2$ $c = 0$ Pure competitive	[a]
α-Chymotrypsin	Nicotinyl-L-tryptophanamide	Indole		[b]
Cholinesterase	Methylaminoethyl acetate	Choline, carbachol, eserine	$k_2 \ll k_3$ $c = 1$ Pure competitive	[c]
Cholinesterase	Acetylcholine	*cis*-2-Dimethyl-amino-cyclo-hexanol	$k_2 > k_3$ $c = 0$ Pure non-competitive	[d]
α-Chymotrypsin	*N*-acetyl-L-tyrosine ethyl ester	Indole		[b]
Cholinesterase	Acetylcholine	Choline, carbachol, eserine	$k_2 > k_3$ $c = 1$ Pure competitive	[c]
α-Chymotrypsin	Methyl hippurate	Indole	$k_2 \approx k_3$ $c = 0$ Mixed non-competitive-competitive	[b], [e]

[a] R. M. Krupka and K. J. Laidler, *Trans. Faraday Soc.*, **56**, 1467 (1960).
[b] H. Kaplan and K. J. Laidler, *Can. J. Chem.*, **45**, 559 (1967).
[c] R. M. Krupka and K. J. Laidler, *J. Am. chem. Soc.*, **83**, 1454 (1961).
[d] R. M. Krupka and K. J. Laidler, *J. Am. chem. Soc.*, **83**, 1445 (1961).
[e] T. H. Applewhite, R. B. Martin, and C. Niemann, *J. Am. chem. Soc.*, **80**, 1457 (1958).

Reactions of two competing substrates

Many enzymes bring about the reaction of more than one substrate, and it is of interest to consider the case in which two such substrates are mixed together in the presence of the enzyme. This case is to be distinguished from the case of two substrates that are reacting with one another, which is discussed later (Chapter 4). An example of the case under consideration is the action of an esterase on a mixture of the D- and L-forms of an ester.

If the two substrates react at different sites on the enzyme there will be no interference between them, and they will react independently of one another; the simple Michaelis–Menten law may then be obeyed by each substrate. If, on the other hand, the two substrates react at the same sites on the enzyme they will compete with one another. The rate equations obtained for this situation will clearly reduce to those for competitive inhibition if one of the 'substrates' does not react at all.

A simple mechanism for this situation, involving two intermediates, is

The steady-state equations are

$$\text{EA}: \quad k_1[\text{E}].[\text{A}] - (k_{-1} + k_2)[\text{EA}] = 0 \qquad (143)$$

$$\text{EA}': \quad k_2[\text{EA}] - k_3[\text{EA}'] = 0 \qquad (144)$$

$$\text{EB}: \quad k_1'[\text{E}].[\text{B}] - (k_{-1}' + k_2')[\text{EB}] = 0 \qquad (145)$$

$$\text{EB}': \quad k_2'[\text{EB}] - k_3'[\text{EB}'] = 0. \qquad (146)$$

The total concentration of enzyme is

$$[\text{E}]_0 = [\text{E}] + [\text{EA}] + [\text{EA}'] + [\text{EB}] + [\text{EB}'] \qquad (147)$$

$$= [\text{EA}]\left[\frac{k_{-1} + k_2}{k_1[\text{A}]} + 1 + \frac{k_2}{k_3} + \frac{k_1'[\text{B}]}{k_{-1}' + k_2'} \cdot \frac{k_{-1} + k_2}{k_1[\text{A}]}\left(1 + \frac{k_2'}{k_3'}\right)\right]. \qquad (148)$$

The rate of disappearances of A, equal to $k_2[\text{EA}]$, is therefore

$$v_\text{A} = \frac{k_2[\text{E}]_0.[\text{A}]}{\left\{\dfrac{(k_{-1} + k_2)}{k_1}\right\} + \left\{1 + \left(\dfrac{k_2}{k_3}\right)\right\}[\text{A}] + \left\{\dfrac{k_1'(k_{-1} + k_2)}{k_1(k_{-1}' + k_2')}\right\}\left\{1 + \left(\dfrac{k_2'}{k_3'}\right)\right\}[\text{B}]}. \qquad (149)$$

If [B] is held constant the rate shows a simple hyperbolic dependence on [A]. When [A] is very large the rate is simply $k_2k_3[\text{E}]_0/(k_2 + k_3)$ and is independent of [B]; the behaviour is therefore competitive.

From eqns (143) and (145) it is to be seen that

$$\frac{[\text{EB}]}{[\text{EA}]} = \frac{k_1'(k_{-1} + k_2)[\text{B}]}{k_1(k_{-1}' + k_2')[\text{A}]}. \qquad (150)$$

The ratio of the rates of disappearance of B and A is therefore

$$\frac{v_B}{v_A} = \frac{k'_1 k'_2 (k_{-1} + k_2)[B]}{k_1 k_2 (k'_{-1} + k'_2)[A]}. \tag{151}$$

Together with (149) this leads to

$$v_B = \frac{k'_2 [E]_0 \cdot [B]}{\left(\dfrac{k'_{-1} + k'_2}{k'_1}\right) + \left\{\dfrac{k_1(k'_{-1} + k'_2)}{k'_1(k_{-1} + k_2)}\right\} \left(1 + \dfrac{k_2}{k_3}\right)[A] + \left(1 + \dfrac{k'_2}{k'_3}\right)[B]} \tag{152}$$

for the rate of disappearance of B.

Equations of the form of (149) and (152) have been shown by Foster and Niemann[1] to apply to the chymotrypsin-catalysed hydrolysis of mixtures of acetyl-L-tryptophanamide and acetyl-L-tyrosinamide.

Various complications could arise; thus A might add on to EB' to form EB'A, and B on to EA' to form EA'B. Equations for this situation can be derived without difficulty. No cases of this seem to have been reported.

[1] R. J. Foster and C. Niemann, *J. Am. chem. Soc.*, **73**, 1552 (1951).

4. Two-Substrate Reactions

THE enzyme-catalysed systems considered in the last chapter involved only a single substrate. It is true that hydrolytic reactions involve the participation of a water molecule in addition to the molecule undergoing hydrolysis. However, from the point of view of the steady-state equations, the step involving the water molecule can be treated by incorporating the water concentration into the rate constant; this was in fact the procedure adopted in previous chapters.

When reaction occurs between two solute species the steady-state equations take a somewhat different, and more complex, form than those derived in the preceding chapter. The present chapter is concerned with the development of such equations, on the basis of various mechanisms that have been found to be common, and with the application of these equations to the experimental results.

One particular type of reaction to which the present treatment will apply comprises those catalysed by enzymes with which coenzymes are associated. Such coenzymes play the role of the second substrate. For example, when lactate dehydrogenase catalyses the oxidation of lactic acid into pyruvic acid the process that occurs is the conversion of nicotinamide adenine dinucleotide (NAD; also known as diphosphopyridine nucleotide, DPN) into its reduced form:

$$CH_3CHOHCOOH + NAD^+ \rightarrow CH_3COCOOH + NADH + H^+$$

General form of rate equations

A number of mechanisms, to be considered below, are possible when an enzyme catalyses a reaction between two substrates, A and B. There may be the formation of binary enzyme–substrate complexes EA and EB, and of the ternary complex EAB. Sometimes this ternary complex is formed exclusively from one of the binary complexes and not from the other; for example, EA + B → EAB may occur but not EB + A → EAB. In this case the mechanism is said to be *ordered*; alternatively the complex EAB may be formed readily from EA and EB, and the mechanism is said to be *random*. Sometimes there is no evidence for the formation of a ternary complex; according to a mechanism suggested by Theorell and Chance[1], for example, EA reacts directly with B to form products. In another mechanism, explored in particular by Koshland[2] and in further detail by Cleland[3], the binary complex EA splits off a product X to form EA', which then

[1] H. Theorell and B. Chance, *Acta chem. scand.*, **5**, 1127 (1951).
[2] D. E. Koshland, *Biol. Revs.*, **28**, 416 (1953); *The Mechanism of Enzyme Action* (ed. McElroy and Glass), The John Hopkins Press, Baltimore (1954), p. 608; *Discuss. Faraday Soc.*, **20**, 142 (1955); *J. cell. comp. Physiol.*, **47**, 217 (1956).
[3] W. W. Cleland, *Biochim. biophys. Acta*, **67**, 104, 173, 188 (1963); *A. Rev. Biochem.*, **36**, 77 (1967).

reacts with B to form the second product Y; such mechanisms have been referred to as *ping-pong* mechanisms. A considerable number of variations on the above basic mechanisms are possible; the steady-state equations for some of the commoner ones are derived later in this chapter.

Before this is done it is of interest to consider the form of the rate equations that can apply to two-substrate systems. Some mechanisms, such as the random ternary mechanism, give rise to very complicated equations, unless certain assumptions and simplifications are made. Many of the other mechanisms, however, give rise to equations of the general form[1]

$$v_1 = \frac{V_1[\text{A}].[\text{B}]}{K_\text{A}'K_\text{B} + K_\text{B}[\text{A}] + K_\text{A}[\text{B}] + [\text{A}].[\text{B}]}. \tag{1}$$

A similar equation applies to the rate v_{-1} in the reverse direction; it involves the product of the concentrations of X and Y. Equation (1) involves four kinetic parameters, V_1, K_A, K_B and K_A'; this is to be contrasted with the two-parameter equation (eqn (9), p. 73) that applies to the single-substrate systems.

When [B] is very large eqn (1) becomes

$$v_1([\text{B}] \rightarrow \infty) = \frac{V_1}{K_\text{A} + [\text{A}]}. \tag{2}$$

This expression is independent of [B]; the rate therefore reaches a limiting value at high B concentrations. The parameter K_A is the Michaelis constant for [A] at high B concentrations. Similarly at high [A],

$$v_1([\text{A}] \rightarrow \infty) = \frac{V_1}{K_\text{B} + [\text{B}]} \tag{3}$$

so that the rate reaches a limiting value at high [A], and K_B is the Michaelis constant for [B] at high [A]. Equation (1) in the Michaelis form is

$$v_1 = \frac{\{V_1[\text{A}].[\text{B}]/(K_\text{B} + [\text{B}])\}}{\{(K_\text{A}'K_\text{B} + K_\text{A}[\text{B}])/(K_\text{B} + [\text{B}]) + [\text{A}]\}} \tag{4}$$

so that at any value of [B] the apparent Michaelis constant for [A] is

$$(K_\text{A}'K_\text{B} + K_\text{A}[\text{B}])/(K_\text{B} + [\text{B}]).$$

In the event that $K_\text{A}' = K_\text{A}$ the apparent Michaelis constant is independent of [B], and equal to K_A; this is actually the case for one mechanism (see p. 121) and applies to certain experimental results. Similarly, the apparent Michaelis constant for [B] is $(K_\text{A}'K_\text{B} + K_\text{B}[\text{A}])/(K_\text{A} + [\text{A}])$; if $K_\text{A}' = K_\text{A}$ this reduces to K_B. Figure 4.1 shows the variations of v_1 with [A] and [B] according to eqn (1).

---

[1] For a comparison of equations see Appendix A, p. 137.

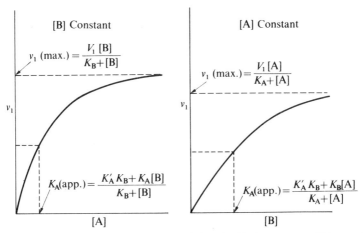

FIG. 4.1. Variations of rate with A at constant [B], and with B at constant [A], according to eqn (1).

It is also instructive to consider eqn (1), and to analyse experimental results, with the use of Lineweaver–Burk plots of $1/v_1$ against $1/[A]$ and of $1/v_1$ against $1/[B]$. Equation (1) may be written as

$$\frac{1}{v_1} = \left(1 + \frac{K_A}{[A]} + \frac{K_B}{[B]} + \frac{K'_A K_B}{[A].[B]}\right)\Big/V_1. \tag{5}$$

A plot of $1/v_1$ against $1/[A]$, at constant [B], will therefore be linear, with a slope of

$$\frac{K_A + \{(K'_A K_B)/[B]\}}{V_1}$$

and the intercept on the $1/v_1$ axis of

$$\frac{1 + K_B/[B]}{V_1}.$$

Both the slope and the intercept will in general decrease with increasing [B]. However, in the event that $K_{A'} = 0$, which is the case for one mechanism (see p. 123), the slope becomes independent of [B], but the intercept decreases with increasing [B]. These relationships are shown schematically in Fig. 4.2. In the general case the lines intersect at

$$\frac{1}{[A]} = -\frac{1}{K_A}. \tag{6}$$

From the plots of slopes and intercepts against [B] it is possible to determine K_A/V_1, $K_A K_B/V_1$, K_B/V_1 and $1/V_1$, and hence the separate kinetic parameters V_1, K_A, K_B and K'_A. Figure 4.3 shows the corresponding relationships for plots of $1/v_1$ against $1/[B]$.

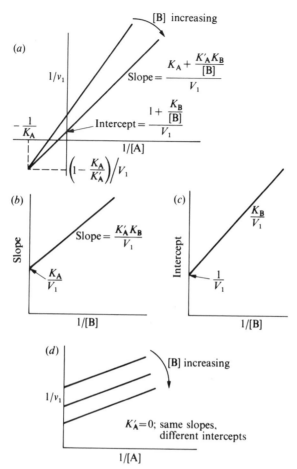

FIG. 4.2. Schematic graphs based on eqn (5); (*a*) shows the plot of $1/v$, against $1/[A]$, and (*b*) and (*c*) show plots against $1/[B]$ of the slopes and intercepts; (*d*) applies to the special case of $K'_A = 0$.

The magnitudes of the kinetic parameters (e.g. the question of whether K'_A is zero or not) provide some information about possible mechanisms; for a further discrimination, recourse has to be had to inhibition studies, as discussed later (pp. 125–132).

Some reaction mechanisms

A large number of alternative mechanisms have been considered for two-substrate systems, particularly by Alberty[1], Dalziel[2], Wong and Hanes[3],

[1] R. A. Alberty, *J. Am. chem. Soc.* **80**, 1777 (1958); V. Bloomfield, L. Peller, and R. A. Alberty, *ibid.*, **84**, 4367 (1962).
[2] K. Dalziel, *Acta chem. scand.*, **11**, 1706 (1957).
[3] J. F. T. Wong and C. S. Hanes, *Can. J. Biochem. Physiol.*, **40**, 763 (1962); *Nature*, **203**, 492 (1964).

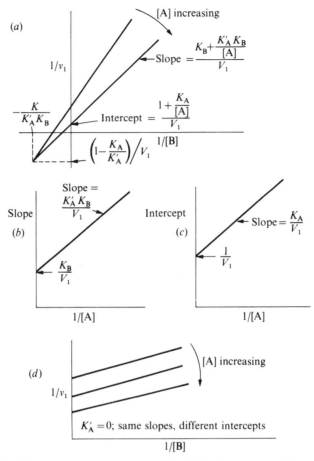

FIG. 4.3.   Schematic graphs, analogous to Fig. 8.2, for plots against $1/[A]$ and $1/[B]$.

and Cleland[1]. Wong and Hanes in particular have shown that several mechanisms can lead to the same form of the overall rate equations; such mechanisms they refer to as *homeomorphs*. As we found with single-substrate systems, it is impossible, by steady-state studies, to distinguish between mechanisms that involve various first-order transformations of intermediates; for example, in a hydrolysis the conversion of an addition complex into an acyl enzyme leads to no change in the form of the equation, and the same is true for similar changes in two-substrate systems. In addition, there are more variations in overall mechanisms for two-substrate ones, so that the number of alternative mechanisms is very great. Considerable ingenuity is therefore required to devise, systematically, alternative mechanisms and to distinguish

[1] W. W. Cleland, *Biochim. biophys. Acta*, **67**, 104, 173, 188 (1963); *A. Rev. Biochem.*, **36**, 77 (1967).

between them on the basis of various lines of experimental evidence. Fortunately the majority of enzyme reactions appear to fall into a few categories, the steady-state equations for which will now be considered.

Random ternary-complex mechanism

In the random ternary complex mechanism the enzyme E can form binary complexes EA and EB and also the ternary complex EAB, with no restriction on the order in which A and B become attached to the enzyme. This mechanism is represented in two different ways in Fig. 4.4; the second is the short-

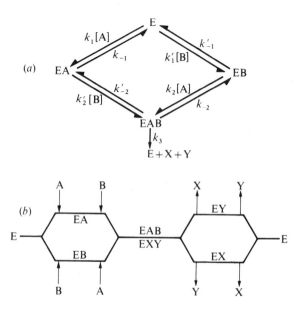

FIG. 4.4.   The random ternary-complex mechanism; two alternative representations.

hand notation introduced by Cleland. The exact steady-state solution for this mechanism was first given by Ingraham and Makower[1] and more generally, using a determinantal method, by Laidler[2]. The resulting rate equation is very complicated involving terms in the numerator in $[E]_0 . [A] . [B]$, $[E]_0[A]^2 . [B]$ and $[E]_0[A] . [B]^2$ and in the denominator in $[A]$, $[B]$, $[A] . [B]$, $[A]^2$, $[B]^2$, $[A]^2 . [B]$ and $[A] . [B]^2$. It would be out of the question to fit experimental data to such an equation, and Laidler[3], making use of the

[1] L. I. Ingraham and B. Makower, *J. phys. Chem.*, **58**, 226 (1954).
[2] K. J. Laidler, *Trans. Faraday Soc.*, **52**, 1374 (1956); see also J. Botts and M. F. Morales, *Trans. Faraday Soc.*, **49**, 696 (1953); K. J. Laidler, *Trans. Faraday Soc.*, **51**, 528 (1955); *Discuss. Faraday Soc.*, **20**, 83 (1955); The method of King and Altman (p. 80) may also be employed to obtain the rate equation.
[3] K. J. Laidler, *Trans. Faraday Soc.*, **52**, 1374 (1956).

properties of determinants, has considered under what conditions the rate equation will reduce to simple forms; there are two, and only two, such conditions:

(1) When there is equilibrium as far as EA, EB and EAB are concerned; i.e. when k_3 is sufficiently small as not to disturb equilibrium.

(2) When the rate of formation of the ternary complex EAB from one of the binary complexes is negligible compared with the rate of its formation from the other. This second condition corresponds to an *ordered* mechanism, and is considered in the next section.

If the first condition is satisfied the problem is readily solved. The equilibrium equations are

$$K'_A[EA] = [E] . [A] \tag{7}$$

$$K'_B[E\overset{\text{B}}{A}] = [E] . [B] \tag{8}$$

$$K'_A K_B[EAB] = K'_B K_A[EAB] = [E] . [A] . [B] \tag{9}$$

where $K'_A = k_{-1}/k_1$, $K'_B = k'_{-1}/k'_1$, $K_A = k_{-2}/k_2$ and $K_B = k'_{-2}/k'_2$. It is to be noted that

$$K'_A K_B = K'_B K_A. \tag{10}$$

The total concentration of enzyme is

$$[E]_0 = [E] + [EA] + [EB] + [EAB] \tag{11}$$

$$= [E]\left(1 + \frac{[A]}{K'_A} + \frac{[B]}{K'_B} + \frac{[A] . [B]}{K'_A K_B}\right). \tag{12}$$

Hence

$$[EAB] = \frac{\dfrac{[A] . [B] . [E]_0}{K'_A K_B}}{1 + \dfrac{[A]}{K'_A} + \dfrac{[B]}{K'_B} + \dfrac{[A] . [B]}{K'_A K_B}} \tag{13}$$

whence

$$v = k_3[EAB] = \frac{\dfrac{k_3[A] . [B] . [E]_0}{K'_A K_B}}{1 + \dfrac{[A]}{K'_A} + \dfrac{[B]}{K'_B} + \dfrac{[A] . [B]}{K'_A K_B}} \tag{14}$$

$$= \frac{V[A] . [B]}{K'_A K_B + K_B[A] + K_A[B] + [A][B]} \tag{15}$$

where $V = k_3[E]_0$; this equation is identical with eqn (1).

A special case of interest arises when $K'_A = K_A$, which requires that $K'_B = K_B$; this means that the attachment of A to E does not affect the ease of attachment of B, and vice versa. Equation (15) then factorizes:

$$v_1 = \frac{V_1[A].[B]}{(K_A + [A])(K_B + [B])}. \tag{16}$$

The Michaelis constant K_A for [A] is then independent of [B] and K_B is independent of [A]. In the Lineweaver–Burk plots (e.g. Figs. 4.2 and 4.3) the lines therefore intersect on the 1/[A] or 1/[B] axis.

The above equations, for the equilibrium case, were first derived by Haldane[1], and on the basis of a somewhat different model by Laidler and Socquet[2].

Ordered ternary-complex mechanism

It was noted above that the general ternary-complex mechanism leads to a simple rate equation if the ternary complex EAB can be formed from only one of the complexes, such as EA. The mechanism is now represented in Fig. 4.5. It is to be noted that, to simplify the equations, B is not allowed to form the complex EB.

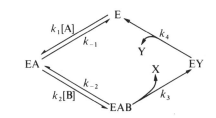

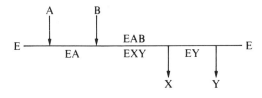

FIG. 4.5. The ordered ternary-complex mechanism; the complex EAB can only be formed from EA.

[1] J. B. S. Haldane, *Enzymes*, Longmans, Green and Co., London (1936), pp. 83–4 (reprinted by M.I.T. Press, 1965).
[2] K. J. Laidler and I. M. Socquet, *J. phys. Colloid Chem.*, **54**, 519 (1950); see R. A. Alberty, *J. Am. chem. Soc.*, **75**, 1928 (1953); I. M. Socquet and K. J. Laidler, *Archs Biochem.*, **25**, 171 (1950); G. W. Schwert and M. T. Hakala, *Archs Biochem. Biophys.*, **38**, 55 (1952).

The steady-state equation corresponding to this scheme is most readily derived by use of the King–Altman method (see p. 80). The terms for E, EA, EY and EAB are:

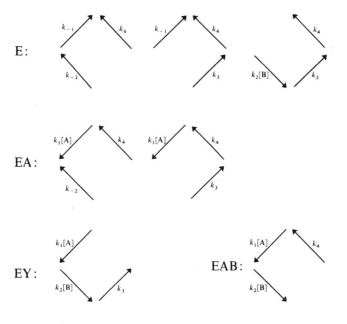

The expressions for the concentrations are therefore

$$[\text{E}] \propto k_{-1}k_{-2}k_4 + k_{-1}k_3k_4 + k_2k_3k_4[\text{B}] \qquad (17)$$

$$[\text{EA}] \propto k_1k_{-2}k_4[\text{A}] + k_1k_3k_4[\text{A}] \qquad (18)$$

$$[\text{EY}] \propto k_1k_2k_3[\text{A}].[\text{B}] \quad [\text{EAB}] \propto k_1k_2k_4[\text{A}].[\text{B}]. \qquad (19)$$

Thus

$$[\text{EAB}] = \frac{k_1k_2k_4[\text{E}]_0.[\text{A}].[\text{B}]}{\begin{array}{c}k_4(k_{-1}k_{-2} + k_{-1}k_3 + k_2k_3[\text{B}]) + \\ + k_1k_4(k_{-2} + k_3)[\text{A}] + k_1k_2(k_3 + k_4)[\text{A}].[\text{B}]\end{array}} \qquad (20)$$

and the rate, equal to $k_3[\text{EAB}]$, is

$$v = \frac{\{k_3k_4/(k_3 + k_4)\}[\text{E}]_0.[\text{A}].[\text{B}]}{\dfrac{k_4(k_{-1}k_{-2} + k_{-1}k_3)}{k_1k_2(k_3 + k_4)} + \dfrac{k_4(k_{-2} + k_3)}{k_2(k_3 + k_4)}[\text{A}] + \dfrac{k_3k_4}{k_1(k_3 + k_4)}[\text{B}] + [\text{A}][\text{B}]}. \qquad (21)$$

This is of the same form as eqn (1).

A variant of this mechanism is when B reacts with EA to form EY + X without the formation of a ternary complex of sufficiently long life to be

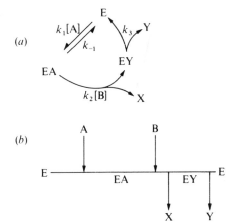

FIG. 4.6. The Theorell–Chance mechanism.

kinetically significant. Such a mechanism was suggested by Theorell and Chance[1], and is shown in Fig. 4.6. The rate equation can readily be derived directly; alternatively, it comes at once from eqn (21) if $k_3 \gg k_{-2}$. The result is

$$v = \frac{k_3[E]_0 \cdot [A] \cdot [B]}{\left(\dfrac{k_{-1}k_3}{k_1k_2}\right) + \left(\dfrac{k_3}{k_1}\right)[B] + \left(\dfrac{k_3}{k_2}\right)[A] + [A] \cdot [B]} \tag{22}$$

$$= \frac{k_1k_2k_3[E]_0 \cdot [A] \cdot [B]}{k_{-1}k_3 + k_1k_3[A] + k_2k_3[B] + k_1k_2[A] \cdot [B]}. \tag{23}$$

Ping-pong mechanism

A third possible mechanism for reaction between two substrates is shown in Fig. 4.7 in two alternative representations. Cleland's full name for this mechanism is *ping-pong bi bi*, the first 'bi' indicating that there are two reactants for the reaction from left to right, the second 'bi' that there are two products. Cleland has derived equations for this mechanism in which the reverse processes are taken into account and for a number of similar mechanisms; the present account will ignore the back reactions of X and Y.

The rate equation for this mechanism will be derived by the King–Altman method, with reference to Fig. 4.7. The terms are

E: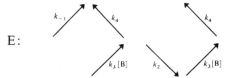

---

[1] H. Theorell and B. Chance, *Acta chem. scand.*, **5**, 1127 (1951).

EA:
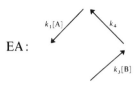

EA':
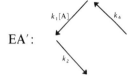

EA'B:
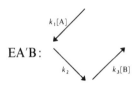

The expressions for the concentrations are therefore

$$[E] \propto k_{-1}k_3k_4[B] + k_2k_3k_4[B] \tag{24}$$

$$[EA] \propto k_1k_3k_4[A] \cdot [B] \tag{25}$$

$$[EA'] \propto k_1k_2k_4[A] \tag{26}$$

$$[EA'B] \propto k_1k_2k_3[A] \cdot [B]. \tag{27}$$

Thus

$$[EA'B] = \frac{k_1k_2k_3[E]_0 \cdot [A] \cdot [B]}{k_1k_2k_4[A] + k_3k_4(k_{-1} + k_2)[B] + k_1k_3(k_2 + k_4)[A] \cdot [B]} \tag{28}$$

and the rate is

$$v = \frac{k_1k_2k_3k_4[E]_0 \cdot [A] \cdot [B]}{k_1k_2k_4[A] + k_3k_4(k_{-1} + k_2)[B] + k_1k_3(k_2 + k_4)[A] \cdot [B]} \tag{29}$$

$$= \frac{\dfrac{k_2k_4}{k_2 + k_4}[E]_0 \cdot [A] \cdot [B]}{\dfrac{k_2k_4}{k_3(k_2 + k_4)}[A] + \dfrac{k_4(k_{-1} + k_2)}{k_1(k_2 + k_4)}[B] + [A] \cdot [B]} \tag{30}$$

This is of the same form as eqn (1), except that the $K'_A K_B$ term is missing.

A variant on this mechanism is when the complex EA'B is of sufficiently short life to have no kinetic significance; this means that k_4 is very large, and in particular $k_4 \gg k_2$. Eqn (30) then reduces to

$$v = \frac{k_2[E]_0 \cdot [A] \cdot [B]}{(k_2/k_3)[A] + \{(k_{-1} + k_2)/k_1\}[B] + [A] \cdot [B]}.$$ (31)

Since this is of the same form as eqn (30), studies of the dependence of rate on A and B obviously cannot lead to conclusions about the existence of EA′B.

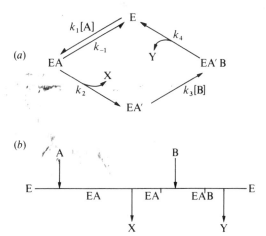

FIG. 4.7. The ping-pong bi bi mechanism.

It is of interest to note that if B is a solvent molecule the latter eqn (31) reduces to that for the double-intermediate single-substrate mechanism (p. 77). The solvent concentration is in that case incorporated in the rate constant k_3; $k_3[B]$ in eqn (21) is thus replaced by k_3, and the result is

$$v = \frac{k_2[E]_0 \cdot [A]}{\{(k_{-1} + k_2)/k_1\} + \{(k_2 + k_3)/k_3\}[A]}$$ (32)

which is identical with eqn (28) on page 79.

Discrimination between mechanisms

Table 4.1 summarizes the rate equations that have been derived for the different mechanisms. Obviously certain conclusions about possible mechanisms may be drawn on the basis of studies of the influence of the concentrations of A and B on the overall rates. It was noted earlier (p. 126) that plots of $1/v_1$ against $1/[A]$ and $1/[B]$, followed by plots of slopes and intercepts against $[B]$ and $[A]$ respectively, can lead to values for the four kinetic parameters V_1, K_A, K_B and K'_A. If, for example, K'_A is found to be zero, the ping-pong mechanism is clearly suggested as a possible mechanism. However, the form of the equations cannot discriminate between the random and ordered ternary-complex mechanism; nor, if the mechanism is an ordered one, can the order of addition be determined on the basis of the equations.

Furthermore, various complications can arise. The equations derived above for the various mechanisms have been based on the simplest of assumptions. If, for example, in the random ternary-complex mechanism there is not complete equilibrium the rate equation is altered and incorrect

TABLE 4.1

Summary of rate equations

Random ternary complex

$$v = \frac{k_3[E]_0 . [A] . [B]}{K'_A K_B + K_B[A] + K_A[B] + [A] . [B]}$$

$$\text{(E)} \qquad \text{(EA)} \qquad \text{(EB)} \qquad \text{(EAB)}$$

Ordered ternary complex (reaction via EA; no EB formed)

$$v = \frac{\dfrac{k_3 k_4}{k_3 + k_4}[E]_0 . [A] . [B]}{\dfrac{k_4(k_{-1}k_{-2} + k_{-1}k_3)}{k_1 k_2(k_3 + k_4)} + \dfrac{k_4(k_{-2} + k_4)}{k_2(k_3 + k_4)}[A] + \dfrac{k_3 k_4}{k_1(k_3 + k_4)}[B] + [A] . [B]}$$

$$\qquad \text{(E)} \qquad\qquad \text{(EA)} \qquad\qquad \text{(E)} \quad \text{(EY and EAB)}$$

Theorell–Chance

$$v = \frac{k_3[E]_0 . [A] . [B]}{(k_{-1}k_3/k_1 k_2) + (k_3/k_1)[B] + (k_3/k_2)[A] + [A] . [B]}$$

$$(\underline{\qquad\qquad E \qquad\qquad}) \qquad \text{(EA)} \qquad \text{(EY)}$$

Ping-pong

$$v = \frac{k_1 k_2 k_3 k_4[E]_0 . [A] . [B]}{k_1 k_2 k_4[A] + k_3 k_4(k_{-1} + k_2)[B] + k_1 k_3 k_4[A] . [B] + k_1 k_2 k_3[A] . [B]}$$

$$\text{(EA')} \qquad\quad \text{(E)} \qquad\qquad \text{(EA)} \qquad\qquad \text{(EA'B)}$$

The species in brackets are the ones that lead to the term printed above.

conclusions might be drawn. If for the ordered mechanism E had been allowed to form an unreactive or 'dead-end' complex with B, the form of the equation is changed, the denominator then having a term in $[B]^2$ as well as in $[B]$; this would again be misleading in the analysis of experimental results. Similarly, a dead-end complex with B also complicates the equation for the ping-pong mechanism.

For all these reasons it is obviously of great importance to have alternative procedures for discriminating between mechanisms. Inhibition studies, and investigations of isotopic exchange, have been used in particular for this purpose, and will now be discussed.

The inhibition of two-substrate reactions

The use of inhibitors to distinguish between different mechanisms depends upon the fact that different inhibition patterns are found with different mechanisms, even when they lead to identical rate laws in the absence of inhibitors. Inhibition equations will be developed for the mechanisms previously considered.

Random ternary-complex mechanism

A simple scheme which may well apply when the mechanism is a random ternary-complex one is illustrated in Fig. 4.8. The inhibitor Q is allowed to attach itself to E, to EA and to EB; it is not allowed to become attached to EAB, since the sites on the enzyme are considered to be filled when EAB has been formed. The inhibitor Q may be a product of reaction; in that case the

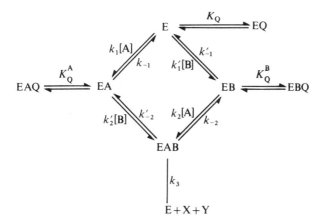

Fig. 4.8.   The random ternary-complex mechanism with inhibitor Q becoming attached to E, EA, and EB but not to EAB.

product X formed from A, and similar to it in structure (e.g. pyruvic acid formed from lactic acid), is likely to become more strongly attached to EB than to EA. This is because the site for attachment of A is free in EB but not in EA. This hypothesis, that X is more likely to become attached to EB than to EA, and that Y is more likely to become attached to EA than to EB, is helpful in the interpretation of experimental results.

The derivation of the rate equation for this scheme, with the assumption of equilibrium between E, EA, EB, EAB, EQ, EAQ and EBQ, involves only a systematic modification to the procedure given on p. 120. The total enzyme concentration is now given not by eqn (11) but by

$$[E]_0 = [E] + [EQ] + [EA] + [EAQ] + [EB] + [EBQ] + [EAB]. \quad (33)$$

However, since K_Q is the dissociation constant of EQ into E + Q,

$$[EQ] = \frac{[E] \cdot [Q]}{K_Q}. \tag{34}$$

Similarly

$$[EAQ] = \frac{[EA] \cdot [Q]}{K_Q^A} \tag{35}$$

and

$$[EBQ] = \frac{[EB] \cdot [Q]}{K_Q^B}. \tag{36}$$

Equation (33) therefore becomes

$$[E]_0 = [E]\left(1 + \frac{[Q]}{K_Q}\right) + [EA]\left(1 + \frac{[Q]}{K_Q^A}\right) + [EB]\left(1 + \frac{[Q]}{K_Q^B}\right) + [EAB]. \tag{37}$$

It is to be noted that each term in eqn (33) is merely multiplied by the appropriate term of the form $(1 + Q/K_Q)$. The final rate equation, instead of eqn (14), is now

$$v = \frac{k_3[A] \cdot [B] \cdot [E]_0 / K_A' K_B}{1 + \dfrac{[Q]}{K_Q} + \dfrac{[A]}{K_A'}\left(1 + \dfrac{[Q]}{K_Q^A}\right) + \dfrac{[B]}{K_B'}\left(1 + \dfrac{[Q]}{K_Q^B}\right) + \dfrac{[A] \cdot [B]}{K_A' K_B}} \tag{38}$$

$$= \frac{V[A] \cdot [B]}{K_A' K_B\left(1 + \dfrac{[Q]}{K_Q}\right) + K_B[A]\left(1 + \dfrac{[Q]}{K_Q^A}\right) + K_A[B]\left(1 + \dfrac{[Q]}{K_Q^B}\right) + [A] \cdot [B]} \tag{39}$$

$$\frac{1}{v} = \frac{1 + \dfrac{K_A}{[A]}\left(1 + \dfrac{[Q]}{K_Q^B}\right) + \dfrac{K_B}{[B]}\left(1 + \dfrac{[Q]}{K_Q^A}\right) + \dfrac{K_A' K_B}{[A][B]}\left(1 + \dfrac{[Q]}{K_A}\right)}{V} \tag{40}$$

A plot of $1/v_1$ against $1/[A]$, at constant [B] and [Q], therefore gives a line of slopes and intercepts indicated in Fig. 4.9. There will be a family of lines for various values of [B], at constant [Q], as discussed earlier with respect to Fig. 8.2. Variation in [Q] will also lead to changes in the slopes and intercepts. As seen earlier, the values of V_1, K_A, K_B and K_A' can be determined by studies in the absence of inhibitor. If the slopes and intercepts of plots of the type shown in Fig. 4.9a, at constant [B], are plotted against [Q], the results are as shown in Fig. 4.9b and 4.9c. The slopes of these plots are $K_A / V_1 K_Q^B + K_A' K_B / K_Q[B]$ and $K_B / V_1 K_Q^B[B]$ respectively, and since V_1, K_A and K_B are known from the studies in the absence of inhibitor K_Q^A and K_Q^B can readily be calculated. A similar procedure can be based on initial plots of $1/v$ against $1/[B]$ at constant $[A]$ and variable $[Q]$.

Alternative methods of plotting may also be used. For example, the primary plot may be of $1/v$ against $[Q]$; the secondary plots of slope and intercept against $1/[A]$ and $1/[B]$.

Table 8.2 summarizes some of the various inhibition patterns that can arise, for a fixed amount of B and variable $[A]$, for different conditions.

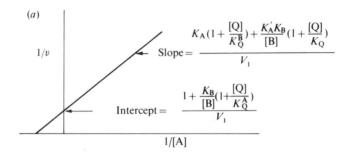

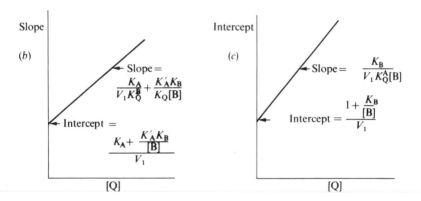

FIG. 4.9. Primary (a) and secondary (b) and (c) plots for the determination of inhibition constants on the basis of the random ternary-complex mechanism.

Ordered ternary-complex mechanism

For the ordered ternary-complex mechanism there are a number of inhibition possibilities. Only one of these will be considered, in order to illustrate the general principles; other cases can be worked out in a similar way.

The case to be treated is the one in which the inhibitor Q can add on to E and EA, but not to EY and EAB; it might be a substance that is structurally analogous to B and to Y. This case is shown in Fig. 4.10. Table 4.1 shows

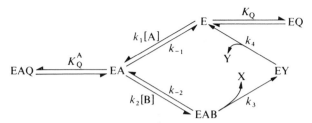

Fig. 4.10. The ordered ternary-complex mechanism, with inhibitor becoming attached to E and EA but not to EY or EAB.

TABLE 4.2

Types of inhibition observed with the random ternary-complex mechanism

Condition	Effect of inhibitor on		Type of inhibition
	Slope of plots of $1/v_1$ against $1/A$	Intercept	
Saturation with B and:			
(1) $K_Q^B \neq \infty$	$1 + \dfrac{Q}{K_Q^B}$	No effect	Competitive with A
(2) $K_Q^B = \infty$	No effect	No effect	No inhibition
No saturation and:			
(1) $K_Q^A \neq \infty$; $K_Q^B \neq \infty$; $K_Q \neq \infty$	$1 + \dfrac{Q}{K_Q'}$	$1 + \dfrac{Q}{K_Q''}$	Mixed; simple non-com- petitive if by accident $K_Q' = K_Q''$
(2) $K_Q^A = \infty$; $K_Q^B \neq \infty$; $K_Q = \infty$	$1 + \dfrac{Q}{K_Q'}$	No effect	Competitive
(3) $K_Q^A \neq \infty$; $K_Q^B = \infty$; $K_Q = \infty$	No effect	$1 + \dfrac{Q}{K_Q^A}$	Anticompetitive

the terms relating to E and EA, and these must now be multiplied by $(1 + [Q]/K_Q)$ and $(1 + [Q]/K_Q^A)$ respectively; the resulting rate equation is therefore

$$v = \frac{k_3[E]_0 \cdot [A] \cdot [B]}{\left\{\dfrac{k_4(k_{-1}k_{-2} + k_{-1}k_3)}{k_1 k_2 (k_3 + k_4)}\right\}\left(1 + \dfrac{[Q]}{K_Q}\right) + \dfrac{k_4(k_{-2} + k_4)}{k_2(k_3 + k_4)}[A] \times \atop \times \left(1 + \dfrac{[Q]}{K_Q^A}\right) + \dfrac{k_3 k_4}{k_1(k_3 + k_4)}[B]\left(1 + \dfrac{[Q]}{K_Q^A}\right) + [A] \cdot [B]}. \quad (41)$$

In Cleland's form this becomes

$$v = \frac{V[A] \cdot [B]}{(K_A' K_B + K_A[B])\left(1 + \dfrac{[Q]}{K_Q}\right) + K_B[A]\left(1 + \dfrac{[Q]}{K_Q^A}\right) + [A] \cdot [B]}. \quad (42)$$

In reciprocal form:

$$\frac{V}{v} = 1 + \frac{K_A}{[A]}\left(1 + \frac{[Q]}{K_Q}\right) + \frac{K_B}{[B]}\left(1 + \frac{[Q]}{K_Q^A}\right) + \frac{K_A'K_B}{[A][B]}\left(1 + \frac{[Q]}{K_Q}\right) \qquad (43)$$

The general equation (40) for the random mechanism is of the same form as this; however, if the inhibitor is analogous to B it will combine with E and EA but not with EB (or EAB); eqn (40) therefore reduces to

$$\frac{V}{v} = 1 + \frac{K_A}{[A]} + \frac{K_B}{[B]}\left(1 + \frac{[Q]}{K_Q^A}\right) + \frac{K_A'K_B}{[A][B]}\left(1 + \frac{[Q]}{K_Q}\right). \qquad (44)$$

The equations now differ in that in (44) the $K_A/[A]$ term is not multiplied by a Q-dependent term, whereas the corresponding term in (43) is multiplied by $(1 + Q/K_Q)$†. Therefore, it is possible by use of such an inhibitor to discriminate between the random and ordered mechanisms; moreover, if the mechanism is found to be ordered it can be concluded that it proceeds via the binary complex formed between the enzyme and the substrate with which the inhibitor competes.

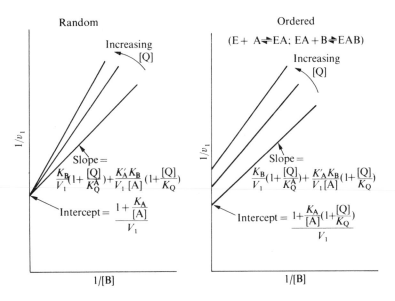

FIG. 4.11. Schematic plots of $1/v$ against $1/[B]$ at constant $[A]$, for reactions proceeding by a random ternary-complex mechanism, and by an ordered mechanism via EA. It is assumed that EB is not formed in the ordered mechanism, and that Q can add on to EA but not to EB.

† This is because for the random mechanism the numerator in the expression for the concentration of free enzyme is independent of B, whereas in the ordered mechanism it is linear in B; in the latter case $(1 + [Q]/K_Q)$ therefore multiplies a term linear in B (eqn (42)).

In order to establish whether eqn (43) or (44) applies, plots could be made of $1/v$ against $1/[B]$, with A held constant; a series of plots at different Q values would be required, as shown in Fig. 4.11. Whether or not the intercepts are dependent on Q distinguishes between the mechanisms. The behaviour with the random mechanism is competitive with respect to B; with the ordered mechanism, on the other hand, the behaviour is mixed.

However, complications can easily arise and may make it difficult to discriminate between mechanisms in this way. Suppose, for example, that the reaction proceeded by an ordered mechanism via EA, but that EB was also formed as a dead-end complex. This could also combine with Q, and the numerator in eqn (44) would then involve an additional term involving $(1 + [Q]/K_Q^B)/[A]$. In these circumstances the equations for the two mechanisms would be of the same form for all inhibitors, so that the inhibitor method could not be satisfactorily employed to distinguish the mechanisms. It must also be borne in mind that the addition of an inhibitor to a system might bring about a change in mechanism.

Ping-pong mechanism

The rate equation for the ping-pong mechanisms is of a different form from the equation for the ternary-complex mechanisms, since it lacks the constant term in the denominator (see Table 4.1). The inhibition patterns for this mechanism may, however, be described briefly since they allow the order of addition to be determined.

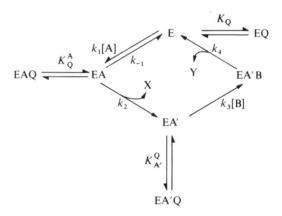

FIG. 4.12.   The ping-pong mechanism, with an inhibitor Q combining with E, EA and EA′ but not with EA′B.

An inhibitor may be able to combine with the three forms E, EA and EA′, with the inhibition constants shown in Fig. 4.12; it is unlikely to combine with EA′B, since both sites are covered. The terms for the concentrations

E, EA and EA′ are shown on p. 124, so that the resulting rate equation is

$$v = \frac{k_1 k_2 k_3 k_4 [E]_0 . [A] . [B]}{k_1 k_2 k_4 [A] \left(1 + \dfrac{[Q]}{K_Q^{A'}}\right) + k_3 k_4 (k_{-1} + k_2)[B] \left(1 + \dfrac{[Q]}{K_Q}\right) + } \tag{45}$$
$$+ k_1 k_3 k_4 [A] . [B] \left(1 + \dfrac{[Q]}{K_Q^{A}}\right) + k_1 k_2 k_3 [A] . [B].$$

This can be written as

$$v = \frac{V[A] . [B]}{K_B [A] \left(1 + \dfrac{[Q]}{K_Q^{A'}}\right) + K_A [B] \left(1 + \dfrac{[Q]}{K_Q}\right) + \dfrac{[A][B]}{k_2 + k_4} \left\{ k_4 \left(1 + \dfrac{[Q]}{K_Q^{A}}\right) + k_2 \right\}} . \tag{46}$$

In reciprocal form

$$\frac{1}{v} = \frac{\dfrac{K_B}{[B]}\left(1 + \dfrac{[Q]}{K_Q^{A'}}\right) + \dfrac{K_A}{[A]}\left(1 + \dfrac{[Q]}{K_Q}\right) + \dfrac{1}{k_2 + k_4} \left\{ k_4 \left(1 + \dfrac{[Q]}{K_Q^{A}}\right) + k_2 \right\}}{V} . \tag{47}$$

Plots of $1/v$ against $1/[A]$ at constant $[B]$ will give slopes from which K_Q can be determined; similarly $K_Q^{A'}$ can be determined from plots of $1/v$ against $1/[B]$. The intercepts in both cases will involve

$$\frac{k_4 \left(1 + \dfrac{[Q]}{K_Q^{A}}\right) + k_2}{V(k_2 + k_4)}$$

but K_Q^{A} cannot be determined unless k_2 and k_4 are known (for example, from transient-phase studies). If the inhibitor combines with E and with EA′ but not with EA the behaviour will be linear non-competitive with respect to both A and B (simple non-competitive if $K_Q^{A'} = K_Q$). If the inhibitor combines only with E the inhibition will be competitive with respect to A and anticompetitive with respect to B.

Isotopic exchange

The previous section has shown how it is possible in some cases to discriminate between different enzyme mechanisms by the use of inhibitors. A second procedure, to be discussed in the present section, is to carry out isotope exchange reactions.

The first application of this technique to an enzyme reaction was made in 1947 by Doudoroff, Barker, and Hassid[1], whose investigation is described

---

[1] M. Doudoroff, H. A. Barker, and W. Z. Hassid, *J. biol. Chem.*, **168**, 725 (1947).

in a later chapter (p. 266). The technique has been later extended by a number of workers[1]. Some of the details are rather complicated and only a very general account can be given here; the reader is referred to the original publications for additional information. One problem is the derivation of the appropriate rate equations for the various mechanisms; use of the method of King and Altman (see p. 80) is very helpful.

A ping-pong mechanism is easily identified using isotopic substitution since exchange can occur between a reactant and product even if the other reactant is not present. This is readily seen with reference to Fig. 4.7. The substrate A interacts with the enzyme to form X even if B is not present, so that if A is labelled the label will appear in X. The rate of this exchange will depend on the concentration of A but be independent of B. In the scheme of Fig. 4.7 as it stands, there can be no exchange between B* and Y in the absence of A. However, if the mechanism is extended to permit B to react with the free enzyme to form Y there will then be exchange between B* and Y.

Some of the complications involved in drawing conclusions from isotope-exchange studies are exemplified by the alcohol dehydrogenase system studied by Silverstein and Boyer[2]. For both yeast and liver alcohol dehydrogenases they proposed the following reaction scheme:

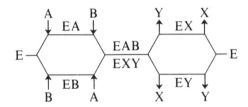

where A = reduced nicotinamide adenine dinucleotide (NADH), B = acetaldehyde, X = nicotinamide adenine dinucleotide (NAD) and Y = ethanol. They found no depression in NAD $\rightleftharpoons$ NADH exchange with increase in concentration of acetaldehyde and ethanol (kept at a constant ratio). They found that acetaldehyde $\rightleftharpoons$ ethanol exchange reached a plateau fairly rapidly as the concentration of NAD was increased (at constant ratio of NAD to NADH). Also, exchange between acetaldehyde dropped again with increase in concentration of ethanol and acetaldehyde, at constant ratio. The authors concluded that these results were in agreement with the above random-ordered mechanism. However, Wong and Hanes[3] pointed out that the results are also consistent with a compulsory-ordered mechanism

[1] P. D. Boyer, *Archs Biochem. Biophys.*, **82**, 387 (1959); R. A. Alberty, V. Bloomfield, L. Peller, and E. L. King, *J. Am. chem. Soc.*, **84**, 4381 (1962); P. D. Boyer and E. Silverstein, *Acta chem. scand.*, **17**, Suppl. 1 (1963); E. Silverstein and P. D. Boyer, *J. biol. Chem.*, **239**, 3908 (1964); I. A. Rose, E. L. O'Connel, and A. H. Mahler, *J. biol. Chem.*, **240**, 1758 (1965); W. W. Cleland, *A. Rev. Biochem.*, **36**, 77 (1967).
[2] E. Silverstein and P. D. Boyer, *J. biol. Chem.*, **239**, 3908 (1964).
[3] J. T. F. Wong and C. S. Hanes, *Nature*, **203**, 492 (1964).

with formation of a dead-end ternary complex. It is possible that NADH has to bind first, the E . NADH complex then being able to bind ethanol, in competition with acetaldehyde, with the formation of a dead-end ternary complex. It is possible that the NADH could dissociate from this ternary E . NADH. ethanol complex but not from the E . NADH acetaldehyde complex; the result would be no block in exchange between NAD and NADH with increase in acetaldehyde concentration.

Single and double exchange

Prior to Cleland's rather detailed analysis of the various two-substrate mechanisms, Koshland[1] had pointed out that such mechanisms could be classified according to whether one or two exchange reactions are involved. His discussion is of importance in relation to isotopic and stereochemical investigations, and will be outlined briefly.

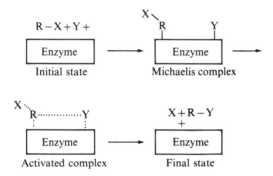

FIG. 4.13. Schematic representation of the single-displacement mechanism. Here the two reactants RX and Y become attached side-by-side on the enzyme and react together directly, the R—Y bond being formed at the same time as the R—X bond is broken.

In the first type of mechanism, illustrated schematically in Fig. 4.13, the two reacting molecules, which will be represented as RX and Y, first become attached side by side on the enzyme molecule, the complex being a ternary one. They then react with one another by what is essentially an S_N2 type of mechanism, and give rise to final products. This type of mechanism has been referred to by Koshland as a *single-displacement* mechanism. An essential feature of it is that the breaking of the R—X bond and the making of the R—Y bond *occur simultaneously*.

The second possibility, referred to as the *double-displacement* mechanism, is illustrated in Fig. 4.14. In this mechanism the substrate RX interacts with the enzyme and a splitting of the R—X bond occurs without the participation of the second substrate, Y. Then, in a second stage, the intermediate

[1] D. E. Koshland, *Biol. Revs.*, **28**, 416 (1953); *The Mechanism of Enzyme Action* (ed. McElroy and Glass), The Johns Hopkins Press, Baltimore (1954), p. 608; *Discuss. Faraday Soc.*, **20**, 142 (1955); *J. cell. comp. Physiol.*, **47**, 217 (1956).

complex formed in this way is attacked by the second substrate Y, and this reaction leads directly to the final products of the reaction. The feature of this type of mechanism that distinguishes it from the single-displacement mechanism is that the R—X bond is broken before the Y—R bond starts to be formed. In this mechanism the substrate Y need not necessarily be attached to the enzyme molecule; it may approach the enzyme-R complex from the aqueous phase.

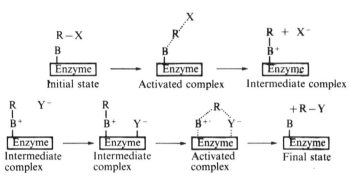

FIG. 4.14. Schematic representation of the double-displacement mechanism. In the first row is shown the formation of an intermediate complex in which the R—X bond has been completely broken. The essential feature of this mechanism is that the R—X bond is broken before R—Y begins to be formed. The stage at which Y becomes attached to the enzyme is immaterial.

The ping-pong mechanism of Cleland is obviously a development of this basic idea.

Some applications of this classification into the single-displacement and double-displacement mechanisms are to be found in Chapter 9, where it will be seen that isotope and stereochemical studies provide ways of distinguishing between them.

During recent years there have also been some studies of three-substrate systems[1].

[1] K. Dalziel, *Biochem. J.*, **114**, 547 (1969); F. B. Rudolph and H. J. Fromm, *Archs Biochem. Biophys.*, **147**, 515 (1971).

Symbols used for two-substrate kinetics

In the present book, the general kinetic equation for two-substrate systems has been written in the form (eqn (1), p. 115).

$$v = \frac{V[A].[B]}{K'_A K_B + K_B[A] + K_A[B] + [A].[B]}. \tag{1}$$

V is the limiting rate at high substrate concentrations when both $[A]$ and $[B]$ are very large. This equation is also conveniently written as

$$\frac{V}{v} = 1 + \frac{K_A}{[A]} + \frac{K_B}{[B]} + \frac{K'_A K_B}{[A].[B]}. \tag{2}$$

The above equations are of the same form as those used by Cleland[1], which are

$$v = \frac{V[A].[B]}{K_{ia}K_b + K_b[A] + K_a[B] + [A].[B]} \tag{3}$$

$$\frac{V}{v} = 1 + \frac{K_a}{[A]} + \frac{K_b}{[B]} + \frac{K_{ia}K_b}{[A].[B]}. \tag{4}$$

Both of the above systems closely resemble those recommended by the Enzyme Commission:

$$\frac{V}{v} = 1 + \frac{K_m^A}{[A]} + \frac{K_m^B}{[B]} + \frac{K_S^A K_m^B}{[A].[B]}. \tag{5}$$

The notation used by Alberty[2] is

$$v = \frac{V[A].[B]}{K_{AB} + K_B[A] + K_A[B] + [A].[B]} \tag{6}$$

$$\frac{V}{v} = 1 + \frac{K_A}{[A]} + \frac{K_B}{[B]} + \frac{K_{AB}}{[A].[B]}. \tag{7}$$

This differs from the above equations only in combining two constants into K_{AB}.

Dalziel[3] has expressed the general equations in the form

$$v = \frac{eS_1 S_2}{\Phi_{12} + \Phi_2 S_1 + \Phi_1 S_2 + \Phi_0 S_1 S_2} \tag{8}$$

or

$$\frac{e}{v} = \Phi_0 + \frac{\Phi_1}{S_1} + \frac{\Phi_2}{S_2} + \frac{\Phi_{12}}{S_1 S_2} \tag{9}$$

where e is the enzyme concentration and S_1 and S_2 the concentrations of the two substrates.

[1] W. W. Cleland, *Biochim. biophys. Acta*, **67**, 104, 173, 188 (1963); *A. Rev. Biochem.*, **36**, 77 (1967).

[2] R. A. Alberty, *J. Am. chem. Soc.*, **80**, 1777 (1958).

[3] K. Dalziel, *Acta chem. scand.*, **11**, 1706 (1957).

Bloomfield, Peller and Alberty[1] have written the equations as

$$v = \frac{V_{AB}[A].[B]}{K_{AB} + \dfrac{K_{AB}}{K_A}[A] + \dfrac{K_{AB}}{K_B}[B] + [A].[B]} \tag{10}$$

$$\frac{V_{AB}}{v} = 1 + \frac{K_{AB}}{K_B[A]} + \frac{K_{AB}}{K_A[B]} + \frac{K_{AB}}{[A].[B]}. \tag{11}$$

The accompanying table (4.3) will help in the translation of one system into another.

TABLE 4.3

Interrelationship of symbols for two-substrate systems

		Present book	Alberty	Dalziel	Bloomfield, *et al.*
Present book,	V		V	e/Φ_0	V_{AB}
Cleland,	K_A, K_a, K_m^A		K_A	Φ_1/Φ_0	K_{AB}/K_B
Enzyme	K_B, K_b, K_m^B		K_B	Φ_2/Φ_0	K_{AB}/K_A
Commission	K_A', K_{ia}, K_S^A		K_{AB}/K_B	Φ_{12}/Φ_2	K_A
Alberty	V	V		e/Φ_0	V
	K_A	K_A		Φ_1/Φ_0	K_{AB}/K_B
	K_B	K_B		Φ_2/Φ_0	K_{AB}/K_A
	K_{AB}	$K_A'K_B$		Φ_{12}/Φ_0	K_{AB}
Dalziel	e	$V\Phi_0$	$V\Phi_0$		$V\Phi_0$
	Φ_0	—	—		—
	Φ_1	$K_A\Phi_0$	$K_A\Phi_0$		$K_{AB}\Phi_0/K_B$
	Φ_2	$K_B\Phi_0$	$K_B\Phi_0$		$K_{AB}\Phi_0/K_A$
	Φ_{12}	$K_A'K_B\Phi_0$	$K_{AB}\Phi_0$		$K_{AB}\Phi_0$
Bloomfield,	V_{AB}	V	V	e/Φ_0	
Peller	K_A	K_A'	K_{AB}/K_B	Φ_{12}/Φ_2	
and	K_B	$K_A'K_B/K_A$	K_{AB}/K_A	Φ_{12}/Φ_1	
Alberty	K_{AB}	$K_A'K_B$	K_{AB}	Φ_{12}/Φ_0	

[1] V. Bloomfield, L. Peller, and R. A. Alberty, *J. Am. chem. Soc.*, **84**, 4367 (1962).

Appendix B

In this Appendix we classify two-substrate enzyme systems in terms of the mechanism by which they appear to occur.

TABLE A

Reactions occurring by a random ternary-complex mechanism

Enzyme	Type of Investigation	Comments	Reference
Yeast alcohol dehydrogenase	Initial velocity and Haldane relationships Isotope effects	See footnote†	1 2
Yeast hexokinase	Initial velocity Isotope exchange (equilibrium) Isotope competition and alternate substrate	— — —	3 4 5
Creatine phosphokinase	Initial velocity and inhibition Isotope exchange at equilibrium	Dead-end binary and ternary complexes	6 7
Phosphoglucomutase	Isotope exchange at equilibrium Initial velocity and product inhibition	Mechanism partly ordered	8 9
Phosphorylase a	Initial velocity, product inhibition and isotope exchange	—	10
Potato phosphorylase	Isotope exchange	—	11
Phosphoribosyl pyro-phosphate synthetase	Product inhibition	—	12
Adenylate kinase	Inhibition with transition state analogue	—	13

† Other workers, however, suggest that the mechanism is ordered; e.g. C. C. Wratten and W. W. Cleland, *Biochemistry*, **2**, 935 (1963) from product inhibition and E. Silverstein and P. D. Boyer, *J. Biol. Chem.*, **239**, 3908 (1964), using equilibrium isotope exchange.

[1] A. P. Mygaard and H. Theorell, *Acta. chem. scand.*, **9**, 1300 (1955); K. Dalziel, *ibid.*, **11**, 1706 (1957).

[2] H. R. Mahler and J. Douglas, *J. Am. chem. Soc.*, **79**, 1159 (1957); V. J. Shiner, H. R. Mahler, R. H. Baker and R. R. Hiatt, *Ann. N.Y. Acad. Sci.*, **84**, 583 (1960); J. F. Thomson, Biochem., **2**, 224 (1963).

[3] H. J. Fromm and V. Zewe, *J. biol. Chem.*, **237**, 3027 (1962).

[4] H. J. Fromm, E. Silverstein, and P. D. Boyer, *J. biol. Chem.*, **239**, 3645 (1964).

[5] F. B. Rudolf and H. J. Fromm, *Biochemistry*, **9**, 4660 (1970).

[6] J. F. Morrison and W. J. O'Sullivan, *Biochem. J.*, **94**, 221 (1965); E. James and J. F. Morrison, *J. biol. Chem.*, **241**, 4758 (1966).

[7] J. F. Morrison and W. W. Cleland, *J. biol. Chem.*, **241**, 673 (1966).

[8] W. Ray, G. Roscelli, and D. S. Kirkpatrick, *J. Biol. Chem.*, **241**, 2603 (1966).

[9] W. Ray and G. Roscelli, *J. biol. Chem.*, **241**, 3499 (1966).

[10] H. D. Engers, S. Schechosky, and N. B. Madsen, *Can. J. Biochem.*, **48**, 746 (1970); H. D. Engers, W. A. Bridger, N. B. Madsen, and B. Neil, *ibid.*, **48**, 755 (1970); see also H. D. Engers, W. A. Bridger, and N. B. Madsen, *J. biol. Chem.*, **244**, 755, 5396 (1969).

[11] A. M. Gold, R. M. Johnson, and G. R. Sanchez, *J. biol. Chem.*, **246**, 3444 (1971).

[12] R. L. Switzer, *J. biol. Chem.*, **246**, 2447 (1971).

[13] D. L. Purlich and H. J. Fromm, *Biochim. biophys. Acta*, **276**, 563 (1972).

TABLE B

Reactions occurring by an ordered ternary-complex mechanism†

Enzyme	Type of investigation	Comments	Reference
Liver alcohol dehydrogenase	Isotope competition and alternate substrate	Random under some conditions	1
	Product inhibition	—	2
	Alternate substrate	—	3
	Isotope exchange	—	4
	Initial velocity and substrate inhibition	—	5
Lactate dehydrogenase (*cf.* also Chapter 10)	Isotope exchange	Ternary complexes and abortive	6
	Kinetic isotope effect	ternary	7
	Fluorescence	complexes	8
	Product inhibition	demonstrated	9
Malate dehydrogenase (pig heart)	Initial velocity and product inhibition	Partly random under some	10
	Isotope exchange	conditions	11
Isocitric dehydrogenase (Neurospora)	Initial velocity and product inhibition	Dead-end ternary complex	12
Shikimate dehydrogenase	Isotope exchange at equilibrium, and product inhibition	—	13
Glyceraldehyde-3-phosphate dehydrogenase	Product inhibition	—	14
Potato acid phosphatase	Product inhibition	—	15
Succinyl-CoA synthetase (*E. coli*)	Initial velocity	Covalent Intermediate	16
Malic enzyme	Product inhibition	Dead-end binary complex	17
Mevalonic kinase (pig liver)	Product and dead-end inhibition	—	18
Isocitric lyase	Product inhibition	—	19
Glutaminase	Product inhibition	—	20

† Many enzymes having a coenzyme (e.g. the dehydrogenases) catalyze by an ordered mechanism, the coenzyme binding first.

[1] F. B. Rudolf and H. J. Fromm, *Biochemistry*, **9**, 4660 (1970).
[2] C. C. Wratten and W. W. Cleland, *Biochemistry*, **2**, 935 (1963).
[3] C. C. Wratten and W. W. Cleland, *ibid.*, **4**, 2442 (1965).
[4] E. Silverstein and P. D. Boyer, *J. biol. Chem.*, **239**, 3908 (1964).
[5] K. Dalziel and F. M. Dickinson, *Biochem. J.*, **100**, 34 (1966).
[6] E. Silverstein and P. D. Boyer, *J. biol. Chem.*, **239**, 3901 (1964).
[7] J. F. Thomson, J. J. Darling, and L. F. Bordner, *Biochim. biophys. Acta*, **85**, 177 (1964).
[8] A. D. Winer and G. W. Schwert, *J. am. chem. Soc.*, **79**, 6571 (1957); H. J. Fromm, *J. biol. Chem.*, **238**, 2938 (1963).
[9] V. Zewe and H. J. Fromm, *J. biol. Chem.*, **237**, 1168 (1962).
[10] D. N. Raval and R. G. Wolfe, *Biochemistry*, **1**, 263, 1112 (1962).
[11] E. Silverstein and G. A. Sulebeles, *Abstracts Am. chem. Soc.*, New York, **124** (1966).
[12] B. D. Sanwal, C. S. Stachow, and R. A. Cook, *Biochemistry*, **4**, 410 (1965).

[13] W. W. Cleland, D. Balinsky, A. W. Dennis, *Biochemistry*, **10**, 1947 (1971).

[14] B. A. Orsi and W. W. Cleland, *Biochemistry*, **11**, 102 (1972).

[15] R. Y. Hsu, W. W. Cleland, and L. Anderson, *Biochemistry*, **5**, 799 (1966).

[16] F. J. Moffet and W. A. Bridger, *J. biol. Chem.*, **245**, 2758 (1970).

[17] R. Y. Hsu, H. A. Lardy, and W. W. Cleland, *J. biol. Chem.*, **242**, 5315 (1967).

[18] E. Beytia, J. K. Dorsey, J. Marr, W. W. Cleland, and J. W. Porter, *J. biol. Chem.*, **245**, 5450 (1970).

[19] B. A. McFadden, J. O. Williams, and T. E. Roche, *Biochemistry*, **10**, 1384 (1971).

[20] B. Treit, G. Svennsky, E. Kvamme, *Eur. J. Biochem.*, **14**, 337 (1970).

TABLE C

Reactions occurring by a ping-pong mechanism

Enzyme	Type of investigation	Comments	Reference
Nucleoside diphosphokinase (yeast)	Initial velocity	—	1
	Product inhibition and isotope exchange	—	2
Phosphoglycerate mutase	Initial velocity and product inhibition	—	3
	Isotope exchange at equilibrium	—	4
Malate–lactate transhydrogenase	Initial velocity Inhibition Isotope effects	Abortive complexes exist	5 6 7
Hexokinase (mammalian)	Initial velocity	—	8, 9
	Inhibition	—	8, 9
	Isotope exchange at equilibrium	—	9
Arginine kinase	Initial velocity, and isotope exchange at equilibrium	—	10

Many reactions having water as the second substrate, and involving covalent intermediates, also occur *via* a ping-pong mechanism; examples are chymotrypsin, trypsin, ribonuclease, subtilisism, and alkaline phosphatase. See also Chapter 10, and Table 9.4, page 268.

[1] N. Mourad and R. E. Parks, *J. biol. Chem.*, **241**, 271 (1966).

[2] E. Garces and W. W. Cleland, *Biochemistry*, **8**, 633 (1969).

[3] S. Grisolia and W. W. Cleland, *Biochemistry*, **7**, 1115 (1968).

[4] M. Cascales and S. Grisolia, *Biochemistry*, **5**, 3116 (1966).

[5] S. H. G. Allen and J. R. Patil, *J. biol. Chem.*, **247**, 909 (1972).

[6] M. I. Dolin, *ibid.*, **244**, 5273 (1969).

[7] S. H. G. Allen, *ibid.*, **241**, 5266 (1966).

[8] H. J. Fromm and V. Zewe, *J. biol. Chem.*, **237**, 1661 (1962).

[9] T. L. Hanson and H. J. Fromm, *ibid.*, **240**, 4133 (1965).

[10] M. L. Uhr, F. Marcus, and J. F. Morrison, *ibid.*, **241**, 5428 (1966).

5. The Influence of Hydrogen Ion Concentration

THE hydrogen ion concentration has a very marked effect on the rates of enzyme reactions, and the situation is a somewhat complicated one. The broad outlines of the problem are now fairly well understood, and much valuable information on the mechanisms of enzyme reactions has been provided by systematic studies of the effects of the pH on the detailed kinetics of the reactions. There is no doubt that many future advances in our knowledge of enzyme mechanisms will come from investigations of this kind.

In many cases the rates of enzyme reactions pass through a maximum as the pH is varied. A typical example of this type of behaviour is shown in Fig. 5.1. The pH corresponding to the maximum rate is known as the *optimum*

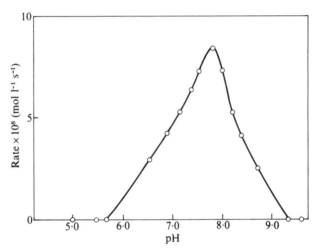

FIG. 5.1. A typical rate–pH curve; results for the α-chymotrypsin–methyl hydrocinnamate system.

pH, but it should be noted that the value of the optimum pH generally varies with the substrate concentration and with the temperature; it cannot therefore be regarded as a characteristic quantity for an enzyme.

Two different types of behaviour are to be distinguished, a *reversible* and an *irreversible* pH effect. If the solution is taken too far to the acid or the alkaline side the enzyme undergoes an irreversible loss of activity: activity cannot be restored by changing the pH back to the neighbourhood of the optimum. This type of behaviour is discussed in some detail in Chapter 13, and is due to the irreversible denaturation of proteins in strongly acid and basic solutions. Such denaturations involve loss of activity of the enzyme.

Irreversible inactivations of this type are not considered in the remainder of this chapter.

The reversible behaviour can be achieved only in a fairly narrow pH region in the neighbourhood of the pH optimum. Within this region the pH can be changed back and forth without any permanent effects ensuing. An explanation for this type of behaviour was first advanced in 1911 by Michaelis and Davidsohn[1], who emphasized the importance of the amphoteric properties of the protein. Although their theory has required modification for the system to which it was originally applied (the inversion of sucrose under the action of sucrase), it may be applicable in certain cases. It is therefore of interest to consider the original explanation, which will now be described from a more modern point of view.

Protein molecules contain a number of ionizing groups, such as $-CO_2H$ which can be ionized to give the negatively charged $-CO_2^-$ ion, and $-NH_2$ which can add on a proton to give $-NH_3^+$. Suppose, for example, that the active part of the enzyme consists of a basic group such as $-CO_2^-$ and an acid group such as $-NH_3^+$; the active enzyme may therefore be represented as

$$\begin{array}{cc} CO_2^- & NH_3^+ \\ | & | \\ \hline \multicolumn{2}{|c|}{\text{Enzyme}} \end{array}$$

The reactions occurring on the addition of acid or base to the active enzyme molecule may be represented as follows:

$$\begin{array}{cc} CO_2H & NH_3^+ \\ | & | \\ \hline \multicolumn{2}{|c|}{\text{Enzyme}} \end{array} \underset{OH^-}{\overset{H^+}{\rightleftharpoons}} \begin{array}{cc} CO_2^- & NH_3^+ \\ | & | \\ \hline \multicolumn{2}{|c|}{\text{Enzyme}} \end{array} \underset{OH^-}{\overset{H^+}{\rightleftharpoons}} \begin{array}{cc} CO_2^- & NH_2 \\ | & | \\ \hline \multicolumn{2}{|c|}{\text{Enzyme}} \end{array}$$

If it is supposed that the species to the left and right, which do not bear the combination $-CO_2^-$ and $-NH_3^+$, are enzymically inactive, it follows that the rate will pass through a maximum as the pH is varied. The optimum pH is the pH at which the concentration of the intermediate form is at a maximum; when the pH is changed from this value by the addition of either acid or base, this form is removed and one of the other forms is produced. The phenomenon of the pH optimum can therefore be explained in terms of three forms of the enzyme, the relationship between which may be represented as follows:

$$EH_2 \underset{OH^-}{\overset{H^+}{\rightleftharpoons}} EH \underset{OH^-}{\overset{H^+}{\rightleftharpoons}} E.$$

In this scheme the form EH_2 obviously bears one more positive charge (or one less negative charge) than EH, and EH has one more positive charge than E. For simplicity the charges will be omitted.

[1] L. Michaelis and H. Davidsohn, *Biochem. Z.*, **35**, 386 (1911).

Subsequent work has fully confirmed the proposal of Michaelis and Davidsohn that the existence of three forms of the enzyme will explain many of the reversible pH effects that have been observed. It is likely, however, that in some instances more than three forms must be postulated in order to explain the results; in such cases it must be that more than two ionizing groups on the enzyme molecule are involved in the enzymic activity.

Michaelis and Davidsohn's proposal was made before the general formulation of the rate equation in terms of the enzyme–substrate complex. It is of interest, however, to consider in the light of later developments what is the form of the rate equation if it is assumed that only the free enzyme molecule undergoes a change of ionization, and that the intermediate form EH is the only one that forms a complex with the substrate A. The scheme of reactions that corresponds to this case is

$$\text{EH}_2 \underset{k_{-b}[\text{H}^+]}{\overset{k_b}{\rightleftharpoons}} \text{EH} \underset{k_{-a}[\text{H}^+]}{\overset{k_a}{\rightleftharpoons}} \text{E}$$

$$k_1[\text{A}] \Big\updownarrow k_{-1} \diagdown k_2 \searrow \text{X}.$$

$$\text{EHA}.$$

The significance of the rate constants is the same as in Chapter 3. The steady-state equation for EHA is

$$k_1[\text{EH}].[\text{A}] - (k_{-1} + k_2)[\text{EHA}] = 0 \tag{1}$$

whence

$$K_{\text{A}}[\text{EHA}] = [\text{EH}].[\text{A}] \tag{2}$$

where K_{A} is equal to $(k_{-1} + k_2)/k_1$. Similarly, for EH_2 and E,

$$[\text{EH}]_2 = \frac{[\text{H}^+].[\text{EH}]}{K_b} \tag{3}$$

and

$$[\text{E}] = \frac{K_a[\text{EH}]}{[\text{H}^+]} \tag{4}$$

where K_a and K_b, equal to k_a/k_{-a} and k_b/k_{-b} respectively, are the two acid dissociation constants.

The total concentration of enzyme is given by

$$[\text{E}]_0 = [\text{EH}] + [\text{E}] + [\text{EH}_2] + [\text{EHA}] \tag{5}$$

$$= [\text{EH}]\left(1 + \frac{K_a}{[\text{H}^+]} + \frac{[\text{H}^+]}{K_b} + \frac{[\text{A}]}{K_{\text{A}}}\right). \tag{6}$$

From eqns (2) and (6)

$$[EHA] = \frac{[E]_0 \cdot [A]}{K_A} \bigg/ \left(1 + \frac{K_a}{[H^+]} + \frac{[H^+]}{K_b} + \frac{[A]}{K_A}\right). \tag{7}$$

The rate is therefore

$$v = k_2[EHA] = \frac{k_2[E]_0 \cdot [A]}{K_A\left(1 + \dfrac{K_a}{[H^+]} + \dfrac{[H^+]}{K_b}\right) + [A]}. \tag{8}$$

Since K_b is necessarily greater than K_a the function

$$1 + \frac{K_a}{[H^+]} + \frac{[H^+]}{K_b}$$

passes through a *minimum* as H^+ is varied, and has its minimum value when $[H^+] = (K_a K_b)^{\frac{1}{2}}$. Consequently the rate, according to eqn (8), will pass through a *maximum* as the pH is varied, and the optimum pH will correspond to a value of $[H^+]$ equal to $(K_a K_b)^{\frac{1}{2}}$. The degree of dependence shown by the rate on the pH will depend on the magnitude of $[A]$ in comparison with the pH-dependent term $K_A(1 + K_a/[H^+] + [H^+]/K_b)$. Indeed, at sufficiently high substrate concentrations eqn (8) predicts that there should be no pH dependence of the rate. The form of eqn (8) is similar to that found with competitive inhibitors, and the present situation is analogous in that the substrate can compete with the hydrogen and hydroxide ions and reduce the extent of formation of the inactive species EH_2 and E.

In 1920 Michaelis and Rothstein[1] proposed an alternative theory, according to which it is the ionization of the enzyme–substrate complex, and not of the enzyme itself, that is responsible for the pH effects. This suggestion can be represented by the following scheme of reactions:

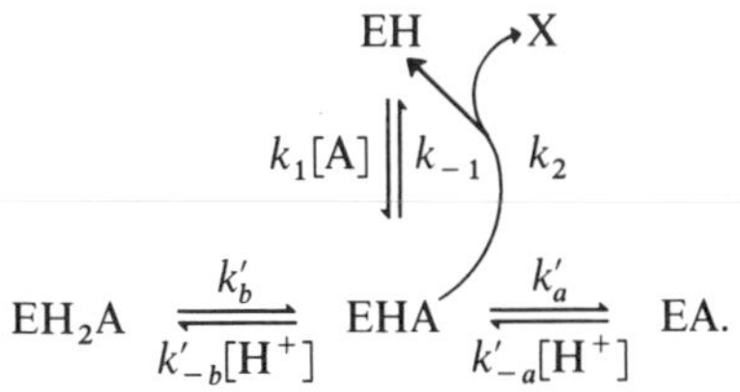

The assumption is now that EH_2A and EA cannot react further to form products, but that EHA can break down to form products with the regeneration of the enzyme EH.

Although this was not done by Michaelis and Rothstein, it is of interest to apply the steady-state method to their scheme of reactions. The result is

$$v = \frac{k_2[E]_0 \cdot [A]}{K_A + [A]\{1 + K_a'/[H^+] + [H^+]/K_b'\}} \tag{9}$$

[1] L. Michaelis and M. Rothstein, *Biochem. Z.*, **110**, 217 (1920).

where $K'_a (= k'_a/k'_{-a})$ and $K'_b (= k'_b/k'_{-b})$ are the two dissociation constants for the enzyme–substrate complex.

This equation, which is somewhat different from eqn (8), also accounts for a maximum in the rate against pH curves. Again the dependence of the rate on the pH is predicted to depend upon the substrate concentration. However, in this case the equation predicts that there will be no variation at sufficiently *low* substrate concentrations; the Michaelis–Davidsohn mechanism, on the other hand, predicted that the dependence would tend to vanish as the substrate concentration was *increased*.

Michaelis and his co-workers applied their theories to the inversion of sucrose, a reaction that is catalysed by the enzyme known variously as sucrase, saccharase, and invertase. According to the Michaelis–Davidsohn ideas the falling-off of the rate on the alkaline side of the pH optimum can be explained adequately on the hypothesis that in this region EH is being converted into E, and that the dissociation constant K_a is equal to $10^{-6 \cdot 6}$ ($pK_a = 6 \cdot 6$). Alternatively, Michaelis and Rothstein explained the falling-off as the solution is made alkaline in terms of a conversion of EHA into EA, with a pK'_a value of $6 \cdot 6$. However, Kuhn[1] pointed out a difficulty with these explanations. Both eqns (8) and (9) predict that the form of the curves should vary with the substrate concentration, but the experimental results fail to show such a variation. Kuhn concluded from this that the pH effects cannot be explained in terms of ionization effects, and suggested that instead the pH influences the rate of breakdown of the enzyme–substrate complex.

In 1924 von Euler, Josephson, and Myrbäck[2] solved the main problem by essentially welding together the ideas of Michaelis and Davidsohn, and of Michaelis and Rothstein. They showed that if the ionizations of *both* the enzyme *and* the enzyme–substrate complex are taken into account the results can be explained satisfactorily. Specifically, they showed that if it is assumed that the values of K_a for the process EH $\rightleftharpoons$ E and of K'_a for EHA $\rightleftharpoons$ EA are both equal to $10^{-6 \cdot 6}$ the lack of dependence of the alkaline branch of the sucrase curve on the substrate concentration can be explained. If, on the other hand, the corresponding dissociation constants for complex and enzyme have different values it is found that there will be a dependence of the curves on substrate concentration. Myrbäck[3] indeed found later that the acid branch of the sucrase curve does depend on the substrate concentration, in the manner predicted by the theory.

The conclusions of the von Euler–Josephson–Myrbäck theory may best be discussed from the standpoint of a much later formulation, worked out by Waley[4] on the basis of the steady-state treatment. The scheme considered

[1] R. Kuhn, *Z. physiol. Chem.*, **125**, 28 (1923).
[2] H. von Euler, K. Josephson, and K. Myrbäck, *Z. physiol. Chem.*, **134**, 39 (1924).
[3] K. Myrbäck, *Z. physiol. Chem.*, **158**, 160 (1926).
[4] S. G. Waley, *Biochim. biophys. Acta*, **10**, 27 (1953).

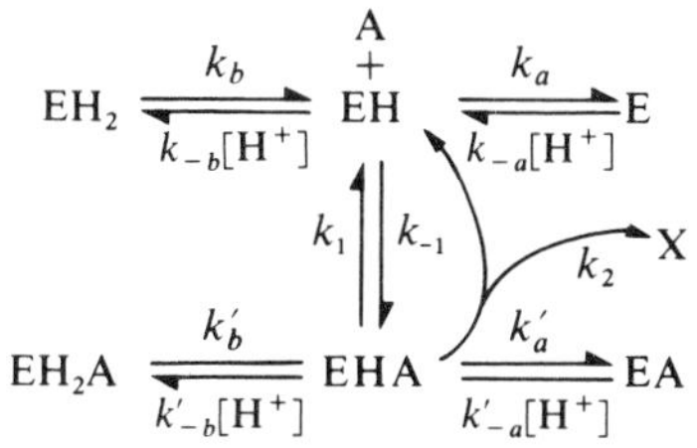

$$\text{EH}_2 \underset{k_{-b}[\text{H}^+]}{\overset{k_b}{\rightleftharpoons}} \overset{\overset{\text{A}}{+}}{\text{EH}} \underset{k_{-a}[\text{H}^+]}{\overset{k_a}{\rightleftharpoons}} \text{E}$$

$$k_1 \Big\| k_{-1} \qquad \overset{k_2}{\searrow}\text{X}$$

$$\text{EH}_2\text{A} \underset{k'_{-b}[\text{H}^+]}{\overset{k'_b}{\rightleftharpoons}} \text{EHA} \underset{k'_{-a}[\text{H}^+]}{\overset{k'_a}{\rightleftharpoons}} \text{EA}$$

FIG. 5.2. Scheme of reactions to explain pH-dependence (von Euler, Josephson, and Myrbäck; Waley).

by Waley, which is essentially that proposed by von Euler and co-workers, is shown in Fig. 5.2. This scheme does not include the reactions

$$\text{EH}_2 + \text{A} \rightleftharpoons \text{EH}_2\text{A} \quad \text{and} \quad \text{E} + \text{A} \rightleftharpoons \text{EA}.$$

This omission limits the applicability of the resulting equation somewhat, but the scheme nevertheless seems to give a satisfactory interpretation of some of the available experimental results.

The steady-state equations for EH_2, E, EH_2A, and EA are as follows:

$$[\text{EH}_2] = \frac{[\text{EH}].[\text{H}^+]}{K_b}, \tag{10}$$

$$[\text{E}] = \frac{K_a[\text{EH}]}{[\text{H}^+]}, \tag{11}$$

$$[\text{EH}_2\text{A}] = \frac{[\text{EHA}].[\text{H}^+]}{K'_b}, \tag{12}$$

$$[\text{EA}] = \frac{K'_a[\text{EHA}]}{[\text{H}^+]}. \tag{13}$$

The steady-state equation for EHA is

$$k_1[\text{EH}].[\text{A}] + k'_b[\text{EH}_2\text{A}] + k'_{-a}[\text{EA}].[\text{H}^+] -$$
$$- (k_{-1} + k_2 + k'_a + k'_{-b}[\text{H}^+]).[\text{EHA}] = 0. \tag{14}$$

Using eqns (12) and (13) this reduces to†

$$k_1[\text{EH}].[\text{A}] = (k_{-1} + k_2)[\text{EHA}], \tag{15}$$

or

$$K_\text{A}[\text{EHA}] = [\text{EH}].[\text{A}]. \tag{16}$$

† Again, we see that in formulating steady-state equations we can ignore reactions leading to dead-ends (see footnote on p. 88).

The total concentration of enzyme is

$$[E]_0 = [EH] + [E] + [EH_2] + [EHA] + [EA] + [EH_2A] \tag{17}$$

$$= [EH]\left(1 + \frac{K_a}{[H^+]} + \frac{[H^+]}{K_b}\right) + [EHA]\left(1 + \frac{K_a'}{[H^+]} + \frac{[H^+]}{K_b'}\right) \tag{18}$$

$$= [EHA]\left\{\frac{K_A}{[A]}\left(1 + \frac{K_a}{[H^+]} + \frac{[H^+]}{K_b}\right) + 1 + \frac{K_a'}{[H^+]} + \frac{[H^+]}{K_b'}\right\}. \tag{19}$$

The rate, $k_2[EHA]$, is therefore

$$v = \frac{k_2[E]_0 \cdot [A]}{K_A\left(1 + \dfrac{K_a}{[H^+]} + \dfrac{[H^+]}{K_b}\right) + [A]\left(1 + \dfrac{K_a'}{[H^+]} + \dfrac{[H^+]}{K_b'}\right)}. \tag{20}$$

This equation may be seen to be to some extent a combination of eqns (8) and (9). It reduces to eqn (8) if K_a' is very small and K_b' very large (which means that the complex will always be in the intermediate form EHA), and it reduces to eqn (9) if K_a is very small and K_b very large.

The way in which this equation interprets the sucrase results is as follows. On the alkaline side the equilibria that are important are $EH \rightleftharpoons E + H^+$ and $EHA \rightleftharpoons EA + H^+$; the terms $[H^+]/K_b$ and $[H^+]/K_b'$ are unimportant in the denominator of eqn (20). If K_a' is equal to K_a eqn (20) may be written in alkaline solution as

$$v = \frac{k_2[E]_0 \cdot [A]}{(K_A + [A])(1 + K_a/[H^+])}. \tag{21}$$

At all substrate concentrations, therefore, the rate is simply proportional to

$$1/(1 + K_a/[H^+]),$$

so that the shape of the rate against pH curve will be the same for all substrate concentrations.

On the acid side the terms involving K_a and K_a' can be neglected but in this case, since the results show a variation in the curve as the substrate concentration is changed, it follows that K_b and K_b' are not equal to one another. The acid results are, as will be seen, satisfactorily explained on the basis of the equation

$$v = \frac{k_2[E]_0 \cdot [A]}{K_A(1 + [H^+]/K_b) + [A]} \tag{22}$$

with a value of K_b equal to $10^{-3 \cdot 0}$. The value of K_b' is apparently extremely large, so that in the range investigated the complex is never in the form of EH_2A.

The above discussion of the sucrase results can all be summed up by the general statement that whenever two dissociation constants for the enzyme and complex are equal, the corresponding branch of the rate-pH curve will be independent of substrate concentration. When they are not equal, on the other hand, some dependence on concentrations is to be expected. The factors that determine whether dissociation constants will be affected by the complexing are discussed later.

A more general formulation of this problem has been given by Laidler[1] with reference to the reaction scheme shown in Fig. 5.3. The vertical reactions

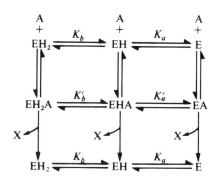

FIG. 5.3.   More general scheme of reactions to explain pH-dependence.

involving the extreme ionized forms of the enzyme are now included. A determinantal procedure was employed for simplicity, since the resulting rate equations are very complicated. The main object of this work was to discover the exact conditions under which it is legitimate to neglect the vertical reactions. These conditions were found to be:

(1) $k_2 \ll k_{-1}$,

or (2) $K'_a = 0$ and $K'_b = \infty$

or (3) The ionization processes are sufficiently rapid compared with the vertical reactions, and in addition the rate of formation of EHA is much greater than the rates of formation of EH_2A and EA.

For further details the reader is referred to the original paper.

A more general formulation

In view of the fact that many enzyme reactions involve a second intermediate, such as an acyl enzyme, it is natural to explore the steady-state equations applicable to this situation. This has been done by various workers[2]; the present account follows the treatment of Kaplan and Laidler[3].

[1] K. J. Laidler, *Trans. Faraday Soc.*, **51**, 528 (1955); see P. Ottolenghi, *Biochem. J.*, **123**, 445 (1971).

[2] L. Peller and R. A. Alberty, *J. Am. chem. Soc.*, **81**, 5907 (1959); R. M. Krupka and K. J. Laidler, *Trans. Faraday Soc.*, **56**, 1467 (1960); R. A. Alberty and V. Bloomfield, *J. biol. Chem.*, **238**, 2804 (1963); J. A. Stewart and H. S. Lee, *J. phys. Chem.*, **71**, 3888 (1967).

[3] H. Kaplan and K. J. Laidler, *Can. J. Chem.*, **45**, 539 (1967).

The scheme of reactions is shown in Fig. 5.4. Six acid dissociation constants are involved; K_a and K_b apply to the free enzyme, K'_a and K'_b to the addition complex, and K''_a and K''_b to the second intermediate. The processes of breakdown of the Michaelis complex and of the second intermediate are considered to be irreversible, which will certainly be true in the earlier stages of reaction, before products have accumulated. The scheme allows for the

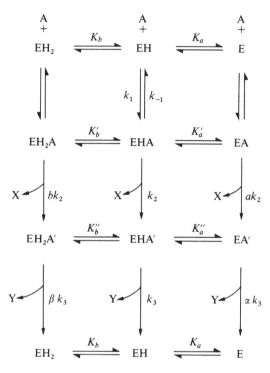

FIG. 5.4. A general scheme for pH-dependence, in which the formation of a second intermediate is considered.

breakdown of different ionization states of the complexes. If, for example, $a = 1$, EA breaks down at the same rate as EHA; this means that the group that has changed its state of ionization when EHA changes to EA is a kinetically *inessential* group. If, on the other hand, $a = 0$ the group is essential. Values of a between 0 and 1 are, of course, possible.

An exact steady-state treatment of this problem leads to exceedingly complex rate equations, which it would be impossible to apply to any experimental results. Kaplan and Laidler therefore employed an approximate procedure which leads to reasonably simple and manageable equations. The steady-state equation for EHA is written as

$$k_1[\text{EH}].[\text{A}] - (k_{-1} + k_2)[\text{EHA}] = 0. \tag{23}$$

The sum of the steady-state equations for EH_2A', EHA' and EA' is

$$k_2[EHA] + ak_2[EA] + bk_2[EH_2A] = k_3[EHA'] +$$
$$+ \alpha k_3[EA'] + \beta k_3[EH_2A']. \tag{24}$$

The ionization equations are assumed to be sufficiently rapid that use can be made of equations such as

$$K_b[EH_2] = [EH].[H^+]. \tag{25}$$

Introduction of such ionization equations into eqns (23) and (24) leads finally to the rate equation

$$v = \frac{\tilde{k}_c[E_0].[A]}{\tilde{K}_A + [A]} \tag{26}$$

where the pH-dependent kinetic parameters $\tilde{k}_c$ and $\tilde{K}_A$ are given by

$$\tilde{k}_c = \cfrac{k_2}{\cfrac{\left(1 + \cfrac{K_a'}{[H^+]} + \cfrac{[H^+]}{K_b'}\right)}{\left(1 + \cfrac{aK_a'}{[H^+]} + \cfrac{b[H^+]}{K_b'}\right)} + \cfrac{k_2}{k_3} \cdot \cfrac{\left(1 + \cfrac{K_a''}{[H^+]} + \cfrac{[H^+]}{K_b''}\right)}{\left(1 + \cfrac{\alpha K_a''}{[H^+]} + \cfrac{\beta[H^+]}{K_b''}\right)}} \tag{27}$$

$$\tilde{K}_A = \cfrac{\left(\cfrac{k_{-1} + k_2}{k_1}\right)\cfrac{\left(1 + \cfrac{K_a}{[H^+]} + \cfrac{[H^+]}{K_b}\right)}{\left(1 + \cfrac{aK_a'}{[H^+]} + \cfrac{b[H^+]}{K_b'}\right)}}{\cfrac{\left(1 + \cfrac{K_a'}{[H^+]} + \cfrac{[H^+]}{K_b'}\right)}{\left(1 + \cfrac{aK_a'}{[H^+]} + \cfrac{b[H^+]}{K_b'}\right)} + \cfrac{k_2}{k_3} \cdot \cfrac{\left(1 + \cfrac{K_a''}{[H^+]} + \cfrac{[H^+]}{K_b''}\right)}{\left(1 + \cfrac{\alpha K_a''}{[H^+]} + \cfrac{\beta[H^+]}{K_b''}\right)}} \tag{28}$$

The ratio of these two parameters is

$$\frac{\tilde{k}_c}{\tilde{K}_A} = \left(\frac{k_1 k_2}{k_{-1} + k_2}\right) \cdot \frac{\left(1 + \cfrac{aK_a'}{[H^+]} + \cfrac{b[H^+]}{K_b'}\right)}{\left(1 + \cfrac{K_a}{[H^+]} + \cfrac{[H^+]}{K_b}\right)} \tag{29}$$

Significance of the $\tilde{k}_c/\tilde{K}_A$ ratio

The ratio $\tilde{k}_c/\tilde{K}_A$ is related to the kinetic behaviour at low substrate concentrations. The way in which this ratio varies with the pH is shown by eqn (29), and the following special cases are of particular interest:

(1) Suppose that the enzyme–substrate complex EA ionizes in the same way as does the free enzyme (i.e. $K_a' = K_a$ and $K_b' = K_b$) and that a and b

are unity. This is the case of *inessential* ionizing groups. Even if these groups are involved in subsequent reactions (e.g. in deacylation), they will not be revealed in studies at low [A].

(2) If a and b are zero the studies at low [A] reveal K_a and K_b for the ionization of the free enzyme. If $a = 0$ but b is not zero there will be pH dependence of $\tilde{k}_c/\tilde{K}_A$ on the basic side, and the results will reveal K_a; if $b = 0$ but a is not zero the pH dependence on the acid side will reveal K_b.

(3) If either $K_a' = 0$ or $K_b' = \infty$, which means that the group is not free to ionize in the addition complex, the studies at low [A] will reveal either K_a or K_b for the ionization of the free enzyme. This is the case of an ionizing group that is *essential* to the binding of the substrate. When K_a and K_b values are obtained from the studies at low [A], one can conclude that the corresponding groups not only ionize in the free enzyme but *are essential to the subsequent reaction of the enzyme–substrate complex.* Even if one is using a substrate for which the final stage (e.g. deacylation) is rate limiting one can still determine the pK values of the ionizing groups involved in the previous stage (e.g. acylation) by studying the pH variation of $\tilde{k}_c/\tilde{K}_A$.

The patterns of pH behaviour

In the scheme under consideration, where there are two intermediates, there is ionization at three stages (E, EA and EA′), and there is the possibility of different rate-limiting steps for different substrates. There are therefore several different types of pH behaviour, and these are classified in Table 5.1. A positive sign (+) indicates that there is pH dependence of the Michaelis parameter indicated, and a zero (0) indicates no pH dependence. The possible combinations that will give rise to the behaviour observed, on the acidic and basic sides are indicated in the last three columns of the table. To simplify matters it has been assumed that if either K_a' or K_a'' is not equal to K_a it is *very* large (so that its ionization is outside the experimental range); similarly if K_b' or K_b'' is not equal to K_b it is *very* small. If these conditions do not hold the modifications are easily made. Table 5.1 is useful in permitting preliminary conclusions to be drawn from experimental pH studies; a decision between alternative possibilities can then often be made on the basis of other evidence—for example, by using substrates for which different steps are rate-determining.

Some examples of the various types of behaviour are listed at the foot of Table 5.1. In some cases the conclusion drawn is a matter of interpretation, and subsequent work might require a change of classification.

The matter of essential and inessential ionizing groups can be looked at in the following manner. A kinetic study can only reveal the presence of those groups for which a change in state of ionization has some effect on the rate of reaction. For any elementary process A → B we may consider three possibilities as far as pH dependence is concerned:

(1) The reactant molecule A can exist in more than one state of ioniza-

TABLE 5.1

Classification of pH behaviour for the case of two intermediates, for uncharged substrates

Case	$\bar{k}_c$	$\bar{K}_A$	$\dfrac{\bar{k}_c}{\bar{K}_A}$	$k_2 \ll k_3$ Acid	$k_2 \ll k_3$ Base	$k_2 \gg k_3$ Acid	$k_2 \gg k_3$ Base	$k_2 \approx k_3$ Acid	$k_2 \approx k_3$ Base
1	+	+	+	N.P.	N.P.	N.P.	N.P.	(i) $b=0, \beta=1$ (ii) $b=0, K_b''=\infty$	(i) $a=0, \alpha=1$ (ii) $a=0, K_a''=0$[a]
2	+	+	0	N.P.	N.P.	$b=1, \beta=0$	$a=1, \alpha=0$	$b=1, \beta=0$	$a=1, \alpha=0$
3	+	0	0	N.P.	N.P.	N.P.	N.P.	N.P.	N.P.
4	0	+	+	$K_b'=\infty$ [b]	$K_a'=0$[b]	(i) $b=0, \beta=1$ (ii) $b=0, K_b''=\infty$ (iii) $K_b'=\infty, \beta=1$ (iv) $K_b'=\infty, K_b''=\infty$	(i) $a=0, \alpha=1$ (ii) $a=0, K_a''=0$[c] (iii) $K_a'=0, \alpha=1$ (iv) $K_a'=0, K_a''=0$	(i) $K_b'=\infty, \beta=1$ (ii) $K_b'=\infty, K_b''=\infty$	(i) $K_a'=0, \alpha=1$ (ii) $K_a'=0, K_a''=0$
5	0	0	+	N.P.	N.P.	N.P.	N.P.	N.P.	N.P.
6	+	0	+	$b=0$[d]	$a=0$[d]	(i) $b=0, \beta=0$[c] (ii) $K_b'=\infty, \beta=0$[e]	(i) $a=0, \alpha=1$ (ii) $K_a'=0, \alpha=0$[e]	(i) $K_b'=\infty, \beta=0$ (iii) $b=0, \beta=0$	(i) $K_a'=0, \alpha=0$ (iii) $a=0, \alpha=0$[e]
7	0	+	0	N.P.	N.P.	N.P.	N.P.	N.P.	N.P.

a α-Chymotrypsin-catalysed hydrolysis of methyl hippurate (G. H. Nielson, J. L. Miles, and W. J. Canady, *Archs Biochem. Biophys.*, **96**, 3043 (1960)).

b Cholinesterase-catalysed hydrolysis of N-methyl amino ethyl acetate (R. M. Krupka and K. J. Laidler, *Trans. Faraday Soc.*, **56**, 1477 (1960)).

c α-Chymotrypsin-catalysed hydrolysis of p-nitrophenyl acetate, N-acetyltyrosine ethyl ester, N-benzoyl-L-alanine methyl ester, N-benzoyl-D-alanine methyl ester; (H. Kaplan and K. J. Laidler, *Can. J. Chem.*, **45**, 547 (1967)); trypsin-catalysed hydrolysis of p-toluene sulphonyl-L-arginine methyl ester (J. J. Bechet, *J. chim. Phys.*, **61**, 584 (1964)).

d α-Chymotrypsin-catalysed hydrolysis of amides (M. L. Bender, G. E. Clement, F. J. Kézdy, and H. d'A. Heck, *J. Am. chem. Soc.*, **86**, 3680 (1964)); α-Chymotrypsin-catalysed hydrolysis of dihydromethyl cinnamate (K. J. Laidler and M. L. Barnard, *Trans. Faraday Soc.*, **52**, 447 (1954)).

e Cholinesterase-catalysed hydrolysis of acetyl choline (R. M. Krupka and K. J. Laidler, *Trans. Faraday Soc.*, **56**, 1477 (1960)).

N.P. indicates that the behaviour specified is not possible.

tion, but each one of these states can equally readily form an activated complex.

For example, for two states of ionization,

$$HA \rightleftharpoons H^+ + A^-$$
$$\downarrow \qquad\qquad\qquad \downarrow$$
$$HA^{\neq} \qquad\qquad (A^-)^{\neq}$$
$$\downarrow \qquad\qquad\qquad \downarrow$$
$$\text{products} \qquad \text{products}$$

and the rate is now unaffected by pH. This is the case of the ionizing group that is inessential to reaction.

(2) If only one ionized state can give rise to an activated complex, the scheme is, for example,

$$HA \rightleftharpoons H^+ + A^-$$
$$\downarrow$$
$$(HA)^{\neq}$$
$$\downarrow$$
$$\text{products}$$

Increase in pH decreases the concentration of HA and lowers the reaction rate. The ionizing group is then *essential*.

(3) The third possibility is that the reactant is not free to ionize, but the activated complex is. For example, ionization in the initial state might be inhibited by hydrogen bonding, but the hydrogen bonding might not exist in the activated state:

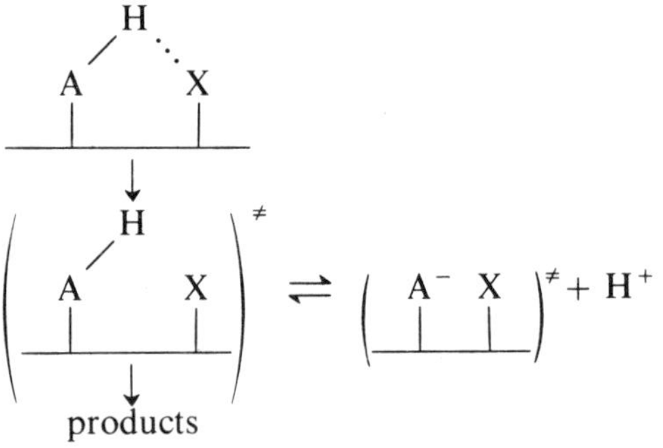

At first sight it might appear that raising the pH would remove activated complexes and reduce the rate. However, activated complexes are not in a state of true equilibrium, and their concentrations cannot be affected in this way; the rate of their decomposition is greater than their rate of ionization. The rate for this scheme is therefore uninfluenced by the pH.

It follows that a kinetic study only reveals the presence of an ionizing group when that group ionizes in the initial state, and when its ionization affects the ease of formation of the activated state.

Analysis of experimental results

Values for ionization constants in enzyme systems are conveniently obtained from plots of the common logarithms of the kinetic parameters against the pH. The procedures will be described and explained with reference to a plot of $\log_{10}(\tilde{k}_c/\tilde{K}_A)$ against pH for the case in which $a = b = 0$ (essential ionizing groups); for this case eqn (29) reduces to

$$\frac{\tilde{k}_C}{\tilde{K}_A} = \frac{k_1 k_2}{(k_{-1} + k_2)\left(1 + \dfrac{K_a}{[H^+]} + \dfrac{[H^+]}{K_b}\right)}. \tag{30}$$

In a sufficiently acid solution the term $[H^+]/K_b$ predominates in the bracketed pH-dependent term, and $\tilde{k}_c/\tilde{K}_A$ is inversely proportional to $[H^+]$:

$$\tilde{k}_c/\tilde{K}_A \approx k_1 k_2 K_b/\{(k_{-1} + k_2)[H^+]\} \tag{31}$$

or

$$\log_{10}(\tilde{k}_c/\tilde{K}_A) = \text{const.} + \text{pH} \tag{32}$$

since $\text{pH} = -\log_{10}[H^+]$. At low pH values the slope of the plot of $\log_{10}(\tilde{k}_c/\tilde{K}_A)$ against pH is therefore $+1$, as shown in Fig. 5.5. As the pH is

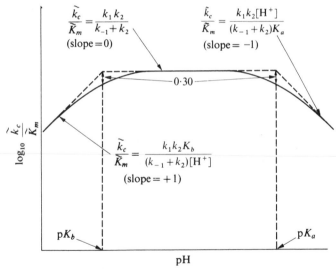

FIG. 5.5. Schematic plot of $\log_{10}(\tilde{k}_c/\tilde{K}_m)$ against pH, for systems to which eqn (30) applies.

raised a region is reached in which unity is predominant in the bracketed pH-dependent term:

$$\tilde{k}_c/\tilde{K}_A \approx k_1 k_2/(k_{-1} + k_2). \tag{33}$$

The slope of the $\log_{10}(\tilde{k}_c/\tilde{K}_A)$-pH plot is now zero. Equation (31) is the equation for the left-hand limb of the plot, and eqn (33) that for the horizontal portion; these two lines therefore intersect when

$$k_1k_2K_b/\{(k_{-1} + k_2)[H^+]\} = k_1k_2/(k_{-1} + k_2) \qquad (34)$$

i.e., when

$$[H^+] = K_b \qquad (35)$$

or

$$pH = pK_b. \qquad (36)$$

It is easy to show by similar arguments that the intersection point on the alkaline side corresponds to

$$pH = pK_a. \qquad (37)$$

The theory does not, of course, predict sharp changes of slope, as shown by the dotted lines in Fig. 5.5; instead there is a rounding off, as shown by the firm line. On the acid side, the value of $\tilde{k}_c/\tilde{K}_A$ is given by

$$\tilde{k}_c/\tilde{K}_A = k_1k_2/\{(k_{-1} + k_2)(1 + [H^+]/K_b)\} \qquad (38)$$

which at the inflection point, when $[H^+] = K_b$, becomes

$$\tilde{k}_c/\tilde{K}_A = k_1k_2/\{2(k_{-1} + k_2)\}. \qquad (39)$$

The curve at $[H^+] = K_b$ therefore lies $\log_{10} 2$, or 0·30, below the intersection point of the straight lines. These relationships are shown in Fig. 5.5.

The above conclusions are only valid if the ionizing groups are *essential*; i.e., if $a = b = 0$. It is evident from eqn (29) that if $b = 1$ there will be no change of slope on the acidic side. Groups that are revealed by this type of plot must therefore be involved in the subsequent reaction of the enzyme–substrate addition complex.

Similar arguments apply to plots of $\log_{10} \hat{k}_c$ and $\log_{10} \tilde{K}_A$ against pH. In these cases the situation is different according to whether $k_2 \gg k_3$ or $k_3 \gg k_2$. In the latter case eqn (27) reduces to

$$k_c = k_2 \bigg/ \left(1 + \frac{K_a'}{[H^+]} + \frac{[H^+]}{K_b'}\right) \qquad (40)$$

if $a = b = 0$ (i.e., if the ionizing groups are essential). The inflection points of plots of $\log_{10} k_c$ against pH will now reveal pK_a' and pK_b', as shown in Fig. 5.6. If $a = 1$, pK_a' will not be found, while if $b = 1$, pK_b' will not be found; in other words, groups will only be revealed by this treatment if they are involved in stage 2 of the reaction.

If, on the other hand, stage 3 is slow and rate-determining (i.e., if $k_2 \gg k_3$), eqn (27) reduces to

$$\hat{k}_c = k_3 \bigg/ \left(1 + \frac{K_a''}{[H^+]} + \frac{[H^+]}{K_b''}\right) \qquad (41)$$

and the inflection points reveal pK_a'' and pK_b'' (see Fig. 5.6).

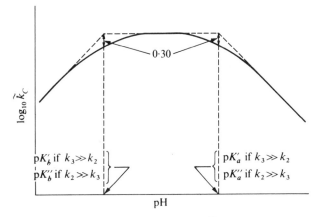

FIG. 5.6.   Schematic plot of $\log_{10} \tilde{k}_c$ against pH.

By choosing for an enzyme two substrates, for one of which $k_3 \gg k_2$ and for the other $k_2 \gg k_3$, it is clearly possible in principle to determine six dissociation constants, K_a, K_a', K_a'' and K_b, K_b' and K_b''. Such a procedure was successfully applied by Krupka and Laidler[1] to the acetylcholinesterase system.

Another plot that can be made in the analysis of experimental results is that of $\log_{10} \tilde{K}_A$ (or $p\tilde{K}_A$, which equals $-\log_{10} \tilde{K}_A$) against pH. Such plots were first suggested by Dixon[2]. Reference to eqn (28) shows that $p\tilde{K}_A$ has a more complex pH dependence than either $\log \tilde{k}_c$ or $\log \tilde{k}_c/\tilde{K}_A$. However, since

$$p\tilde{K}_A = \log_{10} \tilde{k}_c/\tilde{K}_A - \log_{10} \tilde{k}_c, \tag{42}$$

a plot of $p\tilde{K}_A$ is the difference between the plots of $\log_{10} \tilde{k}_c/\tilde{K}_A$ and of $\log_{10} \tilde{k}_c$. This is illustrated schematically in Fig. 5.7, which shows one of the many patterns of behaviour that can be observed. In principle four pK values can be determined from a plot of $p\tilde{K}_A$ alone; however, in the event that, for example, $pK_b' = pK_b$, the two inflection points to the left will obliterate one another. It is therefore best to make all three plots. Dixon has pointed out, with reference to the pK_A plots, that bends that are concave to the pH axis (i.e. ⌒, the slope decreasing with increasing pH) correspond to the ionization of the free enzyme†; conversely, bends convex to the pH axis (i.e. ⌣, the slope increasing with increasing pH) correspond to ionization of a reaction intermediate. It is now realized that this is the reaction intermediate that is the reactant in the slow step (e.g. the acyl enzyme if deacylation is the slow step).

---

† Or of the substrate; it has been assumed in the preceding discussion that substrate ionizations are not involved.

---

[1] R. M. Krupka and K. J. Laidler, *Trans. Faraday Soc.*, **56**, 1477 (1960).
[2] M. Dixon, *Biochem. J.*, **55**, 161 (1953).

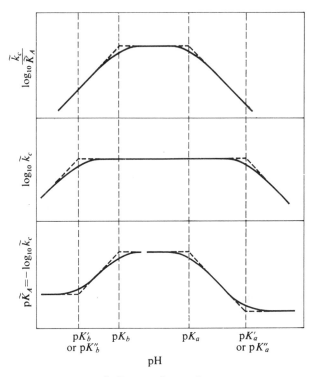

FIG. 5.7. Schematic plots of $\log_{10} \tilde{k}_c/\tilde{K}_m$, $\log_{10} \tilde{k}_c$ and $p\tilde{K}_m$ against pH. If stage 2 in the reaction is rate-determining, the pK values at the outer inflection points pK'_a and pK'_b, corresponding to the ionization of the addition complex; if stage 3 is rate-determining, the quantities are pK''_a and pK''_b, corresponding to the ionization of the second intermediate.

It was noted above that Fig. 5.7 illustrates only one of the several patterns of behaviour that can be observed in a pH study. For this particular case

$$pK'_b \text{ (or } pK''_b) < pK_b < pK_a < pK'_a \text{ (or } pK''_a).$$

If there is a different order of pK values the shapes of the plots will be different. As examples, Fig. 5.8 shows schematically the pH patterns that have been observed for chymotrypsin[1] and trypsin[2] when they act upon substrates for which the deacylation process is rate-determining. It is to be seen that for chymotrypsin

$$pK_b < pK''_b < pK_a < pK''_a$$

(the latter value being outside the range of the experiments), while for trypsin

$$pK''_b < pK_b < pK''_a < pK_a.$$

[1] H. Kaplan and K. J. Laidler, *Can. J. Chem.*, **45**, 547 (1967).
[2] H. P. Kasserra and K. J. Laidler, *Can. J. Chem.*, **47**, 4021 (1969).

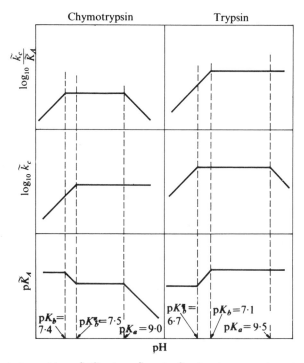

FIG. 5.8.   Variations of $\log_{10} \tilde{k}_c/\tilde{K}_m$, $\log_{10} \tilde{k}_c$ and $p\tilde{K}_m$, for chymotrypsin and trypsin acting upon substrates for which deacylation is rate-determining.

Again, the latter value is too high to be detected in the experiments, which could not be made at pH values higher than 10·0.

At first sight it might be thought that the different pH profiles found for chymotrypsin and trypsin imply that the two enzymes exert their action by fundamentally different mechanisms. This is not, however, the case; in fact, the various lines of evidence for these reactions indicate that the mechanisms are very similar for the two enzymes[1]. The different pH patterns arise from differences in relative magnitudes of pK values for the two cases. Consideration of these differences can lead to valuable conclusions about the details of the mechanisms. For example, there is good reason to believe that in both free enzymes the active group that changes its state of ionization on the alkaline side is the $-NH_2$ group of isoleucine (ile–16 in chymotrypsin, ile–7 in trypsin). However, the pK_a value for chymotrypsin is about 9, while for trypsin it is over 10. A plausible explanation for this is related to the fact that in chymotrypsin there is one $-CO_2^-$ group close to the $-NH_2$ group, whereas in trypsin there are two. This is illustrated in Fig. 5.9. It is

---

[1] For a review of the mechanistic evidence for chymotrypsin and trypsin see H. P. Kasserra and K. J. Laidler, *Can. J. Chem.*, **47**, 4031 (1969); cf. also Chapter 10.

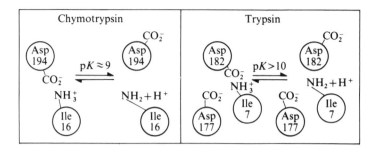

FIG. 5.9. A possible explanation for the higher pK_a value for trypsin than for chymotrypsin.

more difficult, for electrostatic reasons, for the proton to leave in the case of trypsin, and consequently the pK_a value is higher.

Nature of enzyme–substrate interactions

From the differences between pK_a, pK'_a and pK''_a values, and between pK_b, pK'_b and pK''_b values, it is possible to draw tentative conclusions about the nature of the enzyme–substrate complex and of the second intermediate.

The enzyme in its most active form may be represented schematically as

where $-B$ is the basic group and $-A-H$ the acidic group. The formation of an enzyme–substrate complex with a carbonyl compound $R-CO-X$ might occur without any interaction with the ionizing groups; this may be represented as follows:

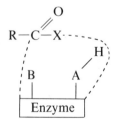

In this case the pK'_a value, which relates to the ionization of the complex, will be very close to pK_a; similarly, pK'_b will be very close to pK_b. On the other hand, there may be interaction with both groups, as follows:

$$
\begin{array}{c}
O^- \\
| \\
R-C-X \\
| \qquad \cdots H \\
| \qquad / \\
B^+ \quad A \\
| \qquad | \\
\boxed{\text{Enzyme}}
\end{array}
$$

The ionizations of this species will be

$$
\begin{array}{ccc}
O & O^- & O^- \\
\| & | & | \\
R-C-X & R-C-X & R-C-X \\
\cdots \quad \overset{pK_b'}{\rightleftharpoons} & \cdots \quad \overset{pK_a'}{\rightleftharpoons} & \\
H \quad \diagdown H & H & \\
| \quad / & | \quad / & \\
B^+ A & B^+ A & B^+ \quad A^- \\
| \quad | & | \quad | & | \qquad | \\
\boxed{\text{Enzyme}} & \boxed{\text{Enzyme}} & \boxed{\text{Enzyme}}
\end{array}
$$

It will be more difficult for the central species to add on a proton than for the free enzyme, so that $pK_b' < pK_b$. Similarly the splitting off of a proton from the central species will be less easy in the complex than in the free enzyme; thus $pK_a' > pK_a$.

Alternatively, if there is binding only to the basic group $-B$, the pK_a' and pK_a values will be much the same, whereas pK_b' and pK_b will be different.

For binding at the acidic group only, $pK_b' \approx pK_b$ whereas $pK_a' > pK_a$.

Similar conclusions can be drawn with regard to the second intermediate, the ionizations of which are related to the pK_a'' and pK_b'' values. The relative values of these pK's thus provide evidence regarding the nature of binding in complexes. The conclusions must be tentative, however, since complications may arise from conformational and other effects.

Influence of dielectric constant[1]

Some evidence as to the nature of the ionizing groups that are involved in enzyme action is provided by a consideration of the magnitudes of the pK values. For example, a pK of about 7 probably corresponds to an imidazole group; other typical pK values are listed on p. 5. However, considerable caution must be taken in drawing conclusions in this way, because the pK values of groups on enzymes are often very different from the values for the same groups in simpler molecules[2].

[1] See H. Kaplan and K. J. Laidler, *Can. J. Chem.*, **45**, 547 (1967); see also D. Findlay, A. P. Mathias, and B. R. Rabin, *Biochem. J.*, **85**, 139 (1963).

[2] See, for example, D. E. Schmidt and F. H. Westheimer, *Biochemistry*, **10**, 1249 (1971).

Further evidence about the nature of ionizing groups on enzymes is provided by studies of the effect of dielectric constants on pK values. Ionizing groups on an enzyme may be divided into two classes:

(a) Neutral groups, such as $-CO_2H$ and $-OH$, which dissociate into a proton and a negative group:

$$-A-H \rightleftharpoons -A^- + H^+$$

(b) Cationic groups, such as $-NH_3^+$, which dissociate into a proton and a neutral group:

$$-B-H^+ \rightleftharpoons -B + H^+.$$

The effect of changing the dielectric constant is very different in the two cases. With groups of class (a), an increase in dielectric constant increases the ease of separation of charges and therefore increases K and decreases pK. In groups of class (b), however, changing the dielectric constant has very little effect on the K value, since cationic species are involved on both sides of the equation. This type of behaviour is well substantiated experimentally for simple substances. For example, the pK of acetic acid is 4·76 in water and 10·14 in a dioxane–water mixture (82 per cent dioxane by weight), which has a much lower dielectric constant[1]. On the other hand, the pK values relating to the $-NH_3^+$ groups of amino acids are only very slightly affected by a change in the dielectric constant[2].

The method should, however, be applied with caution. Variation in dielectric constant can only be effected by making substantial changes in the nature of the solvent: a substance added to water in order to lower the dielectric constant might have a specific inhibitory effect. It is therefore wise to employ more than one mixed solvent system in any investigation of this type. Another complication is that a cationic group on an enzyme might be forming an ion–pair with an anionic group, and therefore behave in mixed solvents as if it were a neutral group; an example is to be found with chymotrypsin[3].

[1] M. Mandel and P. Decroly, *Trans. Faraday Soc.*, **56**, 29 (1960).
[2] E. L. Duggan and C. L. A. Schmidt, *Archs Biochem.*, **1**, 453 (1942).
[3] H. P. Kasserra and K. J. Laidler, *Can. J. Chem.*, **47**, 4031 (1969).

6. The Time Course of Enzyme Reactions

WHEREAS the last three chapters have been concerned with initial reaction rates and their variation with concentrations, the present one deals with changes in concentration during the course of a given reaction. Expressed differently, the last three chapters dealt with the rate equations in their differential forms, while the present deals with the integrated forms. More is involved, however, than a mere integration of the equations already developed, because the equations for the time course of a reaction are not always related in a simple manner to those indicating the variation of initial rate with concentration. This can arise, for example, if a product of reaction is an inhibiting substance; this circumstance will not affect the differential form of the initial rate equation, but will influence the time course of the reaction.

As a result of complexities of this kind investigations of the time course of enzyme reactions frequently cannot be interpreted unambiguously. For example, simple first-order behaviour in the course of a one-substrate reaction might be due to either of the following two causes:

(1) the substrate concentration might be low, so that the true order is the first, or
(2) the true order might be zero, but the reaction might be slowed down during a run owing to the inhibiting action of a product.

Because of such ambiguities it is inadvisable to draw conclusions about kinetics entirely from studies of the time-course of reactions. However, such studies do provide valuable supporting information.

For analysing the kinetic results for an enzyme system the most reliable procedure involves the application of van't Hoff's differential method, which was described on p. 37. After the method has been used to determine the main features of the reaction a further study can be carried out using the integrated equation for the appropriate situation.

The systems treated in the present chapter largely parallel those in Chapter 3. The following situations will be considered:

(1) Uncomplicated one-substrate reactions (p. 164).
(2) Reversible one-substrate reactions (p. 167).
(3) Inhibition by substrate (p. 169).
(4) Inhibition by products (p. 170).
(5) Competing substrates (p. 173).
(6) Enzyme inactivation during the course of reaction (p. 175).
(7) Schutz's law (p. 180).
(8) Kinetics of the transient phase (p. 181).

Many of the integrated solutions have previously been given by Haldane[1].

Uncomplicated one-substrate reactions

When an enzyme reaction involves only one substrate and there are no complications the rate varies with the substrate concentration according to the Michaelis–Menten law (see p. 74, eqn (19)). If a_0 is the initial amount of substrate and x the amount of product after time t this law may be written as

$$\frac{dx}{dt} = \frac{V(a_0 - x)}{K_A + a_0 - x}. \tag{1}$$

Solutions of this will be given for the limiting conditions and for the general case.

If the initial substrate concentration is sufficiently low ($K_A \gg a_0$) the equation reduces to

$$\frac{dx}{dt} = \frac{V}{K_A}(a_0 - x). \tag{2}$$

This integrates to

$$\frac{V}{K_A} = \frac{1}{t} \ln \frac{a_0}{a_0 - x}. \tag{3}$$

The applicability of this law to experimental results can conveniently be tested by plotting $\log a_0/(a_0 - x)$ against the time, or by calculating the values of the right-hand side of the equation at various times. An example of the graphical procedure is shown in Fig. 6.1.

If the initial substrate concentration is sufficiently high ($a_0 \gg K_A$) the rate equation reduces to

$$\frac{dx}{dt} = V, \tag{4}$$

the solution of which is

$$V = \frac{x}{t}. \tag{5}$$

A plot of x against t will therefore give a straight line, of slope V. An example of this type of plot is shown in Fig. 6.2. This law will not be followed throughout the entire course of an enzyme reaction, because as product accumulates $(a_0 - x)$ diminishes, and eventually will become comparable with K_A.

[1] J. B. S. Haldane, *Enzymes*, Longmans, Green and Co., London (1930), (reprinted by the M.I.T. Press, Cambridge, Mass. 1965), chap. v.

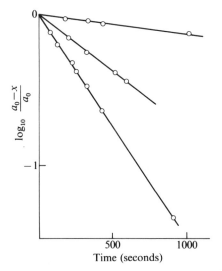

FIG. 6.1. Plot of $\log[(a_0 - x)/a_0]$ against time for a first-order reaction; these results were obtained for the oxidation of lactic acid, catalysed by lactic dehydrogenase (Socquet and Laidler, *Arch. Biochem.*, **25**, 171 (1950)).

The general rate equation (1) integrates as follows:

$$Vt = \int \left\{ \frac{K_A}{(a_0 - x)} + 1 \right\} dx + \text{const},\qquad (6)$$

where the constant is determined by the boundary condition $x = 0$, $t = 0$.

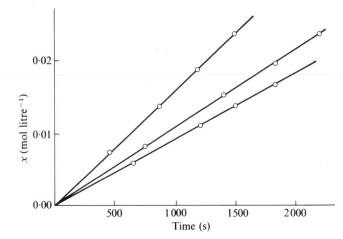

FIG. 6.2. Plot of x against time for a zero-order reaction; these results were obtained for the chymotrypsin-catalysed hydrolysis of methyl hydrocinnamate (Laidler and Barnard, *Trans. Faraday Soc.*, **51**, 499 (1956)).

The solution is

$$Vt = K_A \ln \frac{a_0}{a_0 - x} + x. \tag{7}$$

An equation of this form was first given empirically by Victor Henri[1] in 1903; it is therefore often known as the Henri equation. If K_A is large the equation reduces to eqn (3), while if K_A is small it becomes eqn (5).

The Henri equation actually has a wider applicability than to simple one-substrate reactions. For example, two-substrate reactions obey an equation of the same form as eqn (1) provided that the concentration of one substrate is held constant. It follows, therefore, that if the concentration of one substrate is greatly in excess, and does not vary appreciably during the course of reaction, the reaction should (if uncomplicated by other factors) follow the Henri equation. Also, single-substrate reactions involving additional intermediates have been seen (p. 79) to follow the Michaelis–Menten equation. Similarly, the rate equations for reactions occurring in the presence of constant amounts of inhibitors (competitive or non-competitive) can be written in the same form as the Michaelis–Menten equation, and their course should correspond to the Henri equation. If an inhibitor is produced as a product of reaction, however, the situation is more complicated, and is discussed later.

Analysis of experimental results

A convenient graphical method of testing the applicability of the Henri equation consists of plotting x/t against

$$\frac{1}{t} \ln \frac{a_0}{a_0 - x}.$$

This type of plot, which was apparently first employed by Walker and Schmidt[2], is illustrated schematically in Fig. 6.3. The intercept on the x/t axis is equal to $V (= k_c E)$, that on the other axis is V/K_A, while the slope is equal to $-K_A$. A feature of a plot of this kind, first pointed out by Foster and Niemann[3], is that a straight line drawn through the origin with a slope of a_0 will strike the line at a point that corresponds to the initial conditions of the reaction ($x = 0$). This may be seen from the fact that the slope of the line connecting the origin to the initial point is equal to

$$\lim_{x \to 0} \frac{x/t}{(1/t) \ln \{a_0/(a_0 - x)\}},$$

[1] V. Henri, *Lois generales de l'action des diastases* (1903).
[2] A. C. Walker and C. L. A. Schmidt, *Archs Biochem.*, **5**, 445 (1944); see also G. A. Fleischer, *J. Am. chem. Soc.*, **75**, 4487 (1953), for curve-fitting procedures.
[3] R. J. Foster and C. Niemann, *Proc. natn. Acad. Sci. U.S.A.*, **39**, 999 (1953).

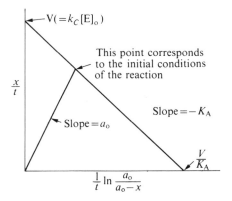

FIG. 6.3. Schematic plot of xt against $(1/t)\ln\{a_0/(a_0 - x)\}$ for a reaction obeying the Michaelis–Menten law.

the value of which is equal to a_0. An extension of this type of plot to reaction inhibited by products is discussed later (p. 172).

Reversible one-substrate reactions

Henri's equation also applies in a special way to the case of reversible unimolecular reactions $A \rightleftharpoons X$. The rate of such a reaction has been seen (p. 82, eqn (48)) to be

$$v = \frac{V_A K_X[A] - V_X K_A[X]}{K_A K_X + K_X[A] + K_A[X]} \tag{8}$$

where the constants are defined on p. 82. If a_0 is the initial concentration of A, equal to the sum of the concentrations of A and X at any time, these concentrations may be written as $a_0 - x$ and x respectively. The rate equation is therefore:

$$\frac{dx}{dt} = \frac{V_A K_X(a_0 - x) - V_X K_A x}{K_A K_X + K_X(a_0 - x) + K_A x}. \tag{9}$$

If x_e is the concentration of X present at equilibrium it follows that

$$V_A K_X(a_0 - x_e) = V_X K_A x_e. \tag{10}$$

The rate equation (9) can be written as

$$\frac{dx}{dt} = \frac{V_A K_X a_0 - (V_A K_X + V_X K_A)x}{K_A K_X + K_X(a_0 - x) + K_A x}. \tag{11}$$

With eqn (10) this becomes

$$\frac{dx}{dt} = \frac{V_A K_X a_0 - V_A K_X a_0 x / x_e}{K_A K_X + K_X a_0 + (K_A - K_X)x} \tag{12}$$

$$= \frac{V_A K_X a_0 (x_e - x)}{(K_A K_X + K_X a_0)x_e + (K_A - K_X)x_e^2 + (K_X - K_A)x_e(x_e - x)}. \tag{13}$$

If z is written for $x_e - x$, the departure from the equilibrium value, eqn (13) becomes

$$-\frac{dz}{dt} = \frac{V_A K_X a_0 z}{K_X(K_A + a_0)x_e + (K_A - K_X)x_e^2 + (K_X - K_A)x_e z} \tag{14}$$

This is of the same form as the simple Michaelis–Menten law, and the system will therefore satisfy Henri's equation as far as the variation of z with t is concerned. In the event that $K_A = K_X$ the variation of z with t will follow the first-order law.

The quantity z may be either positive or negative according to the side from which the equilibrium is being approached; a plot of z against t will therefore consist of two parts approaching the same asymptote. Figure 6.4 shows such plots based on the results of Woolf[1] for the interconversions of

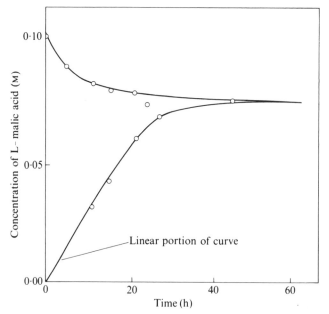

FIG. 6.4.   Results obtained by B. Woolf (*Biochem. J.*, **23**, 472 (1929)) for the interconversion of fumaric and L-malic acids, catalysed by fumarase.

[1] B. Woolf, *Biochem. J.*, **23**, 472 (1929).

fumaric and L-malic acids, catalysed by fumarase present in an *E. coli* preparation. The equilibrium corresponds to nearly 80 per cent of malic acid.

During the early stages of the conversion of fumaric acid into malic acid the term $(K_X - K_A)x_e z$ is much larger than $K_X(K_A + a_0)x_e + (K_A - K_X)x_e^2$, so that the kinetics are zero-order; the curve is seen to be linear in this region. However, the corresponding part of the curve for the reverse reaction is necessarily curved, the kinetics being first-order.

Inhibition by substrate

The rate equations for reactions inhibited by substrate were seen in chapter 3 (p. 85) to be sometimes of the form

$$v = \frac{P[A]}{1 + Q[A] + R[A]^2},$$ (15)

where P, Q, and R are constants. If a_0 is the initial amount of substrate and $a_0 - x$ the amount at time t this equation may be written as

$$\frac{dx}{dt} = \frac{P(a_0 - x)}{1 + Q(a_0 - x) + R(a_0 - x)^2}.$$ (16)

This equation integrates as follows:

$$Pt = \int \left\{ \frac{1}{a_0 - x} + Q + R(a_0 - x) \right\} dx + \text{const},$$ (17)

where the constant is given by the boundary condition $t = 0$, $x = 0$. The result of the integration is

$$Pt = \ln \frac{a_0}{a_0 - x} + Qx + Ra_0 x - \frac{Rx^2}{2}.$$ (18)

If y is the amount of substrate remaining at any time (i.e. $y = a_0 - x$) the equation may be written as

$$Pt = \ln \frac{a_0}{y} + Q(a_0 - y) + Ra_0(a_0 - y) - \frac{R}{2}(a_0 - y)^2.$$ (19)

In reactions of this type there is actually some increase in the rate as the reaction proceeds, after which the rate diminishes. Bamann and Schmeller[1], for example, found an increase of as much as 27 per cent during the course of the hydrolysis of ethyl-L-mandelate under the the action of sheep's liver lipase. If x is plotted against t for a reaction of this kind the curve is at first concave upwards, as in an autocatalytic reaction (Fig. 6.5).

[1] E. Bamann and M. Schmeller, *Z. physiol. Chem.*, **183**, 149 (1929).

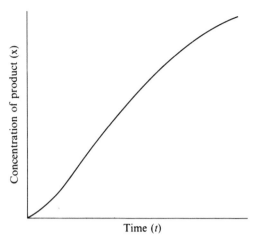

FIG. 6.5.   Schematic plot of x against t for a reaction that is inhibited by the substrate.

Inhibition by products

The products of a reaction may be pure competitive, pure non-competitive, pure anticompetitive or mixed inhibitors. If the inhibition is pure competitive the rate equation is (see p. 92)

$$v = \frac{V[A]}{K_A\left(1 + \frac{[Q]}{K_Q}\right) + [A]}, \qquad (20)$$

where V is the rate in the presence of excess of substrate. If Q is a product of the reaction, of concentration x at time t, the equation may be written as

$$\frac{dx}{dt} = \frac{V(a_0 - x)}{K_A + \frac{K_A}{K_Q}x + (a_0 - x)}. \qquad (21)$$

This integrates as follows:

$$Vt = \int\left\{\frac{K_A}{a_0 - x} + \frac{K_A x}{K_Q(a_0 - x)} + 1\right\} dx + \text{const.} \qquad (22)$$

With the boundary conditions $t = 0$, $x = 0$, the result is

$$Vt = K_A \ln\frac{a_0}{a_0 - x} + x + \frac{K_A}{K_Q}\left(a_0 \ln\frac{a_0}{a_0 - x} - x\right). \qquad (23)$$

The first two terms correspond to Henri's equation, while the last applies to the inhibiting product. It is readily seen that if two products of a reaction are inhibitors the term $1/K_Q$ in the above equation will be replaced by $1/K_Q + 1/K'_Q$, where K'_Q is the inhibition constant for the second product.

Equation (23) has been found to give a satisfactory fit to the experimental curves for the sucrase-catalysed hydrolysis of sucrose[1].

In the event that $K_Q = K_A$, eqn (21) reduces to the form of a simple first-order equation, the term x being lost.

If the inhibition by a product is of the pure non-competitive type the rate equation is (p. 92)

$$v = \frac{V[A]}{(K_A + [A])\left(1 + \dfrac{[Q]}{K_Q}\right)}. \tag{24}$$

This becomes

$$\frac{dx}{dt} = \frac{V(a_0 - x)}{(K_A + a_0 - x)(1 + x/K_Q)}. \tag{25}$$

Integration of this gives rise to

$$Vt = K_A \ln \frac{a_0}{a_0 - x} + x + \frac{K_A}{K_Q}\left(a_0 \ln \frac{a_0}{a_0 - x} - x\right) + \frac{x^2}{2K_Q}. \tag{26}$$

Again, the first two terms correspond to the Henri equation. Other types of product inhibition can be treated by similar methods.

Analysis of experimental results

If the inhibition is of the competitive type, so that eqn (23) applies, a convenient graphical method[2] is available for analysing experimental results. Equation (23) may be written as

$$Vt = K_A\left(1 + \frac{a_0}{K_Q}\right)\ln \frac{a_0}{a_0 - x} + \left(1 - \frac{K_A}{K_Q}\right)x. \tag{27}$$

A plot of x/t against $(1/t)\ln[a_0/(a_0 - x)]$ will therefore give a straight line, with the slopes and intercepts indicated in Fig. 6.6. As in the case of the uninhibited reaction (p. 165) a line drawn through the origin with a slope equal to the initial concentration will again strike the line at a point corresponding to the initial conditions. If runs are made at various initial concentrations the results will therefore be as represented schematically in Fig. 6.7. The points of intersection will fall on a straight line of slope $-K_A$ and of intercepts V and V/K_A, as in the uninhibited case, since the initial points are uninfluenced by the products.

From the line through the initial points V and K_A can be obtained, and from the slopes through the lines for the individual runs $VK_Q/(K_Q - K_A)$

[1] J. M. Nelson and D. J. Hitchcock, *J. Am. chem. Soc.*, **43**, 2632 (1921).
[1] R. J. Foster and C. Niemann, *Proc. natn. Acad. Sci. U.S.A.*, **39**, 999 (1953); see also R. J. Foster and C. Niemann, *J. Am. chem. Soc.*, **77**, 1886 (1955); T. H. Applewhite and C. Niemann, *J. Am. chem. Soc.*, **77**, 4923 (1955); R. R. Jennings and C. Niemann, *J. Am. chem. Soc.*, **77**, 5432 (1955).

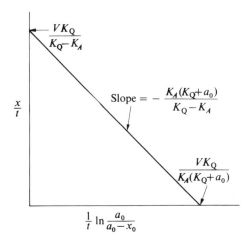

FIG. 6.6.   Schematic plot of x/t against $(1/t) \ln \{a_0/(a_0 - x)\}$ for a reaction inhibited by the products.

and $VK_Q/K_A(K_Q + a_0)$ can be found; it is therefore possible to separate V, K_A, and K_Q. If two reaction products are inhibitors, $1/K_Q$ is replaced by the sum of the $1/K$'s for the two inhibitors; the values for the two inhibitors cannot be obtained separately by this method, but only by independent studies of the effects of the inhibitors.

Figure 6.8 shows a plot of some actual results, obtained by Foster and Niemann for the α-chymotrypsin-catalysed hydrolysis of acetyl-L-tyrosin-

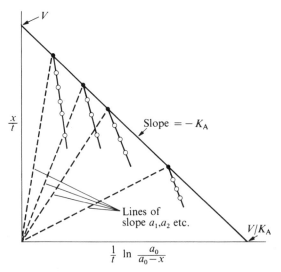

FIG. 6.7.   Foster and Niemann's method of analysis of the results for a reaction inhibited by the products. The experimental points shown are hypothetical.

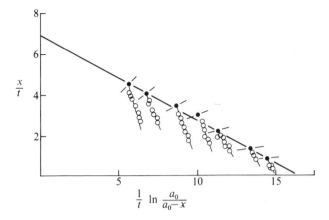

Fig. 6.8. An example of the use of Foster and Niemann's method: the α-chymotrypsin-catalysed hydrolysis of acetyl-L-tyrosinhydroxamide (R. J. Foster and C. Niemann, *Proc. natn. Acad. Sci. U.S.A.*, **39**, 999 (1953)).

hydroxamide. It has been established that only one of the reaction products, acetyl-L-tyrosine, is an inhibitor, so that the K_Q value obtained applies to this substance.

Competing substrates

The differential rate equations for two competing substrates were given on p. 112. Equation (151) may be written as

$$r\frac{1}{a}\frac{da}{dt} = \frac{1}{b}\frac{db}{dt} \tag{28}$$

where a and b are the amounts of the reactants A and B at time t and r is equal to

$$\frac{k_1' k_2'(k_{-1} + k_2)}{k_1 k_2 (k_{-1}' + k_2')}.$$

This equation integrates to

$$r \ln \frac{a}{a_0} = \ln \frac{b}{b_0} \tag{29}$$

where a_0 and b_0 are the amounts of the reactants at $t = 0$. Equation (29) may be written as

$$b = b_0 \left(\frac{a}{a_0}\right)^r \tag{30}$$

so that the rate equation for the disappearance of A (eqn (149) on p. 195) becomes of the form

$$-\frac{da}{dt} = \frac{Pa}{1 + Qa + Ra^r}. \tag{31}$$

Integration of this, with the boundary condition $t = 0$, $a = a_0$, gives

$$Pt = \ln\frac{a_0}{a} + Q/(a_0 - a) + \frac{R}{r}\bigg/(a_0^r - a^r). \tag{32}$$

This may be written in terms of the amount of product $x(= a_0 - a)$ as

$$Pt = \ln\frac{a_0}{a_0 - x} + Qx + \frac{R}{r}[a_0^r - (a_0 - x)^r]. \tag{33}$$

The first two terms of this correspond to Henri's eqn (7), to which eqn (33) reduces if $r = 0$, i.e. if B is unreactive.

Otherwise, eqn (33) differs from Henri's equation by the last term, which is small if r is small and large if r is large. If B is much less reactive than A, A will disappear approximately in accordance with Henri's equation but B will disappear in a different way. Under certain circumstances the rate of reaction of the less reactive substrate will be low at first (because the other is occupying most of the enzyme's sites), but will pass through a maximum and then decrease. Suppose that $r > 1$, i.e. that the substrate A reacts less rapidly than B. Then by differentiation of eqn (31) it can be shown that the rate passes through a maximum when

$$a = \left[\frac{1}{(r-1)R}\right]^{1/r}. \tag{34}$$

Since a/a_0 must be less than unity, a maximum can, therefore, only be observed provided that

$$a_0 > \left[\frac{1}{(r-1)R}\right]^{1/r}. \tag{35}$$

Haldane[1] has applied the above relationships in some detail to the hydrolysis of racemic mixtures, with particular reference to the work of Willstatter, Kuhn, and Bamann[2] and of Weber and Ammon[3] on the hydrolysis of optically active esters by lipases. In a racemic mixture $a_0 = b_0$ and the optical rotation of the products is proportional to $a - b$, or to

$$a - a_0\left(\frac{a}{a_0}\right)^r.$$

[1] J. B. S. Haldane, *Enzymes*, Longmans, Green and Co., London (1930), pp. 85–8.
[2] R. Willstatter, R. Kuhn, and E. Bamann, *Ber. dt. chem. Ges.*, **61**, 886 (1928).
[3] H. H. Weber and R. Ammon, *Biochem. Z.*, **204**, 197 (1929).

This is a maximum when

$$1 - r\left(\frac{a}{a_0}\right)^{r-1} = 0, \tag{36}$$

which is true when

$$a = a_0 r^{1/(1-r)} \tag{37}$$

and

$$b = a_0 r^{r/(1-r)}. \tag{38}$$

The maximum rotation is therefore proportional to

$$r^{1/(1-r)} - r^{r/(1-r)},$$

or to

$$r^{1/(1-r)}\left(1 - \frac{1}{r}\right).$$

When $r = 1$ this is zero; both forms are produced at the same speed so that there is no rotation. When r is either zero or infinity (i.e. the enzyme is absolutely specific with respect to one or other of the substrates) the function is unity. As r either increases or decreases from unity the function gradually increases towards unity, reaching a value of $\frac{1}{2}$ when r is either 4·5 or $\frac{2}{9}$. In order, therefore, for the enzyme to produce during hydrolysis a rotation equal to one-half that which would be achieved were the enzyme absolutely specific, the ratio of rates with the two isomers must be 4·5.

In the case of the hydrolysis of d-ethyl mandelate by pig's liver lipase[1] the value of r is about 3, and the theoretical maximum rotation is 38·5 per cent of that possible with an absolutely specific enzyme. The actual value observed was 35·4 per cent. It can be shown that the maximum rotation should always occur at an extent of hydrolysis of between 50 and 63·2 per cent of the racemic mixture; in the work referred to the maximum was found at about 50 per cent.

Enzyme inactivation during the course of reaction

Under certain circumstances the rate of inactivation of an enzyme may be sufficiently great that it must be taken into account in the study of the time-course of the reaction undergoing catalysis. The kinetics of the inactivation process itself are discussed in some detail in Chapter 13, where it is seen that the reactions are sometimes, although not always, of the first order. In the present section attention will be confined to such first-order inactivations, but the methods are readily extended to other cases.

[1] R. Willstatter, R. Kuhn, and E. Bamann, *loc. cit.*; H. H. Weber and R. Ammon, *loc. cit.*

In some instances[1] it has been observed that a substrate protects an enzyme against inactivation; in other words, the enzyme–substrate complex loses its activity less rapidly than the free enzyme. In other instances, notably with catalase, the enzyme is less stable in the presence of its substrate. With some systems, on the other hand, the substrate has no effect. These three cases will now be considered, the last being taken first. Only uncomplicated single-substrate reactions will be treated.

The general rate equation is

$$\frac{dx}{dt} = \frac{k_c e_0(a_0 - x)}{K_A + (a_0 - x)}, \tag{39}$$

where e_0 is the total concentration of active enzyme at any time, and is a steadily decreasing quantity. The decay law for the first-order inactivation is

$$e_0 = [E]_0 \, e^{-kt} \tag{40}$$

where k is the first-order rate constant. Equation (39) can therefore be written as

$$\frac{dx}{dt} = \frac{V(a_0 - x) \, e^{-kt}}{K_A + (a_0 - x)} \tag{41}$$

where $V(= k_2[E]_0)$ is the limiting initial rate. Integration of this equation, subject to the condition $t = 0$, $x = 0$, proceeds as follows:

$$V \int e^{-kt} \, dt = \int \left(\frac{K_A}{a_0 - x} + 1\right) dx + \text{const} \tag{42}$$

$$-\frac{V e^{-kt}}{k} = -K_A \ln(a_0 - x) + x + \text{const} \tag{43}$$

$$= K_A \ln \frac{a_0}{a_0 - x} + x - \frac{V}{k}, \tag{44}$$

$$\frac{V}{k}(1 - e^{-kt}) = K_A \ln \frac{a_0}{a_0 - x} + x. \tag{45}$$

This differs from the Henri equation (7) only by the factor in brackets on the left-hand side. As k approaches zero $1 - e^{-kt}$ approaches kt, which gives the Henri equation. If the substrate concentration is initially small only the first term on the right-hand side need be considered; if it is large only the second term need be considered, at least during the early stages of reaction.

[1] C. O'Sullivan and F. W. Tompson, *J. chem. Soc.*, **57**, 865 (1890); L. Ouellet, K. J. Laidler, and M. F. Morales, *Archs Biochem. Biophys.*, **39**, 37 (1952).

When t is infinite eqn (45) becomes

$$\frac{V}{k} = K_A \ln \frac{a_0}{a_0 - x_\infty} + x_\infty \qquad (46)$$

where x_∞ is the concentration of substrate at the end of the reaction. This concentration need not necessarily correspond to completion of the reaction. This may readily be seen for the case in which the substrate is initially in very great excess, and saturates the enzyme. In this case the first term on the right-hand side of eqn (46) can be neglected and hence

$$x_\infty = \frac{V}{k}. \qquad (47)$$

In the event that V/k is greater than or equal to a_0 the reaction goes to completion, but if V/k is less than a_0 the final amount of product produced is less than a_0 and is equal to V/k. Since V is proportional to $[E]_0$ the amount of product produced at the end of the reaction is proportional to the initial amount of enzyme employed. This phenomenon was observed by Bach and Chodat[1] for the peroxidase-catalysed reaction between pyrogallol and hydrogen peroxide, and was regarded by Oppenheimer[2] as indicating that peroxidase is not acting catalytically; the above analysis shows, however, that this conclusion is not valid.

The situation in which the substrate completely protects the enzyme from inactivation may be represented by the following reaction scheme:

$$E + A \underset{k_{-1}}{\overset{k_1}{\rightleftharpoons}} EA$$

$$EA \overset{k_2}{\rightarrow} E + X$$

$$E \overset{k_i}{\rightarrow} D.$$

Here D is the inactive enzyme, produced from E in accordance with the first-order law with the rate constant k. EA is assumed not to be inactivated; D does not combine with substrate.

Application of the steady-state treatment to EA gives rise to

$$K_A[EA] = [E] \, . \, [A] \qquad (48)$$

where $[E]$ is the amount of free active enzyme present at any time and K_A is equal to $k_1/(k_{-1} + k_2)$. If e is the total amount of active enzyme present at time t this equation becomes

$$K_A[EA] = (e - [EA])[A]. \qquad (49)$$

[1] A. Bach and R. Chodat, *Ber. dt. chem. Ges.*, **37**, 1342 (1904).
[2] C. Oppenheimer, *Die Fermente und ihre Wirkungen*, Leipzig (fourth edn., 1910).

The amount of free enzyme E present at any time is $e - [\text{EA}]$, so that the decay law is

$$-\frac{de}{dt} = k_i(e - [\text{EA}]). \tag{50}$$

The rate of disappearance of substrate is

$$-\frac{d[\text{A}]}{dt} = k_2[\text{EA}]. \tag{51}$$

The course of the reaction is thus given by a solution of eqns (49) to (51), which for convenience may be written as

$$K_A p = (e - p)a \tag{52}$$

$$-\dot{e} = k_i(e - p), \tag{53}$$

$$-\dot{a} = k_2 p, \tag{54}$$

where $p = [\text{EA}]$ and $a = [\text{A}]$. The boundary condition is $t = 0$, $a = a_0$, $e = e_0$, where a_0 is the initial concentration of substrate and e_0 that of total enzyme.

From eqns (52) and (53),

$$-\dot{e} = \frac{k_i K_A e}{K_A + a} \tag{55}$$

and from eqns (52) and (54),

$$\dot{a} = -\frac{k_2 e a}{K_A + a} \tag{56}$$

$$= \frac{k_2 a}{k_i K_A}\dot{e}. \tag{57}$$

Therefore,

$$de = \frac{k_i K_A}{k_2}\frac{da}{a}. \tag{58}$$

With the boundary condition $a = a_0$, $e = e_0$, this integrates to

$$e = e_0 - \frac{k_i K_A}{k_2}\ln\frac{a_0}{a}. \tag{59}$$

The enzyme has all disappeared and the reaction has stopped when

$$\ln\frac{a_0}{a_\infty} = \frac{k_2 e_0}{k_i K_A} \tag{60}$$

or when

$$a_\infty = a_0 \, e^{-k_2 e_0 / k_i K_A}. \tag{61}$$

The reaction thus stops short of completion by an amount that depends on the magnitude of $k_2 e_0 / k_i$; if this is very small the reaction hardly proceeds at all ($a_\infty \approx a$), while if it is large the reaction essentially goes to completion. If the quantity is very small

$$a_\infty \approx a_0 \left(1 - \frac{k_2 e_0}{k_i K_A} \right), \tag{62}$$

so that the fraction of substrate transformed into product at the end of reaction is

$$\frac{x_\infty}{a_0} = \frac{a_0 - a_\infty}{a_0} \frac{k_2 e_0}{k_i K_A}. \tag{63}$$

In this case too the amount transformed is proportional to the concentration of enzyme. It is also independent of the concentration of substrate.

The course of the reaction is given by combining eqns (56) and (59):

$$\dot{a} = -\frac{k_2 a}{K_A + a} \left(e_0 - \frac{k_i K_A}{k_2} \ln \frac{a_0}{a} \right). \tag{64}$$

This integrates as follows:

$$k_2 t = \int \left[(K_A + a) \Big/ \left\{ a \left(e_0 + \frac{k_i K_A}{k_2} \ln \frac{a}{a_0} \right) \right\} \right] da + \text{const.} \tag{65}$$

Putting

$$z = \frac{a \, e^{k_2 e_0 / k_i K_A}}{a_0} \equiv \frac{a e^q}{a_0}, \tag{66}$$

eqn (65) reduces to

$$k_2 t = \int \frac{K_A + a z e^{-q}}{K_A z \ln z} \, dz + \text{const.} \tag{67}$$

Integration subject to the boundary condition $t = 0$, $z = e^q$, gives

$$k_2 t = \frac{a_0 K_A}{e^q} \left[\text{Ei}(q) - \text{Ei}\left(q \ln \frac{a}{a_0} \right) \right] - \ln \left(1 - \frac{1}{q} \ln \frac{a_0}{x} \right), \tag{68}$$

where

$$\text{Ei}(z) = \int_{-\infty}^{z} \frac{e^u \, du}{u} \tag{69}$$

is a tabulated integral[1].

[1] See, for example, *Tables of Sine, Cosine, and Exponential Integrals*, U.S. Federal Works Agency, vol. ii (1940).

Sensitization by substrate

If the free enzyme is stable but the enzyme–substrate complex is inactivated at an appreciable rate the situation can be represented as:

$$E + A \underset{k_{-1}}{\overset{k_1}{\rightleftharpoons}} EA,$$

$$EA \overset{k_2}{\longrightarrow} E + X$$

$$EA \overset{k_1}{\longrightarrow} D.$$

The treatment of this case is difficult, however, since the steady-state equation cannot now be applied to EA the concentration of which decreases during the course of reaction.

Consecutive reactions

The course, under various conditions, of consecutive reactions, each catalysed by a different enzyme, has been discussed by Haldane[1], and will not be included here.

Schutz's law

An empirical relationship proposed by E. Schutz[2] in 1885 deserves comment in view of its applicability to a variety of enzyme systems. Schutz found that the amount of peptone produced during the hydrolysis of egg albumin by pepsin is proportional to the square root of the enzyme concentration. This result was confirmed by J. Schutz[3] and by Arrhenius[4] for the same system, and for many other reactions such as the action of emulsion on lactose[5], trypsin on gelatin and casein[6], lipase on fats[7], and pepsin on casein[8] and peptone[9]. Deviations from the law are found if measurements are made in the early or later stages of the reactions.

Arrhenius extended this rule to the course of reaction, showing that the amount of substrate transformed is proportional to the square root of the time as well as to the square root of the concentration of protein, i.e.

$$a - x = k\sqrt{([E]_0 t)}. \tag{70}$$

[1] J. B. S. Haldane, *Enzymes*, Longmans, Green and Co., London (1930), pp. 88–91.
[2] E. Schutz, *Hoppe-Seyler's Z. physiol. Chem.*, **9**, 577 (1885).
[3] J. Schutz, *Hoppe-Seyler's Z. physiol. Chem.*, **30**, 1 (1900).
[4] S. Arrhenius, *Immunochem.*, Leipzig (1907), p. 53.
[5] H. E. Armstrong, *Proc. R. Soc.* B, **73**, 507 (1904).
[6] S. Arrhenius, *Quantitative Laws in Biological Chemistry*, London (1915), p. 46; E. A. Moelwyn-Hughes, J. Pace, and W. C. McC. Lewis, *J. gen. Physiol.*, **13**, 323 (1930).
[7] S. Arrhenius, *loc. cit.*, p. 46.
[8] W. van Dam, *Hoppe-Seyler's Z. physiol. Chem.*, **79**, 247 (1912).
[9] J. H. Northrop, *J. gen. Physiol.*, **2**, 471 (1920).

This relationship is consistent with the following law for the velocity of reaction

$$\frac{dx}{dt} = \frac{k}{2}\left(\frac{[E]_0}{t}\right)^{\frac{1}{2}}. \tag{71}$$

The law evidently cannot hold when t is zero (when it predicts an infinite rate) or when t is infinite (when it predicts the amount transformed to be infinite). Northrop[1] made a very careful investigation of the applicability of eqns (70) and (71) and found deviations at high and low times.

Various theoretical explanations of the law have been proposed. Arrhenius[2] gave an explanation in terms of an enzyme-inhibitor complex and Langmuir[3] one on the basis of the replacement of the reactants by the products on the surface of the enzyme. Moelwyn-Hughes[4] showed that the Schutz–Arrhenius eqn (70) can be deduced on the assumption that the rates are controlled by diffusion processes; some support for this type of explanation is provided by the fact that the law applies chiefly to concentrated solutions, where diffusion may certainly be rate-controlling.

In any case, whatever the explanation, the law is evidently not of fundamental significance and cannot throw light on the nature of enzyme action.

Kinetics of the transient phase

The preceding discussions of the time course of reactions under various conditions have in all cases been based on the steady-state treatment. It is also of interest to have a theory of the kinetic behaviour before the steady state is established. This early part of the reaction, which it is convenient to refer to as the *transient phase*, usually occupies a very short period of time, but modern techniques have permitted measurements to be made during this early period.

Several theoretical treatments of transient-phase kinetics have been given. The first of these[5] dealt with the single-intermediate mechanism, but later Ouellet and Laidler[6] extended the treatment to the case of two intermediates. This treatment was further developed by Ouellet and Stewart[7] who were mainly concerned with the kinetics of formation of the product that is released first. In these treatments the assumption was made that the concentration of substrate is greatly in excess of that of the enzyme; the converse case, in which the enzyme is in excess, was treated by Kasserra and Laidler[8] who showed that an experimental study under these conditions can reveal information that is not readily available from studies with excess substrate.

[1] J. H. Northrop, *J. gen. Physiol.*, **6**, 723 (1924).
[2] S. Arrhenius, *Immunochem.*, Leipzig (1907), p. 53.
[3] I. Langmuir, *J. Am. chem. Soc.*, **38**, 2221 (1916).
[4] E. A. Moelwyn-Hughes, *Ergebn. Enzymforsch.*, **6**, 23 (1937); cf. also P. S. Bunting and K. J. Laidler, *Biochem.*, **11**, 4477 (1972) and also Chapter 12.
[5] K. J. Laidler, *Can. J. Chem.*, **33**, 1614 (1955); see H. Gutfreund, *Discuss. Faraday Soc.*, **20**, 167 (1955).
[6] L. Ouellet and K. J. Laidler, *Can. J. Chem.*, **34**, 146 (1956).
[7] L. Ouellet and J. A. Stewart, *Can. J. Chem.*, **37**, 737 (1959).
[8] H. P. Kasserra and K. J. Laidler, *Can. J. Chem.*, **48**, 1793 (1970).

All of the treatments mentioned above ignored the back reactions, which is usually justifiable; Darvey[1] has included these reactions but the resulting equations are rather complex and would be difficult to apply to the experimental results.

In addition, there have been some theoretical treatments which have had as their main object the question of the conditions under which the steady-state assumption is applicable. In particular may be mentioned the publications of Hommes[2], Walter and Morales[3], Walter[4] and Wong[5]. Computer solutions under a variety of experimental conditions have led to the conclusion that the steady state will only be essentially established if the ratio of substrate to enzyme concentration is 10^3 or greater (see also p. 73).

Single-intermediate treatments[6]

The simple Michaelis–Menten case will be considered first; this mechanism probably does not have wide applicability, but some important general principles are revealed by a treatment of this system.

The scheme of reactions applicable to the single-intermediate system, uncomplicated by the effects of activators and inhibitors, is as follows:

$$E \; + \; A \; \underset{k_{-1}}{\overset{k_1}{\rightleftharpoons}} \; EA \overset{k_2}{\longrightarrow} \; E \; + X.$$

$$e_0 - m \quad a_0 - m - x \qquad m \qquad e_0 - m \quad x.$$

If e_0 is the total concentration of enzyme (including that bound to substrate), a_0 the initial total concentration of substrate, x the concentration of product at time t, and m the concentration of complex at time t, the various other concentrations at time t are as represented above. The rate equations are therefore:

$$\frac{dm}{dt} = k_1(e_0 - m)(a_0 - m - x) - \bar{k}m \tag{72}$$

and

$$\frac{dx}{dt} = k_2 m \tag{73}$$

where $\bar{k}$ is equal to $k_{-1} + k_2$.

These equations cannot be solved exactly in explicit form unless one or more of certain conditions are satisfied. One such condition, which is usually realized in practice, is that the substrate concentration is greatly in excess

[1] I. G. Darvey, *J. theor. Biol.*, **19**, 215 (1968); see L. Peller and R. A. Alberty, *Progress in Reaction Kinetics*, **1**, 236 (1961).
[2] F. A. Hommes, *Archs Biochem. Biophys.*, **96**, 28 (1962).
[3] C. Walter and M. F. Morales, *J. biol. Chem.*, **239**, 1277 (1964).
[4] C. Walter, *J. theor. Biol.*, **11**, 181 (1966).
[5] J. T. Wong, *J. Am. chem. Soc.*, **87**, 1788 (1965).
[6] K. J. Laidler, *Can. J. Chem.*, **33**, 1614 (1955).

of that of the enzyme. When this is so the amount of product produced at the end of the transient phase is very much less than the initial substrate concentration. Since, in addition, m must be very much less than a_0, it follows that

$$a_0 - m - x \approx a_0. \tag{74}$$

Equation (72) may therefore be written as

$$\frac{\mathrm{d}m^*}{\mathrm{d}t} = k_1 a_0 (e_0 - m^*) - \bar{k} m^*. \tag{75}$$

Asterisks will now be applied to m and x to indicate that the expressions for them apply only in certain special cases, the most usual of which is $a_0 \gg e_0$.

Equation (75) is readily integrated and gives rise to the equation

$$t = \frac{m_i^*}{k_1 e_0 a_0} \ln \frac{m_i^*}{m_i^* - m^*}, \tag{76}$$

where m_i^* is defined by

$$m_i^* = \frac{k_1 e_0 a_0}{\bar{k} + k_1 a_0}. \tag{77}$$

The quantity m_i^* may be regarded as a hypothetical steady-state concentration at the very beginning of the reaction. Equation (76) may be written as

$$m^* = m_i^* [1 - \exp(-k_1 e_0 a_0 t / m_i^*)], \tag{78}$$

or, using eqn (77), as

$$m^* = \frac{k_1 e_0 a_0}{\bar{k} + k_1 a_0} [1 - \exp\{-(\bar{k} + k_1 a_0)t\}]. \tag{79}$$

The variation of the concentration of a product during the transient phase is found from eqns (73) and (79) to be

$$\frac{\mathrm{d}x^*}{\mathrm{d}t} = \frac{k_1 k_2 e_0 a_0}{\bar{k} + k_1 a_0} [1 - \exp\{-(\bar{k} + k_1 a_0)t\}]. \tag{80}$$

With the boundary condition $t = 0$, $x^* = 0$ this integrates to

$$x^* = \frac{k_1 k_2 e_0 a_0}{\bar{k} + k_1 a_0} t + \frac{k_1 k_2 e_0 a_0}{(\bar{k} + k_1 a_0)^2} [\exp\{-(\bar{k} + k_1 a_0)t\} - 1]. \tag{81}$$

When t is sufficiently small that

$$(\bar{k} + k_1 a_0)t \ll 1 \tag{82}$$

it is permissible to expand the exponential and accept only the first two terms: the result is

$$x^* = \tfrac{1}{2} k_1 k_2 e_0 a_0 t^2. \tag{83}$$

This particular relationship was first given by Roughton[1]. From a study of the kinetics during this very early period of the reaction it is therefore possible to determine the value of $k_1 k_2$. Since k_2 can be determined by conventional methods (from the limiting rate at high substrate concentrations) k_1 can be calculated. Also, since the Michaelis constant K_A is equal to $(k_{-1} + k_2)/k_1$ and can be determined, k_{-1} can be calculated. For a reaction occurring by the single-intermediate mechanism, therefore, the three kinetic constants k_1, k_{-1}, and k_2 can be found by combining a study of the steady-state kinetics with a study of the very early stages of reaction.

The variation of the amount of product x^* with the time, as given by eqn (81), is shown schematically in Fig. 6.9. When t is sufficiently large the

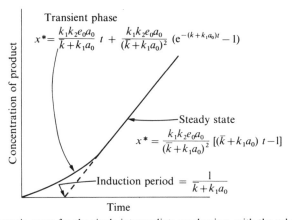

FIG. 6.9. Schematic curve for the single-intermediate mechanism, with the substrate in great excess. The curve shows the variation of the concentration of the products with time during the very early stages of the reaction.

exponential term vanishes and p^* then varies linearly with the time. Sooner or later (depending on the relative magnitude of $k_1 a_0$ and $\bar{k}$) there will be a falling off from linearity owing to the reduction in the amount of substrate. Extrapolation of the linear portion of the curve to the t axis gives rise to an intercept given by

$$\tau = \frac{1}{\bar{k} + k_1 a_0} \tag{84}$$

and this time may conveniently be referred to as the *induction period* of the reaction. Determinations of the induction period at various values of a_0 will allow k_1 and $\bar{k}$ to be determined separately; since k_2 is known k_{-1} can also be calculated.

The variation of the concentration of complex with the time is shown schematically in Fig. 6.10. To a good approximation the kinetic behaviour can be represented by two sets of equations, one referring to the transient

---

[1] F. J. Roughton, *Discuss. Faraday Soc.*, **17**, 116 (1954).

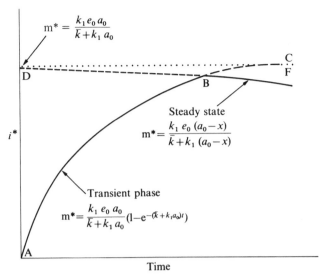

$$m^* = \frac{k_1 e_0 a_0}{\bar{k} + k_1 a_0}$$

Steady state

$$m^* = \frac{k_1 e_0 (a_0 - x)}{\bar{k} + k_1 (a_0 - x)}$$

Transient phase

$$m^* = \frac{k_1 e_0 a_0}{\bar{k} + k_1 a_0} (1 - e^{-(\bar{k} + k_1 a_0)t})$$

Time

FIG. 6.10. Schematic curves for the single-intermediate mechanism with the substrate in great excess. The curves show complex concentration as a function of time. The true course of the reaction is closely represented by ABF.

phase and the other to the steady-state period. The curve ABC corresponds to eqn (79) for the transient phase, the value of m^* approaching m_i^* asymptotically. The curve DBF is that predicted by steady-state theory. The true course of the reaction is closely represented by ABF; this has a discontinuity at B, but the error is nevertheless very slight.

Double-intermediate treatment[1]

When there is a second intermediate, such as an acyl enzyme, the system can be represented as

$$\text{E} \quad + \text{A} \underset{k_{-1}}{\overset{k_1}{\rightleftharpoons}} \text{EA} \overset{k_2}{\rightarrow} \text{EA}' \overset{k_3}{\rightarrow} \quad \text{E} \quad + \text{Y}$$

$$e_0 - m - n \quad a_0 \qquad m \qquad \underset{x}{\text{X}} \; n \qquad e_0 - m - n \quad y$$

If the substrate is assumed to be in excess, the substrate concentration during the transient phase remains close to a_0. The differential equations for the systems are:

$$\dot{m} = k_1 a_0 (e_0 - m - n) - \bar{k}m \tag{85}$$

$$\dot{n} = k_2 m - k_3 n \tag{86}$$

$$\dot{x} = k_2 m \tag{87}$$

$$\dot{y} = k_3 n \tag{88}$$

[1] L. Ouellet and K. J. Laidler, *Can. J. Chem.*, **34**, 146 (1956); L. Ouellet and J. A. Stewart, *Can. J. Chem.*, **37**, 737 (1959).

where $\bar{k} = k_{-1} + k_2$. Elimination of n between eqns (85) and (86) gives

$$\dddot{m} + \ddot{m}(k_1 a_0 + \bar{k} + k_3) + \dot{m}[k_1 a_0(k_2 + k_3) + \bar{k}k_3] - k_1 k_3 e_0 a_0 = 0. \quad (89)$$

This may be written as

$$\dddot{m} + P\ddot{m} + Q\dot{m} - R = 0 \quad (90)$$

where

$$P = k_1 a_0 + \bar{k} + k_3 \quad (91)$$

$$Q = k_1 a_0(k_2 + k_3) + \bar{k}k_3 \quad (92)$$

$$R = k_1 k_3 e_0 a_0. \quad (93)$$

The general solution of eqn (90), subject to the boundary conditions that at zero time $m = 0$ and $\dot{m} = k_1 e_0 a_0 \ (= R/k_3)$ is

$$m = \frac{R}{Q} + M\,e^{-Ft} + N\,e^{-Gt} \quad (94)$$

where R and Q are defined above and

$$F = \tfrac{1}{2}\{P - \sqrt{(P^2 - 4Q)}\} \quad (95)$$

$$G = \tfrac{1}{2}\{P + \sqrt{(P^2 - 4Q)}\} \quad (96)$$

$$M = \frac{R(Q - k_3 G)}{k_3 Q(G - F)} \quad (97)$$

$$N = \frac{R(k_3 F - Q)}{k_3 Q(G - F)}. \quad (98)$$

It is to be noted that of the parameters occurring in eqn (94), R, Q, F and G are necessarily positive. Moreover, G must be greater than F. M and N, on the other hand, may be positive or negative. Thus, if k_3 is sufficiently small that $Q \gg k_3 G$, which requires that $Q \gg k_3 F$, M is positive and approximately equal to $-N$. If k_3 is sufficiently large, M becomes negative and N positive; the magnitude of $-M$ is then much greater than N. A further point, of importance later, relates to the sign of

$$\frac{M}{F} + \frac{N}{G}.$$

This quantity, by use of eqns (91), (92) and (95)–(98), is given by

$$\frac{M}{F} + \frac{N}{G} = \frac{k_1 e_0 a_0(k_1 k_2 a_0 - k_3^2)}{\{k_1 a_0(k_2 + k_3) + \bar{k}k_3\}^2}. \quad (99)$$

It is therefore positive when $k_1 k_2 a_0 > k_3^2$, and negative if $k_3^2 > k_1 k_2 a_0$.

If eqn (94) for m is inserted into eqn (87), and the result integrated, the variation of x with time is given by

$$x = \frac{k_2 R}{Q}t + \frac{k_2 M}{F}(1 - e^{-Ft}) + \frac{k_2 N}{G}(1 - e^{-Gt}). \qquad (100)$$

When t is sufficiently large the exponentials become negligible; eqn (100) then reduces to

$$x_s = \frac{k_2 R}{Q}t + k_2\left(\frac{M}{F} + \frac{N}{G}\right). \qquad (101)$$

This linear variation of x with t corresponds to the steady state. Figures 6.11 and 6.12 show two possible types of behaviour; in Fig. 6.11 the quantity

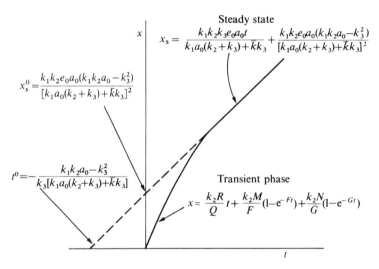

FIG. 6.11. The variation with time of the concentration x of the first product released in a double-intermediate mechanism, with the substrate in great excess. This is the behaviour found when $k_1 k_2 a_0 > k_3^2$ (contrast Fig. 6.12).

$(M/F) + (N/G)$ is positive (i.e., $k_1 k_2 a_0 > k_3^2$), while in Fig. 6.12, the quantity is negative.

In the former case the values of x during the transient phase are less than the steady-state values (Fig. 6.11), the exponential terms in eqn (100) being subtracted. In the latter case the values of x during the transient phase are higher than the steady-state values; since $(M/F) + (N/G)$ is negative there is now an addition of positive exponential terms. The reason for this difference between the two kinds of behaviour is that in the second case the steady-state value of m is reduced by the fact that step 3 is rapid, and this means a reduction in the steady-state rate of formation of the product X.

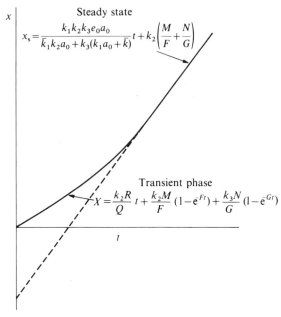

Fig. 6.12. The variation with time of the concentration x of the first product released in a double-intermediate mechanism, with the substrate in great excess. This figure, to be contrasted with the preceding one, shows the behaviour when $k_3^2 > k_1 k_2 a_0$. The general expressions shown in Fig. 6.11 apply to this case also.

The expressions shown on Fig. 6.11 for the intercepts suggest a possible way of analysing experimental results. If it is known that the substrate concentration is sufficiently high that

$$k_1 a_0 (k_2 + k_3) \gg \bar{k} k_3, \tag{102}$$

and

$$k_1 k_2 a_0 \gg k_3^2, \tag{103}$$

and if $k_2 \gg k_3$, the expressions for the intercepts reduce to

$$x_s^0 = e_0 \tag{104}$$

and

$$t^0 = -\frac{1}{k_3}. \tag{105}$$

Since for the double-intermediate mechanism the Michaelis constant K_A is $\bar{k} k_3 / k_1 (k_2 + k_3)$ (see eqn (29) on p. 79), the first of these conditions implies that†

$$a_0 \gg K_A. \tag{106}$$

† As will be discussed later, this condition is often hard to satisfy experimentally.

Provided that the necessary conditions are satisfied, k_3 can therefore be determined by extrapolating the steady-state curve to zero concentration of product X. If t^0 values are determined for a range of substrate concentrations, a plot of $1/t^0$ against $1/a_0$ will allow k_3 to be determined, as may be seen from the expression for t^0 shown on Fig. 6.11.

Moreover, the intercept of the steady-state line on the x axis can give a value for e_0; again, if conditions (102) and (103) are not satisfied an extrapolation procedure may be employed. This method is useful in establishing the number of active centres on an enzyme, since the e_0 obtained is the concentration of active centres; comparison with the molar concentration of enzyme gives the number of active centres. This procedure was employed by Hartley and Kilby[1] for the hydrolysis of *p*-nitrophenyl acetate by chymotrypsin. The value of x_s^0 extrapolated to high substrate concentrations was equal to the molar concentration of the enzyme, which confirms that there is only one active site on the enzyme.

Experimental results can also be analysed in terms of the shape of the transient-phase curve and of its deviation from the hypothetical steady-state line. This deviation, Δx, is given by (compare eqns (100) and (101))

$$\Delta x = \frac{k_2 M}{F} e^{-Ft} + \frac{k_2 N}{G} e^{-Gt}. \tag{107}$$

G is always greater than F, sometimes considerably so, so that e^{-Gt} approaches zero much more rapidly than e^{-Ft} as the steady state is approached. It has been seen that if k_3 is sufficiently small M is positive and approximately equal to $-N$; if k_3 is large M becomes negative and N positive, the magnitude of $-M$ being much greater than N. It follows that almost inevitably the approach to the steady state will be dominated by the term in e^{-Ft}, so that we can write

$$\Delta x \approx \frac{k_2 M}{F} e^{-Ft}. \tag{108}$$

A plot of $\ln \Delta x$ against t will then allow

$$k_2 M \quad \text{and} \quad F$$

to be determined. In the event that $a_0 \gg K_A$ these quantities are given by

$$k_2 M = e_0(k_2 - k_3) \quad \text{and} \quad F = k_2 + k_3. \tag{109}$$

Thus in principle, from extrapolations up to high substrate concentrations, the constants k_2 and k_3 can be separated.

A similar treatment can be given for the formation of the second product, Y. Elimination of m between eqns (85) and (86) leads to

$$\ddot{n} + \dot{n}(k_1 a_0 + \bar{k} + k_3) + n\{k_1 a_0(k_2 + k_3) + \bar{k}k_3\} - k_1 k_2 e_0 a_0 = 0 \tag{110}$$

[1] B. S. Hartley and B. A. Kilby, *Biochem. J.*, **65**, 288 (1954).

which may be written as

$$\ddot{n} + P\dot{n} + Qn - R^* = 0 \tag{111}$$

(note that the constants P and Q are the same as in the preceding treatment, but that R^* is different from R; $R^* = k_2 R/k_3$). The solution of eqn (111), subject to the boundary conditions $t = 0$, $n = 0$, and $t = 0$, $\dot{n} = 0$, is

$$n = \frac{k_2 R}{k_3 Q} + M^* e^{-Ft} + N^* e^{-Gt} \tag{112}$$

where

$$F = \tfrac{1}{2}\{P - \sqrt{(P^2 - 4Q)}\} \tag{113}$$

$$G = \tfrac{1}{2}\{P + \sqrt{(P^2 - 4Q)}\} \tag{114}$$

$$M^* = \frac{GR^*}{Q(G - F)} = -\frac{k_2 GR}{k_3 Q(G - F)} \tag{115}$$

$$N^* = \frac{FR^*}{Q(G - F)} = \frac{k_2 FR}{Q(G - F)}. \tag{116}$$

Insertion of (112) into (88) followed by integration, subject to the boundary condition $t = 0$, $y = 0$, then leads to an expression for y as a function of time:

$$y = \frac{k_2 R}{Q} t + \frac{k_3 M^*}{F}(1 - e^{-Ft}) + \frac{k_3 N^*}{G}(1 - e^{-Gt}). \tag{117}$$

The general form of this relationship is shown in Fig. 6.13. When t is sufficiently large the exponential terms vanish and the equation reduces to the steady-state equation

$$y_s = \frac{k_2 R}{Q} t + \frac{k_3 M^*}{F} + \frac{k_3 N^*}{G}. \tag{118}$$

The intercepts of this line on the axes are

$$y_s^0 = k_3 \left(\frac{M^*}{F} + \frac{N^*}{G} \right) \tag{119}$$

and

$$t^0 = -\frac{k_3 Q}{k_2 R} \left(\frac{M^*}{F} + \frac{N^*}{G} \right). \tag{120}$$

Since M^* is necessarily negative, and G is greater than F, y_s^0 must be negative and t^0 positive. The expressions to which y_s^0 and t^0 reduce are shown in the diagram.

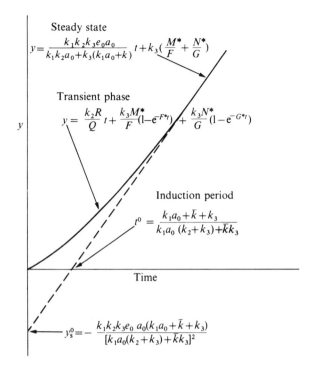

Steady state

$$y = \frac{k_1 k_2 k_3 e_0 a_0}{k_1 k_2 a_0 + k_3 (k_1 a_0 + k)} t + k_3 \left(\frac{M^*}{F} + \frac{N^*}{G} \right)$$

Transient phase

$$y = \frac{k_2 R}{Q} t + \frac{k_3 M^*}{F} (1 - e^{-Ft}) + \frac{k_3 N^*}{G} (1 - e^{-Gt})$$

y

Induction period

$$t^0 = \frac{k_1 a_0 + \bar{k} + k_3}{k_1 a_0 (k_2 + k_3) + \bar{k} k_3}$$

Time

$$y_s^0 = - \frac{k_1 k_2 k_3 e_0 \, a_0 (k_1 a_0 + \bar{k} + k_3)}{[k_1 a_0 (k_2 + k_3) + \bar{k} k_3]^2}$$

FIG. 6.13. The variation with time of the concentration y of the second product released in a double-intermediate mechanism.

Various methods are available for analysing experimental y–t curves. If the intercepts of the steady-state line on the y and t axes are extrapolated to high values the results are

$$y_s^0(a_0 \to \infty) = -\frac{k_2 k_3 e_0}{(k_2 + k_3)^2} \tag{121}$$

$$t^0(a_0 \to \infty) = \frac{1}{k_2 + k_3}. \tag{122}$$

In principle, if e_0 is known, $k_2 + k_3$ and $k_2 k_3$, and hence the individual constants k_2 and k_3 can be determined. In practice, however, the extrapolation to high a_0 cannot always be made in a reliable manner.

Alternatively, information can be obtained from an analysis of the y–t curve prior to the establishment of the steady state. G is larger than F, so that as t increases e^{-Gt} approaches zero more rapidly than does e^{-Ft}. The deviation between y and its steady state value is thus approximately

$$\Delta y \approx \frac{k_3 M^*}{F} e^{-Ft} \tag{123}$$

A plot of $\ln \Delta y$ against t therefore gives F and $k_3 M^*$. These quantities are, to a good approximation,

$$F \approx \frac{k_1 a_0 (k_2 + k_3) + \bar{k} k_3}{k_1 a_0 + \bar{k} + k_3} \tag{124}$$

and

$$k_3 M^* \approx -\frac{k_1 k_2 k_3 e_0 a_0}{k_1 a_0 (k_2 + k_3) + \bar{k} k_3}. \tag{125}$$

If the results can be satisfactorily extrapolated to high a_0 they can be identified with

$$F(a_0 \to \infty) = k_2 + k_3 \tag{126}$$

and

$$k_3 M^*(a_0 \to \infty) = -\frac{k_2 k_3 e_0}{k_2 + k_3}. \tag{127}$$

In principle k_2 and k_3 can be determined in this way, if e_0 is known. However, in practice it is frequently difficult to achieve the condition of sufficiently high substrate concentrations.

Kinetics at high enzyme concentrations

The foregoing treatment has shown that the analysis of experimental x–t or y–t curves, for the usual case in which the substrate is in excess of the enzyme, is a matter of some difficulty. The analysis is particularly troublesome unless one can work at rather high substrate concentrations, in which case the expressions take a simpler form. It is frequently impossible to work at substrate concentrations that are high enough to permit a reliable extrapolation.

These difficulties are avoided if one works at enzyme concentrations that are much higher than the substrate concentrations. The theoretical treatment of this case has been given by Kasserra and Laidler[1], and proceeds as follows. The reaction scheme is

$$\text{E} + \quad \text{A} \quad \underset{k_{-1}}{\overset{k_1}{\rightleftharpoons}} \text{EA} \overset{k_2}{\searrow} \text{EA}' \overset{k_3}{\to} \quad \text{E} \quad + \text{Y}$$

$$e_0 \quad a_0 - m - n - y \qquad m \quad \underset{x}{X} \ n \qquad a_0 - m - n - y \quad y.$$

Since the enzyme is in great excess its concentration remains essentially at e_0 throughout the reaction. The differential equations describing the system

[1] H. P. Kasserra and K. J. Laidler, *Can. J. Chem.*, **48**, 1793 (1970).

are:

$$\dot{m} = k_1 e_0 (a_0 - m - n - y) - \bar{k}m \tag{128}$$

$$\dot{n} = k_2 m - k_3 n \tag{129}$$

$$\dot{x} = k_2 m \tag{130}$$

$$\dot{y} = k_3 n. \tag{131}$$

Differentiation of eqn (128) and elimination of $\dot{n}$ and $\dot{y}$ by use of eqns (129) and (131) gives

$$\ddot{m} + \dot{m}(k_1 e_0 + \bar{k}) + m(k_1 k_2 e_0) = 0. \tag{132}$$

It is to be noted that this equation is simpler than either eqn (89) or (110); in particular, it does not involve k_3, and the constant term has disappeared.

Equation (132) can be written as

$$\ddot{m} + P'\dot{m} + Q'm = 0 \tag{133}$$

where

$$P' = k_1 e_0 + \bar{k} \tag{134}$$

and

$$Q' = k_1 k_2 e_0. \tag{135}$$

The general solution of eqn (133), subject to the boundary conditions $t = 0, m = 0$, and $t = 0, \dot{m} = k_1 e_0 a_0$, is

$$m = M'(e^{-F't} - e^{-G't}) \tag{136}$$

where

$$F' = \tfrac{1}{2}\{P' - \sqrt{(P'^2 - 4Q')}\} \tag{137}$$

$$G' = \tfrac{1}{2}\{P' + \sqrt{(P'^2 - 4Q')}\} \tag{138}$$

and

$$M' = \frac{k_1 e_0 a_0}{G' - F'}. \tag{139}$$

The variation of x with time is obtained by substituting (136) into (130) and integrating with the boundary condition that at $t = 0, x = 0$; the result is

$$x = \frac{k_2 M'}{F'}(1 - e^{-F't}) - \frac{k_2 M'}{G'}(1 - e^{-G't}). \tag{140}$$

The variation of this function is shown schematically in Fig. 6.14. When t is large the exponentials vanish and the value of x becomes

$$x(t \to \infty) = k_2 M'\left(\frac{1}{F'} - \frac{1}{G'}\right). \tag{141}$$

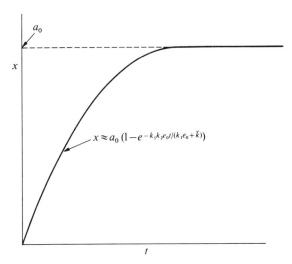

FIG. 6.14. The variation with time of x, the concentration of the first product released, for a reaction in which the enzyme concentration is in great excess of the substrate concentration.

Introduction of the expressions for F', G' and M' shows that in fact

$$x(t \rightarrow \infty) = a_0. \tag{142}$$

In this case, there is no steady state established when enzyme is in excess; the concentration of product rises exponentially until it becomes equal to the initial substrate concentration, i.e. until the substrate is all used up. There is really no transient phase under these circumstances; however, the reaction may be completed in a very short period of time, and the usual experimental techniques for transient-phase kinetics may have to be employed.

Since G' is greater than F' (under most conditions it is considerably greater) the time-course of the reaction is dominated by the first term in eqn (140):

$$x = \frac{k_2 M'}{F'}(1 - e^{-F't}). \tag{143}$$

Analysis of x–t curves can be carried out by plotting $\ln x$ against t, which according to eqn (143) will give

$$k_2 M' \quad \text{and} \quad F'.$$

Since k_1 is large, and the enzyme concentration is high, P'^2 is usually very much greater than $4Q'$; if this is the case the expressions for F' and G' reduce as follows:

$$F' = \tfrac{1}{2}\{P' - P'(1 - 4Q'/P'^2)^{\frac{1}{2}}\} \approx Q'/P' \tag{144}$$

and

$$G' = \tfrac{1}{2}\{P' + P'(1 - 4Q'/P'^2)^{\frac{1}{2}}\} \approx P'. \tag{145}$$

With these approximations, and use of eqns (134), (135) and (139), eqn (143) becomes

$$x \approx a_0\{1 - e^{-k_1 k_2 e_0 t/(k_1 e_0 + \bar{k})}\}. \tag{146}$$

From the plot of ln x against t the quantity

$$\frac{k_1 k_2 e_0}{k_1 e_0 + \bar{k}}$$

can therefore be determined. Extrapolation to high enzyme concentrations thus yields the rate constant k_2.

The above treatment may be extended[1] to the kinetics of formation of the second intermediate Y. Elimination of m from eqns (128) and (129) yields a differential equation for which no solution can be obtained. It is therefore necessary to introduce the solution for m into eqn (129) and solve for n. The result is

$$n = \frac{k_2 M'}{k_3 - F'} e^{-Ft} - \frac{k_2 M'}{k_3 - G'} e^{-Gt} - k_2 M'\left(\frac{1}{k_3 - F'} - \frac{1}{k_3 - G'}\right) e^{-k_3 t}. \tag{147}$$

Substitution of this into (131) gives, after integration with the boundary condition $t = 0, y = 0$,

$$y = \frac{k_2 k_3 M'}{F'(k_3 - F')}(1 - e^{-F't}) - \frac{k_2 k_3 M'}{G'(k_3 - G')}(1 - e^{-G't}) -$$
$$- k_2 M'\left(\frac{1}{k_3 - F'} - \frac{1}{k_3 - G'}\right)(1 - e^{-k_3 t}). \tag{148}$$

It can further be shown that, to a good approximation in most cases,

$$y = a_0(1 - e^{-k_3 t}) \tag{149}$$

irrespective of the relative magnitudes of k_2 and k_3. A kinetic study of the rate of formation of Y thus allows k_3 to be determined. It is to be noted that y simply rises exponentially to the limiting value a_0; again, as with the first product X, there is no steady state.

Recently, some additional systems have been treated, including inhibited systems and two-substrate systems.[2]

[1] N. H. Hijazi and K. J. Laidler, *Can. J. Chem.*, **50**, 1440 (1972).
[2] N. H. Hijazi and K. J. Laidler, *Can. J. Biochem.*, **51**, (1973).

7. Molecular Kinetics

KINETIC studies on enzyme systems can conveniently be divided into two classes. The first of these comprises investigations of the phenomenological rate laws that are obeyed—of the way in which rates vary with concentrations of substrates, activators, inhibitors, and hydrogen ions. Such studies have been dealt with in Chapters 3–5, and lead to values for certain rate constants and equilibrium constants. Investigations of the effects of pH on the rate, for example, have been seen in Chapter 5 to lead to dissociation constants for enzymes and enzyme–substrate complexes, and the numerical values of these allow conclusions to be drawn about the nature of active centres of enzymes, and about the way in which these centres interact with substrates, activators, and inhibitors. The phenomenological rate laws also involve certain purely kinetic rate constants, such as the first-order rate constants for the subsequent reactions of the enzyme–substrate complex.

The investigations of the second main class, which constitute the field known as *molecular kinetics*, are dependent upon those of the first, and comprise more detailed studies of the thermodynamic and kinetic constants. Up to now most of the work of this kind has consisted in studying the effects of temperature on these constants, and evaluating the corresponding heats and entropies. The magnitudes of these constants, taken in conjunction with other evidence, can throw a good deal of light on the details of the enzyme–substrate interaction and on the mechanism of enzyme action. Other investigations of this type that have been carried out are related to the effects of high hydrostatic pressures, the dielectric constant, and the ionic strength on the kinetic and thermodynamic constants. Some of the investigations of this second main class are discussed in the present chapter.

The influence of temperature on enzyme reaction rates

The effects of temperature on the rates of enzyme reactions are matters of considerable interest and importance, and have been studied for many years. The subject is a fairly complicated one, and it is only recently that a reasonable degree of understanding has been achieved. Some of the earlier studies were of too superficial a nature to be useful; it is only when experiments are carried out in the light of a precise theoretical treatment that satisfactory results can be obtained.

It has long been known that if an enzyme-catalysed reaction is studied over a range of temperature the overall rate passes through a maximum: this type of behaviour is shown schematically in Fig. 7.1. The temperature at which the rate is a maximum is known as the *optimum temperature*, and was at one time thought to be characteristic of the enzyme system; it is now known, however, to be an ill-defined quantity, and to vary with concentrations and with other factors such as the pH.

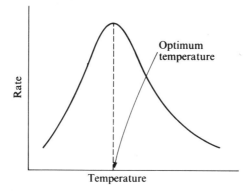

Fig. 7.1. Schematic representation of the variation of the rate of an enzyme reaction with temperature.

The correct explanation of this type of behaviour, first given by Tammann[1], is that raising the temperature affects two independent processes, the catalysed reaction itself and the thermal inactivation of the enzyme. In the lower temperature range, up to 30°C or so for a typical enzyme, inactivation is very slow and has no appreciable effect on the rate of the catalysed reaction; the overall rate therefore increases with rise in temperature, as with ordinary chemical reactions. At higher temperatures inactivation becomes more and more important, so that the concentration of active enzyme falls during the course of reaction. It is an essential feature of the explanation that the temperature coefficient of the rate of inactivation must be greater than that of the rate of the catalysed reaction; in the low-temperature range the rate of inactivation may be negligible compared with the rate of the catalysed reaction, whereas in the high-temperature range it may be much higher. In the extreme case, if the temperature is quite high (say 50° to 60°C for most enzymes), the enzyme is inactivated extremely rapidly, so that the rate of the catalysed reaction is close to zero. It will be seen later that temperature coefficients for enzyme inactivations are generally much higher than for enzyme-catalysed reactions, so that the phenomenon of the optimum temperature seems to be a universal one.

As soon as it was realized that temperature affects the two separate processes, enzyme inactivation and the enzyme-catalysed reaction, efforts were devoted to studying the two problems separately. The influence of temperature on the inactivation process is considered in Chapter 13; the present chapter will be concerned solely with enzyme-catalysed reactions, and will treat not only the question of the temperature coefficients of these reactions but also certain other matters (e.g. entropies of activation), information about which is provided by the temperature studies. In the remainder of this chapter, it will be assumed that the catalysed reactions have been

[1] G. Tammann, *Z. phys. Chem.*, **18**, 426 (1895).

investigated under such conditions that the enzyme inactivation is unimportant (in the low-temperature range) or that a suitable correction has been made for the inactivation.

Even when the effects of enzyme inactivation have been eliminated it is still necessary to carry out a careful preliminary analysis of the kinetics before attempting to study temperature effects. This is because the rate law for an enzyme-catalysed reaction always involves at least three kinetic constants, and temperature may be expected to affect these individually. The simplest case is when the simple Michaelis–Menten mechanism applies: the kinetic law then involves the three rate constants k_1, k_{-1} and k_2. These individual constants vary with temperature according to the Arrhenius law. The variation of the rate v with the temperature is, however, more complicated, and in order for a satisfactory study to be made it is necessary for the constants to be separated as far as possible. This has not always been done; some confusion has resulted from attempts to analyse the effects of temperature on the reaction velocity itself. Nevertheless, the Arrhenius law is frequently found to apply to complex processes that are certainly controlled by more than one enzyme system.[1]

It was seen in Chapter 2 that the Arrhenius law takes the form

$$k = A\,e^{-E/RT} \tag{1}$$

where E is the energy of activation for the reaction and A is the frequency factor. In terms of activated-complex theory the equation, in the case of reactions in solution, can be expressed in the alternative form (see p. 46)

$$k = e\frac{kT}{h}e^{\Delta S^{\neq}/R}\,e^{-E/RT}. \tag{2}$$

Here $\Delta S^{\neq}$ is the entropy of activation, which is the increase in entropy when the system passes from the initial to the activated state. Energies of activation are usually calculated from plots of $\log_{10} k$ against $1/T$, and the entropies of activation can then be calculated from the value of E and from k at any one temperature.

This type of procedure has been applied to a considerable number of enzyme systems, and in all cases the Arrhenius law has been found to be obeyed satisfactorily; an example of an Arrhenius plot is shown in Fig. 7.2. Statements have occasionally appeared in the literature to the effect that the law does not apply to biological processes, but this belief has arisen largely from an improper application of the law to the rates of reactions rather than to the individual rate constants.

In the treatment that follows of the application of the Arrhenius law to enzyme systems attention will be confined to work done with reasonably

[1] For example, the creeping of ants (*Proc. natn. Acad. Sci. U.S.A.*, **6**, 204 (1920)) and the frequency of heart beat (*Am. J. Physiol.*, **11**, 383 (1904)). See K. J. Laidler, *J. chem. Educ.*, **49**, 343 (1972).

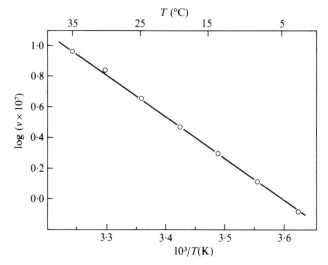

Fig. 7.2. An Arrhenius plot for the myosin-catalysed hydrolysis of adenosine triphosphate; $\log v$ is plotted against the reciprocal of the absolute temperature (Ouellet, Laidler, and Morales, *Archs Biochem. Biophys.*, **39**, 37 (1952)). Under the conditions of the experiments (high substrate concentrations) the velocity v is proportional to the rate-constant k_c.

pure enzyme preparations. The pH employed will in each case be close to the optimum, so that no complications will result from changes in the degree of ionization with the temperature.

The second and third stages of reaction

It has been seen in previous chapters that the evidence, kinetic and otherwise, points to the conclusion that many enzymes act according to the general scheme

$$E + A \underset{k_{-1}}{\overset{k_1}{\rightleftharpoons}} EA \overset{k_2}{\rightsquigarrow} EA' \overset{k_3}{\rightarrow} E + Y.$$
$$\searrow$$
$$X$$

The second stage in this scheme is the reaction of the enzyme–substrate complex to give a product X and a second intermediate EA', which in the case of the proteolytic enzymes is frequently an acyl-enzyme. Stage 3 is the conversion of the acyl-enzyme into the enzyme and the second product Y. The present section is concerned with the kinetic parameters (i.e., energy and entropy of activation) associated with the rate constants k_2 and k_3 for those two stages of reaction.

Separation of rate constants k_2 and k_3

A steady-state investigation of the kinetics of an enzyme-catalysed reaction does not itself permit the evaluation of the individual constants

k_2 and k_3. The first-order rate constant at high substrate concentrations, k_c, is

$$k_c = \frac{k_2 k_3}{k_2 + k_3}. \tag{3}$$

The two individual constants, or at least one of them, can, however, be derived from a steady-state study if other information is available. In addition, studies of the transient phase enable k_2 and k_3 to be determined.

The most important methods which lead to individual values of k_2 and k_3 are as follows:

(1) In several investigations measurements have been made of k_c values for a series of substrates which give rise to the same intermediate EA′. Schwert and Eisenberg[1], for example, studied the hydrolysis by trypsin of a series of esters of benzoyl-L-arginine, of general structure

$$C_6H_5CONH \diagdown \quad \diagup NH$$
$$C$$
$$|$$
$$NH$$
$$|$$
$$(CH_2)_2$$
$$|$$
$$H_2N-CH-COOR.$$

The acyl-enzyme formed from any member of this series will be the same, irrespective of the group R. The hydrolysis of EA′, i.e., the deacylation reaction, will therefore be the same for all of the substrates. It was found that the k_c values were the same for all members of the series, and it is a reasonable inference that the slow process is the deacylation reaction; that is, it is concluded that $k_2 \gg k_3$, so that $k_c \approx k_3$. The acylation process will be different for the various substrates, so that if step 2 were other than fast compared with step 3 one would expect to find different k_c values for the different substrates.

A similar argument was employed by Wilson and Cabib[2] for the enzyme acetylcholinesterase. They worked with the following four substrates:

acetylcholine,	$CH_3CO_2CH_2CH_2N^+(CH_3)_3$
dimethylaminoethyl acetate,	$CH_3CO_2CH_2CH_2N^+H(CH_3)_2$
methylaminoethyl acetate,	$CH_3CO_2CH_2CH_2N^+H_2CH_3$
aminoethyl acetate,	$CH_3CO_2CH_2CH_2NH_2$

Again there is a common EA′ species, the acetyl (CH_3CO-) enzyme. The k_c values are now different, the order being

acetylcholine > dimethylaminoethyl acetate > methylaminoethyl
acetate > aminoethyl acetate.

[1] G. W. Schwert and M. A. Eisenberg, *J. biol. Chem.*, **179**, 665 (1949).
[2] I. B. Wilson and E. Cabib, *J. Am. chem. Soc.*, **78**, 202 (1956).

Step 3 cannot be rate-controlling, as the variations between the substrates are due to variations in k_2, which must vary in the order indicated above. Wilson and Cabib extended their argument by studying the k_c values at various concentrations and making Arrhenius plots; the results are shown in Fig. 7.3. The curvature of the plots with acetylcholine and dimethyl-aminoethyl acetate is taken to indicate that for these substrates the k_2 and

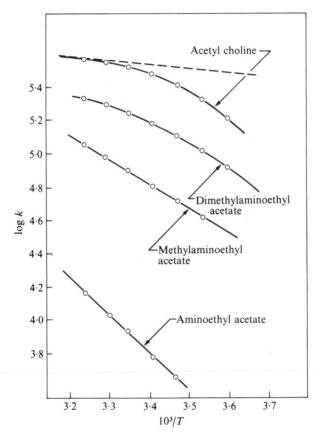

FIG. 7.3. Plots of $\log k$ against $1/T$ for the acetylcholinesterase-catalyzed reactions of various substrates (Wilson and Cabib, *loc. cit.*).

k_3 values are comparable, so that k_c does not follow the Arrhenius law. For the other two substrates, however, there is a linear plot, indicating that one of the k's is significantly more important than the other; in view of the order of k_c values, and of the fact that k_3 is the same for all of the substrates, it follows that $k_2 \ll k_3$ for methylaminoethyl acetate and aminoethyl acetate.

(2) Variation of the species that reacts with EA′ in stage 3 can also lead to conclusions about relative values of k_2 and k_3. Normally EA′ is hydrolysed by water to give E + Y; thus if EA′ is an acyl enzyme, RCO_2-enzyme, step 3 is

$$RCO_2\text{-enzyme} + H_2O \xrightarrow{k_3} RCO_2H \; + \; HO\text{-enzyme.}$$

However, H_2O may be replaced by another species, such as an alcohol R′OH:

$$RCO_2\text{-enzyme} + R'OH \xrightarrow{k'_3} RCO_2R' + HO\text{-enzyme.}$$

Measurements of k_c values with water and a series of alcohols can then lead to conclusions about the relative magnitudes of k_2 and k_3.

The theory of the method is as follows[1]. If one works at high substrate concentrations, so that the rate is zero-order in substrate, the enzyme is essentially all converted into either EA or EA′. The reaction scheme now simplifies to†

$$\text{EA} \underset{X}{\overset{k_2}{\rightleftharpoons}} \text{EA}' \overset{k_3[W]}{\underset{k'_3[Z]}{\rightrightarrows}} \begin{matrix} Y \\ Y' \end{matrix}$$

where W and Z are two species which react with EA′ (for example, W may be H_2O and Z an alcohol). The steady-state equation is

$$k_2[\text{EA}] - (k_3[\text{W}] + k'_3[\text{Z}])[\text{EA}'] = 0. \tag{4}$$

The total enzyme concentration is

$$[\text{E}]_0 = [\text{EA}] + [\text{EA}'] \tag{5}$$

(the amount of free enzyme being negligible at high A), so that

$$[\text{E}]_0 = [\text{EA}']\left(\frac{k_3[\text{W}] + k'_3[\text{Z}]}{k_2} + 1\right). \tag{6}$$

The rate of formation of the product Y is therefore

$$v_Y = k_3[\text{W}] \cdot [\text{EA}'] = \frac{k_2 k_3[\text{W}] \cdot [\text{E}]_0}{k_3[\text{W}] + k'_3[\text{Z}] + k_2}. \tag{7}$$

The rate of formation of the product Y in the absence of Z is

$$v_Y(Z = 0) = \frac{k_2 k_3[\text{W}] \cdot [\text{E}]_0}{k_3[\text{W}] + k_2}. \tag{8}$$

† Note that k_3 and k'_3 are now second-order rate constants for the reaction of EA′ with W and Z respectively.

[1] R. M. Epand and I. B. Wilson, *J. biol. Chem.*, **238**, 1718 (1963); cf. I. Hinberg and K. J. Laidler, *Can. J. Biochem.*, **50**, 1334 (1972).

Thus

$$\frac{v_Y}{v_Y(Z = 0)} = \frac{k_3[W] + k_2}{k_3[W] + k_3'[Z] + k_2}. \tag{9}$$

The two limiting cases are now:
  (i) $k_2 \gg k_3[W] + k_3'[Z]$, when

$$\frac{v_Y}{v_Y(Z = 0)} = 1 \tag{10}$$

and
  (ii) $k_3[W] + k_3'[Z] \gg k_2$, when

$$\frac{v_Y}{v_Y(Z = 0)} = \frac{k_3[W]}{k_3[W] + k_3'[Z]}. \tag{11}$$

If, therefore, it is found that the addition of Z leads to a significant diminution in the rate of formation of Y, as predicted by eqn (11), it can be concluded that step 2 is rate limiting. Conversely, if addition of Z has no detectable effect on the rate of formation of Y, the conclusion is that step 3 is rate-limiting.

The reason for the different behaviour in the two extreme cases is as follows. If step 3 is slow and step 2 fast, the enzyme is essentially all present as EA', the supply of which is rapidly replenished as it reacts with W or Z. There is therefore always plenty of EA' available to react with W, irrespective of the reaction of EA' with Z. When, on the other hand, step 2 is slow and step 3 fast, the enzyme is present mainly as EA, and the concentration of EA' is low; there is then competition between W and Z for EA', so that the rate of reaction of W with EA' is reduced by the addition of Z.

An example of the use of this method is to be found in a study by Wilson, *et al.*[1] of the hydrolysis of 2-naphthol phosphate by alkaline phosphatase; the reaction scheme is

$$\text{E} + \text{A} \underset{k_{-1}}{\overset{k_1}{\rightleftharpoons}} \text{EA} \overset{k_2}{\searrow} \text{EA}' \xrightarrow{k_3[\text{H}_2\text{O}]} \text{E} + \text{phosphate.}$$

$$\text{2-naphthol}$$

EA' is phosphoryl-enzyme, which is hydrolysed by water in the final stage to enzyme and phosphate. Ethanolamine also reacts with EA', with the formation of enzyme and O-phosphorethanolamine. It was found that addition of ethanolamine increased the rate of formation of 2-naphthol but had no effect on the rate of formation of phosphate. This indicates that dephosphorylation (stage 3) is the rate-controlling step, in agreement with previous conclusions[2].

[1] D. Levine, T. W. Reid, and I. B. Wilson, *Biochemistry*, **8**, 2374 (1969).
[2] H. N. Fernley and P. G. Walker, *Nature*, **212**, 1435 (1966); W. N. Aldridge, T. E. Barman, and H. G. Gutfreund, *Biochem. J.*, **92**, 23c (1964); D. R. Trentham and H. Gutfreund, *Biochem. J.*, **106**, 455 (1968).

(3) Measurements of concentrations of EA and EA′ during the steady state can also provide information about the relative magnitudes of k_2 and k_3. Such concentrations of intermediates can be estimated by spectrophotometric methods. The steady-state treatment for the general scheme shown on p. 78 leads to the result that at high substrate concentrations

$$\frac{[EA]}{[E]_0} = \frac{k_3}{k_2 + k_3} \tag{12}$$

and

$$\frac{[EA']}{[E]_0} = \frac{k_2}{k_2 + k_3}. \tag{13}$$

Thus if $k_3 \gg k_2$,

$$[EA] \approx [E]_0 \quad \text{and} \quad [EA'] \approx 0$$

while if $k_2 \gg k_3$

$$[EA] \approx 0 \quad \text{and} \quad [EA'] \approx [E]_0.$$

An example of the use of this method is to be found in a study of the chymotrypsin-catalysed hydrolysis of *p*-nitrophenyl trimethylacetate[1]. The acyl-enzyme intermediate was present in high concentrations, showing that the acylation process is rapid and the deacylation slow ($k_2 \gg k_3$). For this system, in fact, it was possible to isolate and crystallize the acyl-enzyme intermediate.

(4) As described in the last chapter, transient-phase studies of the rates of formation of products can lead to individual values of the rate constants k_2 and k_3.

Table 7.1 gives systems for which it appears that $k_3 \gg k_2$, and Table 7.2 gives systems for which $k_2 \gg k_3$.

Energies and entropies of activation

A few investigations have given rise to values of the rate constants k_2 and k_3 over a range of temperatures. From the results, as explained in Chapter 2, it is possible to calculate energies and entropies of activation. Some of the results are collected in Tables 7.3 and 7.4. It is of interest that the $\Delta S_2^{\neq}$ and $\Delta S_3^{\neq}$ values are all negative.

In considering the magnitudes of the entropies of activation listed in Tables 7.3 and 7.4 it is necessary to take into account two types of effect, which it is convenient to refer to as *solvent* and *structural* effects.† By the

† The present discussion of the magnitudes of entropies of activation of enzyme-catalysed reactions is based on one by K. J. Laidler, *Discuss. Faraday Soc.*, **20**, 83 (1955).

[1] A. K. Balls, C. E. McDonald, and A. S. Breecher, *Proc. Int. Symposium on Enzyme Chemistry*, Marizan Co., Tokyo, 1957, p. 392; M. L. Bender, G. R. Schonbaum, and B. Zerner, *J. Am. chem. Soc.*, **84**, 2562 (1962).

TABLE 7.1

Systems for which $k_2 \ll k_3$

Enzyme	Substrate	Ref.
Chymotrypsin	N-benzoyl-L-tyrosinamide	[a]
	furylacryloyltryptophanamide	[b]
Cholinesterase	methylaminoethyl acetate	[c]
	3-acetoxypyridine-N-oxide	[d]
	isoamyl acetate	[e]
	substituted aminoethyl acetates	[f]
Carbonic anhydrase	p-nitrophenylacetate	[g]
Papain	N-benzoyl-L-arginine ethyl ester	[h]

[a] M. L. Bender, G. R. Schonbaum, and B. Zerner, *J. Am. chem. Soc.*, **84**, 2540 (1962).
[b] G. P. Hess, J. McConn, E. Ku, and G. McConkey, *Phil. Trans. Roy. Soc.*, **257B**, 89 (1970).
[c] I. B. Wilson and E. Cabib, *J. Am. chem. Soc.*, **78**, 202 (1956).
[d] D. E. Lenz and G. E. Hein, *Biochim. biophys. Acta*, **220**, 617 (1970).
[e] R. M. Krupka, *Biochemistry*, **5**, 1983 (1966).
[f] R. M. Krupka, *ibid.*, **5**, 1988 (1966).
[g] Y. Pocker and D. R. Storm, *Biochemistry*, **7**, 1202 (1968).
[h] L. A. A. E. Sluyterman, *Biochim. biophys. Acta*, **151**, 178 (1968).

former is meant the interaction between the solvent and the reaction system, an interaction that may change during the course of reaction. By structural effects is meant the possibility that the enzyme itself actually undergoes some reversible change in conformation during the process of reaction.

The possibility of such structural changes was first proposed[1] in an attempt to interpret the effects of pressure on enzyme reactions, an aspect that is discussed later in this chapter. It was suggested that in some cases the enzyme molecule assumes a more open conformation when it forms a complex, and that when it undergoes subsequent reactions there is a refolding of the enzyme molecule. Such a picture is consistent with the results for a number of enzyme systems, but cannot be said to have been firmly established. The idea does, however, seem reasonable since it brings the enzyme reactions into line with what takes place during enzyme inactivation, in which it seems clear that unfolding occurs.

The theory that enzymes become partly unfolded during complex formation and become refolded in the subsequent reactions leads to the conclusion that the entropy of activation will be negative for the subsequent reactions. Tables 7.3 and 7.4 show that this latter prediction is correct in all cases.

Before the solvent effects are considered, reference may be made to another type of conformational change for which there is some evidence, particularly with myosin. In the above discussion of unfolding it has been assumed that bonds (perhaps hydrogen bonds) are broken, so that the unfolded molecule possesses more freedom of movement, and therefore more entropy, than the

[1] K. J. Laidler, *Archs Biochem.*, **30**, 226 (1951).

TABLE 7.2

Systems for which $k_3 \ll k_2$

Enzyme	Substrate	Ref.
Chymotrypsin	p-nitrophenyl acetate	[a], [b]
Chymotrypsin	o- and p-nitrophenyl cinnamate	[c]
Chymotrypsin	N-acetyl-L-tyrosine ethyl ester	[c]
Chymotrypsin	N-benzoyl-L-tyrosine ethyl ester	[c]
Chymotrypsin	N-trans-cinnamoyl imidazole	[d]
Trypsin	p-nitrophenyl acetate	[e]
Trypsin	α-N-carbobenzoxy-L-lysine-p-nitrophenyl ester hydrochloride	[f]
Trypsin	N-carbobenzoxy-L-alanine-p-nitrophenyl ester	[g]
Trypsin	N-trans-cinnamoyl-imidazole	[d]
Trypsin	benzoyl-L-arginine esters	[h]
Alkaline phosphatase (*E. coli*)	2,4-dinitrophenylphosphate	[i]
Alkaline phosphatase (*E. coli*)	4-methyl umbelliferyl phosphate	[j]
Alkaline phosphatase (*E. coli*)	2-naphthol phosphate	[k]
Papain	benzoyl glycine ethyl ester	[l]
Papain	methylthiohippurate	[m]
Papain	cinnamoylimidazole	[n]
Papain	BOC-glycine-p-nitrophenyl phosphate	[o]
Papain	p-nitrophenyl-α-D-BOC-p-norleucinate	[p]
Papain	N-BOC-L-tyrosine-o-nitrophenol	[q]
Papain	furylacryloylimidazole	[r]
Cholinesterase	m-(trimethylammonium)-phenylacetate iodide	[s]
Cholinesterase	3-acetoxypyridine-N-methiodide	[s]
Cholinesterase	phenylacetate	[t]
Cholinesterase	acetylcholine	[t]
Cholinesterase	acetylthiocholine	[t]
Cholinesterase	p-nitrophenyldimethylcarbamate	[q]
Subtilisin	N-trans-cinnamoylimidazole	[u]
Elastase	p-nitrophenyldiethylphosphate	[q]

[a] B. S. Hartley and B. A. Kilby, *Biochem. J.*, **50**, 672 (1952); **56**, 288 (1954).
[b] H. Gutfreund and J. M. Sturtevant, *Proc. natn. Acad. Sci. U.S.A.*, **42**, 719 (1956).
[c] M. L. Bender, G. R. Schonbaum, and B. Zerner, *J. Am. chem. Soc.*, **84**, 2540 (1962).
[d] M. L. Bender and E. T. Kaiser, *J. Am. chem. Soc.*, **84**, 2556 (1962).
[e] J. A. Stewart and L. Ouellet, *Can. J. Chem.*, **37**, 751 (1959).
[f] M. L. Bender, F. J. Kézdy, and J. Feder, *J. Am. chem. Soc.*, **81**, 1874 (1959).
[g] H. P. Kasserra and K. J. Laidler, *Can. J. Chem.*, **48**, 1793 (1970).
[h] G. W. Schwert and M. A. Eisenberg, *J. biol. Chem.*, **179**, 665 (1949).
[i] S. H. G. Ko and F. J. Kézdy, *J. Am. chem. Soc.*, **89**, 7139 (1967).
[j] H. N. Fernley and P. G. Walker, *Biochem. J.*, **111**, 187 (1969).
[k] D. Levine, T. W. Reid, and I. B. Wilson, *Biochemistry*, **8**, 2374 (1969).
[l] L. A. A. E. Sluyterman, *Biochem. biophys. Acta*, **151**, 178 (1968); see also K. Brocklehurst, E. M. Crook and W. C. Wharton, *FEBS Lett.*, **2**, 69 (1968).
[m] G. Lowe and A. Williams, *Biochem. J.*, **96**, 189 (1965).
[n] L. J. Brubacher and M. L. Bender, *J. Am. chem. Soc.*, **88**, 5971 (1966).
[o] D. C. Williams and J. R. Whitaker, *Biochemistry*, **6**, 3711 (1967).
[p] A. Williams and E. C. Lucas, *Analyt. Chem.*, **42**, 1491 (1970).
[q] M. L. Bender and coworkers, *J. Am. chem. Soc.*, **88**, 5890 (1966).
[r] P. M. Hinkle and J. F. Kirsch, *Biochemistry*, **9**, 4633 (1970).
[s] D. E. Lenz and G. E. Hein, *Biochim. biophys. Acta*, **220**, 617 (1970).
[t] R. M. Krupka, *Biochemistry*, **4**, 429 (1965).
[u] S. A. Bernhard, S. J. Lau, and H. Noller, *Biochemistry*, **4**, 1108 (1965); and M. L. Bender and coworkers, *loc. cit.*

TABLE 7.3

Kinetic parameters relating to the conversion of the enzyme–substrate complex into the second intermediate and the first product

Enzyme	Substrate	Temperature (0°C)	pH	k_2 (sec^{-1})	E_2 (kcal mol^{-1})	$\Delta S_2^{\neq}$ (cal deg^{-1} mol^{-1})	Ref.
Chymotrypsin	benzoyl-L-tyrosyl-glycinamide	25·0	7·5	37	11·5	−19·8	[a]
Chymotrypsin	benzoyl-L-tyrosinamide	25·0	7·8	0·625	14·6	−13·0	[b]
Trypsin	benzoyl-L-argininamide	25·0	7·8	270	14·9	−8·2	[a]

[a] J. A. V. Butler, *J. Am. chem. Soc.*, **63**, 2971 (1941).
[b] S. Kaufman, H. Neurath, and G. W. Schwert, *J. biol. Chem.*, **177**, 792 (1949).

TABLE 7.4

Kinetic parameters relating to the conversion of the second intermediate
into the enzyme and the second product

Enzyme	Substrate	Temperature (0°C)	pH	k_3 (sec^{-1})	E_3 (kcal mol^{-1})	$\Delta S_3^{\neq}$ (cal deg^{-1} mol^{-1})	Ref.
Chymotrypsin	methyl-L-β-phenyl-lactate	25·0	7·8	1·38	11·1	−23·4	[a]
Chymotrypsin	N-acetyl-L-tyrosine ethyl ester	25·0	8·7	0·0683	10·9	−13·4	[b]
Chymotrypsin	benzoyl-L-tyrosine ethyl ester	25·0	7·8	78·0	9·2	−21·4	[c]
Chymotrypsin	benzoyl-L-phenyllanine ethyl ester	25·0	7·8	37·4	12·5	−11·0	[a]
Chymotrypsin	N-trans-cinnamoyl imidazole	25·0	8·7	0·0125	11·8	−29·6	[b]
Alkaline phosphatase (E. coli)	p-nitrophenyl phosphate	25·0	8·5	28·0	9·4	−22·8	[d]
Papain	furylacryloylimidazole	25·0	7·0	0·02	14·9	−22·7	[e]

[a] J. E. Snoke and H. Neurath, *J. biol. Chem.*, **182**, 577 (1950).
[b] M. L. Bender, F. J. Kézdy, and C. R. Gunter, *J. Am. chem. Soc.*, **86**, 3714 (1964).
[c] S. Kaufman, H. Neurath, and G. W. Schwert, *J. biol. Chem.*, **177**, 792 (1949).
[d] C. Lazdunski and M. Lazdunski, *Biochim. biophys. Acta*, **113**, 551 (1966).
[e] P. M. Hinkle and J. F. Kirsch, *Biochemistry*, **9**, 4633 (1970).

folded molecule. A process of a different kind could also occur during complex formation, and give rise to an entropy increase. When the fully extended (β) form of a protein molecule becomes converted into the less extended (α) form, without the formation or breaking of bonds, there is an increase of entropy;[1] such a process may conveniently be referred to as a 'rubber-like coiling', since a similar change occurs during the shortening of stretched rubber. A structural change of this type has in fact been considered by Morales and Botts[2] for the myosin–ATP system. It is supposed that myosin is normally maintained in an elongated form by charges on the protein molecule, and that when these charges are neutralized on complex formation the myosin contracts with an increase of entropy. Such an interpretation is supported by the kinetic results for this system.

The way in which entropies of activation may be interpreted in terms of solvent effect is as follows (see Chapter 2). During the course of reaction there may be polarity changes that will result in either an increase or a decrease in solvent binding. If charges are formed during a reaction, for example, the solvent will be bound more firmly, and there will be a loss of entropy. If, on the other hand, charges are neutralized during the reaction there will be a release of solvent molecules and a corresponding gain of entropy.

The values in Tables 7.3 and 7.4 for $\Delta S_2^{\neq}$ and $\Delta S_3^{\neq}$ are all negative, and could therefore be explained on the hypothesis that there are increases in polarity during the processes $EA \rightarrow EA' + X$ and $EA' \rightarrow E + Y$. Such increases are not unreasonable; in reactions involving solvent molecules such as water it is commonly found that there are increases in polarity when the activated complex is formed[3].

Further comments on the significance of entropies of activation will be made in the section on solvent effects.

Interaction between enzyme and substrate

There are only a few cases for which the kinetic constant k_1 for the bi-molecular interaction between enzyme and substrate has been obtained at more than one temperature. This was done for catalase and peroxides, for which Chance has shown, using rapid-flow techniques, that the overall rate constants correspond to k_1 for the initial step. Using a rapid titration method Bonnichsen, Chance, and Theorell[4] have determined rate constants at two temperatures; their results for catalase are summarized in Table 7.5. Rates at two temperatures do not, of course, permit a test of the Arrhenius law,

[1] J. Frenkel, *Kinetic Theory of Liquids*, Clarendon Press, Oxford (1946), p. 485.

[2] M. F. Morales and J. Botts, *Archs Biochem. Biophys.*, **37**, 283 (1952); see also L. Ouellet, K. J. Laidler, and M. F. Morales, *Archs Biochem. Biophys.*, **39**, 37 (1952); K. J. Laidler and M. Ethier, *Archs Biochem. Biophys.*, **44**, 338 (1953); K. J. Laidler and A. J. Beardell, *Archs Biochem. Biophys.*, **55**, 138 (1955).

[3] See, for example, K. J. Laidler and P. A. Landskroener, *Trans. Faraday Soc.*, **52**, 200 (1956); K. J. Laidler and D. T. Y. Chen, *Trans. Faraday Soc.*, **54**, 1026 (1958).

[4] R. K. Bonnichsen, B. Chance, and H. Theorell, *Acta chem., scand.*, **1**, 685 (1947).

TABLE 7.5

Activation energy data for the decomposition of hydrogen peroxide catalysed by catalase

Type of catalase	T_2 (°C)	T_1 (°C)	$k_1 \times 10^{-7}$ $M^{-1} s^{-1}$ T_2	T_1	E_1 (kcal mol^{-1})
Horse blood	25·5	1·5	3·50	2·66	1·9
Horse blood	23·5	7·2	3·50	3·11	1·5
Horse blood	22·0	2·0	3·54	2·84	1·8
Horse blood	23·5	3·0	3·50	2·86	1·8
Horse liver	22·0	2·0	3·00	2·55	1·3

but if the validity of the law is assumed, the figures in the last column can be calculated. For horse-blood catalase, the average value of E is 1·7 kcal, and the corresponding value of $\Delta H_1^{\ddagger}$, equal to $E - RT$, is 1·1 kcal. This is an unusually low activation energy for any reaction and, as shown in Table 7.6, is much lower than for the uncatalysed hydrogen peroxide decomposition and for the reaction with other catalysts. The entropies of activation are about -20 cal deg^{-1} mol^{-1}.

TABLE 7.6

Comparison of activation energies for hydrogen peroxide decompositions

Catalyst	E (kcal)	Ref.
Horse-blood catalase	1·7	[a]
Platinum	11·7	[b]
Fe^{2+}	10·1	[c]
I^-	13·5	[d]
No catalyst	17–18	[e]

[a] R. K. Bonnichsen, B. Chance, and H. Theorell, *Acta chem. scand.*, **1**, 685 (1947).
[b] G. Bredig and R. M. von Berneck, *Z. phys. Chem.*, **31**, 258 (1899).
[c] J. H. Baxendale, M. G. Evans, and G. S. Park, *Trans. Faraday Soc.*, **42**, 155 (1946).
[d] J. H. Walton, *Z. phys. Chem.*, **47**, 185 (1904).
[e] J. Williams, *Trans. Faraday Soc.*, **24**, 245 (1928); C. Pana, *Trans. Faraday Soc.*, **24**, 486.

The corresponding results for the formation of the peroxidase–hydrogen peroxide complex are shown in Table 7.7. There is apparently no appreciable effect of temperature, and the activation energy is therefore zero. The heat of activation in this case is $-0·6$ kcal, and the entropy of activation is about -28 cal deg^{-1} mol^{-1}.

One point of significance with regard to these results is that in spite of the high rate constants the reactions cannot be diffusion controlled; this follows

TABLE 7.7

Effect of temperature on k_1 for the formation of the peroxidase–hydrogen peroxide complex

Temperature (°C)	$k_1 \times 10^{-6}$ ($M^{-1} s^{-1}$)
3	8
25	9
42	9

from the very low activation energies, which are much lower than those corresponding to diffusion in aqueous solution. Because of the low entropies of activation the rates are not great enough for diffusion to be rate controlling.

A second point relates to the magnitudes of the entropies of activation. The large negative values are consistent with an increase in polarity during reaction, or with some tightening of the enzyme structure.

The steady-state treatment of the double-intermediate mechanism, discussed in Chapter 2 (see eqn (29), p. 79), leads to the conclusion that the limiting rate at low substrate concentrations is

$$v_0 = \frac{k_1 k_2}{k_{-1} + k_2} [E]_0 \cdot [A]. \tag{14}$$

The second-order rate constant k_0, equal to $k_1 k_2/(k_{-1} + k_2)$, can readily be determined from a Lineweaver–Burk plot if $[E]_0$ is known. Since this rate constant k_0 is composite it is not necessarily the case that the Arrhenius law will apply to it directly. There are, however, two special cases under which this composite constant will obey the law, as follows:

(a) When $k_2 \gg k_{-1}$. If this is the case k_{-1} may be neglected in comparison with k_2, so that k_0 is equal to k_1. This means that the rate constant for the overall reaction at low substrate concentrations is simply that for the first step, the formation of the complex. If that is so the Arrhenius law should certainly apply to k_0, and the corresponding activation energy will be E_1, that for the initial complex formation.

(b) When $k_{-1} \gg k_2$. This case corresponds to the original assumption of Michaelis and Menten. The constant k_0 is now equal to $k_2 k_1/k_{-1}$. The ratio k_1/k_{-1} is simply the equilibrium constant for the complex formation:

$$E + A \underset{k_{-1}}{\overset{k_1}{\rightleftharpoons}} EA.$$

Its variation with temperature is given by

$$\frac{k_1}{k_{-1}} = \frac{A_1 e^{-E_1/RT}}{A_{-1} e^{-E_{-1}/RT}} = \frac{A_1}{A_{-1}} e^{-\Delta E/RT}, \tag{15}$$

where E_1 and E_{-1} are the activation energies for the forward and reverse reactions, A_1 and A_{-1} are constants which are practically temperature-independent, and ΔE is the increase in energy per mole for the change from E + A to EA. Since k_2 varies exponentially with temperature according to the Arrhenius law,

$$k_2 = A_2 \, e^{-E_2/RT}, \tag{16}$$

it follows that

$$\frac{k_2 k_1}{k_{-1}} = \frac{A_2 A_1}{A_{-1}} e^{-(E_2 + E_1 - E_{-1})/RT} \tag{17}$$

$$= \frac{A_2 A_1}{A_{-1}} e^{-(E_2 + \Delta E)/RT}. \tag{18}$$

In other words, the Arrhenius law should apply to this composite constant $k_1 k_2 / k_{-1}$, but the activation energy does not correspond to a single elementary step; it is the sum of the activation energy for the second step (EA → EA′ + X) and the total energy increase for the first step (E + A → EA).

These relationships are illustrated schematically by the energy diagrams shown in Fig. 7.4. In case (a), which corresponds to $k_2 \gg k_{-1}$ and $k_0 = k_1$,

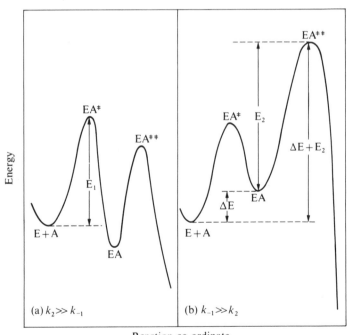

FIG. 7.4.   Schematic energy-diagrams for enzyme systems.

the highest energy barrier over which the system must pass corresponds to $EA^{\neq}$, the activated complex for the reaction $E + A \rightarrow EA$. The measured activation energy corresponding to k_0, the overall rate constant at low substrate concentrations, is thus E_1, corresponding to this initial step. In case (b), on the other hand, the highest energy barrier corresponds to $EA^{\neq\neq}$, the activated complex for the second step, $EA \rightarrow EA' + X$. In order to reach this level the system must first reach the level of EA, where the energy is ΔE higher than that of $E + A$, and then acquire an additional E_2; the activation energy corresponding to k_0 is thus $\Delta E + E_2$. In this case the overall rate is in no way dependent on the rate of the initial step, so that the data can give no information about k_1 or E_1; these quantities can then only be obtained by the use of special techniques such as those of transient-phase investigations.

It follows from what has been said that if either of the conditions $k_2 \gg k_{-1}$ or $k_{-1} \gg k_2$ is satisfied, the Arrhenius law will apply to the composite constant k_0, and that in either case the resultant activation energy will correspond to the increase in energy in going from the initial state $E + A$ to one or other of the activated states $EA^{\neq}$ or $EA^{\neq\neq}$.

When the situation corresponds to neither of these special cases the plots of $\log_{10} k_0$ against $1/T$ will not necessarily be linear, but energies of activation can still be calculated from the slope at any one temperature. The resultant value is related to the individual values by the equation:

$$E_0 = \frac{k_{-1}(E_1 + E_2 - E_{-1}) + k_2 E_1}{k_{-1} + k_2}. \tag{19}$$

The proof of this is as follows. The activation energy E at any temperature is defined by

$$E_0 = RT^2 \frac{\mathrm{d} \ln k_0}{\mathrm{d}T}. \tag{20}$$

By eqn (14)

$$\ln k_0 = \ln k_1 + \ln k_2 - \ln (k_{-1} + k_2) \tag{21}$$

so that

$$\frac{E}{RT^2} = \frac{\mathrm{d} \ln k_0}{\mathrm{d}T} = \frac{\mathrm{d} \ln k_1}{\mathrm{d}T} + \frac{1}{k_2} \frac{\mathrm{d}k_2}{\mathrm{d}T} - \frac{1}{k_{-1} + k_2} \frac{\mathrm{d}(k_{-1} + k_2)}{\mathrm{d}T} \tag{22}$$

$$= \frac{\mathrm{d} \ln k_1}{\mathrm{d}T} + \frac{1}{k_2} \frac{\mathrm{d}k_2}{\mathrm{d}T} - \frac{1}{k_{-1} + k_2} \left(\frac{\mathrm{d}k_{-1}}{\mathrm{d}T} + \frac{\mathrm{d}k_2}{\mathrm{d}T} \right) \tag{23}$$

$$= \frac{\mathrm{d} \ln k_1}{\mathrm{d}T} + \frac{k_{-1}}{k_{-1} + k_2} \frac{\mathrm{d} \ln k_2}{\mathrm{d}T} - \frac{k_{-1}}{k_{-1} + k_2} \frac{\mathrm{d} \ln k_1}{\mathrm{d}T}. \tag{24}$$

The individual activation energies are defined by

$$E_1 = RT^2\frac{\mathrm{d}\ln k_1}{\mathrm{d}T}, \qquad E_{-1} = RT^2\frac{\mathrm{d}\ln k_{-1}}{\mathrm{d}T}, \qquad E_2 = RT^2\frac{\mathrm{d}\ln k_2}{\mathrm{d}T}. \qquad (25)$$

It then follows from (24) and (25), with a little rearrangement, that

$$E_0 = \frac{k_{-1}(E_1 + E_2 - E_{-1}) + k_2 E_1}{k_{-1} + k_2}. \qquad (26)$$

The overall activation energy E_0 is thus the weighted mean of the values E_1 and $E_1 + E_2 - E_{-1}$ for the extreme cases (a) and (b), the weighting factors being

$$\frac{k_2}{k_{-1} + k_2} \quad \text{and} \quad \frac{k_{-1}}{k_{-1} + k_2}.$$

Values of E_0 and $\Delta S_0^{\neq}$ corresponding to k_0 are shown in Table 7.8. It is to be seen that the entropy of activation values fall into two main groups[†]. Thus with pepsin, trypsin, and myosin the values are positive, while with chymotrypsin, carboxypeptidase, and urease they are all negative. Some light is thrown on these differences by a consideration of the types of substrates that are hydrolysed by these six enzymes. Chymotrypsin and urease invariably act upon electrically neutral molecules. Adenosine triphosphatase, on the other hand, is itself positively charged, and it acts upon a negatively charged substrate. Pepsin will only act upon substrates that contain a free $-CO_2H$ group, and at pH 4 this exists at least in part as a negatively charged group. Carboxypeptidase also acts only on substrates containing the $-CO_2H$ group, which will be ionized. The specificity requirement of trypsin is for a free $-NH_2$ group, and this will exist as the positively charged $-NH_3^+$ group. It is thus likely that with ATP-ase, pepsin, carboxypeptidase, and trypsin the interaction between enzyme and substrate involves a charge neutralization, whereas with chymotrypsin and urease there can be no such neutralization.

During the reaction between an enzyme and an uncharged substrate there will result certain electron shifts as a result of which the activated complex will be more polar than the reactants; there is therefore an increase in electrostriction and a corresponding negative entropy of activation. With ATP-ase, pepsin, and trypsin, on the other hand, this effect is presumably counteracted by the charge neutralization that occurs when the enzyme and substrate come together. This neutralization will lead to a release of water molecules, and there will be a corresponding increase of entropy. This explanation of the signs of the $\Delta S^{\neq}$ values derives some support from

[†] It may be noted that in any bimolecular reaction in solution there is always a small entropy contribution resulting from the fact that a solute species disappears when the activated complex is formed; there is therefore an entropy of the 'unmixing' of this with the solvent. For an aqueous solution the entropy loss due to this is -7.9 e.u.

TABLE 7.8

Kinetic values relating to the formation of the enzyme–substrate complex

Enzyme	Substrate	Temperature (°C)	pH	$k°$ (M^{-1} s^{-1})	E_0	$\Delta S_0^{\neq}$	Ref.
Pepsin	carbobenzoxy-L-glutamyl-L-tyrosine ethyl ester	31·6	4·0	0·57	23·1	14·1	[a]
Pepsin	carbobenzoxy-L-glutamyl-L-tyrosine	31·6	4·0	0·79	20·2	2·6	[a]
Trypsin	chymotrypsinogen	19·6	7·5	2900	16·3	8·5	[b]
Chymotrypsin	methyl hydrocinnamate	25·0	7·8	6·66	11·5	−23·2	[c]
Chymotrypsin	methyl-DL-α-chloro-β-phenylpropionate	25·0	7·8	11·2	6·9	−33·0	[c]
Chymotrypsin	methyl-D-β-phenyllactate	25·0	7·8	4·0	3·1	−47·2	[c]
Chymotrypsin	methyl-L-β-phenyllactate	25·0	7·8	138	3·8	−38·5	[c]
Chymotrypsin	benzoyl-L-tyrosine ethyl ester	25·0	7·8	19 500	0·8	−38·5	[d]
Chymotrypsin	benzoyl-L-tyrosinamide	25·0	7·8	14·9	3·7	−43·0	[d]
Carboxypeptidase	carbobenzoxyglycyl-L-tryptophan	25·0	7·5	17 444	9·9	−8·5	[c]
Carboxypeptidase	carbobenzoxyglycyl-L-phenylalanine	25·0	7·5	27 900	9·6	−8·5	[c]
Carboxypeptidase	carbobenzoxyglycyl-L-leucine	25·0	7·5	3920	11·0	−7·5	[e]
Urease	urea	20·8	7·1	$5·0 \times 10^6$	6·8	−6·8	[f]
Adenosine-triphosphatase (myosin)	adenosine triphosphate	25·0	7·0	$8·2 \times 10^6$	21·0	44·0	[g]

[a] E. J. Casey and K. J. Laidler, *J. Am. chem. Soc.*, **72**, 2159 (1950).
[b] J. A. V. Butler, *J. Am. chem. Soc.*, **63**, 2971 (1941).
[c] J. E. Snoke and H. Neurath, *J. biol. Chem.*, **182**, 577 (1950); *Archs Biochem.*, **21**, 351 (1949).
[d] S. Kaufman, H. Neurath, and G. W. Schwert, *J. biol. Chem.*, **177**, 792 (1949).
[e] R. Lumry, E. I. Smith, and R. R. Glantz, *J. Am. chem. Soc.*, **73**, 4330 (1951).
[f] M. C. Wall and K. J. Laidler, *Archs Biochem. Biophys.*, **43**, 299 (1953).
[g] L. Ouellet, K. J. Laidler, and M. F. Morales, *Archs Biochem. Biophys.*, **39**, 37 (1952).

the general correlation between the sign of the entropy change and whether the substrate is charged or not. Carboxypeptidase, however, appears to be an exception, in that the substrates are charged but the entropies of activation are negative; a possible explanation is that the $-CO_2^-$ group on the substrate does not come into contact with a positive group on the enzyme, and is therefore not neutralized. The ionic strength effects for this enzyme do in fact tend to suggest that there is an approach of *like* (negative) charges.

Some of the entropy changes shown in Table 7.8 seem to be too large to be explained in terms of electrostatic effects alone†. For chymotrypsin, where there is no charge neutralization, the entropies of activation are much more negative than can be accounted for on the basis of 'unmixing' with solvent (see footnote on p. 214). With ATP-ase there is seen to be a large positive entropy change and here the results indicate an important structural effect[1].

Influence of dielectric constant (ε_r)

In view of this uncertainty in the interpretation of the entropies of activation for enzyme systems, it is useful to have a method of separating the solvent and structural effects from one another. An attempt to do this in a simple manner has been made by Barnard and Laidler[2] with reference to some proteolytic enzyme systems, and has been also employed on the myosin–ATP system[3]. The method involves making a study of the effect of the dielectric constant on the magnitudes of k_c and k_0.

The theory of the method is as follows. If a process involves a movement of charges‡ $z_A e$ and $z_B e$ from an initial separation of r_i to a final separation of r_f the increase in electrostatic free energy per mole is given by

$$\Delta G_{\text{e.s.}} = \frac{z_A z_B e^2 N}{\varepsilon_r}\left(\frac{1}{r_f} - \frac{1}{r_i}\right). \tag{27}$$

This equation can be applied to the two separate processes, (1) the bimolecular interaction between the enzyme and the substrate, and (2) the subsequent breakdown of the complex.

Case 1. Interaction between enzyme and substrate. The electrostatic contribution to the free-energy increase in forming $EA^{\neq}$ or $EA^{\neq\neq}$ from $E + A$ is given by

$$\Delta G_{\text{e.s.}}^{\neq} = \frac{z_A z_B e^2 N}{\varepsilon_r r^{\neq}} \tag{28}$$

† As a rough rule it can be said that the entropy change due to neutralization on complex formation is equal to $-10 z_A z_B$, where z_A and z_B are the valencies of the ions.

‡ The values of z_A and z_B are positive or negative according as the charges are positive or negative.

[1] J. J. Blum and M. F. Morales, *Archs Biochem. Biophys.*, **43**, 208 (1953).
[2] M. L. Barnard and K. J. Laidler, *J. Am. chem. Soc.*, **74**, 6099 (1952).
[3] K. J. Laidler and M. C. Ethier, *Archs Biochem. Biophys.*, **44**, 338 (1953).

since r_i in eqn (27) is now infinity and r_f is $r^{\neq}$, the separation between the charges in the activated complex. The total free-energy increase $\Delta G^{\neq}$ in forming the activated complex is $\Delta G^{\neq}_{\text{e.s.}} + \Delta G^{\neq}_{\text{n.e.s.}}$, where $\Delta G^{\neq}_{\text{n.e.s.}}$ is the nonelectrostatic contribution. The equilibrium constant $K^{\neq}$ for the formation of the activated complex is thus given by

$$\ln K^{\neq} = -\frac{\Delta G^{\neq}}{RT} = -\frac{\Delta G^{\neq}_{\text{n.e.s.}}}{RT} - \frac{z_A z_B e^2}{\varepsilon_r k T r^{\neq}}. \tag{29}$$

The rate constant k_0 for this process is proportional to this equilibrium constant $K^{\neq}$ so that

$$\ln k_0 = \text{const.} - \frac{\Delta G^{\neq}_{\text{n.e.s.}}}{RT} - \frac{z_A z_B e^2}{\varepsilon_r k T r^{\neq}}. \tag{30}$$

This may be written as

$$\ln k_0 = \ln k_0^0 - \frac{z_A z_B e^2}{\varepsilon_r k T r^{\neq}} \tag{31}$$

where k_0^0 is the value of k_0 when ε is equal to infinity. Equation (31) may be written as

$$\ln k_0 = \ln k_0^0 + \frac{A}{\varepsilon_r T}, \tag{32}$$

where A is equal to $-z_A z_B e^2/k r^{\neq}$.

From the relationship between entropy and equilibrium constant it is found that the electrostatic contribution to the entropy of activation is given by

$$(\Delta S_0^{\neq})_{\text{e.s.}} = R\left(\frac{\partial(T \ln K^{\neq})}{T}\right)_p \tag{33}$$

$$= \frac{z_A z_B e^2 N}{r^{\neq} \varepsilon_r^2}\left(\frac{\partial \varepsilon}{\partial T}\right)_p \tag{34}$$

$$= -\frac{RA}{\varepsilon_r^2}\left(\frac{\partial \varepsilon_r}{\partial T}\right)_p. \tag{35}$$

In aqueous solution ε_r is about 80 and $\partial \ln \varepsilon_r/\partial T$ is about -0.0046; hence

$$(\Delta S_0^{\neq})_{\text{e.s.}} \approx 1.13 \times 10^{-4} A, \tag{36}$$

or

$$(\Delta S_0^{\neq})_{\text{e.s.}} \approx 2.0 \times 10^{-7} z_A z_B/r^{\neq}. \tag{37}$$

A can be obtained by plotting $\ln k_0$ against $1/\varepsilon_r$ and using eqn (32); $(\Delta S_0^{\neq})_{\text{e.s.}}$ can then be calculated using eqn (36) and $z_A z_B/r^{\neq}$ calculated using eqn (37).

Case 2. Subsequent reaction of the enzyme–substrate complex. In this case r_i is equal to r_c, the separation of the charges in the enzyme–substrate complex, and r_f is equal to $r_{\neq}$, the separation in the activated state. The equations that are now obtained are

$$(\Delta G_c^{\neq})_{\text{e.s.}} = \frac{z_A z_B e^2 N}{\varepsilon_r} \left(\frac{1}{r_{\neq}} - \frac{1}{r_c} \right) \tag{38}$$

$$\ln k_c = \ln k_c^0 + \frac{A'}{\varepsilon_r T}, \tag{39}$$

$$(\Delta S_c^{\neq})_{\text{e.s.}} = -\frac{RA'}{\varepsilon_r^2} \left(\frac{\partial \varepsilon_r}{\partial T} \right)_p, \tag{40}$$

$$(\Delta S_c^{\neq})_{\text{e.s.}} = 1\cdot13 \times 10^{-4} A^{\neq}, \tag{41}$$

$$(\Delta S_c^{\neq})_{\text{e.s.}} = 2\cdot0 \times 10^{-7} z_A z_B \left(\frac{1}{r_{\neq}} - \frac{1}{r_c} \right). \tag{42}$$

A' is now given by

$$A' = -\frac{z_A z_B e^2}{k} \left(\frac{1}{r_{\neq}} - \frac{1}{r_c} \right). \tag{43}$$

Application to enzyme data

This treatment has been applied to a number of enzyme systems, rates being measured at different substrate concentrations and in mixed solvents. Methanol–water mixtures have been used for the most part, since in low concentrations methanol has no specific effect on proteins. Mixtures of up to 25 per cent methanol have been found to be satisfactory in the systems investigated, but at higher concentrations there is inactivation of the enzyme.

The procedure is to determine, by the usual extrapolation procedures, the values of k_0 and k_c at various dielectric constants. The logarithms of k_0 and of k_c are then plotted against $1/\varepsilon_r$ and the value of A or A' determined. The electrostatic entropies $\Delta S_{\text{e.s.}}^{\neq}$ and $(\Delta S_c^{\neq})_{\text{e.s.}}$ are then obtained, and the non-electrostatic entropies obtained by subtraction. An example of a plot of $\log k_c$ against $1/\varepsilon_r$ is shown in Fig. 7.5, in which it is seen that the points lie satisfactorily on a straight line.

The procedure has been applied to some reactions of α-chymotrypsin and to the myosin–ATP system. The results for the hydrolysis of methyl hydrocinnamate[1] are as follows, the overall entropies being those of Snoke and Neurath[2]:

Complex formation:

$$\Delta S_0^{\neq} = -23 \qquad (\Delta S_0^{\neq})_{\text{e.s.}} = -38 \qquad (\Delta S_0^{\neq})_{\text{n.e.s.}} = 15.$$

[1] M. L. Barnard and K. J. Laidler, *J. Am. chem. Soc.*, **74**, 6099 (1952).
[2] J. E. Snoke and H. Neurath, *Archs Biochem.*, **31**, 351 (1949); *J. biol. Chem.*, **182**, 577 (1950).

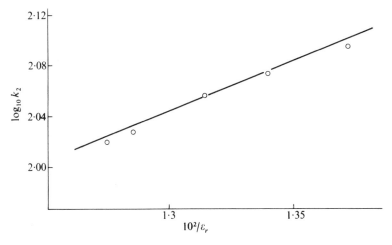

FIG. 7.5. Plot of $\log_{10} k_2$ against $1/\varepsilon_r$ for the myosin-catalysed hydrolysis of ATP (K. J. Laidler and M. C. Ethier, *Archs Biochem. Biophys.*, **44**, 338 (1953)).

Subsequent reaction of complex:

$$\Delta S_c^{\neq} = -12 \qquad (\Delta S_c^{\neq})_{\text{e.s.}} = -20 \qquad (\Delta S_c^{\neq})_{\text{n.e.s.}} = 8.$$

The value of -38 e.u. for $(\Delta S^{\neq})_{\text{e.s.}}$ is of special interest in suggesting that there is considerable charge separation during the formation of the activated complex, and the value of -20 for $(\Delta S_c^{\neq})_{\text{e.s.}}$ is a normal one for a unimolecular reaction, but the value of 15 e.u. for $(\Delta S^{\neq})_{\text{n.e.s.}}$ implies some structural change (perhaps unfolding) during complex formation.

Data are also available in the literature for making a similar, but less complete, analysis of the α-chymotrypsin-catalysed hydrolysis of benzoyl-L-tyrosine ethyl ester, a substrate that is also uncharged. The data are given by Kaufman, Neurath, and Schwert[1] in the form of k_0 at three different methanol concentrations. The results obtained are

$$\Delta S_0^{\neq} = -38 \qquad (\Delta S_0^{\neq})_{\text{e.s.}} = -27 \qquad (\Delta S_0^{\neq})_{\text{n.e.s.}} = -11.$$

The value of -11 is reasonable for bimolecular interaction, and it is not necessary to suggest structural change. The negative electrostatic term again implies charge separation.

The hydrolysis of acetyl-L-tyrosinamide has also been studied at different methanol concentrations[2], and analysis of the results gives -30 e.u. for $(\Delta S_0^{\neq})_{\text{e.s.}}$ and 0 e.u. for $(\Delta S_c^{\neq})_{\text{e.s.}}$. For the closely analogous substrate benzoyl-L-tyrosinamide the entropy values[3] are $\Delta S^{\neq} = -43.0$ and $\Delta S_c^{\neq} = -13.0$. If these values can be accepted tentatively for the acetyl compound the

---

[1] S. Kaufman, H. Neurath, and G. W. Schwert, *J. biol. Chem.*, **177**, 793 (1949).
[2] S. Kaufman and H. Neurath, *J. biol. Chem.*, **180**, 181 (1949).
[3] S. Kaufman, H. Neurath, and G. W. Schwert, *loc. cit.*

non-electrostatic contributions are found to be $(\Delta S^{\neq})_{\text{n.e.s.}} = 13$ and $(\Delta S_c^{\neq})_{\text{n.e.s.}} = -13$. The value of -30 for $(\Delta S^{\neq})_{\text{e.s.}}$ suggests that, as with hydrocinnamic ester, there is considerable charge separation during complex formation; this is rather to be expected for an uncharged substrate. There appears to be little further separation during the reaction of the complex. The non-electrostatic entropy values indicate some structural change in this system, the enzyme unfolding during the complex formation and folding during the subsequent process.

The work on the myosin–ATP system[1] was done at 0·6 M potassium chloride and 0·001 M calcium chloride. The results are:

$$\Delta S_0^{\neq} = 41 \qquad (\Delta S_0^{\neq})_{\text{e.s.}} = 31 \qquad (\Delta S_0^{\neq})_{\text{n.e.s.}} = 10.$$

Subsequent breakdown of complex:

$$\Delta S_c^{\neq} = -8 \qquad (\Delta S_c^{\neq})_{\text{e.s.}} = 6 \qquad (\Delta S_c^{\neq})_{\text{n.e.s}} = -14.$$

The value of 31 e.u. for the formation of the enzyme–substrate complex suggests that positive charges on the enzyme molecule are being neutralized by the negative charges on the ATP; this is consistent with other evidence. The value of 6 for $(\Delta S_c^{\neq})_{\text{e.s.}}$ is consistent with there being a separation of *like* charges; this is presumably to be correlated with the fact that the hydrolysis of ATP involves a splitting of the negatively charged terminal phosphate group from the remainder of the molecule, which is also negatively charged.

The values of 10 and -14 for $\Delta S_{\text{n.e.s.}}^{\neq}$ and $(\Delta S_c^{\neq})_{\text{n.e.s.}}$ are consistent with structural changes occurring during the course of reaction.

Although this method of varying the dielectric constant has led to some interesting conclusions, it must be admitted that it is not as satisfactory as one would have wished[2]. A considerable source of uncertainty arises from the simplification of regarding the solvent as a continuous dielectric. A second difficulty is that specific effects may result from the use of mixed solvents; methanol molecules may, for example, undergo some interaction with part of the enzyme molecule. In view of this it is probably best to regard the results of the method as only semi-quantitative, and to seek confirmation from other sources.

Influence of hydrostatic pressure[3]

Extremely valuable information about the details of chemical reactions in solution is provided by measurement of their rates over a range of hydrostatic pressures. Such investigations provide at least as much insight into mechanisms as do temperature studies, and there is need for much more work in this field. Even among non-biological reactions only a very few

[1] K. J. Laidler and M. C. Ethier, *Archs Biochem. Biophys.*, **44**, 338 (1953).
[2] See J. M. Sturtevant, *Discuss. Faraday Soc.*, **20**, 254 (1955); D. J. R. Lawrence, *Discuss. Faraday Soc.*, **20**, 255 (1955).
[3] For a general review see S. D. Hamann, *Physico-chemical Effects of Pressure*, Chap. 9, Butterworth Scientific Publications, London, 1957.

systems have been studied at higher pressures. The present account outlines the general theory of pressure effects, describes the main conclusions drawn from studies of non-biological systems, and summarizes and interprets some of the results obtained for enzyme reactions.

The theory of hydrostatic-pressure effects on reaction rates was first formulated in 1901 by van't Hoff; his treatment will be given here in terms of more modern concepts. The theory is based on the equation for the effect of pressure on the equilibrium constant. The equilibrium constant for a reaction such as

$$A + B \underset{k_{-1}}{\overset{k_1}{\rightleftharpoons}} X + Y$$

is related to the standard Gibbs free-energy change $\Delta G^{\ominus}$ by the equation

$$\Delta G^{\ominus} = -RT \ln K. \tag{44}$$

The thermodynamical relationship between the volume, the Gibbs free energy G, and the hydrostatic pressure is

$$V = \left(\frac{\partial G}{\partial P}\right)_T \tag{45}$$

so that

$$\Delta V = \left(\frac{\partial \Delta G^{\ominus}}{\partial P}\right)_T. \tag{46}$$

From (44) and (46),

$$\left(\frac{\partial \ln K}{\partial P}\right)_T = -\frac{\Delta V}{RT}. \tag{47}$$

If a reaction occurs with an increase in volume (ΔV positive), the equilibrium constant therefore decreases with increasing pressure; conversely, if ΔV is negative, the equilibrium constant increases with increasing pressure. The proportion of products at equilibrium is therefore decreased or increased by pressure according as the volume change is positive or negative.

In order to obtain a corresponding equation for the variation with pressure of the rate constant, van't Hoff employed an argument which is analogous to the one that he used in connection with temperature effects. The volume change ΔV is equal to the difference between the volume of the products V_p and that of the reactants V_r,

$$\Delta V = V_p - V_r. \tag{48}$$

If $V^{\neq}$ is the volume of the activated complex, this can be written as

$$\Delta V = (V^{\neq} - V_r) - (V^{\neq} - V_p) \tag{49}$$

$$= \Delta V_1^{\neq} - \Delta V_{-1}^{\neq} \tag{50}$$

In this equation $\Delta V_1^{\ddagger}$ is the increase in volume in passing from the initial state to the activated state, and is known as the *volume of activation* for the forward reaction; $\Delta V_{-1}^{\ddagger}$ is the volume of activation for the reverse reaction. Since K is equal to k_1/k_{-1}, eqn (47) may be written as

$$\left(\frac{\partial \ln k_1}{\partial P}\right)_T - \left(\frac{\partial \ln k_{-1}}{\partial P}\right)_T = -\frac{\Delta V}{RT} \tag{51}$$

and together with eqn (50) this gives

$$\left(\frac{\partial \ln k_1}{\partial P}\right)_T - \left(\frac{\partial \ln k_{-1}}{\partial P}\right)_T = -\frac{\Delta V_1^{\ddagger}}{RT} + \frac{\Delta V_{-1}^{\ddagger}}{RT}. \tag{52}$$

The assumption was then made that the behaviour of the forward reaction depends only on the volume change in going from the initial state to the activated state, which means that eqn (52) can be split into the two equations

$$\left(\frac{\partial \ln k_1}{\partial P}\right)_T = -\frac{\Delta V_1^{\ddagger}}{RT} \tag{53}$$

and

$$\left(\frac{\partial \ln k_{-1}}{\partial P}\right)_T = -\frac{\Delta V_{-1}^{\ddagger}}{RT}. \tag{54}$$

These equations can also be derived by making use of the fact that, according to activated-complex theory, the rate constant k is proportional to the equilibrium constant $K^{\ddagger}$ between initial and activated states. The variation of the equilibrium constant $K^{\ddagger}$ with pressure is given by (see eqn (47))

$$\left(\frac{\partial \ln K^{\ddagger}}{\partial P}\right)_T = -\frac{\Delta V^{\ddagger}}{RT} \tag{55}$$

where $\Delta V^{\ddagger}$ is the volume of activation. Since k is proportional to $K^{\ddagger}$ it follows that

$$\left(\frac{\partial \ln k}{\partial P}\right)_T = -\frac{\Delta V^{\ddagger}}{RT}. \tag{56}$$

According to this equation, the rate constant of a reaction increases with increasing pressure if $\Delta V^{\ddagger}$ is negative, i.e., if the activated state has a smaller volume than the initial state. Conversely, pressure has an adverse effect on rates if there is a volume increase when the activated complex is formed. By using eqn (56), values of $\Delta V^{\ddagger}$ can be determined from experimental measurements of rates at different pressures. In practice, it is necessary to use fairly high pressures for this purpose (of the order of thousands of pounds per square inch), since otherwise the changes in rate are too small for accurate $\Delta V^{\ddagger}$ values to be obtained. The procedure involves plotting the logarithm of the rate constant against the pressure; according to eqn (56), the slope at any pressure is equal to $-\Delta V^{\neq}/RT$ (or $-\Delta V^{\neq}/2\cdot303\,RT$ if common logarithms are used). In some cases such plots are straight lines, which means that the volume of activation is independent of the pressure. If this is so, eqn (56) can be integrated to give

$$\ln k = \ln k_0 - \frac{\Delta V^{\neq}}{RT}P \qquad (57)$$

where k_0 is the rate constant at zero pressure (this is always very close to the value at atmospheric pressure). If eqn (57) is obeyed, a plot of $\ln(k/k_0)$ against P will be a straight line through the origin. A number of reactions of widely different types obey eqn (57), so that for them it can be concluded that $\Delta V^{\neq}$ is independent of pressure over the range investigated, which usually covers from atmospheric pressure up to about 20 000 lb per sq in. Table 7.9 shows some values of $\Delta V^{\neq}$ for a number of non-enzymic reactions in various solvents, and also gives the entropies of activation $\Delta S^{\neq}$.

TABLE 7.9
Volumes and entropies of activation[a]

Reaction	Solvent	$\Delta V^{\neq}$ cc mol^{-1}	$\Delta S^{\neq}$ cal deg^{-1} mol^{-1}
$Co(NH_3)_5Br^{2+} + OH^- \rightarrow Co(NH_3)_5OH^{2+} + Br^-$	H_2O	8·5	22
$(CH_3)(C_2H_5)(C_6H_5)(C_6H_5CH_2)N^+Br^- \rightarrow$ $(CH_3)(C_6H_5)(C_6H_5CH_2)N + C_2H_5Br$	H_2O	3·3	15
$CH_2BrCOOCH_3 + S_2O_3^{2-} \rightarrow$ $CH_2(S_2O_3)COOCH_3 + Br^-$	H_2O	3·2	6
Sucrose + $H_2O \xrightarrow{H^+}$ glucose + fructose	H_2O	2·5	8
$C_2H_5O^- + C_2H_5I \rightarrow C_2H_5OC_2H_5 + I^-$	C_2H_5OH	$-4\cdot1$	-10
$CH_2ClCOO^- + OH^- \rightarrow CH_2OHCOO^- + Cl^-$	H_2O	$-6\cdot1$	-12
$CH_2BrCOO^- + S_2O_3^{2-} \rightarrow$ $CH_2(S_2O_3)COO^- + Br^-$	H_2O	$-4\cdot8$	-17
$CH_3COOCH_3 + H_2O \xrightarrow{H^+} CH_3COOH + CH_3OH$	H_2O	$-8\cdot7$	-10
$CH_3CONH_2 + H_2O \xrightarrow{OH^-} CH_3COOH + NH_3$	H_2O	$-14\cdot2$	-34
$C_5H_5N + C_2H_5I \rightarrow C_5H_5(C_2H_5)N^+I^-$	CH_3COCH_3	$-16\cdot8$	-35
$C_6H_5CCl_3 \rightarrow C_6H_5CCl_2^+ + Cl^-$	80% C_2H_5OH	$-14\cdot5$	-35

[a] For references to the original literature see C. T. Burris and K. J. Laidler, *Trans. Faraday Soc.*, **51**, 1497 (1955); D. T. Y. Chen and K. J. Laidler, *Can. J. Chem.*, **37**, 599 (1959).

Significance of volumes of activation

A number of theoretical interpretations of the magnitudes of the volumes of activation, $\Delta V^{\neq}$, have been put forward. It was pointed out in 1938 by Perrin[1] that the reactions studied up to that time fell into three broad classes, and that these classes differ from one another in other kinetic characteristics.

[1] M. W. Perrin, *Trans. Faraday Soc.*, **34**, 144 (1938).

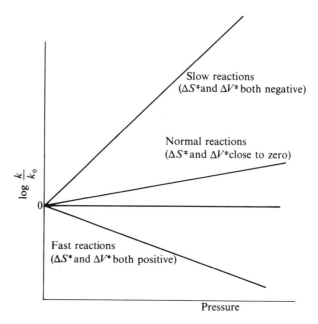

Fig. 7.6. Three classes of reactions, as indicated by the pressure effects.

These classes are illustrated schematically in Fig. 7.6, and in terms of more modern concepts may be described as follows:

1. *'Slow' reactions*, which are bimolecular reactions having abnormally low frequency factors and therefore negative entropies of activation. These reactions are markedly accelerated by pressure, which means, according to eqn (56) that their volumes of activation are negative.

2. *'Normal' reactions*, which are bimolecular reactions whose frequency factors are normal ($\sim 10^{11}$ litres $\text{mol}^{-1}\,\text{s}^{-1}$) so that their entropies of activation are close to zero. Such reactions are usually very slightly accelerated by pressure, which means that they have slightly negative volumes of activation.

3. *'Fast' reactions*, for which the entropies of activation are positive. These reactions are retarded by pressure, so that the $\Delta V^{\neq}$ values are positive.

Perrin's conclusions can all be summarized by the statement that entropies of activation and volumes of activation tend to fall in line with each other. A plot of the volumes of activation against the entropies of activation for a number of reactions in aqueous solution is shown in Fig. 7.7, and there is seen to be a fairly good linear correlation.

Evans and Polanyi[1] pointed out that two distinct effects must be considered in connection with the interpretation of volumes of activation.

[1] M. G. Evans and M. Polanyi, *Trans. Faraday Soc.*, **31**, 875 (1935); **32**, 1333 (1936); M. G. Evans, *ibid.*, **34**, 49 (1938).

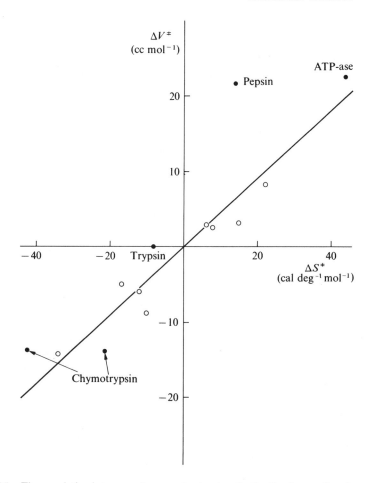

FIG. 7.7.   The correlation between volumes and entropies of activation for reactions in aqueous solution. The open circles are for non-enzymic reactions (cf. K. J. Laidler, *Chemical Kinetics*, McGraw-Hill, New York, 1965, Table 33, p. 234). The filled circles are for enzyme systems (cf. Tables 7.12 and 7.13), but the substrates are unfortunately different in all cases for the $\Delta V^{\ddagger}$ and $\Delta S^{\ddagger}$ values.

In the first place, there may be a change, due to structural factors, in the volume of the reactant molecules themselves as they pass into the activated state; for a bimolecular process this always leads to a volume decrease, while for a unimolecular process there is a volume increase. Secondly, there may be a volume change resulting from reorganization of the solvent molecules. Studies of a variety of reactions[1] have led to the conclusion that for reactions in which ions or fairly strong dipoles are concerned the solvent effects are generally more important than the structural ones.

[1] J. Buchanan and S. D. Hamann, *Trans. Faraday Soc.*, **49**, 1425 (1953); C. T. Burris and K. J. Laidler, *ibid.*, **51**, 1497 (1955).

These effects of solvent on volumes are explained in a manner similar to the effects of solvent on entropies of activation, a fact that explains the correlation between volumes and entropies of activation (Fig. 7.7). Thus if a reaction occurs with the approach of ions of the same sign, or a separation of ions of the opposite sign, there is an intensification of the electric field, and therefore an increase in electrostriction and a resulting decrease in volume; there is also a decrease in entropy owing to the loss of freedom of the solvent molecules. Conversely, if the electric field is weakened when the activated complex is formed (as when two ions of opposite signs come together), there will be some release of bound solvent molecules, and the volumes and entropies of activation will be positive. Reference to the reactions listed in Table 7.9 shows that all the results can be qualitatively explained on this basis. The above conclusions are summarized in Table 7.10, which shows the three main types of reactions.

TABLE 7.10
Summary of pressure effects on reactions

Classification	Ionic character	Examples	Volume of activation	Entropy of activation
'Slow'	Formation of opposite charges, or approach of like charges	Reactions between ions of same sign, ester hydrolyses, esterifications, Menschutkin reactions, unimolecular solvolyses	Large negative	Large negative
'Normal'	Electrostatic effects unimportant	Negative-ion replacements	Small negative	Small negative
'Fast'	Approach of opposite charges, or spreading of charges	Reverse Menschutkin reactions, reactions between ions of opposite sign	Positive	Positive

Whalley[1] has made use of the results of pressure studies in order to arrive at conclusions about reaction mechanisms. He has shown that volumes of activation are more reliable than entropies of activation in suggesting the type of reaction that is taking place. This is the case because entropies of activation are quite sensitive to factors such as the loosening or strengthening of chemical bonds, whereas volumes depend to a much greater extent on electrostriction effects than on any other effects. Volume changes are therefore much more constant for reactions of a given type than are entropies, and therefore lead to more clear-cut conclusions about the processes taking place.

[1] E. Whalley, *Trans. Faraday Soc.*, **55**, 798 (1959).

Pressure and enzyme reactions

In studying the effects of high hydrostatic pressures on the rates of enzyme reactions it is necessary for a number of factors to be borne in mind. In the first place, if the pressures are extremely high (around 100 000 lb/sq. in.) the enzyme may undergo irreversible inactivation even at low temperatures[1]. Secondly, even at moderate pressures (around 10 000 lb/sq. in.) the behaviour under pressure may, if the temperature is high enough, be complicated by the fact that the pressure may affect a reversible equilibrium between active and inactive forms of the enzyme[2]. These effects on the enzyme will be discussed later (p. 445); consideration will first be given to investigations in which the complications involving inactivation have been avoided by working at moderately high pressures (up to 10 000 lb/sq. in.) and at sufficiently low temperatures.

The influence of pressure on enzyme reactions can only be studied satisfactorily by making measurements over a wide range of substrate concentrations, so that the rate constants k_c and k_0 can be determined separately. If the reaction involves a single substrate and two intermediates the rate constant k_c is given by

$$k_c = \frac{k_2 k_3}{k_2 + k_3}.$$ (59)

If step 2 is rate-determining, i.e. if $k_3 \gg k_2$, k_c is equal to k_2, so that the pressure study gives a value for $\Delta V_2^{\neq}$. If step 3 is rate-determining ($k_2 \gg k_3$) the volume of activation that is measured is $\Delta V_3^{\neq}$. By an argument similar to that on p. 213, for the energy of activation, it can be shown that in general the volume of activation measured at high substrate concentrations, $\Delta V_c^{\neq}$, is given by

$$\Delta V_c^{\neq} = \frac{k_3 \Delta V_2^{\neq} + k_2 \Delta V_3^{\neq}}{k_2 + k_3}.$$ (60)

That is, it is the weighted mean of the values $\Delta V_2^{\neq}$ and $\Delta V_3^{\neq}$, the weighting factors being $k_3/(k_2 + k_3)$ and $k_2/(k_2 + k_3)$. The significance of this result is shown in Fig. 7.8.

Similarly, for the rate constant k_0 (equal to $k_1 k_2/(k_{-1} + k_2)$) determined at low substrate concentrations the relationships are

$$k_{-1} \gg k_2 \nearrow \qquad k_0 = \frac{k_1 k_2}{k_{-1}}; \qquad \Delta V_0^{\neq} = \Delta V_1^{\neq} - \Delta V_{-1}^{\neq} + \Delta V_2^{\neq}$$ (61)

$$k_0 \searrow \qquad$$

$$k_2 \gg k_{-1} \qquad k_0 = k_1; \qquad \Delta V_0^{\neq} = \Delta V_1^{\neq}.$$ (62)

[1] J. E. Matthews, R. B. Dow, and A. K. Anderson, *J. biol. Chem.*, **135**, 697 (1940); A. L. Curl and E. F. Jansen, *J. biol. Chem.*, **184**, 45 (1950); **185**, 713 (1950).
[2] H. Eyring and J. L. Magee, *J. cell. comp. Physiol.*, **20**, 169 (1942).

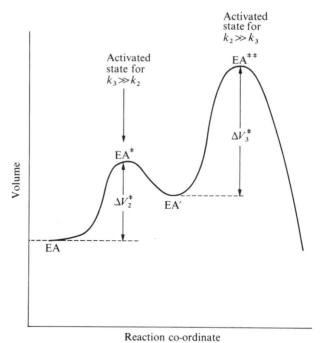

FIG. 7.8. Volume diagram for the pressure results obtained at high substrate concentrations. If $k_3 \gg k_2$, the enzyme exists mainly as EA and the volume of activation is $\Delta V_2^{\ddagger}$; if $k_2 \gg k_3$ the enzyme is present mainly as EA' and the volume of activation is $\Delta V_3^{\ddagger}$.

In general,

$$\Delta V_0^{\ne} = \frac{k_{-1}(\Delta V_1^{\ne} - \Delta V_{-1}^{\ne} + \Delta V_2^{\ne}) + k_2 \Delta V_1^{\ne}}{k_{-1} + k_2}.$$ (63)

The value is now the weighted mean of the values for the extreme cases, the weighting factors being $k_{-1}/(k_{-1} + k_2)$ and $k_2/(k_{-1} + k_2)$. These conclusions are illustrated in Fig. 7.9.

Analysis of experimental results

Unfortunately, the experiments on pressure effects in enzyme systems have rarely been carried out from the standpoint of a theoretical treatment such as the above, and in particular the distinction between the high- and low-concentration regions has not always been appreciated. It is possible, however, to make a fairly satisfactory interpretation of some of the earlier results, as follows.

Brown, Johnson, and Marsland[1] investigated the luciferase–luciferin system, and their data have been analysed by Eyring and Magee[2], who find

[1] D. E. S. Brown, F. H. Johnson, and D. A. Marsland, *J. cell. comp. Physiol.*, **20**, 151 (1942).
[2] H. Eyring and J. L. Magee, *J. cell. comp. Physiol.*, **20**, 169 (1942).

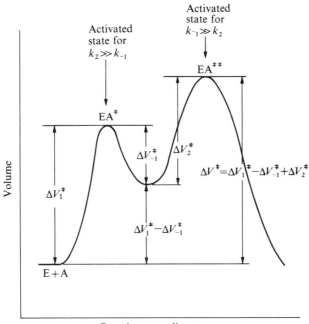

Reaction co-ordinate

Fig. 7.9. Volume diagram for the pressure results obtained at low substrate concentrations.

a volume change of 546·4–1·813T cc. This is equal to 11·6 cc at 22°C and to 51·5 cc at 0°C. Under biological conditions substrate concentrations are generally low, and we may consider tentatively that the values refer to $\Delta V_0^{\ddagger}$.

Eyring, Johnson, and Gensler[1] found that high pressure increased the rate of hydrolysis of sucrose by yeast invertase. Since the substrate concentration was 10 per cent, or 0·29 M, and the Michaelis constant K_A has been given[2] as 0·016 M and as 0·04 M, it follows that in either case $[A] \gg K_A$; the calculated volume change therefore corresponds to $\Delta V_c^{\ddagger}$. At 22°C the data indicate a value of about -8 cc at pH 7·03 to 7·07 and of about -3 cc at pH 4·5.

Benthaus[3] published pressure data for four different reactions, and stated the substrate concentration; the Michaelis constant K_A is known for each reaction or for an analogous one, so that it is possible to decide whether the volume change corresponds to $\Delta V_c^{\ddagger}$ or to $\Delta V_0^{\ddagger}$. Unfortunately, rate constants are not given in the paper, but on the basis of the curves given a rough estimate of v/v_0 can be made in each case. The data are summarized in Table 7.11, the v/v_0 ratios referring to a pressure of 1500 atm; the Michaelis constants are taken from Haldane's book[2]. It is to be seen that in the first

[1] H. Eyring, F. H. Johnson, and R. L. Gensler, *J. phys. Collord Chem.*, **50**, 453 (1946).
[2] J. B. S. Haldane, *Enzymes*, Longmans, Green and Co. (1930), M.I.T. Press (1965).
[3] J. Benthaus, *Biochem. Z.*, **311**, 108 (1941–2).

two cases [A] is greater than K_A; the volume changes calculated therefore correspond approximately to $\Delta V_c^{\neq}$. In the third and fourth reactions [A] is less than K_A, so that the volume changes are $\Delta V_0^{\neq}$. The actual volume values calculated from Benthaus's data are included in columns 6 and 7 of Table 7.11.

TABLE 7.11

Results of Benthaus on the influence of pressure on enzyme-catalysed reactions

Enzyme	Substrate	K_A	[A]	v/v_0	$\Delta V_0^{\neq}$ (cc mol^{-1})	$\Delta V_c^{\neq}$ (cc mol^{-1})
Salivary amylase	starch	0·4, 0·8%	1%	4·0	—	−22
Pancreatic amylase	starch	0·25%	2%	5·5	—	−28
Pancreatic lipase	tributyrin	0·03 M†	very low	0·45	13	—
Pepsin	gelatin	4·5%‡	0·83%	0·25	22	—

† This value actually applies to ethyl butyrate.
‡ This value actually applies to egg albumin.

The effects of pressure on a number of reactions of chymotrypsin[1] and trypsin[2] have been investigated by Werbin and McLaren. The substrates used with chymotrypsin were casein and tyrosine ethyl ester. With casein the order of the reaction with respect to time was the first, and this means that the condition $[A] \ll K_A$ is probably satisfied; the volume change of -14 cc mol^{-1} is therefore probably $\Delta V_0^{\neq}$. With tyrosine ethyl ester, on the other hand, the reaction was zero-order with respect to time, so that the volume change of $-13·5$ cc mol^{-1} is assigned to $\Delta V_c^{\neq}$.

The work with trypsin was done with β-lactoglobulin, α-benzoyl-L-argininamide, α-benzoyl-L-arginine isopropyl ester, and L-arginine methyl ester. The β-lactoglobulin reaction was first-order with respect to time, and the resulting value of -36 cc mol^{-1} is thus $\Delta V_0^{\neq}$. The kinetics of the benzoyl-L-argininamide hydrolysis correspond to a true order of zero in the higher concentration range. Werbin and McLaren's value of 0 cc mol^{-1} (no effect of pressure) must therefore be assigned to $\Delta V_c^{\neq}$. The kinetics of the benzoyl-L-arginine isopropyl ester and L-arginine methyl ester reaction also suggest that $[A] \gg K_A$, so that the resulting values of 0 and $-5·5$ cc mol^{-1}, respectively, are to be regarded as corresponding to $\Delta V_c^{\neq}$.

The effects of pressure on the kinetics of the myosin-catalysed hydrolysis of adenosine triphosphate (ATP) have been investigated by Laidler and Beardell[3]. By working over a range of substrate concentrations as well as hydrostatic pressures it was possible to separate the constants k_0 and k_c and

[1] H. Werbin and A. D. McLaren, *Archs Biochem. Biophys.*, **31**, 285 (1951).
[2] H. Werbin and A. D. McLaren, *Archs Biochem. Biophys.*, **32**, 325 (1951).
[3] K. J. Laidler and A. J. Beardell, *Archs Biochem. Biophys.*, **55**, 138 (1955).

TABLE 7.12
Volume and entropy data for $[A] \ll K_A$

Enzyme	Substrate	$\Delta V_0^{\neq}$	$\Delta S_0^{\neq}$
Pepsin	cbz-L-glutamyl-L-tyrosine ethyl ester	—	14.1
Pepsin	gelatin	22	—
Trypsin	β-lactoglobulin	-36	—
Trypsin	chymotrypsinogen	—	6·5
Chymotrypsin	casein	$-13·8$	—
Chymotrypsin	benzoyl-L-tyrosinamide	—	$-43·0$
Carboxypeptidase	cbz-glycyl-L-phenylalanine	—	$-8·5$
Pancreatic lipase	tributyrin	13	—
Myosin	ATP	8 to 23†	44·0
Luciferase	luciferin	11·6	—

† Depending on KCl concentration.

to determine the influence of pressure on each one. The resulting volumes are $\Delta V_0^{\neq}$ and $\Delta V_c^{\neq}$ respectively.

The results of the various high-pressure investigations are summarized in Tables 7.12 and 7.13. Table 7.12 includes the data for which the condition $[A] \ll K_A$ is satisfied, while Table 7.13 applies to the case of $[A] \gg K_A$. It is unfortunate that for few of the reactions studied at high hydrostatic pressures are there any entropy of activation data. There have, however, been included in Tables 7.12 and 7.13 some entropies of activation for related reactions, and it will be noted that these are generally of the same sign as the volumes of activation for similar systems.

The theoretical interpretation of these volumes of activation is a matter of considerable interest and importance. In a discussion of some of the earlier data emphasis has been placed on the importance of structural changes in

TABLE 7.13
Volume and entropy data for $[A] \gg K_A$

Enzyme	Substrate	$\Delta V_c^{\neq}$	$\Delta S_c^{\neq}$
Pepsin	cbz-L-glutamyl-L-tyrosine	—	$-21·8$
Trypsin	benzoyl-L-argininamide	0	$-8·2$
Trypsin	benzoyl-L-arginine isopropyl ester	0	$-16·5$
Trypsin	L-arginine methyl ester	$-5·5$	—
Chymotrypsin	L-tyrosine ethyl ester	$-13·5$	—
Chymotrypsin	benzoyl-L-tyrosine ethyl ester	—	$-21·4$
Carboxypeptidase	cbz-glycyl-L-phenylalanine	—	$-18·5$
Invertase	sucrose	-8	—
Salivary amylase	starch	-22	—
Pancreatic amylase	starch	-28	—
Myosin	ATP	-32 to -25†	$-8·0$

† Depending on KCl concentration.

enzyme systems[1]. In particular, it has been noted that the data are consistent with the view that enzymes sometimes unfold during the process of complex formation and refold during the subsequent breakdown of the complex. That such effects occur is supported by certain lines of evidence, but on the basis of the more recent work in both biological and non-biological systems it would seem that the earlier explanation was incomplete in that it disregarded the importance of electrostatic effects. The review of the non-biological data given earlier in this chapter did, in particular, show that such electrostatic effects are at least as important as the structural ones.

The entropies of activation have been interpreted earlier in this chapter according to a certain pattern. For example, when enzyme substrate interaction involves a charge neutralization (as with pepsin, trypsin, and myosin) the $\Delta S_0^{\neq}$ values are generally positive, a fact that is attributed to release of bound water molecules. When, on the other hand, there is no charge neutralization (as with urease and chymotrypsin) the data show the expected negative values of $\Delta S_0^{\neq}$; such values may be attributed partly to structural factors and partly to polarity increases. The values of $\Delta S_c^{\neq}$ for enzyme systems are invariably negative, a fact that is consistent with increases in polarity during the process $EA \rightarrow EA^{\neq}$ or $EA^{\neq\neq}$.

Electrostriction factors that give rise to an increase in entropy of activation should also produce an increase in volume, and vice versa. If electrostatic effects are important one would therefore expect that the entropies and volumes of activation would fall somewhat in line with one another. Inspection of Tables 7.12 and 7.13 shows that, with minor exceptions, this is the case. In Table 7.12, for example, the positive $\Delta V_0^{\neq}$ for pepsin and the negative one for chymotrypsin are understandable on the basis of the fact that there is charge neutralization in the former but not in the latter. The large negative volume for the trypsin-β-lactoglobulin reaction is anomalous and possibly, as suggested by Werbin and McLaren, is due to preliminary denaturation of the protein. The $\Delta V_c^{\neq}$ values in Table 7.13 are all negative and support the view that the reaction $EA \rightarrow EA'$ is generally accompanied by an increase in polarity.

A detailed study of the effect of pressure on the activity of ribonuclease has been carried out[2]. The conclusion is that a neutralization of like univalent charges occurs on formation of the enzyme–substrate complex.

[1] K. J. Laidler, *Archs Biochem.*, **30**, 226 (1951).
[2] R. K. Williams and C. Shen, *Archs Biochem. Biophys.*, **152**, 606 (1972).

8. Kinetic Isotope Effects

THE kinetic effect of making an isotopic substitution, such as the replacement of a hydrogen atom by a deuterium atom, can provide valuable information about the nature of the process that is occurring. Thus, for an elementary process in which an H or D atom is being transferred from one molecule to another it is frequently found that the H atom is transferred some 6 to 10 times more rapidly than the D atom. A kinetic isotope effect of this magnitude can hardly be explained unless the H or D atom is directly involved in bond making or breaking during the reaction, so that such an isotope effect provides good evidence for this type of mechanism. However, it has been realized for the past few years that the theory of kinetic isotope effects is considerably more complicated than had previously been supposed, and that great care must be taken in drawing conclusions. Thus, even if H or D transfer is involved in the process the kinetic isotope effect may be very much smaller than indicated above.

The present chapter will give an outline of the basic theory of isotope effects[1] and will indicate how the theory is applicable to enzyme systems. Since the theory of kinetic isotope effects is substantially more complicated than that of isotope effects on equilibrium constants, and is closely related to it, the equilibrium case will be considered first.

Isotope effects on equilibrium constants

Consider the very simple equilibrium

$$H_2 \rightleftharpoons 2H$$

occurring in the gas phase. According to statistical mechanics, the equilibrium constant is given by the ratio of partition functions for products and reactants, multiplied by a Boltzmann term which involves the difference between the zero-point levels of the reactant and product molecules:

$$K_{H_2} = \frac{Q_H^2}{Q_{H_2}} e^{-E_0/RT}. \tag{1}$$

Here the Q's are the partition functions and E_0 is the energy difference shown in Fig. 8.1; its value is 103·2 kcal per mol. The partition function for a hydrogen atom relates only to translational freedom, and involves the mass of the atom m_H:

$$Q_H = \frac{(2\pi m_H kT)^{\frac{3}{2}}}{h^3}. \tag{2}$$

---

[1] For further information on the general theory see L. Melander, *Isotope Effects on Reaction Rates*, Ronald Press Co., New York, 1960; K. J. Laidler, *Theories of Chemical Reaction Rates*, McGraw-Hill Book Co., New York, 1969, pp. 86–98; M. Wolfsberg, *Acct. chem. Res.*, **5**, 225 (1972).

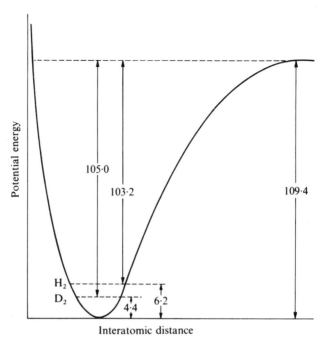

FIG. 8.1. Schematic potential-energy curves for H_2 and D_2, showing the zero-point levels. The energy values are in kcal. mol^{-1}.

That for the hydrogen molecule also has a translational factor, which is now multiplied by a rotational factor involving the moment of inertia I of the molecule, and by a vibrational factor which will be written as q_v:

$$Q_{H_2} = \frac{(2\pi m_{H_2}\mathbf{k}T)^{\frac{3}{2}}}{h^3} \frac{8\pi^2 I \mathbf{k}T}{2h^2} q_v. \tag{3}$$

The moment of inertia I for a homonuclear diatomic molecule is related to the masses as follows:

$$I = \frac{m_H}{2} d_{H_2}^2 \tag{4}$$

where d_{H_2} is the distance between the atoms. In dealing with kinetic isotope effects we are interested only in the dependence on mass of the pre-exponential part of the expression for K_{H_2}, and can write

$$K_{H_2} = \frac{\dfrac{(2\pi m_H \mathbf{k}T)^3}{h^6} e^{-103\,200/RT}}{\dfrac{(2\pi m_{H_2}\mathbf{k}T)^{\frac{3}{2}}}{h^3} \dfrac{8\pi^2 I \mathbf{k}T}{2h^2} q_v} \tag{5}$$

$$\propto \frac{m_{\mathrm{H}}^{3}}{m_{\mathrm{H_2}}^{\frac{3}{2}} m_{\mathrm{H}}} \, \mathrm{e}^{-103\,200/RT} \tag{6}$$

$$\propto m_{\mathrm{H}}^{\frac{1}{2}} \, \mathrm{e}^{-103\,200/RT} \tag{7}$$

since $m_{\mathrm{H_2}} = 2m_{\mathrm{H}}$. The vibrational factor q_{v} is practically independent of mass at ordinary temperatures.

Application of the same treatment to the dissociation of the D_2 molecule

$$D_2 \rightleftharpoons 2D$$

leads to the analogous result

$$K_{\mathrm{D_2}} \propto m_{\mathrm{D}}^{\frac{1}{2}} \, \mathrm{e}^{-105\,000/RT}. \tag{8}$$

The exponential term is now different, the zero-point energy for the D_2 molecule being below that for the H_2 molecule (see Fig. 8.1). The ratio of the equilibrium constants is thus

$$\frac{K_{\mathrm{H_2}}}{K_{\mathrm{D_2}}} = \left(\frac{m_{\mathrm{H}}}{m_{\mathrm{D}}}\right)^{\frac{1}{2}} \mathrm{e}^{1800/RT}. \tag{9}$$

At 300 K this ratio is

$$\frac{K_{\mathrm{H_2}}}{K_{\mathrm{D_2}}} = \left(\frac{1}{2}\right)^{\frac{1}{2}} 10^{1\cdot315} = 0\cdot71 \times 20\cdot7$$

$$= 14\cdot7.$$

The isotope effect is seen to be largely due to the difference between the zero-point energies of H_2 and D_2; H_2 has not as far to go to reach the activated state, and this alone leads to a factor of 20·7 in the equilibrium constant at 300 K.

The isotope effect will be less if only one atom is changed, for example if we compare

$$R{-}H \rightleftharpoons R + H$$

with

$$R{-}D \rightleftharpoons R + D$$

where R might be an organic radical. If R is much heavier than H or D the isotope effect on the preexponential factor is negligible, so that the main effect is due to the difference in zero-point levels. A typical zero-point energy difference for the C—H and C—D bonds is $1\cdot15\ \mathrm{kcal\ mol^{-1}}$, so that

$$\frac{K_{\mathrm{RH}}}{K_{\mathrm{RD}}} \propto \mathrm{e}^{1150/RT} = 6\cdot9 \text{ at } 300\ \mathrm{K}.$$

The above considerations lead at once to an important principle, namely that *substitution with a heavier atom will always favour the formation of the stronger bond.* The point is that the stronger the bond the greater is the vibrational frequency v, so that there is a wider separation between the zero-point levels for the two isotopes, these levels being $\frac{1}{2}hv$ higher than the classical ground states. The situation will therefore be as represented in Fig. 8.2, which shows two possible cases. In case (*a*), E_{heavy} is seen to be greater than E_{light}, so that the equilibrium for the light system will be more in favour of the weak bond than for the heavy system. In (*b*), $E_{light} > E_{heavy}$, but the weak bond is still more favoured by the light system.

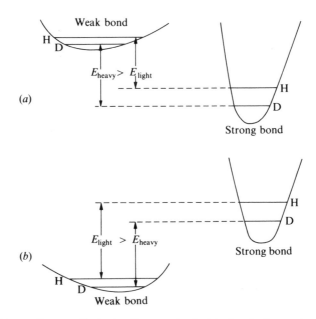

FIG. 8.2.   Energy-diagrams illustrating the general principle that the heavy atom favours the strong bond.

Equilibria in H_2O and D_2O

The above treatment is relevant to the problem of the dissociations of acids in H_2O and D_2O. Experimentally, it is frequently found that the rates of acid-catalysed reactions are 2–3 times faster in D_2O than in H_2O, and this can be explained if they occur by the following type of mechanism:

$$A + H^+ \underset{}{\overset{fast}{\rightleftharpoons}} AH^+$$

$$AH^+ \overset{slow}{\rightarrow} products.$$

This result can be explained if the effect of replacing H by D causes the first equilibrium to lie more over to the right, and if the effect on the second process is small. This implies that acid dissociation constants are smaller in D_2O than in H_2O; i.e., for the dissociations

$$HA \rightleftharpoons A^- + H^+ \qquad K_H = \frac{[A^-].[H^+]}{[AH]} \qquad (10)$$

$$DA \rightleftharpoons A^- + D^+ \qquad K_D = \frac{[A^-].[D^+]}{[AD]} \qquad (11)$$

K_H is greater than K_D. It is to be noted that if an acid HA is dissolved in D_2O there is rapid exchange so that the acid exists largely as DA.

It is actually found that acids are stronger in H_2O than D_2O, the pK difference, pK_D − pK_H, being approximately 0·6 in many cases. A somewhat oversimplified explanation for this is as follows[1]. These acid dissociations actually involve a transfer of a proton from HA to a water molecule

$$HA + H_2O \rightleftharpoons H_3O^+ + A^-.$$

Usually HA is a much weaker acid than H_3O^+, and will have the proton bound more tightly; thus in terms of Fig. 8.2, replacement of H by D will favour the more strongly bonded state DA as compared with D_3O^+, and the dissociation constant is reduced. The weakness of this explanation is that it is too simple, in that it neglects the numerous other interactions between HA, H_2O and the surrounding water molecules. A much more detailed treatment, which takes account of these interactions, has been given by Bunton and Shiner[2].

Kinetic isotope effects[3]

It is useful to consider first the very simple reaction

$$H + H{-}H \rightarrow H{-}H + H$$

and to compare it with

$$H + D{-}H \rightarrow H{-}D + H.$$

The advantage of this comparison is that it is one of the very few in which (if quantum-mechanical tunnelling is ignored) a rigorous treatment of the isotope effect is possible. The potential-energy surface is the same for both processes, only the zero-point levels being different; the surface is sketched

[1] R. P. Bell, *The Proton in Chemistry*, Cornell University Press, 1959, p. 187.
[2] C. A. Bunton and V. J. Shiner, *J. Am. chem. Soc.*, **83**, 42 (1961).
[3] For more detailed treatments see L. Melander, *Isotope Effects on Reaction Rates*, Ronald Press Co., New York, 1960; J. W. Bigeleisen and M. Wolfsberg, *Adv. Chem. Phys.*, **1**, 15 (1958). Briefer accounts are given by K. J. Laidler, *Theories of Chemical Reaction Rates*, McGraw-Hill Book Co., New York, 1969, pp. 86–98; W. P. Jencks, *Catalysis in Chemistry and Biochemistry*, McGraw-Hill Book Co., New York, 1969, Chapter 4.

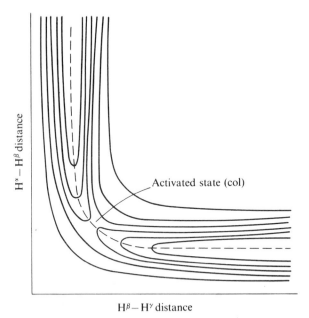

FIG. 8.3. Schematic potential-energy surface for the $H + H_2 \rightarrow H_2 + H$ system. The atoms are labelled α, β, and γ, and the dotted curve shows the course of reaction when H^{α} reacts with H^{β}—H^{γ}.

in Fig. 8.3. According to activated-complex theory the rate constant for the $H + H_2$ reaction is given by

$$k_{H+H_2} = \frac{kT}{h} \frac{Q_{\neq}}{Q_H Q_{H_2}} e^{-E_0/RT} \qquad (12)$$

where the Q's are the partition functions and E_0 is the difference between the zero-point levels for initial and activated states. A similar expression applies to the $H + D$—H process:

$$k_{H+DH} = \frac{kT}{h} \frac{Q_{\neq}}{Q_H Q_{HD}} e^{-E_0'/RT}. \qquad (13)$$

The magnitudes of E_0 and E_0' are uncertain (they lie between 8 and 10 kcal mol^{-1}), but their difference is known very precisely. As shown in Fig. 8.4, the zero-point level for H_2 lies 6·2 kcal mol^{-1} above the minimum in the potential-energy curve (the classical ground state), while the zero-point energy for HD lies 5·4 kcal above the ground state. In the activated state the only effective vibrations† are the symmetric stretch

$$\leftarrow H \cdots H \cdots H \rightarrow \qquad \leftarrow H \cdots D \cdots H \rightarrow$$

† The remaining vibration, the antisymmetric stretch, corresponds to passage of the system through the activated state, and is unquantized.

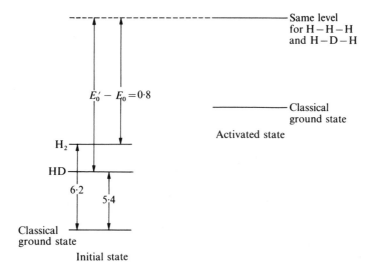

FIG. 8.4. Zero-point levels in the initial and activated states for the systems H—H—H and H—D—H.

and a pair of degenerate bending vibrations

$$
\begin{array}{cc}
\uparrow \quad \uparrow & \uparrow \qquad \uparrow \\
\text{H} \cdots \text{H} \cdots \text{H} & \text{H} \cdots \text{D} \cdots \text{H.} \\
\downarrow & \downarrow
\end{array}
$$

In none of these normal modes of vibration will there be any motion of the central atom; the frequency of the symmetric stretch will therefore be the same for H—H—H as for H—D—H, and the same is true of the degenerate bending vibrations. It follows that the zero-point level for the activated complexes H—H—H and H—D—H are the same. The difference $E'_0 - E_0$ is therefore determined solely by the difference between the zero-point levels in the initial states, and is thus equal to $6\cdot2 - 5\cdot4 = 0\cdot8 \text{ kcal mol}^{-1}$.

The rate-constant ratio is thus given by

$$
\frac{k_{\text{H}+\text{H}_2}}{k_{\text{H}+\text{DH}}} = \frac{Q_{\neq}Q_{\text{HD}}}{Q'_{\neq}Q_{\text{H}_2}} e^{800/RT}.
\tag{14}
$$

The dependencies of the partition functions on the masses are as follows:

$$
Q_{\text{H}_2} \propto m_{\text{H}_2}^{\frac{3}{2}}/m_{\text{H}} \qquad Q_{\text{HD}} \propto m_{\text{HD}}^{\frac{3}{2}} \frac{m_{\text{H}} m_{\text{D}}}{m_{\text{H}} + m_{\text{D}}}
$$

whence

$$
\frac{Q_{\text{HD}}}{Q_{\text{H}_2}} = \left(\frac{3}{2}\right)^{\frac{3}{2}} \frac{2}{3} = \left(\frac{3}{2}\right)^{\frac{1}{2}}
$$

and

$$Q_{\neq} \propto m_{H_3}^{\frac{3}{2}} \qquad Q'_{\neq} \propto m_{HDH}^{\frac{3}{2}}$$

whence

$$\frac{Q_{\neq}}{Q'_{\neq}} = \left(\frac{3}{4}\right)^{\frac{3}{2}}.$$

Thus

$$\frac{k_{H+H_2}}{k_{H+DH}} = \left(\frac{3}{2}\right)^{\frac{1}{2}}\left(\frac{3}{4}\right)^{\frac{3}{2}} e^{800/RT}$$

$$= 0.8 \times 3.35 = 2.68 \text{ at } 300 \text{ K}.$$

It is to be noted that the isotope effect arises mainly from the difference in zero-point levels of the reactants. The pre-exponential factor introduces only a minor effect.

The case just considered is a particularly simple one, the zero-point levels being the same in the two activated states. This arose from the symmetry of the activated state; the central H or D atom lies at the centre of gravity of the activated complex and therefore does not influence the frequencies of the symmetric vibrations. Asymmetry in an activated state can arise in

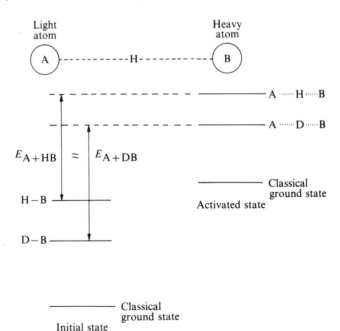

FIG. 8.5. Reaction between A and HB, in which A is a light atom or group and B is a heavy one. The kinetic isotope effect is reduced because of the asymmetry of the activated state.

two ways; the force constants of the two bonds may be different, or the atom being transferred may be attached to groups of different masses. Suppose, for example, that the activated complex were as represented in Fig. 8.5; the central atom is attached to a light atom and a heavy atom, but the force constants are the same. It is then easy to show, by solving the dynamical equations, that the frequency of the stretching vibration is reduced if H is replaced by a D atom. This is shown in the lower part of Fig. 8.5. The effect of the asymmetry is to decrease the difference between E_{A+HB} and E_{A+DB}, and therefore to diminish the kinetic isotope effect. A similar result is obtained if asymmetry results from unequal force constants for the A—H and H—B bonds.

The situation is different again if the two atoms A and B are *both* significantly heavier than the central atom H or D (Fig. 8.6). The solution of the dynamical equation now leads to the result that the frequency of the symmetric vibration is practically independent of the mass of the central atom. This is the case whatever the relative magnitudes of the force constants

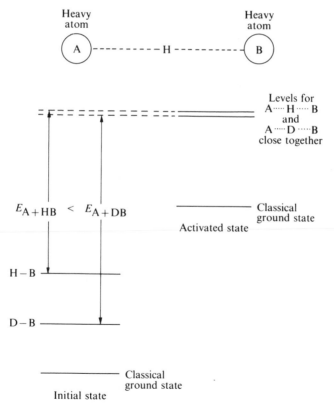

FIG. 8.6. Reaction between A and HB, in which both A and B are much heavier than H. There is now a substantial isotope effect.

involved. This situation is of particular interest for enzyme and other organic reactions, since the atoms A and B are frequently carbon, oxygen or nitrogen atoms, all of which are a good deal heavier than the H or D atoms. For such systems it is therefore expected that the zero-point levels in the activated state will be very close together for H and D atoms, so that the situation will be controlled largely by the zero-point energies of the reactants. The situation will therefore be somewhat as represented in Fig. 8.6, and there will be a substantial kinetic isotope effect.

The general conclusion is therefore that the rate constant for an H-atom transfer in an enzyme system should be markedly greater than that for a D-atom transfer. The difference between the zero-point levels for an organic reactant R—H as compared with R—D is approximately $1 \cdot 15 \, \text{kcal mol}^{-1}$ (note that this is greater than the difference between H—H and H—D), so that the estimated isotope effect is

$$\frac{k_H}{k_D} \approx e^{1150/RT} = 6 \cdot 9 \text{ at } 300 \text{ K}.$$

Ratios of this magnitude are, in fact, frequently observed. Much smaller values suggest that the process being studied may not involve an H-atom or D-atom transfer.

Quantum-mechanical tunnelling[1]

The preceding discussion has neglected the possibility of quantum-mechanical tunnelling through the potential-energy barrier; it has involved the assumption that, as required by classical mechanics, the system passes over the top of the barrier. Quantum-mechanical theory admits the possibility that a system having less energy than required to surmount the barrier may nevertheless pass from the initial to the final state; it is said to *tunnel* or *leak* through the barrier. Tunnelling is most important for a particle of small mass, and when the barrier is low and narrow. The theoretical treatments indicate that tunnelling may, for certain barriers, be quite important for the transfer of H atoms or of H^+ and H^- ions; it is probably of negligible importance for D, D^+ and D^-, and is certainly negligible for heavier atoms. As a result of this possibility that tunnelling may occur with H but not with D it follows that the kinetic isotope ratio can be very much greater than the figure of about 7 that was indicated above; values of over 20 have in fact been observed.

The quantitative theory of tunnelling is a matter of very considerable difficulty, even for a reaction as simple as $H + H_2$. The problem is that the rate of tunnelling depends rather strongly on the exact shape of the potential-energy barrier, and this is still not reliably known for any system. Even if the exact shape of the potential-energy surface were known, there still

[1] See K. J. Laidler, *Theories of Chemical Reaction Rates*, McGraw-Hill Book Co., New York, 1969.

remains uncertainty about the theoretical treatment of tunnelling. Consequently, it is necessary to obtain information about tunnelling largely from the experimental evidence.

It is not even easy to obtain reliable experimental evidence for tunnelling. Theory predicts that when tunnelling occurs certain effects will be observed; however, these effects may be due to other causes, so that the evidence may be ambiguous. The main effects predicted if tunnelling occurs are:

(1) There will be a deviation from linearity in an Arrhenius plot, as shown schematically in Fig. 8.7. The reason is that tunnelling is fairly insensitive

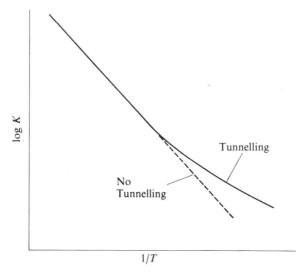

FIG. 8.7. Schematic Arrhenius plot for a reaction in which quantum-mechanical tunnelling occurs.

to temperature, so that it becomes relatively more important as the temperature is lowered. Behaviour of the type shown in Fig. 8.7 has been observed for a number of reactions, e.g. for the reactions

$$H + H_2 \rightarrow H_2 + H \quad \text{and} \quad D + H_2 \rightarrow DH + H.$$

A number of reactions in solution have also been found to show nonlinear Arrhenius behaviour. The most thoroughly investigated example is the base-catalysed bromination, in heavy water, of the cyclic ketone 2-carbethoxy-cyclopentanone; this reaction is a base-catalysed reaction and is undoubtedly controlled by proton transfer from the substrate to the base. For this reaction catalysed by fluoride ions and other bases Bell and coworkers[1] found a very marked deviation from linearity in the Arrhenius plot, an effect that cannot

[1] R. P. Bell, J. A. Fendley, and J. R. Hulett, *Proc. R. Soc., Lond.*, **A235**: 453 (1956); J. R. Hulett, *Proc. R. Soc., Lond.*, **A251**:274 (1959).

be explained in other ways. At 20°C, for example, the rate was 75 per cent faster than that calculated by extrapolating the linear Arrhenius plot obtained at higher temperatures. The experiments were interpreted in terms of the equations for the penetration of a parabolic barrier; the width of this at its base was calculated as 1·17 Å, and the height is about 20 per cent greater than the observed activation energy. Similar deviation from the Arrhenius behaviour has been found by Caldin and coworkers[1] for the reactions of the trinitrobenzyl ion with acetic acid and with hydrofluoric acid, and for the reaction of 4-nitrobenzyl cyanide with ethoxide ions. The work on these three reactions was done down to -115, -90 and -124°C; the corresponding deviations of the rate constants at the lowest temperatures, from the values expected from extrapolation of the Arrhenius plots, were 45, 120, and 100 per cent. These deviations are far outside the experimental error, and no explanation other than tunnelling is satisfactory.

It is important to note that this type of deviation from Arrhenius behaviour is also obtained if two reactions are occurring simultaneously. The reaction of higher activation energy is unimportant at low temperatures, and the activation energy observed is therefore that of the reaction having the lower activation energy. As the temperature is raised the reaction of higher activation energy becomes relatively more important and finally predominates. The observed activation energy therefore increases with increasing temperature as the temperature is raised, and the behaviour will be very similar to that shown in Fig. 8.7. Care must therefore be taken to eliminate this possibility of parallel reactions, as well as of consecutive reactions.

(2) The second effect predicted to occur when there is tunnelling is related to the first; this is that frequency factors will be abnormally small. It can be seen from Fig. 8.7 that, since $\log A$ is $\log k$ extrapolated to a zero value of $1/T$, the frequency factor at low temperatures, where tunnelling is important, will be low. The result is that, for example, the frequency factor of a proton transfer may be considerably less than that for the corresponding deuteron transfer; this would be difficult to explain on other grounds. In the work on the bromination of 2-carbethoxycyclopentanone, A_D/A_H was found to be 24 at low temperatures; this is strong evidence for tunnelling, especially taken in conjunction with the nonlinear Arrhenius behaviour.

(3) Apart from these frequency-factor differences, abnormally large isotope effects are predicted where tunnelling occurs, the lighter isotope reacting relatively more rapidly than can be explained without taking tunnelling into account. This kind of evidence is never very clear-cut, since, as has been seen, the treatment of isotope effects presents some difficulty. However, a value of k_H/k_D which is substantially greater than 10 is good evidence for tunnelling.

[1] E. F. Caldin and E. Harbron, *J. chem. Soc.*, 3454 (1962); E. F. Caldin, M. Kasparian, and G. Tomalin, *Trans. Faraday Soc.*, **64**: 2802 (1968); E. F. Caldin and G. Tomalin, *Trans. Faraday Soc.*, **64**: 2814, 2823 (1968); E. F. Caldin, *Fast Reactions in Solution*, p. 273, Blackwell, Oxford, 1964.

(4) A fourth indication of tunnelling can be obtained if experiments are done with tritium (T) as well as D and H. According to the theory of kinetic isotope effects in which tunnelling is neglected, the effects are due largely to differences in zero-point levels, which depend on the masses of the atoms involved. It can readily be shown that, if there is no tunnelling, k_H, k_D and k_T must be related by

$$\log \frac{k_H}{k_T} = 1 \cdot 44 \log \frac{k_H}{k_D}. \qquad (15)$$

However, if there is tunnelling for H the k_H value will be abnormally high, and the effect of this is that the factor on the right-hand side will be less than 1·44. The existence of a low value for this factor is thus good evidence for tunnelling.

Isotope effects in solution

When hydrogen–deuterium isotope effects are studied in solution there is one problem to which careful attention must be given. Hydrogen atoms attached to oxygen, nitrogen and certain other atoms are rapidly exchanged with the solvent because of the rapidity of the ionization processes that occur; for example

$$-O-H \overset{\text{fast}}{\rightleftharpoons} -O^- + H^+.$$

If, therefore, an alcoholic H atom is replaced by D, and the substance is dissolved in ordinary water, there will be rapid exchange and the compound will become largely undeuterated. In order to study the kinetic isotope effect it is therefore necessary to dissolve the undeuterated compound in ordinary H_2O, and the deuterated compound in D_2O.

Unfortunately, this procedure introduces additional factors, and the results are less easy to interpret. The change of solvent alone may have a significant effect on the rate, in a manner that is difficult to predict. For example, the viscosity of D_2O at 25°C is 23 per cent higher than that of H_2O, the latent heat of vaporization is higher[1] (1·50 as compared with 1·44 kcal mol^{-1} at the freezing points) and the dielectric constant is slightly lower[2] (77·9 at 25°C, as compared with 78·3 for H_2O). The evidence is that there is rather more structure in liquid D_2O than in H_2O; this is consistent with the principle that the heavy atom favours the strong bond. As a result of these differences between the two solvents caution must be taken in interpreting the kinetic isotope effects involving exchangeable hydrogen atoms, where the solvent is necessarily changed. However, this solvent

---

[1] F. D. Rossini, *et al.*, *Chemical Thermodynamic Properties*, National Bureau of Standards Circular 500 (1952).
[2] C. G. Malmberg, *J. Res. natn. Bur. Stand.*, **60**, 609 (1958).

effect is certainly fairly small; Bender, Pollock and Neveu[1] have suggested a k_H/k_D ratio of 1·0 to 1·5 resulting from this cause.

These difficulties do not arise if the hydrogen is attached to a carbon atom, in which case it is generally not exchangeable. Unfortunately, however, such hydrogen atoms are not as commonly involved in enzymic processes.

Transfer of H^+, H and H^-

An interesting question that arises is whether, by studying the magnitude of an H-D kinetic isotope effect, one can decide whether the process involves the proton H^+, the hydrogen atom H or the hydride ion H^-.

It is certainly true that, in view of the complexities of the situation, it is very difficult to discriminate between these possibilities. Jencks[2], in fact, takes the view that there is no evidence that one can draw any valid conclusions on the basis of kinetic isotope effects. On the other hand, Swain, Wiles, and Bader[3] have considered the types of activated states which are expected to arise for H^+, H and H^- transfers, and suggest that some significant differences are to be expected. Thus, when H^- is transferred the bonds in the activated state are expected to be shorter and stronger than for H^+ transfer; consequently one would expect for H^- transfer a smaller k_H/k_D ratio, and even possibly an inverse isotope effect (i.e., k_H/k_D less than unity). The behaviour with H transfer should lie in between that for H^- and H^+ transfer. Also, the isotope effect with H^- should be less sensitive to structural changes than those with H^+ (and again H will lie in between) because the activated state is less polarized. These criteria have been applied to enzyme systems by Belleau and Moran[4].

A decision as to whether or not the kinetic isotope effect in fact permits a reliable discrimination between H^+, H and H^- transfer mechanisms must await further studies in which alternative evidence is used in addition to the kinetic isotope work. In the meantime Jencks's scepticism seems to be justified.

Isotope effects with heavier atoms

The discussion so far in this chapter has been confined to hydrogen isotope effects. A few investigations, some of them for enzyme systems (see p. 252), have been concerned with isotopes of heavier elements such as carbon; the most abundant isotope of carbon is C^{12}, and studies have been made with C^{13} and C^{14}.

The effects to be expected are naturally much smaller for the heavier elements, since the ratio of masses is much less. Table 8.1 shows some values

[1] M. L. Bender, E. J. Pollock, and M. C. Neveu, *J. Am. chem. Soc.*, **84**, 595 (1962).
[2] W. P. Jencks, *Catalysis in Chemistry and Enzymology*, McGraw-Hill Book Co., New York, 1969, p. 273.
[3] C. G. Swain, R. A. Wiles, and R. F. W. Bader, *J. Am. chem. Soc.*, **83**, 1945 (1961).
[4] B. Belleau and J. Moran, *Ann. N.Y. Acad. Sci.*, **107**, 822 (1963).

TABLE 8.1
Typical kinetic isotope ratios at 25°C

Isotopic forms	Ratios
H, D	7†
H, T	17†
C^{12}, C^{13}	1·04
C^{12}, C^{14}	1·08
N^{14}, N^{16}	1·04
O^{16}, O^{18}	1·04

† These values are significantly greater if tunnelling occurs.

that are more or less typical, but it must be remembered that, for the reasons discussed above, wide variations are possible.

Secondary isotope effects

The kinetic isotope effects that have been considered so far are those in which the isotopic substitution involves an atom that is at one end of a bond that is formed or broken during the reaction. Such effects are known as *primary* kinetic isotope effects. It is also found that effects arise when there is substitution to an atom that is not directly at the seat of reaction—i.e. is not involved in a bond that is being formed or broken. Such effects are known as *secondary* isotope effects. Only a brief account will be given, since such effects are by no means as useful in leading to conclusions about reaction mechanisms.

Essentially, secondary isotope effects are explained in the same way as the primary effects, in terms of changes in vibrational frequencies and hence zero-point levels. For example, the secondary effects found in S_N^1 substitution reactions can be explained in terms of the frequency changes resulting from a change of hybridization. The process is essentially

$$\overset{\diagdown}{\underset{\diagup}{-}}C{-}X \;\rightarrow\; \overset{\diagup\diagdown}{C}\!\cdots X \;\rightarrow\; \overset{\diagup\diagdown}{C^{+}} + X^{-}.$$

Initially there is sp^3 hybridization and finally there is sp^2 hybridization; in the activated state the hybridization is between sp^3 and sp^2. There is a decrease in the bending frequencies when the change $sp^3 \rightarrow sp^2$ occurs, and therefore a decrease when the activated complex is formed. The H and D levels for the activated state are therefore closer together than in the initial state; the energy diagram is somewhat as in Fig. 8.6 (but for quite different reasons), and there is a positive kinetic isotope effect. A value for k_H/k_D of 1·3 to 1·4 is predicted on this basis.

In some cases it has proved more expedient to interpret secondary kinetic isotope effects in terms of inductive and steric effects, and hyperconjugation:

(1) *Inductive effects.* For the ionizations

$$HCO_2H \rightleftharpoons HCO_2^- + H^+ \quad K_H$$

$$DCO_2H \rightleftharpoons DCO_2^- + H^+ \quad K_D$$

it is found that

$$K_H \approx 1 \cdot 12 K_D; \quad \Delta pK \approx 0 \cdot 035.$$

This can be explained by saying that D is slightly more electron-donating than H. This effect can also be invoked to explain differences in rates.

(2) *Steric effects.* On the average the C—D bond is slightly shorter than the C—H bond; this arises from the anharmonicity of the vibration and the fact that the amplitude of vibration of the C—D bond is less than that of the C—H bond. As a result, the —CH_3 group, for example, shows a little less steric hindrance than —CD_3.

(3) *Hyperconjugation.* In a solvolysis reaction of the type

$$-\overset{\displaystyle H}{\underset{\displaystyle |}{\underset{|}{C}}}-\overset{\displaystyle |}{\underset{\displaystyle |}{C}}-X \rightarrow -\overset{\displaystyle H}{\underset{\displaystyle |}{\underset{|}{C}}}-\overset{\displaystyle |}{\underset{\displaystyle |}{C^+}} + X^-$$

it is found that replacement of H by D gives a decrease in rate. This is explained in terms of hyperconjugation in the initial state—i.e. resonance between the structures

$$-\overset{\displaystyle H}{\underset{\displaystyle |}{\underset{|}{C}}}-\overset{\displaystyle |}{\underset{\displaystyle |}{C}}-X \leftrightarrow -\overset{\displaystyle H^+}{\underset{\displaystyle |}{C}}=\overset{\displaystyle |}{\underset{\displaystyle |}{C}} \quad X^-$$

a process which aids the reaction. This resonance will be less important for D than for H because of the loss of zero-point energy in hyperconjugation.

Conclusions

It will be evident from what has been said that the theoretical interpretation of kinetic isotope effects is a very difficult matter, in view of lack of knowledge of the potential-energy surfaces. As a result, one should be very careful in drawing conclusions from the magnitudes of the effects. It is essential that a careful kinetic study be made so that one is sure that one can separate out the rate constants for individual reaction steps; much confusion will result if the kinetic constants are composite. For example, in an enzyme study the results at low substrate concentrations are difficult to interpret, since several rate constants are involved. It is important in kinetic studies to determine whether there are any pre-equilibria, such as ionizations, since these can have a considerable effect on the results.

If the rate constant for an elementary process shows a normal isotope effect (e.g. if $k_H/k_D \approx 7$), this is good evidence that there is transfer of H^+, H, or H^-. A lower ratio is more difficult to interpret. It might mean that the process does not involve an H^+, H, or H^- transfer, the effect observed being a secondary one. Alternatively, it might arise from transfer of one of these species with asymmetry in the activated state (see Fig. 8.5) or with fairly tight bonding (e.g. if H^- is being transferred).

Kinetic isotope effects in enzyme systems

In view of the complexity of the situation regarding kinetic isotope effects it need hardly be said that great care must be taken in applying the technique to enzyme reactions. It is obviously not satisfactory to study the effect under a single set of conditions and draw any satisfactory conclusions. It is necessary to make a complete kinetic study, covering a variety of conditions. The following are the main points to which consideration must be given:

(1) As with non-enzymic systems, it is essential to distinguish between effects involving exchangeable and non-exchangeable atoms. If, for example, one is concerned with the transfer of an H or D atom which is exchangeable with the solvent, the work with D must be done using D_2O as solvent. Care must then be taken with the interpretation, since there may be important effects arising from the change of solvent.

(2) Kinetic isotope effects must always be related to rate constants and not simply to rates, which may involve a number of rate constants in complicated ways. The methods available for separating rate constants in enzyme systems have been considered in Chapter 7. For a one-substrate system, for example, a steady-state study provides the constants K_A and k_c, where the latter may under certain conditions be a simple rate constant (e.g. for acylation or deacylation). The ratio k_c/K_A is not, in general, a simple rate constant, but it relates to the second-order conversion of enzyme and substrate into an intermediate activated complex. It is best to make a more complete analysis of kinetic-isotope effects by using stopped-flow techniques, which reveal individual rate constants; however, even a steady-state investigation can reveal significant results if care is taken with the interpretation.

(3) The study must not be carried out at a single pH, but must cover a significant pH range so that one obtains information about the relevant ionizations. It has been seen that pK values are frequently increased by substitution of D for H, so that studies over a pH range might give results such as shown schematically in Fig. 8.8. It is to be seen that as far as the active centre of the enzyme is concerned there is a positive isotope effect $(k_H/k_D > 1)$. However, because of the shift of the pK values to the right as H is replaced by D, the apparent isotope effect at higher pH values is a negative one $(k_H/k_D < 1)$. It is evident that unreliable conclusions might be drawn from a study at any one pH value.

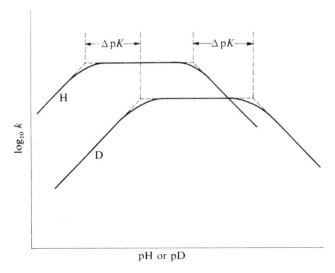

FIG. 8.8. Schematic pH (pD) profiles for an enzyme-catalysed system in which there is a significant positive isotope effect.

After a kinetic isotope study has been made, with the above points dealt with, the interpretation of the results must take account of the following factors:

(1) The process being investigated may involve an H or D transfer. In this case a k_H/k_D ratio of ~ 7 may well be found; however, as has been discussed, considerably higher or lower values are consistent with H or D transfer. An isotope effect of this magnitude may only be taken as evidence for H or D transfer if other possibilities (see below) have been eliminated.

(2) If the work with D is done in D_2O (as is necessary if the D is exchangeable with the solvent) it is possible that the enzyme has a different conformation in D_2O and in H_2O. Such differences in conformations, by changing the structure at the active centre, may lead to a significant isotope effect. The possibility of different conformations is very considerable in view of the very different hydrogen bonding properties of H_2O and D_2O; there is, in fact, direct physical evidence for different conformations with ribonuclease[1] and proline racemase[2].

(3) If the process under study involves a nucleophilic attack (e.g. a deacylation reaction), there will be an isotope effect due to the different nucleophilicities of H_2O and D_2O. Since D is more electron-donating than H (see p. 248), D_2O is somewhat more nucleophilic than H_2O. However, this effect is a minor one compared with that expected if there is transfer of H or D.

[1] J. Hermans and H. A. Scheraga, *Biochim. biophys. Acta*, **36**, 634 (1959).
[2] G. J. Cardinale and R. H. Abeles, *Biochemistry*, **7**, 3970 (1968).

Chymotrypsin

Detailed kinetic isotope studies on chymotrypsin-catalysed reactions have been carried out by Bender and coworkers[1]. The substrates they used were N-acetyl-L-tryptophanamide, for which it is known that acylation is rate-limiting ($k_2 \ll k_3$), and N-acetyl-L-tryptophan ethyl ester, for which deacylation is rate limiting ($k_3 \ll k_2$). Since the H atoms involved are rapidly exchangeable with the solvent, the studies were made in H_2O and D_2O.

The results at high substrate concentrations are shown in Fig. 8.9. For the amide, two pK values are revealed and both are increased when H is

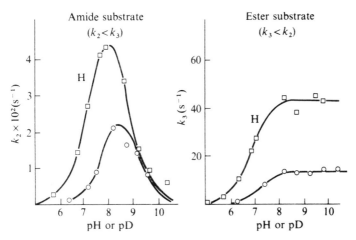

FIG. 8.9. Kinetic isotope effects in the chymotrypsin-catalysed hydrolysis of N-acetyl-L-tryptophanamide and of N-acetyl-G-tryptophan ethyl ester (results of Bender *et al.*).

replaced by D. For the ester there is no falling off up to pH 10; the lower pK value is also increased when H is replaced by D. In all cases ΔpK is approximately 0·5. The k_H/k_D ratios at the maximum rates are 2·0–2·5 for both substrates, the value being slightly but significantly greater for the ester substrate.

There are several lines of evidence which point to the conclusion that there are no important conformational changes in chymotrypsin when H_2O is replaced by D_2O:

(1) D_2O brings about no irreversible change in the behaviour of the enzyme. This is not a strong argument, since D_2O might well bring about a conformational change which is reversed when the D_2O is replaced by H_2O.

(2) The kinetic isotope effect is the same for chymotrypsin as for imidazole, which cannot undergo any substantial conformational change, and which is involved in the catalytic action of chymotrypsin.

[1] M. L. Bender, G. E. Clement, F. J. Kézdy, and H. d'A. Heck, *J. Am. chem. Soc.*, **86**, 3680 (1964).

(3) The kinetic isotope effect is the same for chymotrypsin as for trypsin, which is mechanistically very similar to chymotrypsin. If conformational changes accounted for the kinetic isotope effect they would therefore have to be the same for the two enzymes. Again, this is not a strong argument since the enzymes are structurally rather similar.

(4) Varying kinetic isotope effects were found with chymotrypsin and a group of anilides. If the effect were due to conformational changes one would expect less variation with the different substrates.

The results obtained with chymotrypsin are also incompatible with solvation effects, which can hardly give a k_H/k_D ratio of greater than 1·5. D_2O does have a different nucleophilicity from H_2O, but this explanation appears incapable of explaining the results. Thus, the same k_H/k_D ratio is found in both acylation (with the amide substrate) and in deacylation (with the ester). In acylation H_2O (or D_2O) does not act as a nucleophile, while in deacylation it does. There is in fact a slightly higher k_H/k_D ratio for deacylation, and this may be a small effect of the differing nucleophilicities. However, the main effect cannot be explained on this basis.

These various possibilities having been eliminated on the basis of fairly strong arguments, it seems necessary to conclude, for these chymotrypsin reactions, that the results are to be explained on the hypothesis that the processes of acylation and deacylation both involve transfer of H or D (see also Chapter 10).

Decarboxylases

A few investigations have also been made in enzyme systems of the kinetic isotope effects obtained when C^{12} is replaced by C^{13}. According to theory, a value of 1·04 is expected for the ratio k^{12}/k^{13} if the carbon atom is involved in bond making or breaking.

A study of glutamic acid decarboxylase (from *E. coli*) has been made by O'Leary and coworkers[1]; the glutamic acid was labelled with C^{13} in its carboxyl group. At low substrate concentrations they found a k^{12}/k^{13} value of 1·02. This result may be interpreted in terms of the Michaelis–Menten scheme, according to which the second-order rate constant is given by (see p. 211)

$$k = \frac{k_1 k_2^*}{k_{-1} + k_2^*}. \tag{16}$$

The k_2 is starred because a kinetic isotope effect is expected for it; k_1 and k_{-1}, on the other hand, are not expected to exhibit an isotope effect, since complex formation is presumably not associated with the breaking of a bond involving the carbon atom. Two special cases are to be distinguished:

(1) If $k_2^* \gg k_{-1}$, k is equal to k_1, and no kinetic isotope effect is expected.
(2) If $k_{-1} \gg k_2^*$, $k = k_1 k_2^*/k_{-1}$ and a kinetic isotope effect is expected.

[1] M. H. O'Leary, *J. Am. chem. Soc.*, **91**, 6886 (1969); M. H. O'Leary, D. T. Richards, and D. W. Hendrickson, *ibid.*, **92**, 4435 (1970).

Since an effect was observed, the first case is excluded, and it appears that step 2 in the reaction is more or less rate limiting.

On the other hand, a study of the decarboxylation of oxalacetic acid showed no kinetic isotope effect for the enzyme-catalysed process[1]; the reaction catalysed by magnesium ions gave a k^{12}/k^{13} value of 1·06. It thus appears that in this enzyme reaction stage 2 is not rate-limiting, in contrast to the situation with glutamic acid decarboxylase.

Some further examples of kinetic isotope effects in enzyme systems have been given by Richards[2].

[1] S. Seltzer, E. A. Hamilton, and F. H. Westheimer, *J. Am. chem. Soc.*, **81**, 4018 (1959).
[2] J. H. Richards, in *The Enzymes* (ed. P. D. Boyer, H. Lardy, and K. Myrbäch), Academic Press, New York, 1970, Vol. 11, p. 321.

9. Some Deductions About Mechanisms

THE ultimate aim of the kinetic work that is carried out on enzyme systems is the elucidation of the mechanisms by which enzymes catalyse the reactions that are undergone by their substrates. There has now accumulated a considerable body of evidence which suggests plausible hypotheses about the mechanisms of action of certain enzymes. Some of this evidence has been referred to in previous chapters. In Chapter 3, for example, we have seen that the variations of rate with the concentrations of substrates and other substances lead to some rather definite conclusions about the nature of the complexes that the enzymes form during the course of reaction. Chapter 5 has shown that a study of pH dependence leads to more specific conclusions about mechanisms, and Chapters 7 and 8 have shown that further detail is provided by studies of temperature solvent, pressure and isotope effects. In the present chapter an attempt will be made to collect together the main lines of evidence that lead to an understanding of mechanisms†. Some of this evidence will be of an essentially non-kinetic nature—as, for example, evidence from stereochemistry, from isotope studies, and from X-ray investigations.

It is convenient to begin a discussion of enzyme mechanisms by considering two outstanding features in which enzymes show marked differences from other catalysts, and which therefore call for special explanation; first of these is that enzymes, when compared on a molecular basis with other catalysts, are always very much more effective. Comparisons of this type are sometimes made in terms of the 'turnover number' of the enzyme, which is the number of substrate molecules transformed by a single molecule of the enzyme in one minute under standard conditions. Since the turnover number is not a precisely defined quantity, being dependent on the substrate concentration, a more suitable comparison can be made in terms of the rate constants for the reactions. As seen in Chapter 3, many enzyme reactions follow the law

$$v = \frac{k_c [E]_0 \cdot [A]}{K_A + [A]} \tag{1}$$

so that, at substrate concentrations sufficiently low that $K_A \gg [A]$, the rate is equal to

$$v = k_c [E]_0 \cdot [A]/K_A. \tag{2}$$

---

† For reviews on various aspects of mechanisms, see D. E. Koshland and K. E. Neet, *A. Rev. Biochem.*, **37**, 359 (1968); H. Gutfreund and J. R. Knowles, *Essays in Biochemistry*, **3**, 25 (1967); W. P. Jencks, *Catalysis in Chemistry and Enzymology*, McGraw-Hill, New York, 1969.

This may be written as

$$v = k_0[E]_0 . [A] \qquad (3)$$

where k_0, equal to k_c/K_A, is a second-order rate constant for the reaction at low substrate concentrations. This second-order rate constant may be compared with second-order rate constants for the same (or a similar) substrate when it is acted upon by a catalyst other than an enzyme.

Such a comparison is shown in Table 9.1, which includes results for enzymes of various types. In all cases the enzyme can bring about the reaction of a given substrate with a very much higher rate constant than can any other catalyst. Some of the differences are particularly striking: thus urease can catalyse the hydrolysis of urea 10^{14} times as rapidly as can hydrogen ions at a temperature 40° higher. The rate constant quoted for the catalase-hydrogen peroxide system is one of the highest known for any reaction, and exceeds by a factor of 10^6 the rate constant for the reaction catalysed by ferrous ions, which are generally regarded as good catalysts for this reaction. Several other comparisons of this type could have been given, and there seems to be no case where a non-biological substance is a more effective catalyst than the appropriate enzyme.

Enzyme specificity

The second outstanding characteristic of the enzymes is their high specificity. This is in marked contrast to the much lower degree of specificity exhibited by most non-enzymic catalysts; examples of such are the hydrogen and hydroxide ions, which catalyse a wide variety of reactions such as ester hydrolyses, amide hydrolyses, and isomerizations. The only inorganic catalysts which show any marked specificity are surface catalysts, but even with them one never meets the situation, so often found with the enzymes, that the substitution of one group for another completely eliminates the catalytic activity.

The different types of specificity exhibited by enzymes have been mentioned briefly in Chapter 1. The original theory of specificity was due to Emil Fischer, whose famous 'lock-and-key' hypothesis[1] forms the basis of most modern concepts of specificity. According to this hypothesis, in order for an enzyme to act upon a substrate it is necessary for the enzyme and substrate molecules to fit together in somewhat the way that a key fits into a lock. Normally a lock can only be operated by a given key; a slight modification to the key, or to the lock, may mean that the key can no longer open or close the lock. All types of specificity can be explained by suitable versions of this idea; in reaction specificity, for example, it must be supposed that only the reactive part of the substrate molecule is critical, so that changes in other parts of the molecule are analogous to changes in non-essential parts of the lock (or key).

[1] E. H. Fischer, *Ber. dtsch. chem. Ges.*, **27**, 2987 (1894).

TABLE 9.1

Comparison of kinetic constants

Substrate	Catalyst	T (°C)	k_0 (M^{-1} s^{-1})	A (M^{-1} s^{-1})	E (kcal mol^{-1})	Ref.
Hydrolysis of peptides						
Glycyl-glycine	H_3O^+	54.5	1.11×10^{-6}	8.8×10^7	20.9	[a]
Carbobenzoxy-glycyl-L-phenylalanine	carboxypeptidase	25.0	2.8×10^4	3.5×10^{11}	9.6	[b]
Carbobenzoxy-L-glutamyl-L-tyrosine	pepsin	31.6	0.79	2.2×10^{14}	20.2	[c]
Hydrolysis of amides						
Benzamide	H_3O^+	52.0	2.4×10^{-6}	1.3×10^{10}	23.3	[d]
Benzamide	OH^-	53.0	8.5×10^{-6}	2.8×10^7	18.7	[d]
Benzoyl-L-tyrosinamide	chymotrypsin	25.0	14.9	6.8×10^3	3.7	[e]
Hydrolysis of esters						
Ethyl benzoate	H_3O^+	100.0	9.0×10^{-5}	2.9×10^7	19.6	[f]
Ethyl benzoate	OH^-	25.0	5.5×10^{-4}	5.3×10^9	17.7	[g]
Benzoyl-L-tyrosine ethyl ester	chymotrypsin	25.0	1.95×10^4	7.5×10^4	0.8	[e]
Methyl hydrocinnamate	chymotrypsin	25.0	6.7	8.4×10^7	9.6	[h]
Hydrolysis of urea						
Urea	H_3O^+	62.0	7.4×10^{-7}	1.8×10^{10}	24.6	[i]
Urea	urease	20.8	5.0×10^6	1.7×10^{13}	6.8	[j]
Hydrolysis of adenosine triphosphate						
ATP	H_3O^+	40.0	4.7×10^{-6}	2.35×10^9	21.2	[k]
ATP	myosin	25.0	8.2×10^6	1.64×10^{22}	21.1	[l]
Decomposition of hydrogen peroxide						
H_2O_2	Fe^{++}	22.0	56.0	1.78×10^9	10.1	[m]
H_2O_2	catalase	22.0	3.5×10^7	6.38×10^8	1.7	[n]

[a] L. Lawrence and W. J. Moore, *J. Am. chem. Soc.*, **73**, 3973 (1951).
[b] R. Lumry, E. L. Smith, and R. R. Glantz, *J. Am. chem. Soc.*, **73**, 4330 (1952).
[c] E. J. Casey and K. J. Laidler, *J. Am. chem. Soc.*, **72**, 2159 (1950).
[d] I. Meloche and K. J. Laidler, *J. Am. chem. Soc.*, **73**, 1712 (1951).
[e] S. Kaufman, H. Neurath, and G. W. Schwert, *J. biol. Chem.*, **177**, 792 (1949).
[f] E. W. Timm and C. N. Hinshelwood, *J. chem. Soc.*, 862 (1938).
[g] C. K. Ingold and W. S. Nathan, *J. chem. Soc.*, 222 (1936).
[h] J. E. Snoke and H. Neurath, *J. biol. Chem.*, **182**, 577 (1950); *Archs Biochem.*, **21**, 351 (1949).
[i] K. J. Laidler and J. P. Hoare, *J. Am. chem. Soc.*, **72**, 2489 (1950).
[j] M. C. Wall and K. J. Laidler, *Archs Biochem. Biophys.*, **43**, 299 (1953).
[k] S. L. Friess, *J. Am. chem. Soc.*, **75**, 323 (1953).
[l] L. Ouellet, K. J. Laidler, and M. F. Morales, *Archs Biochem. Biophys.*, **39**, 37 (1952).
[m] J. H. Baxendale, M. G. Evans, and G. S. Park, *Trans. Faraday Soc.*, **42**, 155 (1946).
[n] R. K. Bonnichsen, B. Chance, and H. Theorell, *Acta chem. scand.*, **1**, 685 (1947).

The more recent investigations of enzyme kinetics have indicated that whereas this lock-and-key analogy is essentially correct, it requires development in various directions. One of these is related to the more detailed nature of the enzyme–substrate interactions. The lock-and-key hypothesis implies that specificity is governed by purely steric effects, the effect of substituent groups being simply that they get in the way and prevent intimate contact between the two molecules. It is now clear, however, that electrical interactions involving charged groups and dipoles are also of great importance.

A second line of development of the lock-and-key hypothesis is required in connection with our more detailed knowledge of the mechanism of enzyme action. Enzyme reactions usually involve the initial formation of one or more complexes. The question of specificity can apply either to complex formation or to the subsequent reactions of the complex. If a modification to a substrate leads to failure to react this may mean either that complexing does not occur or that the complex formed is unreactive. A decision between these two possibilities can readily be made by finding out whether or not the modified substrate acts as an inhibitor of the reaction of the normal substrate; if it does, failure to react is not due to failure to complex with the enzyme.

In connection with specificity two distinct factors have therefore to be taken into account: the first is the actual complexing (the fitting of the key into the lock), and the second is the subsequent reaction (the turning of the key). A failure of either of them to occur will lead to lack of reaction. With regard to complexing it is clear that certain steric requirements must be satisfied and that in some cases charged groups must be suitably placed on the substrate molecule. The simplest view would be that the substrate molecule must be exactly complementary in form to the normal enzyme molecule, so that there is a mere fitting together of the two molecules. However, such a picture is too simple. The increased reactivity of the complexed substrate molecule must be due to its having been in some way modified by the enzyme, and the enzyme in turn must suffer some change of structure during complex formation; evidence for such changes have been considered in Chapter 7. According to this point of view, the question of whether a molecule will complex or not depends not so much upon whether it fits the enzyme when both it and the enzyme are in their normal configurations, as upon whether both can easily be modified to allow such a fit. This theory has been developed extensively by Koshland in his theory of induced fit; details are given later in this chapter.

Stereochemical specificity[1]

One point of some special interest arises when stereochemical specificity occurs. When such is the case there must be attachment of the enzyme to the

[1] For a general review of enzyme stereochemistry see I. A. Rose, *Crit. Rev. Biochem.*, **1**, 33 (1971).

substrate at three positions at least[1], as shown schematically in Fig. 9.1. In this figure the three groups R_2, R_3, and R_4 on the asymmetric substrate molecule are shown attached to the enzyme at the points a, b, and c, and it must be assumed that R_2 can only become attached at a, R_3 only at b, and R_4 only at c. The optical enantiomorphs of the substrate cannot undergo the right type of attachment, as shown in Fig. 9.1b. If there were only two points of attachment, as shown in Fig. 9.1c and d, both enantiomorphs could become attached to the enzyme molecule.

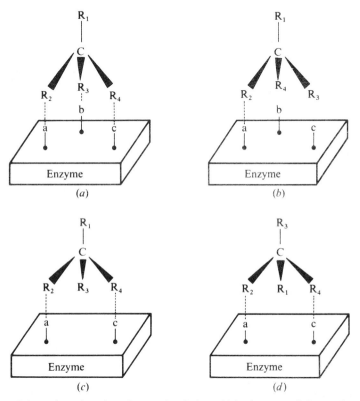

FIG. 9.1.   Schematic explanation of stereochemical specificity in terms of three-point attachment; (a) and (b) show that three-point attachment leads to stereochemical specificity; (c) and (d) that two-point attachment does not.

The ability of enzymes to distinguish between optical enantiomers has an interesting extension: enzymes also sometimes distinguish between what would normally be regarded as two identical atoms or groups on a substrate molecule. This was first discovered by Westheimer and his coworkers[2] with

[1] A. G. Ogston, *Nature*, **164**, 180 (1949).
[2] F. A. Loewus, P. Ofner, H. F. Fisher, F. H. Westheimer, and B. Vennesland, *J. biol. Chem.*, **202**, 699 (1953).

muscle lactate dehydrogenase. The enzyme has associated with it a coenzyme, Coenzyme I or nicotinamide adenine dinucleotide (NAD), of structure

The process catalysed is

Westheimer *et al.* prepared reduced NAD containing one deuterium atom, and caused it to reduce pyruvate in the presence of the enzyme. They found that if the deuterio-reduced-NAD had been prepared by the enzymic reduction of NAD, by use of alcohol dehydrogenase and CH_3CD_2OH, the reaction with pyruvate produced lactate containing one deuterium atom per molecule, as indicated by the equation

When, on the other hand, mono-deuterio-reduced-NAD was prepared by chemical means, the lactate that was formed contained only one half of a deuterium atom per molecule. The chemical reduction with D_2 will produce two forms of the reduced NAD, as follows:

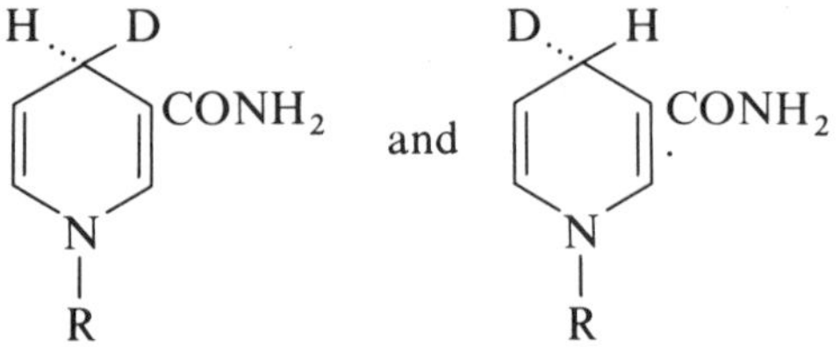

The lactic dehydrogenase exerts stereochemical specificity, transferring only the H or D atom that is on the correct side of the molecule. Also, the alcohol dehydrogenase exerts stereochemical specificity, placing the deuterium atom on the side that is correct for the lactic dehydrogenase; i.e., both enzymes are stereospecific in the same way, removing H or D atoms from the same side of the nicotinamide ring.

A similar example is found with the enzyme aconitase, which catalyses the interconversion of citrate to isocitrate:

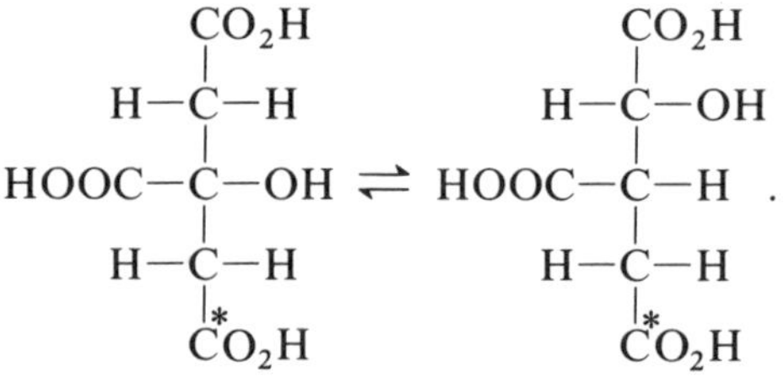

citrate   isocitrate

Citrate itself has a plane of symmetry, but can be prepared with one of the end carbon atoms labelled with C^{14}, as indicated above. When this is done it is found that the enzyme acts upon the $-CH_2-$ group that is furthest away from the labelled $-CO_2H$ group, as in the equation above.

It is to be noted that a similar discrimination between like groups has been observed in several non-enzymic systems. An example is the reaction of a bulky amine with 3-phenyl-glutaric anhydride[1]:

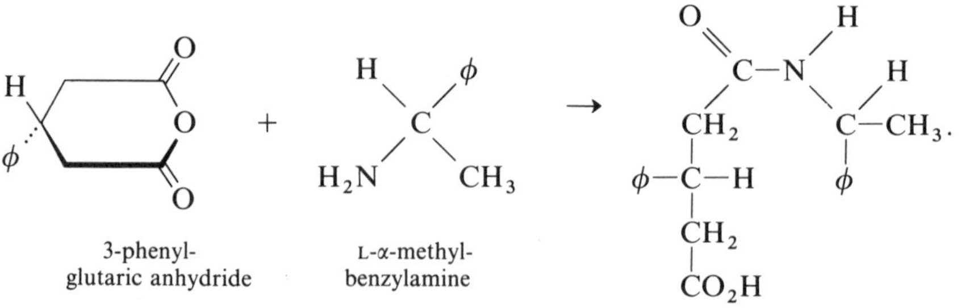

3-phenyl-glutaric anhydride   L-α-methyl-benzylamine

[1] P. Schwartz and H. E. Carter, *Proc. natn. Acad. Sci. U.S.A.*, **40**, 499 (1954).

The process occurs as shown above, the amine approaching only at the top of the ring; approach from below is hindered by the bulky phenyl group in the 3-position. Similar examples are to be found in the chemistry of steroids, where the geometry of the molecule is also fixed.

It is therefore possible that in some enzymic reactions, particularly those where there is only weak specificity for binding, steric considerations of this kind may be important. This matter has been discussed in some detail by Hirschmann[1].

Detailed specificity studies

The specificity of a number of enzymes has been studied in considerable detail, and the results have been helpful in mapping the active centre of the enzyme. Recent studies have been made on the proteolytic enzyme papain. Earlier studies revealed that this enzyme exhibits fairly broad specificity; it hydrolyses peptide bonds involving most of the amino acids, but these must be in their L configurations. Schechter and Berger[2] carried out experiments with two series of peptides: (i) the first series contained phenylalanine at different positions, P, in the substrate (Fig. 9.2, where bond breakage occurs

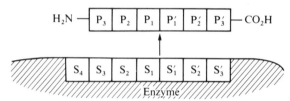

FIG. 9.2. A schematic representation of the binding of a substrate, comprising peptides P_3 to P_3', at the active centre of the enzyme papain, comprising subsites S_4 to S_3'. The bond broken by the enzyme is P_1-P_1', as indicated by the arrow (after Schechter and Berger, *loc. cit.*).

between P_1 and P_1'); (ii) the second series contained L-alanine residues, with D-alanine being substituted at various positions in the peptides. The relative rates of hydrolysis of these two series of peptides revealed that: (i) papain exhibits a strong preference for phenylalanine (or other aromatic residue) at subsite S_2 of the active centre, a point removed from the position of cleavage; (ii) the subsites S_1' and S_2' are highly stereospecific for L-residues—more so than other subsites. These two facts are readily interpreted in terms of data available from other sources on a variety of substrates for papain.

Similar studies have been carried out with chymotrypsin[3], carboxypeptidase[4], and elastase[5]. In all cases, the investigations revealed specificity sites some distance from the bond to be cleaved.

[1] H. Hirschmann, *Compr. Biochem.*, **12**, 236 (1964).
[2] I. Schechter and A. Berger, *Biochem. biophys. Res. Commun.*, **27**, 157 (1967); A. Berger and I. Schechter, *Phil. Trans. R. Soc.*, **B257**, 249 (1970).
[3] G. L. Neil, C. Niemann, and G. E. Hein, *Nature*, **210**, 903 (1966); W. K. Baumann, S. A. Bizzazaro, and H. Dutler, *FEBS Lett.*, **8**, 257 (1970).
[4] I. Schechter, *Eur. J. Biochem.*, **14**, 516 (1970).
[5] D. Atlas, S. Levit, I. Schechter, and A. Berger, *FEBS Lett.*, **11**, 281 (1970).

A slightly different approach has been used to study the specificity of the subtilopeptidases (subtilisins). These proteolytic enzymes were allowed to react with insulin[1] and with oxidized lysozyme[2]. The frequency with which the various amino acids occurred as the N- and C-terminal groups in the resulting small peptides was taken to indicate the specificity of these enzymes. They were found to hydrolyse aliphatic non-polar residues preferentially, although exhibiting a broad specificity.

A rather less detailed idea of the polarity of the active site may be obtained by other studies with substrates and inhibitors. For example, subtilopeptidase A (Carlsberg) reacts more slowly than subtilopeptidase B (Novo) with several esters containing non-polar (benzyl) groups[3]. This is an indication of the more polar nature of the active site of the former enzyme. Similar results were obtained with measurements of binding constants of several inhibitors.

X-Ray studies

The development of X-ray crystallography as a tool for the determination of enzyme structure has thrown considerable light on enzyme specificity. Some of the work done with chymotrypsin and trypsin, considered in some detail in the next chapter, will illustrate this point. Chymotrypsin exhibits a specificity for aromatic amino-acid residues, and the X-ray studies show that there is a non-polar cavity large enough to allow the entry of fairly bulky aromatic residues (see Fig. 10.11, p. 309). Trypsin prefers peptide bonds adjacent to which there is a positively-charged amino-acid residue such as arginine or lysine; the X-ray studies indicate that trypsin has a negatively charged $-COO^-$ group correctly situated for interaction with the positive charge on the substrate. The enzyme elastase presents an interesting contrast to chymotrypsin and trypsin; the X-ray work indicates two fairly bulky groups (threonine and valine) in its active site cavity, in place of two glycine residues for chymotrypsin. These bulky groups prevent larger amino acid residues from entering the cavity and as a result elastase is specific towards smaller residues.

Detailed information on the structures of enzyme–substrate and enzyme–inhibitor complexes is also available for many enzymes from X-ray studies[4]. Results with chymotrypsin are described in the next chapter (p. 308). Another example is carboxypeptidase A, which hydrolyses proteins residue by residue from the carboxyl end. The X-ray studies[5] have revealed that the carboxyl-terminal residue fits into a cavity in the enzyme (see Fig. 9.3). The negatively-charged carboxyl group interacts with the positively-charged arginine–145 at the active site; it is this that produces the characteristic specificity. Various

[1] B. Meedom, *C. r. Trav. Lab. Carlsberg*, **29**, 403 (1956).

[2] K. Okunuki, H. Matsubara, S. Nishimura, and B. Hagihawa, *J. Biochem.*, **43**, 857 (1956).

[3] O. A. Barel and A. N. Glazer, *J. biol. Chem.*, **243**, 1344 (1968).

[4] See R. E. Dickerson and I. Geis, *The Structure and Action of Proteins*, Harper and Row, New York, 1969.

[5] W. N. Lipscomb, G. N. Reeke, J. A. Hartsuck, F. A. Quiocho, and P. H. Bethge, *Phil. Trans. R. Soc.*, **B251**, 177 (1970).

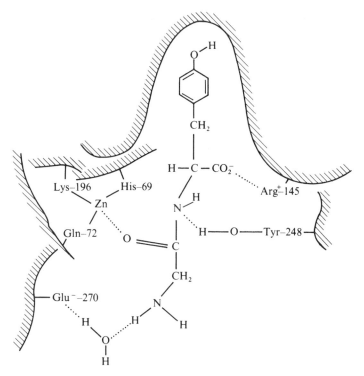

F$_{IG}$. 9.3. The interaction between carboxypeptidase A and a dipeptide, as revealed by X-ray studies. The above does not necessarily represent the action of the enzyme on a true substrate.

other interactions of the substrate with the enzyme are shown in the figure. A number of different types of binding, such as hydrogen bonding, electrostatic interactions, and hydrophobic bonding are involved, as well as steric effects.

Mechanistic evidence from non-kinetic sources

Much valuable information about the mechanisms of enzyme reactions has been derived from investigations that do not strictly speaking lie in the field of chemical kinetics. The most important and useful of these lines of investigation are isotopic tracer studies, studies of the isotope exchange reactions, and stereochemical investigations. These may now be considered.

Isotopic tracer studies

Investigations of reactions in which atoms in the substrate molecules have been isotopically labelled are of great value in determining which bonds are broken during the course of reaction. This technique has been applied particularly to a number of reactions which occur in the following manner:

$$ROR' + R''OH \rightarrow ROR'' + R'OH.$$

This class of reactions includes hydrolyses (in which R″ is H, i.e. HOR″ is H_2O), transglycosidations (in which R, R′, and R″ are sugar residues), and transphosphorylations (in which R is PO_3). These reactions are catalysed by the hydrolases, the transglycosidases, and the transphosphorylases, respectively. It is clear that the same products are obtained irrespective of whether the reaction occurs by one or other of the following mechanisms:

(1)
$$R \dot{\div} O{-}R'$$
$$R''{-}O\dot{\div}H$$
$$\rightarrow R{-}O{-}R'' + R'{-}O{-}H$$

(2)
$$R{-}O\dot{\div}R'$$
$$R''\dot{\div}O{-}H$$
$$\rightarrow R{-}O{-}R'' + R'{-}O{-}H.$$

A study of the products of the reaction, or of the kinetics, can give no indication of whether there is cleavage of the R—O bond or of the R′—O bond. This point of cleavage is of course of interest from the standpoint of the detailed mechanism of the reaction.

The most direct method of determining the point of cleavage is to label the oxygen atom in either the donor molecule (ROR′) or the acceptor (R″OH), and then to determine the presence or absence of the label in one or both of the products. Thus, if the oxygen atom in ROR′ is labelled and the label appears in ROR″ and not in R′OH, the reaction occurred by the second of the two mechanisms indicated above. The oxygen isotope of mass number 18 is most conveniently used for labelling an oxygen atom.

Table 9.2 summarizes the results for several reactions with which this technique has been employed. In this table the substrates ROR′ are shown

TABLE 9.2

Results of isotope tracer studies, showing cleavage points for reaction of the type ROR′ + R″OH → ROR″ + R′OH

Enzyme	ROR′	R″OH	Ref.
Muscle phosphorylase	$G{-}OPO_2$	glycogen	[a]
Sucrose phosphorylase	$G{-}OPO_3$	fructose	[a]
Alkaline phosphatase	$O_3P{-}OG$	H_2O	[a]
Acid phosphatase	$O_3O{-}OG$	H_2O	[a]
Adenosinetriphosphatase	$O_3P{-}OPO_3PO_3A$	H_2O	[b]
Sucrase	fructosyl-OG	H_2O	[c]
Acetylphosphatase	$O_3P{-}OCOCH_3$	H_2O	[d]
Acetylcholinesterase	$CH_3CO{-}OCH_2CH_2NMe_3^+$	H_2O	[e]

[a] M. Cohn, *J. biol. Chem.*, **180**, 771 (1949).
[b] E. Clarke and D. E. Koshland, *Nature*, **171**, 1223 (1953); D. E. Koshland and E. Clarke, *J. biol. Chem.*, **205**, 917 (1953).
[c] D. E. Koshland and S. S. Stein, *Fedn Proc.*, **12**, 233 (1953).
[d] R. Bentley, *J. Am. chem. Soc.*, **71**, 2765 (1949).
[e] S. S. Stein and D. E. Koshland, *Archs Biochem. Biophys.*, **45**, 467 (1953).

either as R—OR′ or as RO—R′ in order to indicate the bond that is broken, i.e. to show whether the reaction occurs by mechanism (1) or by mechanism (2). It is of interest to note that glycosyl phosphate undergoes one type of splitting (G—OPO₃) when undergoing reaction with glycogen and fructose, and the other type (GO—PO₃) when undergoing hydrolysis under the action of the phosphatases.

The question of whether, on the basis of results such as those given in Table 9.2, one can arrive at any generalizations regarding the cleavage point has been considered by Koshland and Stein[1]. Their conclusion is that two general rules appear to apply as follows:

(1) Other things being equal, the weaker bond is the one that will be cleaved; in other words if, in the ordinary reaction of the substrates, the R—O bond is more readily split than the O—R′, the same will be true in the enzyme processes.

(2) Other things being equal, if the enzyme exhibits severe specificity requirements for one group (e.g. R) and wide tolerance for structural changes in the other (e.g. R′), then R—O cleavage will be observed.

These two rules have been found to be useful guides for making predictions.

The first rule may be illustrated with respect to the hydrolysis of acetylcholine by the enzyme acetylcholinesterase. The choice is here between cleavage of the acyl-oxygen bond

$$\overset{\displaystyle O}{\underset{\uparrow}{-\overset{\|}{C}-O-}}$$

and the methylene oxygen bond

$$-CH_2 \underset{\uparrow}{-} O-.$$

Chemical studies have shown that the acyl-oxygen bond is readily split, but that the methylene oxygen bond is split only with difficulty. One would therefore expect the CO—O bond to be the one that is broken, and this is found to be the case (see Table 9.2). Similar considerations lead to the prediction that the esterases split the CO—O bond, and the transglycosidases the bond between the oxygen of the bridge and the carbon of the group being transferred. This rule should only be applied when the differences between the two bonds are very great.

The second rule may be applied in such cases as invertase (β-fructosidase), which is quite specific with respect to the β-fructoside ring, but will allow wide variations in the rest of the molecule. The rule would therefore predict that the fructosyl—O bond is the one broken, and this is the case.

[1] D. E. Koshland and S. S. Stein, *Fedn Proc.*, **12**, 233 (1953).

It is of interest that generally the two rules lead to the same conclusion. It appears that the enzyme requires the group at which the reaction occurs to obey more stringent requirements than the group that is being split off.

With the aid of these two rules it is possible to construct an additional table showing cleavage points as predicted in this way. Such a table is shown as Table 9.3.

TABLE 9.3

Cleavage points as predicted by the rules of Koshland and Stein

Enzyme	ROR'	R"OH
α-Amylase	α-maltosyl-O polysaccharide	H_2O
α-Clucosidase	α-glucosyl-OCH_3	H_2O
β-Glucosidase	β-glucosyl-OCH_3	H_2O
trans-N-glycosidase	β-ribosyl-O hypoxanthine	adenine
Amylosucrase	α-glycosyl-O fructosyl (sucrose)	polysaccharide
Dextransucrase	α-glycosyl-O fructosyl	dextran
Dextrandextrinase	α-glucosyl-O polysaccharide	dextran
Q enzyme	α-glucosyl-O polysaccharide	polysaccharide

Isotope exchange reactions

It has been seen in Chapter 4 that the various mechanisms for enzyme reactions can be broadly classified as *single-displacement* and as *double-displacement* mechanisms. The two main techniques that may be used for distinguishing between the two possibilities are exchange studies and stereochemical studies. The latter, which are limited to compounds having an asymmetric carbon atom at the seat of reaction, are discussed in the next section; the former will now be considered.

The first application of the exchange technique to an enzymic reaction was made in 1947 by Doudoroff, Barker, and Hassid[1]. The enzyme they used was sucrose phosphorylase, the normal action of which is to catalyse the phosphorolysis of sucrose or the reverse reaction between glucose-1-phosphate and fructose:

$$\text{glucosyl}-OPO_3H + \text{fructose} \rightleftharpoons \underset{\text{(sucrose)}}{\text{glucosyl}-O-\text{fructosyl}} + H_3PO_4.$$

Doudoroff *et al.* allowed a solution of sucrose phosphorylase to stand in the presence of glucose-1-phosphate and of $KH_2P^{32}O_4$, but in the absence of the acceptor fructose. They found an exchange of the isotopic phosphorus; that is, the glucose-1-phosphate was found to contain some of the labelled phosphorus. This exchange does not occur in the absence of the enzyme, and it must be concluded that some kind of glucosyl-enzyme intermediate is produced.

[1] M. Doudoroff, H. A. Barker, and W. Z. Hassid, *J. biol. Chem.*, **168**, 725 (1947).

The simplest explanation for an exchange of this kind can be given in terms of the double-displacement mechanism. The first stage of such a mechanism, shown in the upper part of Fig. 4.14 (p. 136), involves the formation of a complex between R and the enzyme, and a complete splitting of the R—X bond, a splitting that has occurred without the participation of Y (fructose in the case under consideration). If these reactions shown in the first line of Fig. 4.14 are reversible it follows that if labelled X ($KH_2P^{32}O_4$ in the sucrose phosphorylase case) is present in the system there will be RX* in the final products.

Further consideration, however, shows that the occurrence of exchange between RX and X* does not necessarily exclude the single-displacement mechanism. Exchange might in fact occur by the single-diplacement mechanism if it were possible for X or X* to become attached to the enzyme at the site normally occupied by Y. If this were so it would be possible for exchange to occur according to the mechanism shown in Fig. 4.13. The conditions for exchange by this single-displacement mechanism are: (1) it must be possible for X* to be attached to the Y site, (2) the enzyme —RX—X* complex formed must be of such a nature that exchange can occur, i.e. that it can change over into enzyme—RX*—X, and (3) the enzyme—RX*—X complex must be capable of splitting up into enzyme + RX* + X.

The double-displacement mechanism permits a more ready interpretation of the exchange reactions, since not as many conditions need be satisfied. The single-displacement explanation in unlikely if the specificity requirements of the enzyme are rather rigid, in which case the possibility that X* would be attached to the Y site is fairly remote. In the case under consideration the specificity requirements are such as to make it very unlikely that the phosphate ion could become attached to the fructose site.

Following the original work of Doudoroff *et al.* a number of additional systems have been studied, and in many of these exchange has been observed. In only a few such cases has it been possible, however, to conclude with reasonable certainty that the double-displacement mechanism is the correct one. The evidence in such cases has been mainly based on the specificity studies. The reactions for which the double-displacement mechanisms has been established with reasonable certainty are shown in Table 9.4. Other reactions in which positive exchange results were obtained, but where there is insufficient supplementary evidence to conclude that the double-displacement mechanism applies, are shown in Table 9.5. Table 9.6 shows some reactions for which no exchange could be detected.

For the cases in which no exchange appears to occur it is important to realize that no conclusions can be drawn with regard to the mechanisms of these reactions. Even if the double-displacement mechanisms hold there will be no observable exchange if the equilibrium shown in the first line of Fig. 4.14 is well over to the right; there will then be little reaction of X* with the complex, and little exchange will occur. It is, however, possible to draw

TABLE 9.4

Systems in which isotopic exchange occurs, and for which
the evidence favours the double-displacement mechanism

Enzyme	Substrate	Labelled compound	Ref.
Sucrose phosphorylase	glucose-1-phosphate	P^{32}-phosphate	[a]
Sucrose phosphorylase	sucrose	C^{14}-fructose	[b]
Triosephosphate dehydrogenase	acetyl phosphate	P^{32}-phosphate	[c]
Glutathione synthetase	ATP	ADP^{32}	[d]
Chymotrypsin	benzoyltyrosyl-glycinamide	C^{14}-glycinamide	[e]
	phenylalanine	H_2O^{18}	[f]

[a] M. Doudoroff, H. A. Barker, and W. Z. Hassid, *J. biol. Chem.*, **168**, 725 (1947).

[b] H. Wolochow, F. W. Putnam, M. Doudoroff, W. Z. Hassid, and H. A. Barker, *J. biol. Chem.*, **180**, 1237 (1949).

[c] J. Harting and S. F. Velick, *Fedn Proc.*, **11**, 226 (1952); P. Oesper, *J. biol. Chem.*, **207**, 421 (1954).

[d] J. E. Snoke, *J. Am. chem. Soc.*, **75**, 4872 (1953).

[e] R. B. Johnston, M. J. Mycek, and J. S. Fruton, *J. biol. Chem.*, **185**, 629 (1950).

[f] D. B. Sprinson and D. Rittenberg, *Nature*, **167**, 484 (1951); D. G. Doherty and F. Vaslow, *J. Am. chem. Soc.*, **74**, 931 (1952).

some conclusions about mechanisms if exchange experiments are made on a reaction in the forward and reverse directions. Such a procedure has been applied by Koshland[1] to the 5'-nucleotidase system. The experiments on this were done as follows:

(1) adenylic acid was partially hydrolysed by the enzyme in the presence of C^{14} adenosine, and

(2) phosphoric acid was allowed to stand in the presence of the enzyme and H_2O^{18}. In neither case was any exchange observed. The overall

TABLE 9.5

Other systems in which isotope exchange occurs

Enzyme	Substrate	Labelled compound	Ref.
Alkaline Phosphatase	KH_2PO_4	H_2O^{18}	[a]
Spleen NAD-ase	NAD	C^{14}-nicotinamide	[b]
β-Glucosidase	glucose	H_2O^{18}	[c]
Papain	benzoylglycinamide	$N^{15}H_3$	[d]
Acetylcholinesterase	acetic acid	H_2O^{18}	[e]
Lipase A	butyric acid	H_2O^{18}	[e]

[a] S. S. Stein and D. E. Koshland, *Archs Biochem. Biophys.*, **39**, 279 (1954).

[b] L. J. Zatman, N. O. Kaplan, and S. P. Colowick, *J. biol. Chem.*, **200**, 197 (1953).

[c] D. E. Koshland, unpublished results (quoted by Koshland in *Discuss. Faraday Soc.*, **20**, 142 (1955)).

[d] R. B. Johnston, M. J. Mycek, and J. S. Fruton, *J. biol. Chem.*, **185**, 629 (1950).

[e] R. Bentley and D. Rittenberg, *J. Am. chem. Soc.*, **76**, 4883 (1954).

[1] D. E. Koshland, *Discuss. Faraday Soc.*, **20**, 142 (1955).

TABLE 9.6

Systems in which no isotope exchange occurs

Enzyme	Substrate	Labelled compound	Ref.
Muscle phosphorylase	glucose-1-phosphate	P^{32}-phosphate	[a]
Thymidine phosphorylase	desoxyribose-1-phosphate	P^{32}-phosphate	[b]
Maltose phosphorylase	glucose-1-phosphate	P^{32}-phosphate	[c]
NAD synthetase	ATP	P^{32}-pyrophosphate	[d]
Myosin	ATP	ADP^{32}	[e]
Myosin	KH_2PO_4	H_2O^{18}	[e]
5'-Nucleotidase	AMP	C^{14}-adenosine	[f]
5'-Nucleotidase	KH_2PO_4	H_2O^{18}	[f]
Neurospora NAD-ase	DPN	C^{14}-nicotinamide	[g]
Acetyl kinase	ATP	ADP^{32}	[h]
Acetyl kinase	acetylphosphate	C^{14}-acetate	[h]

[a] M. Cohn and G. T. Cori, *J. biol. Chem.*, **175**, 89 (1948).
[b] M. Friedkin and D. Roberts, *J. biol. Chem.*, **207**, 245 (1954).
[c] C. Fitting and M. Doudoroff, *J. biol. Chem.*, **199**, 193 (1952).
[d] A. Kornberg and W. E. Pricer, *J. biol. Chem.*, **191**, 535 (1951).
[e] D. E. Koshland, Z. Budenstein, and A. Kowalski, *J. biol. Chem.*, **211**, 279 (1954).
[f] D. E. Koshland and S. S. Springhorn, *J. biol. Chem.*, **221**, 469 (1956).
[g] L. J. Zatman, N. O. Kaplan, and S. P. Colowick, *J. biol. Chem.*, **200**, 197 (1953).
[h] I. A. Rose, M. Grunberg-Manago, S. R. Korey, and S. Ochoa, *J. biol. Chem.*, **211**, 737 (1954).

reaction for this enzyme is

$$AdOPO_3^{--} + H_2O \rightarrow AdOH + HPO_4^{--}$$

where Ad stands for adenosine. If the reaction occurred by a double-displacement mechanism the steps would be:

(1)
$$AdOPO_3^{--} + EH \underset{k_{-1}}{\overset{k_1}{\rightleftharpoons}} AdOH + EPO_3^{--}$$

and

(2)
$$EPO_3^{--} + H_2O \underset{k_{-2}}{\overset{k_2}{\rightleftharpoons}} EH + HOPO_3^{--}$$

where EH is the enzyme. The fact that adenylic acid and C^{14}-adenosine do not undergo exchange excludes the possibility that reaction (1) occurs rapidly and reversibly and that (2) is slow; similarly the lack of exchange of H_2O^{18} and $HOPO_3^{--}$ excludes a rapid step (2) and a low k_{-1}. The failure of the first exchange to take place actually imposes a lower limit on k_2/k_{-1}, and the value calculated for this limit is 1·4. Similarly, the failure of the second exchange to occur together with the overall rate of hydrolysis allowed a lower limit of $1·9 \times 10^3$ to be calculated for k_1/k_{-2}. Combination of these two values leads to the conclusion that

$$\frac{k_1 k_2}{k_{-1} k_{-2}} > 2·6 \times 10^3.$$

This ratio is, however, the equilibrium constant of the overall reaction, and this is known to have a value of about 0·7. The discrepancy between the two values is convincing proof that the double-displacement mechanism does not apply to this system.

It is clear that considerations of this kind, since they provide a lower limit to the overall equilibrium constant for the reaction, will not lead to useful conclusions if the equilibrium constant is in fact very high.

An entirely different type of exchange study may also be helpful in distinguishing between the single-displacement and the double-displacement mechanisms. The essential difference between these two types of mechanisms is that in the double-displacement mechanisms the R—X bond is broken before the Y—R bond starts to form. Evidence that the R—X bond is intact in the initial complex formed by the enzyme and the RX molecule would therefore exclude the double-displacement mechanism. Such evidence is given if it can be shown that the enzyme brings about exchange between the unreacted substrate molecule and the solvent. This criterion is particularly easy to apply in the case of ester hydrolyses, since H_2O^{18} can be added to the solvent and the unreacted ester tested for O^{18}. In the case of ester hydrolyses catalysed by hydrogen and hydroxide ions Bender[1] has shown that such exchange occurs, and these results provide evidence for an additive intermediate in such reactions.

Koshland[2] has applied the same procedure to the hydrolysis of adenylic acid, catalysed by 5'-nucleotidase. No detectable exchange occurred in this case. If it had occurred the conclusion would have been that the hydrolysis took place by a mechanism such as the following:

(1)
$$\text{Ad}-\text{O}-\underset{\underset{\text{OH}}{|}}{\overset{\overset{\text{O}}{\uparrow}}{\text{P}}}-\text{OH} + \text{H}_2\text{O} \rightleftharpoons \text{Ad}-\text{O}-\underset{\underset{\text{OH}}{|}}{\overset{\overset{\text{H}\diagdown\text{O}\diagdown\text{O}\diagup\text{H}}{|}}{\text{P}}}-\text{OH}$$

(2)
$$\text{Ad}-\text{O}-\underset{\underset{\text{OH}}{|}}{\overset{\overset{\text{H}\diagdown\text{O}\diagdown\text{O}\diagup\text{H}}{|}}{\text{P}}}-\text{OH} \rightarrow \text{AdOH} + \text{HO}-\underset{\underset{\text{OH}}{|}}{\overset{\overset{\text{O}}{\uparrow}}{\text{P}}}-\text{OH}.$$

If reaction (1) were reversible the addition of H_2O^{18} to the water would clearly give rise to adenylic acid into which O^{18} had entered. The type of intermediate shown above is consistent with the single-displacement mechanism and not the double-displacement mechanism. The failure of

[1] M. L. Bender, *J. Am. chem. Soc.*, **73**, 1626 (1951).
[2] D. E. Koshland, *Discuss. Faraday Soc.*, **20**, 142 (1955).

exchange to occur of course means that no conclusions can be drawn in this particular case.

A similar procedure has been applied to alkaline phosphatase, acetylcholinesterase, ATP-ase, and chymotrypsin, and in no case was this type of exchange observed. No conclusions can therefore be drawn, but the criterion may be valuable for other systems. If exchange were observed it would, of course, be necessary to exclude the possibility that it occurred by the reversal of the overall reaction.

Stereochemical investigations[1]

A third type of non-kinetic study that has provided much valuable information comprises investigation of whether changes of stereochemical configuration occur during the course of reaction. Since displacement reactions are known to proceed almost always with inversion of configuration (Walden inversion), a single-displacement mechanism should be associated with such an inversion. In the case of double-displacement mechanisms, on the other hand, there will be two inversions, one when the complex is formed between enzyme and RX (with breaking of the R—X bond) and one in the subsequent reaction with Y. The net result will therefore be retention of stereochemical configuration.

TABLE 9.7

Enzyme reactions proceeding with inversion of configuration

Enzyme	Substrate	Product
Maltose phosphorylase	maltose	β-glucose-1-phosphate
β-Amylase	amylose	β-maltose
Galactowaldenase	uridine diphosphoglucose	uridine diphosphogalactose
Alanine	L-alanine	D-alanine

Only a few reactions are found to occur with inversion of configuration. They are listed in Table 9.7; for further details, particularly with regard to how the configurations are determined, reference should be made to Koshland's reviews. In the last two reactions it is possible that simple displacement reactions are not involved, but in the first two the evidence points to direct displacements on the asymmetric carbon atom. In these reactions it seems likely that the two substrate molecules become attached side by side on the surface of the enzyme molecule, during which process they are rendered more reactive than in their original states. They then react together by a single-displacement mechanism, and give rise to final products.

Table 9.8 shows a number of reactions for which retention of configuration has been observed. This list excludes those reactions for which, as revealed

[1] For a detailed review see G. Popjak, in *The Enzymes* (ed. P. D. Boyer), Academic Press, N.Y., and London, 1970, p. 115.

TABLE 9.8

Enzyme reactions proceeding with retention of configuration

Enzyme	Substrate	Product
Sucrose phosphorylase	α-glucose-1-phosphate	α-glucose-1-fructose (sucrose)
α-Amylase	starch	α-maltose
α-Glycosidase	α-methyl glucoside	α-glucose
β-Glucosidase	β-methyl glucoside	β-glucose
trans-N-Glycosidase	inosine	adenosine
Amylosucrase	sucrose	glycogenic polysaccharide
trans-Glucosidase	maltose	panose
Dextran sucrase	sucrose	dextran
Dextran dextrinase	dextrin	dextran
Q enzyme	amylose	amylopectin
B. macerans amylase	amylose	cycloamylose

by isotope studies, the bond broken is such that inversion would not be expected on any mechanism. In other words, the reactions in Table 9.8 are limited to those in which a single displacement would be expected to give inversion. The fact that there is retention therefore supports the double-displacement type of mechanism, as illustrated in Fig. 4.14.

X-Ray studies

Besides providing information about enzyme specificity, X-ray crystallography has contributed a great deal to our understanding of enzyme mechanisms. Until recently, enzyme mechanisms had to be deduced without knowledge of the structure of the enzyme; now the structures of a few enzymes are known (see p. 10) and also the structures of a few enzyme–substrates and enzyme–inhibitor complexes have been determined, so that much more detail can be inferred about the mechanisms. This aspect of the X-ray studies has been well covered in a review by Blow and Steitz[1]. Some particular enzymes are dealt with in the next chapter.

Modes of binding in enzyme systems

A number of different kinds of force are involved in the binding of substrates to enzymes. The theoretical treatment of these forces is a matter of very considerable difficulty, particularly in view of the complexities of the molecules involved and of the fact that the environment is partly aqueous and partly non-aqueous. Only a very brief account is given in the present section[2].

In the first place, covalent bonding may be involved in the binding of substrates to enzymes. An example of this is a possible mode of binding

[1] D. M. Blow and T. A. Steitz, *A. Rev. Biochem.*, **39**, 63 (1970).
[2] For a much more detailed account see W. P. Jencks, *Catalysis in Chemistry and Enzymology*, McGraw-Hill, New York, 1969, Chapters 6–9.

between a substrate containing a carbonyl group and a basic nitrogen atom on an enzyme molecule, for example:

$$\underset{\displaystyle \underset{\diagup N \diagdown}{\diagup\diagdown}}{\overset{\overset{\displaystyle O}{\parallel}}{-\underset{}{C}-}} \longrightarrow \underset{\displaystyle \underset{\diagup N^{+} \diagdown}{\diagup\diagdown}}{\overset{\overset{\displaystyle O^{-}}{\mid}}{-\underset{}{C}-}}.$$

This process can be envisaged as involving an ionization of the carbonyl group, to give $-\overset{\overset{\displaystyle O^{-}}{\mid}}{C}{}^{+}-$, followed by the formation of a dative bond between the nitrogen atom and the carbon atom. The arrangement of bonds round the carbon atom has become tetrahedral, and intermediates of this type are often referred to as *tetrahedral intermediates*. It is difficult to make a reliable estimate of the extent to which the solvent water molecules have access to this region.

TABLE 9.9

Energies of attraction for different types of intermolecular force

Type of force	Example	Separation (Å)	Attractive energy at equilibrium in vacuum (kcal mol^{-1})
Ion-ion	$Na^{+}\cdots F^{-}$	2·3	160
Ion–dipole	$Na^{+}\cdots OH_2$	2·4	20
Hydrogen bond	$\overset{\displaystyle H\diagdown}{\underset{\displaystyle H\diagup}{O}}\cdots H-O\diagup{}^{H}$	2·8	5
Van der Waals (dispersion)	$Ne\cdots Ne$	3·3	0·06
Charge-transfer	$C_6H_6\cdots I_2$	3·4	1·3
Hydrophobic	$\diagdown\!\underset{\diagup}{C}H_2\ H_2C\!\overset{\diagup}{\diagdown}$	—	†

† Very variable; the attraction of non-polar groups depends in a complicated way on enthalpy and entropy factors.

Apart from covalent bonds, there are also a number of intramolecular forces that may be involved in bonding. The most important of these are listed in Table 9.9, which includes an example of each type and gives an indication of the strength of the bonding. However, great care must be taken in relating these binding energies to enzyme systems. For example, the separation of 2·3 Å for the $Na^{+}\cdots F^{-}$ ion–ion electrostatic attraction is the separation in the sodium fluoride crystal, and the value of 160 kcal mol^{-1} is the attractive energy between the ions in a vacuum. In an enzyme–substrate system there might be binding between a cationic group such as $-NH_3^+$ and an anionic group such as $-CO_2^-$,

$$-NH_3^+ \cdots {}^-OOC-$$

and the separation may be not far from 2·3 Å; however the binding energy may be very substantially less than 160 kcal mol^{-1} because of the effect of the environment. If the region were completely surrounded by water, behaving as if it had a dielectric constant of 80, the effect would be to reduce the binding energy by a factor of 80, the result being 2 kcal mol^{-1}. This, however, is much too simple a way of looking at the situation, for two main reasons:

(1) the environment is in fact only partly aqueous, some of the surrounding space being occupied by organic structures which have a much lower dielectric constant than water.

(2) the dielectric constant of water is very strongly dependent on the electric field strength, the value falling as low as 2 in the very near neighbourhood of ions.

Both of these effects result in a higher binding energy than the 2 kcal mol^{-1} estimated above; a value of 10–30 kcal mol^{-1} might be a reasonable estimate.

Similar comments apply to the ion–dipole and dipole–dipole forces. The hydrogen bond is the strongest of the bonds which result from dipole–dipole forces, the reason being that the small size of the hydrogen atom allows the atoms to come much closer together.

Van der Waals (dispersion) and charge-transfer forces play a very unimportant role in the binding of substrates to enzymes, being largely overshadowed by stronger forces. Considerable bonding can, however, result when non-polar groups come close together in aqueous solutions, and it is customary to speak of *hydrophobic bonds* in this connection. Hydrophobic bonding is really an indirect bonding, resulting from the existence of hydrogen bonding between the surrounding water molecules. Consider, for example, two isolated methane molecules in water. In the neighbourhood of these molecules there is considerable disruption of the hydrogen-bonded structure of the water. If, however, the molecules come together the external surface area of the pair of molecules is less than when they were apart, and as a result there can be more hydrogen bonding in the water. The structure in which the molecules have come together is therefore more stable (i.e. has a lower standard free energy) than that in which they are apart. This effect explains the low solubility of hydrocarbons in water. It has already been noted in Chapter 1 (p. 20) that hydrophobic bonding plays a very important role in connection with protein structure, there being a tendency, in an aqueous environment, for hydrophobic side chains to stick together in the interior of the protein molecule. An equally important role is played by these bonds in the binding of substrates to enzymes; examples are to be mentioned later (Chapter 10).

General features of enzyme mechanisms

The remainder of this chapter will be concerned with some of the general features of enzyme mechanisms that have been proposed with a view to explaining the high effectiveness of enzymes and their specificities. Individual enzymes and the mechanisms of their action are considered in the next chapter.

The features that will be considered are:

(1) The *proximity effect*; the suggestion that enzymes owe their effectiveness in part to bringing the reactant molecules together.
(2) *Sequential reactions*; the enzyme allows the reaction to occur in several stages, having lower barriers than for the uncatalysed reaction.
(3) *Chain reactions*; involving the participation of free radicals.
(4) *Bifunctional acid-base catalysis*; special features of this that have been suggested are *facilitated proton transfer*, and the *neighbouring group effect*.
(5) *Induced fit*; the enzyme undergoing a conformational change in order to accommodate the substrate.
(6) *Strain*; the enzyme causing a change in shape and size (i.e. a strain) in the substrate molecule.
(7) *Non-productive binding*; this is the suggestion put forward to explain certain specificity results.
(8) *Orientation effects*

Proximity effects

It was first suggested in 1825 by Michael Faraday that the main effect of a catalyst is to cause the reactant molecules to come close together. It was soon realized, however, that this *proximity* effect cannot make an important contribution to catalytic action. Solid surfaces, for example, are somewhat specific in their action, and this cannot be explained on the proximity theory.

It has more recently been suggested that proximity effects may be of some significance in connection with enzyme action. It is now fully realized that other effects must be important also, but the suggestion is that proximity may make some contribution to the high efficiencies of enzymes. The matter has been discussed in some detail by Koshland[1] and by Jencks[2].

A very simple argument, based on free energy changes, shows that proximity cannot have a favourable effect on rates, and that in general it leads to reductions of rate. Curve (*a*) in Fig. 9.4 shows schematically the free-energy changes during the course of a reaction between A and B, the curve passing

---

[1] D. E. Koshland, *J. theor. Biol.*, **2**, 75 (1962); *J. cell. comp. Physiol.*, **54**, suppl. 1, 245 (1959).
[2] W. P. Jencks, *Catalysis in Chemistry and Enzymology*, McGraw-Hill Book Co., New York, 1969, Chapter 1; M. I. Page and W. P. Jencks, *Proc. natn. Acad. Sci. U.S.A.*, **68**, 1678 (1971).

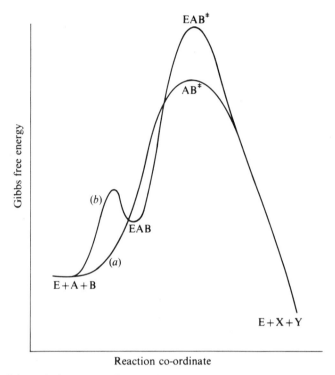

FIG. 9.4.   Schematic free-energy diagram for a reaction between two reactants A and B, showing the expected variations without (*a*) and with (*b*) the catalyst.

through a maximum at the activated state $AB^{\neq}$. If the enzyme E is introduced there may be the initial formation of the complex EAB, which may undergo further reaction by passage through the activated state $EAB^{\neq}$. Now $EAB^{\neq}$ necessarily has a lower entropy than $AB^{\neq}$, and if no energy changes are involved (i.e. a pure proximity effect is being assumed) $EAB^{\neq}$ must have a higher free energy than $AB^{\neq}$ (see Curve (*b*) of Fig. 9.4). Since the rate at low substrate concentrations is determined by the energy of the activated complex with reference to that of A + B(+E) it follows that the proximity effect will lead to a lowering of the rate at low substrate concentrations. The situation at higher substrate concentrations is different, the rate now being determined by the free-energy difference between EAB and $EAB^{\neq}$. The former will have a higher free energy than E + A + B, because of the loss in entropy. The situation is therefore less unfavourable at high substrate concentrations. However, under these circumstances very high substrate concentrations would be required to give saturation, because of the positive free energy; in normal adsorption processes the entropy losses have to be offset by binding forces, leading to negative ΔH values.

Sequential reactions

Catalysts frequently cause reactions to occur by a sequence of reactions, each one of which is more rapid than the overall reaction that occurs in the absence of the catalyst. A very simple case of this is illustrated in Fig. 9.5. A reaction between two species A and B may occur in the absence of catalysts by passage over a high free-energy barrier. In the presence of an enzyme E

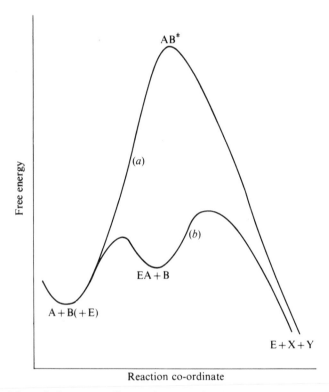

FIG. 9.5.   Schematic free-energy diagram for a reaction between A + B occurring (*a*) un-catalysed, (*b*) by a sequence of two reactions involving an enzyme E.

it might occur by the primary formation of a complex EA, followed by reaction of EA with B. Each of the barriers may be low, so that the rate by this mechanism will be faster than that for the uncatalysed reaction. Solid surfaces frequently catalyse reactions in this manner.

A simple example is provided by the action of the enzyme rhodanese on the reaction

$$SSO_3^{2-} + CN^- \rightarrow SO_3^{2-} + SCN^-.$$

In the absence of catalyst the reaction is very slow because of the approach of charges of like sign. The mechanism in the presence of the enzyme has been shown[1] to be

$$E + SSO_3^{2-} \rightarrow E-S + SO_3^{2-}$$
$$E-S + CN^- \rightarrow SCN^-.$$

There is no electrostatic hindrance for either of these processes, each of which is therefore much more rapid than the uncatalysed reaction.

It seems likely that this factor is involved in many enzymic mechanisms. It has been seen that many enzyme reactions occur in at least two stages, for example, with the initial formation of an enzyme–substrate complex; many occur in three stages, with an acyl enzyme formed as well as the enzyme-substrate complex.

Chain reactions

The occurrence of many non-biological reactions by chain mechanisms has led some workers[2] to believe that free-radical chains may be involved in the reactions of certain enzymes, such as the oxidative ones. Suggestions of this kind have been made in part on the basis of analogy with the non-enzymic reactions, and in part to explain the very high effectiveness of enzymes. If a chain occurs the production by an enzyme molecule of a single radical may lead to a considerable amount of reaction, and the turnover number may be correspondingly high. On the whole, however, the evidence does not appear to support the idea of free-radical chains. There are some general objections, and in some particular cases there are results that are quite incompatible with the idea.

The general objections are as follows. In the first place, there are some very reliable methods for the detection of free radicals in solution, and these have always given negative results when applied to enzyme systems. A second argument is that it is now well known that free radicals have a very destructive effect on enzymes and indeed on all protein systems; the denaturation of proteins by ultraviolet light and by ionizing radiations is in fact brought about by the free radicals produced. Consequently, if free radicals were present in enzyme reactions, the enzymes themselves would be rapidly inactivated. A third objection to the idea of free-radical chains is based on fact that enzymes are usually very specific; oxidative enzymes, for example, are specific both with respect to the molecule being oxidized and to the molecule being reduced. It is difficult, if not impossible, to account for such dual specificity if chain reactions are involved. An enzyme might be specific with respect to

[1] J. R. Green and J. Wesley, *J. biol. Chem.*, **236**, 3047 (1961); J. Wesley and T. Nakamoto, *J. biol. Chem.*, **237**, 547 (1962).

[2] F. Haber and R. Willstätter, *Ber. dt. chem. Ges.*, **64B**, 2844 (1931); E. A. Moelwyn-Hughes, *Ergebn. Enzymforsh.*, **6**, 23 (1937); J. Weiss, *Trans. Faraday Soc.*, **42**, 16, 133 (1946); S. J. Lench, *Adv. Enzymol.*, **15**, 1 (1954); W. A. Waters, *The Chemistry of Free Radicals*, Clarendon Press, Oxford (1946).

one reactant and set up free-radical chains with it, but the radicals so produced would necessarily be very indiscriminate in their action and would bring about reaction with a variety of other substances.

For certain enzymes the results of Westheimer, Vennesland, and co-workers using deuterium-substituted compounds (see p. 259) have provided strong evidence against the idea of free-radical chains. This work, discussed earlier for two particular enzyme systems, has been carried out with a variety of different enzymes, and in each case the result has been that the deuterium or hydrogen atom transferred is non-exchangeable with the solvent; it therefore proves that the transfer is a direct one and cannot involve the occurrence of a chain of processes. The enzymes for which this has been shown to be the case[1] are alcohol dehydrogenase[2], lactic dehydrogenase[3], malic dehydrogenase[4], a testosterone dehydrogenase from *Pseudomonas*[5], a transhydrogenase from *Pseudomonas*[6], glucose dehydrogenase[7], and triose phosphate dehydrogenase[7].

The above evidence renders rather unlikely the possibility that free-radical chains are involved in enzyme action. It does not, however, exclude the possibility that bound radicals play a role in the reactions of oxidative enzymes. Such bound radicals, analogous to adsorbed hydrogen atoms in the hydrogenation reactions on metals, would have no destructive action on proteins and would not be detected by the tests for free radicals referred to above. These bound radicals could not be involved in reaction chains, so that there is no difficulty in explaining the specificity with respect to both reactants. Finally, such bound radicals (or hydrogen atoms) could be non-exchangeable, and therefore could effectively be transferred directly from one substrate molecule to another.

The participation of bound radicals in oxidative enzyme systems has been proposed in particular by Michaelis[8], and appears to represent a much more likely possibility than the existence of free-radical chains. These ideas are based on conclusions, also arrived at by Michaelis[9], regarding the mechanisms of certain organic oxidations in solution.

Acid-base catalysis and enzyme action

When hydrolyses are catalysed by acids and bases the catalysing species are hydrogen ions and hydroxide ions, and also certain other species that

[1] For a review see B. Vennesland, *Discuss. Faraday Soc.*, **20**, 240 (1955).

[2] F. H. Westheimer, H. F. Fisher, E. E. Conn, and B. Vennesland, *J. Am. chem. Soc.*, **73**, 2403 (1951); H. F. Fisher, E. E. Conn, B. Vennesland, and F. H. Westheimer, *J. biol. Chem.*, **202**, 687 (1953). J.-F. Biellmann and M. J. Jung, *Eur. J. Biochem.*, **19**, 130 (1971).

[3] F. A. Loewus, P. Ofner, H. F. Fisher, F. H. Westheimer, and B. Vennesland, *J. biol. Chem.*, **202**, 699 (1953).

[4] F. A. Loewus, T. T. Tchen, and B. Vennesland, *J. biol. Chem.*, **212**, 787 (1955).

[5] P. Talalay, F. A. Loewus, and B. Vennesland, *J. biol. Chem.*, **212**, 801 (1955).

[6] A. San Pietro, N. O. Kaplan, and S. P. Colowick, *J. biol. Chem.*, **212**, 941 (1955).

[7] F. A. Loewus, H. R. Levy, and B. Vennesland, *Fedn Proc.*, **14**, 245 (1955).

[8] L. Michaelis and C. V. Smythe, *A. Rev. Biochem.*, **7**, 1 (1938).

[9] L. Michaelis, *J. biol. Chem.*, **36**, 703 (1932); *Chem. Rev.*, **16**, 243 (1935); L. Michaelis and M. P. Schubert, *Chem. Rev.*, **22**, 437 (1938).

have a tendency to donate or accept protons. According to modern views (see p. 61) an acid can be defined as any substance that can give up protons, and a base as a substance that can accept protons. The $-COO^-$ group of a protein is therefore a basic group, while the $-NH_3^+$ and $-NH^+=$ groups (the latter occurring in the imidazole ring of histidine) are acidic groups. Not only hydrogen and hydroxide ions, but all other acidic and basic molecules and groups, have the power to catalyse reactions, particularly those of the hydrolytic class. As seen on p. 61, the catalytic powers of such acids and bases depends on their strengths as acids and bases.

Considerable evidence has now accumulated about the actual mechanisms of reactions catalysed by acids and bases. Such mechanisms have been reviewed in detail in various places[1], and the evidence for them can only be outlined briefly here. At the outset, however, it is worth noting that the kinetic laws obeyed by acid-catalysed and base-catalysed reactions are not completely analogous to those obeyed by enzyme-catalysed reactions. There is nothing, for example, resembling a Michaelis–Menten type of law for acid-base catalysis, the rates of which increase indefinitely with increasing substrate concentration. This is due to the establishment, in acid-base catalysis, of solution equilibria involving the solvent and the various acidic and basic species present. This difference between acid-base and enzyme catalysis is of particular significance in showing that enzymes cannot be regarded as acidic or basic catalysts *in a simple sense*; their reactions must involve some special feature, the possible nature of which is considered later.

Kinetic studies on reactions catalysed by acids and bases have been concerned mainly with the dependence of the rate on the concentrations of catalyst and of substrate, and on the temperature, with the effects of added salts and of changing the solvent, and with substituent effects. Work of this kind has led to certain conclusions about the details of the reaction mechanisms for such reactions. Thus the acid hydrolysis of ester, RCO_2R', has been regarded[2] as occurring according to the following mechanism:

(Initial state)   (Activated state)   (Final state)

[1] See, for example, A. E. Remick, *Electronic Interpretations of Organic Chemistry*, John Wiley and Sons, New York (1943); C. K. Ingold, *Structure and Mechanism in Organic Chemistry*, Cornell University Press, Ithaca, New York (1953); K. J. Laidler, *Chemical Kinetics*, McGraw-Hill, New York, 2nd ed., 1965; W. P. Jencks, *Catalysis in Chemistry and Enzymology*, McGraw-Hill, New York, 1969.
[2] K. J. Laidler and P. A. Landskroener, *Trans. Faraday Soc.*, **52**, 200 (1956).

Here it is supposed that there is a simultaneous attack of an oxonium ion (H_3O^+) on the alcoholic oxygen atom and of a water molecule on the carbonyl carbon atom. This attack is seen to lead to a complete ionization of the carbonyl group. There is a transfer of a proton from the H_3O^+ ion to the alcoholic oxygen atom, and a transfer of a hydroxide ion from a water molecule to the carbonyl carbon atom. It is possible that during the course of the reaction a tetrahedral intermediate such as

$$
\begin{array}{c}
O^- \\
| \\
R-C-O^+-R' \\
| \\
O\ \ H \\
/ \\
H
\end{array}
$$

is formed; since this satisfies the valence rules it may be expected to have a certain stability, although by a flow of electrons from the carbonyl oxygen atom to the alcoholic oxygen atom via the carbonyl carbon atom it will decompose into the acid and the alcohol.

In the case of the hydrolysis catalysed by hydroxide ions the evidence suggests that the hydroxide ions attack at the carbonyl carbon atom, whereas a proton is transferred from water to the alcoholic oxygen atom:

| (Initial state) | (Activated state) | (Final state) |

Again it is possible that essentially the same stable intermediate would be formed as in the case of acid hydrolysis.

It is to be noted that in both acidic and basic catalysis the direct participation of a water molecule is required; in the former case the water molecule provides an OH^- ion and therefore acts as a base, while in the latter case the water molecule provides a proton and thus acts as an acid. In both acid and base catalysis there is therefore the simultaneous action of an acid and a base, in each case a water molecule playing the role of one or the other. In both cases reaction therefore occurs by what has been referred to as a 'push–pull' mechanism, an acid species pushing a proton on to the substrate and a basic species pulling a proton from another part of the molecule. Alternatively, the basic species can be regarded as pushing electrons towards the substrate molecule, and the acidic species as pulling them out.

An important point about this type of catalysis is that the ability of a water molecule to act as an acid or a base is a strictly limited one; water has amphoteric properties (i.e. it is both acidic and basic), but is neither a strong acid nor a strong base. In view of Brønsted's relationship between acidic or basic strength and catalytic ability (see p. 61) it follows that the efficiencies of both acidic and basic catalysts are limited by the necessity for the participation of the relatively inefficient water molecules. If the part played by these water molecules could instead be played by more strongly basic or acidic species (in the case of acid or base catalysis, respectively) the efficiency of the catalytic process would be increased. An effective catalysis of ester hydrolysis would, for example, be achieved if there could be simultaneous attacks by hydrogen ions (at the alcoholic oxygen atom) and hydroxide ions (at the carbonyl carbon atom). These ions cannot, however, exist together in solution to any extent, owing to the recombination process

$$H_3O^+ + OH^- \rightleftharpoons 2H_2O,$$

the equilibrium for which lies well over to the right.

In the case of enzymes, however, the pH studies reveal that acidic and basic groups are both involved in the catalytic action, and it is possible that this fact provides an explanation for the very high effectiveness of enzymes. It seems very reasonable to suppose that the simultaneous attack of the substrate molecule by acidic and basic groups of average strength might be much more effective than the simultaneous attack by a very strong acid (or base) and the very weak water molecule. Suppose, for example, that an enzyme has, at its active centre, two groups with pK values of 6·0 and 8·0. At a pH of 7·0 the group of pK 6·0 will be in its basic form, and will be about as strong a base as can exist at that pH; the group of pK 8·0 will be protonated and will be one of the strongest acids that can exist at pH 7·0. It is not unreasonable that the simultaneous attack of these two groups, situated at positions which allow them to approach the ester in a suitable manner, may be more effective than the simultaneous attack of a strong acid (or base) and the very weak base (or acid) water, which has a pK of 14.

Support for this type of explanation for the high effectiveness of enzymes has been provided by the discovery by Swain and Brown[1] that other catalysts containing both acidic and basic groups are also very effective catalysts. Swain and Brown, for example, have investigated the effectiveness of 2-hydroxypyridine as a catalyst for mutarotation reactions. These reactions are catalysed by both acids and bases, and Swain and Brown studied the action of the weak acid phenol

OH

[1] C. G. Swain and J. F. Brown, *J. Am. chem. Soc.*, **74**, 2538 (1952).

and the weak base pyridine

is a much weaker acid than phenol, and a much weaker base than pyridine; consequently, if it were to act as a catalyst in exactly the same way as phenol or as pyridine, it would be expected to be a very much less effective catalyst than either of them. It was found, on the contrary, that 2-hydroxypyridine is a very much more effective catalyst than either phenol or pyridine. In one study, in fact, Swain and Brown found a solution of 2-hydroxypyridine to be about 7000 times as effective a catalyst as a mixture of equivalent amounts of phenol and pyridine. This clearly indicates that 2-hydroxypyridine is not acting as a simple acidic or basic catalyst, and the only reasonable explanation is that it is acting as a 'bifuctional' catalyst, the acidic and basic groups acting simultaneously on the substrate.

Facilitated proton transfer

A suggestion that proton transfers may be facilitated by the hydrogen-bonded network in the enzyme has been put forward by Wang[1], with special reference to trypsin, chymotrypsin, the subtilopeptidases (subtilisins), and other hydrolytic enzymes. This idea is related to theories[2] put forward to explain the high rates of proton transfers in water and ice. Protons are suppose to jump from an H_3O^+ ion to a neighbouring water molecule, and the resulting ion, after some rotation, can transfer a proton to another water molecule.

Wang's suggestion is that a similar process can occur in the enzyme molecule itself; the X-ray evidence has indicated a hydrogen-bonded structure in certain enzyme–substrate systems.

Evidence for Wang's viewpoint is provided by studies in deuterium oxide, D_2O. For subtilisin and thiol-subtilisin the rate of deacylation of the acyl enzyme in D_2O was about a third as great as in H_2O. For the acetylation, however, this effect was found in subtilisin only, indicating that the thiol-subtilisin had undergone some change which prevented the proton transfer from being involved in the rate-controlling step.

[1] J. H. Wang, *Science*, **161**, 328 (1968); *Proc. natn. Acad. Sci. U.S.A.*, **66**, 874 (1970).
[2] M. Eigen, *Proc. R. Soc.*, **A247**, 505 (1958).

Another example of facilitated proton transfer is to be found in a sugges-tion[1] relating to the mechanisms of action of trypsin and chymotrypsin. It was postulated that the transfer of a proton during the acylation stages occurs as shown in Fig. 9.6. This suggested scheme makes use of the remarkable properties of the imidazole group as a proton-transfer agent. Eigen and his

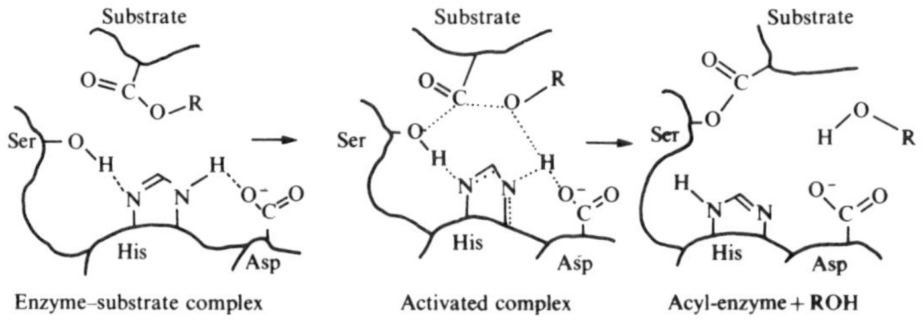

Fig. 9.6.   Possible proton-transfer processes during acylation of trypsin and chymotrypsin.

coworkers[2] have shown that the tautomeric character of the unprotonated imidazole ring makes it possible for a proton to be accepted and donated at the same time. The suggestion is therefore that the efficiency of acylation in these enzymes is due to the fact that the imidazole group is at the same time interacting with the carboxyl group and assisting in the removal of the proton from the serine hydroxyl group. The transfer of the proton of the imidazole ring to the alcoholic oxygen atom of the substrate is aided by the carboxyl group.

Neighbouring-group effects

It is well established that the rates of proton transfer can be strongly influenced by neighbouring groups. This effect has been invoked to explain the high efficiencies of enzymes. Thus ribonuclease catalyses the hydrolyses of polyribonucleotides forming a cyclic phospho-ester intermediate. The presence of the carbonyl group is found to be essential for catalytic activity, and it has been postulated[3] that the oxygen atom of this group forms a hydrogen bond with a hydroxyl group of the sugar ring, as shown in Fig. 9.7. As a result, the oxygen atom assists in the polarization of the —OH group which aids the formation of the cyclic ester.

Induced fit

It has been seen in the last chapter that entropies and volumes of activation for a number of enzyme reactions cannot be explained satisfactorily without

[1] D. M. Blow, J. J. Birktoft, and B. S. Hartley, *Nature*, 221, 337 (1969); H. P. Kasserra and K. J. Laidler, *Can. J. Chem.*, 47, 403 (1969).
[2] M. Eigen and G. G. Hammes, *Adv. Enzymol.*, 25, 1 (1963); M. Eigen, C. G. Hammes, and K. J. Kustind, *J. Am. chem. Soc.*, 87, 126 (1966).
[3] H. Witzel, *Prog. Nucl. Acid Res.*, 2, 221 (1963); E. N. Ramsden and K. J. Laidler, *Can. J. Chem.*, 44, 2197 (1966).

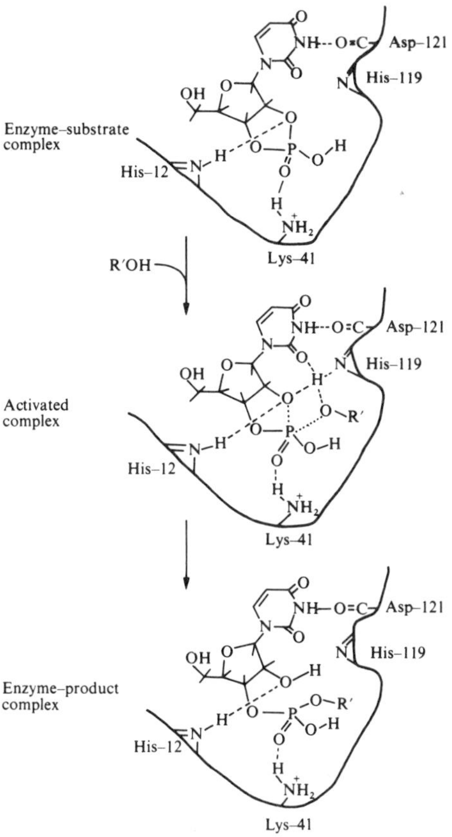

FIG. 9.7. The mechanism of Ramsden and Laidler for ribonuclease action. This is an example of a neighbouring group effect.

invoking the possibility of conformational changes. The occurrence of such conformational changes was first suggested with reference to an analysis of pressure effects; it was postulated[1] that during complexing there is sometimes an unfolding of the enzyme, during which process the active sites are made more available to the substrate. During the subsequent reaction the enzyme returns to its original conformation. Similar suggestions were put forward to explain the entropy values.

These ideas have been developed further by Koshland[2] in his *induced-fit* theory. According to this, the catalytic groups at the active centre of a free enzyme molecule are not in precisely the right positions to exert effective catalytic action. When, however, a substrate is bound to the enzyme, the

---

[1] K. J. Laidler, *Archs Biochem.*, **30**, 226 (1951); see *Chemical Kinetics of Enzyme Action*, 1st edition, pp. 223–4, 1958.

[2] D. E. Koshland, *Proc. natn. Acad. Sci.*, **44**, 98 (1958); *Adv. Enzymol.*, **22**, 45 (1960); D. E. Koshland and K. E. Neet, *A. Rev. Biochem.*, **37**, 359 (1968).

binding forces between the enzyme and the substrate force the enzyme into a conformation which is catalytically more active. A poor substrate, or a substrate analogue which is an inhibitor, may also bind to the active site, but will not have the necessary structural features to bring about the appropriate conformational change. The theory is represented schematically in Fig. 9.8.

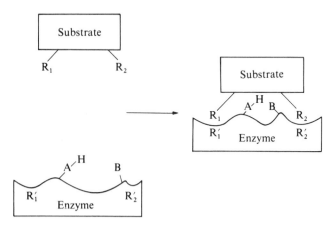

FIG. 9.8. Schematic representation of Koshland's induced-fit mechanism. The binding of substrate through the groups R_1 and R_2 causes a conformational change in the enzyme so that the active groups —A—H and —B, required for catalysis, are now in more effective positions.

As has been discussed by Bender *et al.*[1], the induced-fit theory can provide explanations for the very large positive and negative entropies of activation which are not readily interpreted on other grounds. It also accounts for the fact that substrate specificity is frequently observed to be more important at high substrate concentrations than at low. This may be seen in terms of the free-energy diagram shown in Fig. 9.9. Path 1 shows the hypothetical course of the reaction if it occurred without any conformational change in the enzyme. The normal state of the enzyme is represented as E_i, and this can in principle form the complex E_iA, which undergoes further reaction on passing through the activated state $EA^{\neq}$. If, on the other hand, the enzyme undergoes a conformational change on binding the substrate, as postulated in the induced-fit theory, it reaches the state E_aA, of higher energy. This complex undergoes reaction via the same activated complex $EA^{\neq}$. At low substrate concentrations this alternative route is of no advantage, since the enzyme is largely present as E_i, and $E_i + A$ has to acquire a free-energy increase of $\Delta G_0^{\neq}$ to reach the activated state. At high substrate concentrations, however, the enzyme is largely present as E_aA, and the free energy of activation is now $\Delta G_c^{\neq}$; this is less than the free energy of activation for the reaction via E_iA. It follows that when induced fit is involved the

[1] M. L. Bender, F. J. Kézdy, and C. R. Gunter, *J. Am. chem. Soc.*, **86**, 3714 (1964).

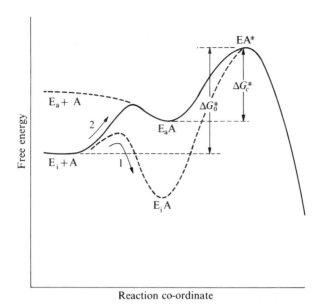

FIG. 9.9. Schematic energy-diagram which shows that the induced-fit theory will lead to an enhancement in rate, particularly at high substrate concentrations.

Michaelis constant K_A should be abnormally high, the strength of binding of substrate to enzyme being reduced.

One enzyme to which the induced-fit hypothesis has been applied is hexokinase; this catalyses the transfer of a phosphate group from ATP to water as well as to the hydroxyl group of glucose, but the rate of the reaction with water is 5×10^{-6} times that with glucose[1]; this difference corresponds to about 7 kcal of free energy. The suggestion is that the water molecule lacks the necessary side groups to bring about a distortion of the enzyme, but that the glucose molecule can do so; it will form a less stable, and more reactive, complex with the enzyme.

Strain or distortion

An alternative theory, put forward many years ago by Haldane[2] and later developed by Jencks[3], involves the postulate that the binding forces between the substrate and the enzyme are directly utilized to bring about the strain or distortion in the substrate molecule. On the simplest view the enzyme is assumed to be absolutely rigid, so that the substrate must become distorted when binding occurs. This theory is shown schematically in Fig. 9.10. The free energy diagram of Fig. 9.9 applies also to this theory, if E_aA is taken to be

[1] K. A. Trayser and S. P. Colowick, *Archs Biochem. Biophys.*, **94**, 161 (1961).
[2] J. B. S. Haldane, *Enzymes*, Longmans, Green and Co., London (1930); reproduced by the M.I.T. Press, 1965, p. 182.
[3] W. P. Jencks, *Catalysis in Chemistry and Enzymology*, McGraw-Hill Book Co., 1969, Chapter 5.

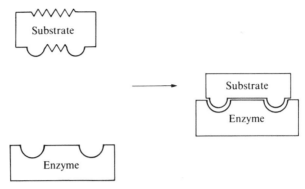

F<small>IG</small>. 9.10.   Schematic representation of the strain or distortion theory. The enzyme is assumed to be rigid; the substrate is distorted in forming the complex.

the distorted substrate bound to the enzyme, and E_iA the hypothetical complex formed from an undistorted substrate. Jencks discusses various lines of evidence in support of the strain theory; for example, strained cyclic compounds are frequently hydrolysed more rapidly than less strained cyclic ones and the corresponding straight-chain compounds. Further evidence for the strain theory derives from the X-ray studies of lysozyme[1], which show that the sugar ring of the inhibitor adjacent to the bond to be broken is in a strained, planar conformation, whereas all of the other sugar rings are in the unstrained chair conformation.

The induced-fit and strain theories can obviously be blended into one theory, in which *both* enzyme and substrate become distorted during complex formation; there is a corresponding raising of the free energy of the complex, with enhancement of the rate at high substrate concentrations. An alternative strain theory is that the free enzyme is strained, the strain being released on substrate binding[2].

Non-productive binding

Bernhard and Gutfruend[3] have put forward a hypothesis which is particularly designed to explain certain specificity results. Their suggestion is that a substrate may become bound to an enzyme in a catalytically unfavourable position. A good substrate may have several binding sites which are complementary to sites on the enzyme, as shown in the upper part of Fig. 9.11. A substance which lacks one of these binding sites, or has them in the wrong

[1] C. C. F. Blake, D. F. Koenig, G. A. Meir, A. C. T. North, D. C. Phillips, and V. R. Sarna, *Nature*, **206**, 757 (1965); C. C. F. Blake, G. A. Meir, A. C. T. North, D. C. Phillips, and V. R. Sarna, *Proc. R. Soc.*, **B167**, 365 (1967).
[2] A. Wishnia, *Biochemistry*, **8**, 5064, 5070 (1969).
[3] S. A. Bernhard and H. Gutfreund, *Proc. Int. Symp. Enzyme Chem.*, Tokyo, Kyoto, 124 (1958); see also S. A. Bernhard, *J. cell. comp. Physiol.*, **54**, Supp. 1, 256 (1959); T. Spencer and J. M. Sturtevant, *J. Am. chem. Soc.*, **81**, 1874 (1959); G. E. Hein and C. Niemann, *J. Am. chem. Soc.*, **84**, 4495 (1962); C. Niemann, *Science*, **143**, 1287 (1964); C. L. Hamilton, C. Niemann, and G. S. Hammond, *Proc. natn. Acad. Sci. U.S.A.*, **55**, 664 (1966).

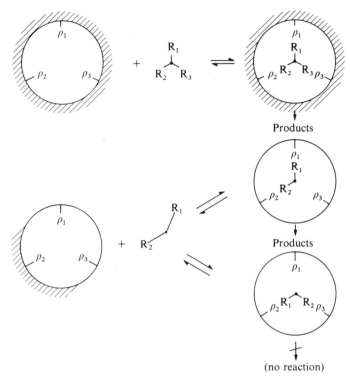

FIG. 9.11. Productive and non-productive binding of substrates.

position, may bind correctly, and give rise to reaction products, but it may also bind incorrectly to give a catalytically inactive complex. This is shown at the bottom of Fig. 9.11. Such a substance will be a less effective substrate; the observed k_c' value will be less than that for the productively-bound substrate, k_c, by the fraction of the substrate molecules that are bound productively (apart from other factors).

This hypothesis has been applied successfully to reactions catalysed by chymotrypsin[1]. For substrates of the type

$$\text{R}_1\text{CONHCHCOR}_3$$
$$\overset{\displaystyle \text{R}_2}{\underset{\displaystyle |}{}}$$

it is postulated that there are three sites on the enzyme, $\rho_1, \rho_2,$ and ρ_3, which are binding sites for $\text{R}_1, \text{R}_2,$ and R_3 respectively. With a good substrate there is correct matching of the substrate groups with the enzyme groups; for a substrate with incorrect matching, there is non-productive binding and low

[1] G. E. Hein and C. Niemann, *loc. cit.*; C. Niemann, *loc. cit.*; C. L. Hamilton, C. Niemann, and G. S. Hammond, *loc. cit.*

reactivity. This treatment has given a satisfactory interpretation of the reactivities of various substrates with chymotrypsin.

The fact that derivatives of D-amino acids are hydrolysed very slowly or not at all by chymotrypsin, although they are bound to the enzyme approximately as strongly as the L-substrates, can be explained in terms of non-productive binding. It has been observed[1] that acyl-enzymes formed from D-substrates are hydrolysed more slowly than those from non-specific substrates (such as acetyl-chymotrypsin), and that the faster the hydrolysis of the acyl enzyme formed from the L-substrate the slower is the hydrolysis of the acyl enzyme formed from the corresponding D-substrate. The non-specific substrate can presumably become bound to the enzyme in several ways, only a few of which lead to hydrolysis. A specific L-substrate is bound correctly and is rapidly hydrolysed; the corresponding D-substrate, however, is bound incorrectly and is not hydrolysed.

This hypothesis is capable of explaining many results, but other results seem to require a combination of various hypotheses. Thus, the fact referred to earlier, that hexokinase acts with glucose 2×10^5 times more rapidly than with water can hardly be explained in terms of the unproductive binding of water; induced fit or strain must also be involved in order to explain this result.

Orientation effects

Koshland[2] has compared the possible contributions of both proximity and orientation effects to enhancement of the rates of enzymic reactions. He has suggested that proper orientation of the substrate on the enzyme surface provides for good alignment of the reacting groups on enzyme and substrate, and hence for a rapid reaction. In free solution such correct alignment or orientation is not possible to the same extent. This effect was calculated to be particularly large where several substrates or reacting groups on the enzyme are involved. (In some cases, however, this effect led to a predicted reduction in rate.)

More recently one particular type of orientation has been examined in some detail, namely *orbital steering*[3]. Most chemical reactions are more likely to occur if the orbitals of the electrons involved overlap as much as possible. Storm and Koshland envisage such orbital overlap as being greatly increased by correct orientation of reacting groups on the surface of an enzyme. Experiments carried out by them with non-enzyme reactions indicated rate enhancements by factors of up to 10^4 as a result of such orbital steering effects. Different orbitals will have different angular requirements, thus making for more or less stringent orientation factors.

[1] D. W. Ingles and J. R. Knowles, *Biochem. J.*, **104**, 369 (1967).
[2] D. E. Koshland, *J. theor. Biol.*, **2**, 75 (1962).
[3] D. R. Storm and D. E. Koshland, *Proc. natn. Acad. Sci. U.S.A.*, **66**, 445 (1970).

Objections have, however, been raised to this theory. Port and Richards[1] carried out calculations which indicated that the large rate enhancements observed (and predicted) by Storm and Koshland cannot be explained by orbital overlap effects alone. Furthermore, orbital overlap requirements seemed to be less stringent than the theory. Other workers[2] pointed out that severe orientation effects would require very large force constants in the activated complex—larger even than those for a covalent bond.

Page and Jencks[3] adopt the view that orientation effects are likely to be of secondary importance in enzyme reactions. They suggest that translational and rotational motions provide an important entropic driving force for reactions.

[1] G. N. J. Port and W. G. Richards, *Nature*, **231**, 312 (1971).
[2] T. C. Bruice, A. Brown, and D. O. Harris, *Proc. natn. Acad. Sci. U.S.A.*, **68**, 658 (1971).
[3] M. I. Page and W. P. Jencks, *Proc. natn. Acad. Sci. U.S.A.*, **68**, 1678 (1971).

10. Some Reaction Mechanisms

DURING the past few years a considerable number of kinetic and mechanistic investigations have been carried out on individual enzyme systems. In a few cases these have been complemented by X-ray structural studies, which have contributed greatly to an understanding of the mechanisms. In the present chapter some examples of these investigations are given, with special reference to the purely kinetic aspects.

Lysozyme (E.C.3.2.1.17)

The enzyme lysozyme is found in most bodily secretions and in large quantities in the whites of eggs. The enzyme from hen eggs is a globular protein of molecular weight 14 600, having 129 amino acids in a single polypeptide chain with four disulfide bridges. It was the second protein, and the first enzyme, to have its detailed molecular structure worked out by X-ray analysis[1].

Lysozyme attacks many bacteria by dissolving (lysing) the polysaccharide structure of the cell wall. The basic unit of the polysaccharide, a hexose sugar, is shown in Fig. 10.1a. This sugar unit can be polymerized by linking its C_1 and C_4 carbons: the OH group attached to the C_1 atom is shown in the β configuration, which is the form in which it exists in the bacterial polysaccharides. If the R group in the sugar unit is —OH the unit is N-acetylglucosamine (NAG); if it is —$CH(CH_3)CO_2H$ the unit is N-acetylmuramic acid (NAM). The bacterial cell wall polymer is an alternating polymer of NAG and NAM. These chains are cross linked by short polypeptides that are attached to the —OR side chains of NAM by peptide bonds. The action of lysozyme is to catalyse the hydrolysis of the polysaccharide chain at a specific place, as shown in Fig. 10.1b; the cut is always at the far side of a linking oxygen atom that is attached to the C_4 of NAG. As well as the alternating—NAG-NAM-NAG-NAM—polymer, a polymer—NAG-NAG-NAG-NAG—made up entirely of NAG units is also a substrate, but poly-NAM is not. The trimer NAG-NAG-NAG is not a substrate but becomes attached to the enzyme and acts as an inhibitor; some of the X-ray studies were made with this lysozyme-tri-NAG complex.

A very schematic view of the enzyme, as revealed by X-ray studies, is shown in Fig. 10.2. A hexasaccharide substrate, containing units represented as A, B, C, D, E and F, is shown attached to the enzyme. The enzyme is seen to be egg-shaped, and its most striking feature is a crevice running across it. This crevice is largely lined with hydrophobic groups, so that substrate and inhibitor binding is to a considerable extent due to hydrophobic interactions.

It has been noted that the trimer NAG-NAG-NAG can become attached to the enzyme; the X-ray studies show that it is attached at the A, B and C

[1] For references see p. 10.

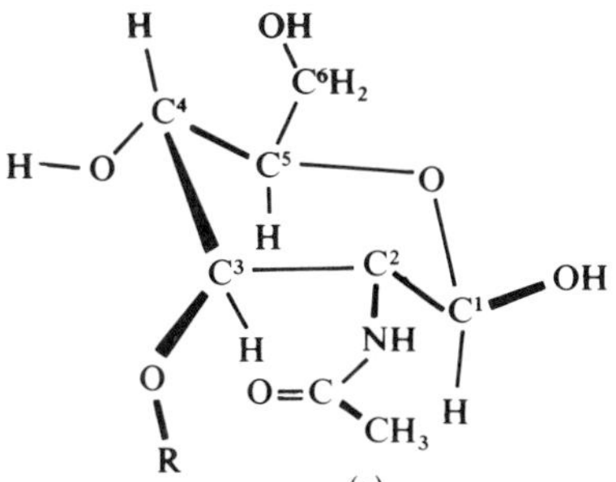

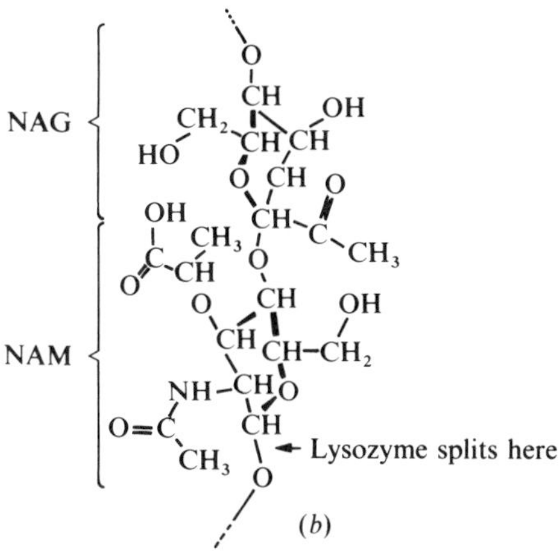

FIG. 10.1. (*a*) The basic unit of the polysaccharides found in the cell wall; if OR is OH the unit is *N*-acetylglucosamine; if OR is OCH(CH$_3$)CO$_2$H the unit is *N*-acetylmuramic acid. (*b*) Part of an alternating polymer of NAG and NAM.

positions shown in Fig. 10.2. Since this trimer is not split, it follows that the site of catalysis cannot be in this A, B, C region. Furthermore, the NAG-NAM-NAG-NAM-NAG-NAM hexamer is split on the far side of a linking oxygen atom that is attached to the C$_4$ of NAG. It therefore follows that the site of catalysis must be between units D and E, as shown in Fig. 10.2.

When consideration is given to possible catalytic groups near the linkage between D and E, two groups are found to be likely: aspartic acid–52, which lies in a polar environment so that the carbonyl group will exist as —CO$_2^-$, and glutamic acid–35, which lies in a hydrophobic environment and may well remain unionized.

On the basis of his X-ray studies Phillips[1] has proposed the mechanism shown in Fig. 10.3. The proton of glu–35 first attacks the linking oxygen atom and weakens the C$_1$—O bond with the formation of a carbonium ion; the formation of this ion is aided by the neighbouring charged asp–52. The

---

[1] For references see p. 10; see also D. M. Chipman and N. Sharon, *Science*, **165**, 454 (1969).

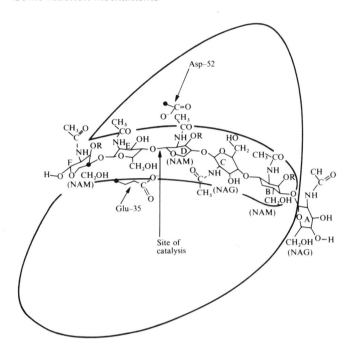

FIG. 10.2. Schematic representation of the lysozyme molecule, showing the crevice with hexamer fitting into it.

glu–35 proton is replaced by another from an ionizing water molecule; the resulting hydroxide ion then attacks the carbonium ion (stage 2).

This mechanism of the lysozyme reaction is one to which the X-ray structural studies have made the primary contribution. Kinetic and other chemical studies[1] are now providing evidence in support of Phillips' proposals.

Ribonuclease (E.C.2.7.7.16)

Ribonuclease† is also an enzyme which catalyses the cutting of long chains, and it resembles lysozyme in a number of ways. It is also a small egg-shaped or kidney-shaped molecule, having a molecular weight of 13 700 with 124 amino acids. Its action is to cut the links between the ribonucleotide units of ribonucleic acid (RNA). The ribonucleotides consist of ribose and phosphate with one of four kinds of pure base attached to each ribose. As shown in

† For a detailed review of the work on this enzyme see F. M. Richards and H. W. Wyekoff in *The Enzymes* (3rd edn.), ed. P. D. Boyer, Vol. 4, p. 647, 1971, Academic Press, New York and London.

[1] R. W. Rupley and V. Gates, *Proc. natn. Acad. Sci. U.S.A.*, **57**, 496 (1967); F. W. Dahlquish, T. Rand-Meir, and M. A. Raftery, *ibid.*, **61**, 1194 (1968); T. Y. Lin and D. E. Koshland, *J. biol. Chem.*, **244**, 505 (1969); S. M. Parsons and M. A. Raftery, *Biochemistry*, **8**, 701 (1969).

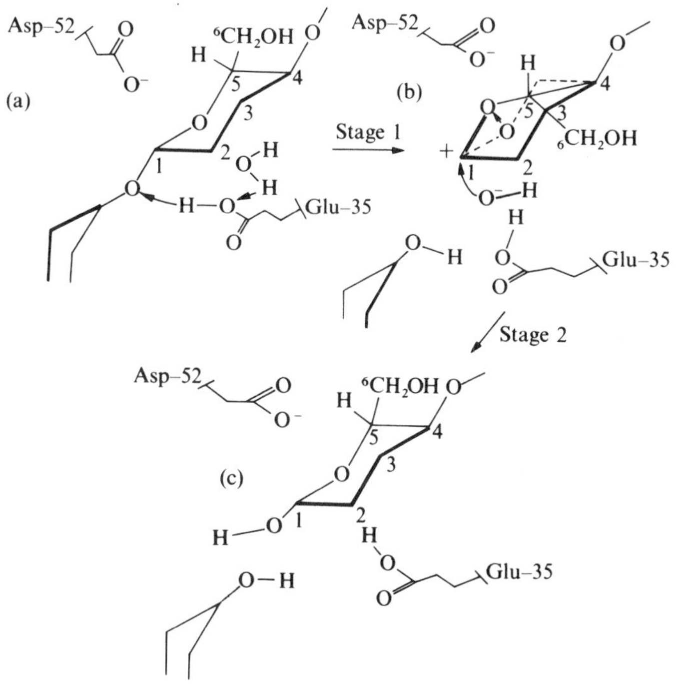

FIG. 10.3.   Suggested mechanism of hydrolysis by lysozyme.

Fig. 10.4, the ribonuclease first produces a cyclic phosphate (a cyclic 2′,3′-diester), and then catalyses the opening of this ring[1]. Some of the kinetic work on ribonuclease has been done with nucleic acid fragments rather than the nucleic acids themselves. The cleavage takes place only at those nucleotides containing pyrimidine bases (uracyl and cytosine) and not at purine bases (adenine and guanine). This specificity assumes considerable importance in most of the mechanisms which have been proposed, as will be seen below.

One of the first stages in any mechanistic investigation usually involves an attempt to identify those groups which are essential for catalysis. In the case of ribonuclease, a number of groups have been implicated in catalysis by means of a variety of techniques. Some of the evidence for this will now be discussed.

Histidines

A study of the effect of pH on the rate of hydrolysis, at high substrate concentrations, and on the Michaelis constant K_A, is a simple means of obtaining some indication of the nature of the groups involved, for example whether they are acidic or basic. For ribonuclease, the pH dependence of k_c

[1] L. Vandendriessche, *Acta chem. scand.*, **7**, 699 (1963); R. Markham and J. D. Smith, *Biochem. J.* **52**, 552 (1952); D. M. Brown and A. R. Todd, *J. chem. Soc.*, **2040** (1953).

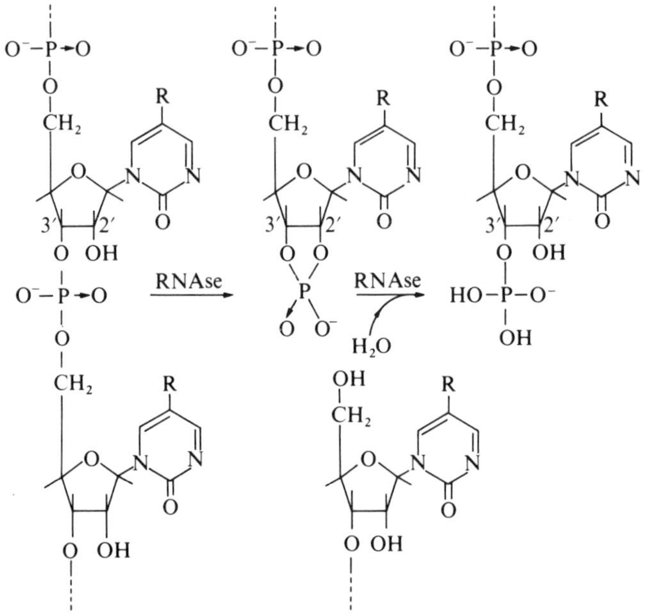

Fig. 10.4.   The mode of action of ribonuclease (RNAse).

and K_A for a variety of synthetic substrates has been observed to be bell-shaped[1]. (Synthetic substrates are normally used since the breakdown products of RNA produce an inhomogeneous solution, rendering a study of the kinetics by conventional techniques very difficult.) This is an indication that at least two groups are involved in both catalysis and binding. This conclusion, if reached purely on the basis of such a bell-shaped curve, is not always valid. For chymotrypsin, for example (see p. 314) the alkaline limb of the pH-rate (k_c) curve is the result of a pH-dependent conformational change, and the group involved does not participate directly in the catalytic reaction. The pK's of the groups involved in ribonuclease catalysis are about 5·5 to 6·5 for the acidic group, and about 7·5 to 8·0 for the basic group, depending on the particular substrate used, and on experimental conditions. The acidic group is almost certainly a histidine residue. The basic group, however, is consistent with the ionization both of histidine and lysine. This has complicated the elucidation of the mechanism, as will be seen below.

Numerous chemical modification studies have provided additional evidence for the presence of histidine residues in the active centre of this enzyme. Photooxidation leads to inactivation of the enzyme as a result of modification of a single histidine residue[2]. Reaction of the enzyme with

[1] D. G. Herries, A. P. Mathias, and B. R. Rabin, *Biochem. J.*, **85**, 127 (1962); E. N. Ramsden and K. J. Laidler, *Can. J. Chem.*, **44**, 2597 (1966); P. Wigler, *J. biol. Chem.*, **243**, 3466 (1968).
[2] F. M. Richards and P. J. Vithayathil, *Biochemistry*, **13**, 115 (1960).

bromoacetate[1] has been shown to result in alkylation of two histidines, numbers 12 and 119 in the sequence[2], yielding N^1-carboxymethylhistidine–119 and N^3-carboxymethylhistidine–12 as alternate modified residues in the enzyme. These products are both enzymically inactive, which indicates that both of these residues are necessary for catalysis. The ability of the enzyme to bind the inhibitor, phosphate, is reduced by such alkylation, indicating that these histidines are probably close to the phosphate-binding site[3].

Nuclear magnetic resonance studies[4] have also been used in an elegant manner to confirm the conclusion that, of the four histidines in the enzyme, it is his–12 and his–119 that are essential for catalysis (cf. p. 12).

Evidence that histidine–48 is necessary for catalysis can be found in n.m.r. studies[5] which showed that the peak for histidine–48 was significantly altered when the inhibitor 3′-cytidine monophosphate became bound to the enzyme, showing that this residue may be involved in binding this inhibitor. It is equally likely, however, that this peak shift is the result of some other conformational change, affecting histidine–48 indirectly. The other evidence for this residue comes from relaxation studies (see below). One of the relaxation processes observed for RNAse A is not seen at all for RNAse S†. Since there is a difference in the environment of histidine–48, this could be the residue responsible. The pK of the group involved in this relaxation process (shown to be a conformation change) was found to be about 6, implying that the residue is likely to be a histidine. It could, however, be another histidine residue. Further n.m.r. evidence was that the n.m.r. peak of histidine–48 was considerably different in sodium acetate compared with sodium chloride[6]. Since a similar difference has been noted for binding of certain inhibitors, histidine–48 might be implicated in this binding. The above evidence is, however, rather circumstantial.

Lysine–41

There is abundant evidence that this one lysine residue, out of a total of 10 in the enzyme, is essential for catalysis. Lysine–41 is resistant to reaction with

---

† Ribonuclease A is the normal form of the enzyme; ribonuclease S is a more easily crystallized form in which the polypeptide chain has been cleaved by subtilisin between residues 20 and 21, leaving the small S peptide which contains histidine–12.

[1] A. M. Crestfield, W. H. Stein, and S. Moore, *J. biol. Chem.*, **238**, 2413 (1963).
[2] R. R. Redfield and C. B. Anfinsen, *J. biol. Chem.*, **221**, 385 (1956); C. H. W. Hirs, S. Moore, and W. H. Stein, *J. biol. Chem.*, **235**, 633 (1960); D. G. Smyth, W. H. Stein, and S. Moore, *J. biol. Chem.*, **238**, 227 (1963).
[3] R. G. Fruchter and A. M. Crestfield, *J. biol. Chem.*, **240**, 3868 (1965).
[4] D. H. Meadows, O. Jardetzky, R. M. Epand, H. Ruterjans, and H. A. Scheraga, *Proc. natn. Acad. Sci. U.S.A.*, **60**, 766 (1968); D. H. Meadows and O. Jardetzky, *Proc. natn. Acad. Sci.*, **61**, 406 (1968).
[5] D. H. Meadows, J. L. Marbley, J. S. Cohen, and O. Jardetzky, *Proc. natn. Acad. Sci. U.S.A.*, **58**, 1307 (1967).
[6] G. C. K. Roberts, D. H. Meadows, and O. Jardetzky, *Biochemistry*, **8**, 2053 (1969).

N-carboxy-DL-alanine anhydride in phosphate buffer, but it is very reactive in bicarbonate buffer (both at neutral pH)[1]. The loss in enzymic activity after alanination was correlated with the extent of reaction at this residue. Further, dinitrobenzylation of ribonuclease at lysine–41 was shown to destroy enzyme activity[2]. The loss of activity in this case was reduced in the presence of the inhibitor phosphate, indicating that lysine–41 is near the binding site for phosphate. Similar experiments were carried out by Hirs[3] and by Klee and Richards[4].

X-ray evidence

Very convincing confirmation of all the above conclusions as to the nature of the groups present in the active centre of ribonuclease has been provided by X-ray crystallography. These studies have been carried out on the phosphate-inhibited RNAse A[5], and on RNAse S inhibited by uridine–2′, 3′-cyclic phosphate[6]. More recently, some work has been carried out on RNAse S complexed with cytidine 3′-monophosphate[7].

The overall picture of the enzyme is as a kidney-shaped molecule with a fairly deep crevice in the middle. Histidines–12 and –119 are situated on opposite sides of this crevice, both in the neighbourhood of the bound inhibitors in all cases, and near the phosphate group in particular. Lysine–41 is also near the active site, with histidine–48 not far away. The X-ray work also indicates that there is very little difference in structure between the enzymes RNAse A and RNAse S, except in the region of the S-peptide. There is a slight opening up of the structure around histidine–48 compared with RNAse A, for example.

Role of active groups in binding and catalysis

As has been mentioned, ribonuclease is specific for those bonds which have a neighbouring pyrimidine ring as the base in the substrate. Evidence has been obtained[8] that varying the substituents in the pyrimidine ring leads to considerable variation in k_c, but very little variation in the K_A; this indicates that the pyrimidine ring participates in catalysis, but not in binding. Addi-

[1] J. P. Cooke, A. B. Anfinsen, and M. Sela, *J. biol. Chem.*, **238**, 2034 (1963).

[2] C. B. Anfinsen, *Brookhaven Symp. Biol.*, **15**, 184 (1962).

[3] C. H. W. Hirs, *Brookhaven Symp. Biol.*, **15**, 154 (1962); see also R. L. Heinricksen, *J. biol. Chem.*, **241**, 1393 (1966).

[4] W. R. Klee and F. M. Richards, *J. biol. Chem.*, **229**, 489 (1957).

[5] G. Kartha, J. Bellow, and D. Harker, *Nature*, **213**, 862 (1967); and in *Structural, Chemical and Molecular Biology*, A. Rich (ed.), p. 29 (Freeman, 1968).

[6] H. W. Wyckoff, K. D. Hardman, N. M. Allewell, T. Inagami, L. N. Johnson, and F. M. Richards, *J. biol. Chem.*, **242**, 3984 (1967).

[7] H. W. Wyckoff, K. D. Hardman, N. M. Allewell, T. Inagami, D. Tsernoglou, L. N. Johnson, and F. M. Richards, in *Abstracts*, 155th meeting, American Chemical Society, April 1–7 (1968).

[8] H. Witzel, *Ann. Chem.* **635**, 191 (1960); *Prog. nucl. Acid Res.* (& mol. Brol), **2**, 221 (1963); H. Witzel, and E. A. Barnard, *Biochem. biophys. Res. Commun.*, **7**, 295 (1962); E. N. Ramsden and K. J. Laidler, *Can. J. Chem.*, **44**, 2597 (1966).

tional evidence for this was obtained by other workers[1] and is supported by X-ray work[2].

The inhibitor studies of Ukita *et al.*[3] led to the following three conclusions: (1) there is a site for binding to the $-CO-NH-CO-$ part of the pyrimidine ring; (2) the 2'-oxygen atom of the ribose sugar ring on the substrate binds to the enzyme, and (3) there is a site for the binding of phosphate. This implies that a total of three groups are required for binding of the substrate alone. The pH studies discussed above indicate that at least two groups must be involved in the catalytic reaction as well.

The effect of organic solvent on ribonuclease catalysis was studied by Rabin and coworkers[4]. Their results showed that the two groups involved in catalysis are both cationic in their acid forms. This means that one must be protonated, and the other not, and suggests an acid–base mechanism. Ramsden and Laidler[5] observed an increase in the pK of the basic group on complex formation, which indicates that this group is probably not involved in binding the substrate, since a decrease in pK would be expected as a result of interaction of the positive group on the enzyme with the negatively-charged substrate. The basic group is probably involved in catalysis. The acidic pK was also observed to increase on complex formation, indicating the possible formation of a hydrogen bond between this group and the substrate. Further evidence for such a hydrogen bond is given by Hammes *et al.*†. Very similar results for both acidic and basic groups were obtained by Rabin *et al.*[6]. These workers further suggested[7] that the 2'-oxygen atom on the substrate may interact with a positive group on the enzyme.

Several attempts have been made to assign specific groups to the various roles which have been discussed above. Unfortunately, the available evidence is ambiguous at best, and often circumstantial. Alkylation of histidine–119 destroys the enzyme activity, but nevertheless allows the enzyme to combine with substrate and inhibitors[8]. This would appear to indicate that histidine–12 is involved in binding; alternatively, it suggests that histidine–119 is involved in catalysis and not in binding. However, as has been discussed, the basic group of the enzyme is thought to be involved in catalysis. Nuclear magnetic resonance studies[9] indicate that histidine–119 has a more acidic pK (around

---

† However, n.m.r. work suggests that there may be no hydrogen bond. [G. C. K. Roberts, D. H. Meadows, and O. Jardetzky, *Biochemistry*, **8**, 2053 (1969).]

---

[1] D. G. Anderson, G. G. Hammes, and F. G. Watz, *Biochemistry*, **7**, 1637 (1968); M. Irie and F. Sawada, *J. Biochem.*, **62**, 282 (1967).
[2] H. W. Wyckoff, *et al.*, *loc. cit.*
[3] T. Ukita, K. Waku, M. Irie, and O. Hoshino, *J. Biochem.*, **50**, 405 (1961).
[4] D. Findlay, A. P. Mathias, and B. R. Rabin, *Biochem. J.*, **85**, 136 (1962).
[5] E. Ramsden and K. J. Laidler, *loc. cit.*
[6] A. Deavin, A. P. Mathias, and B. R. Rabin, *Nature*, **211**, 252 (1966).
[7] D. G. Herries, A. P. Mathias, and B. R. Rabin, *Biochem. J.*, **85**, 127 (1962).
[8] C. A. Ross, A. P. Mathias, and B. R. Rabin, *Biochem. J.*, **85**, 145 (1962).
[9] D. H. Meadows, G. C. K. Roberts, and O. Jardetzky, unpublished results, quoted in G. C. K. Roberts, E. A. Dennis, D. H. Meadows, J. S. Cohen, and O. Jardetzky, *Proc. natn. Acad. Sci. U.S.A.*, **62**, 1151 (1969).

5·8) than histidine–12 (around 6·4). This suggests that the catalytic group is histidine–12, contrary to the above. The n.m.r. experiments revealed that the pK of histidine–119 is sensitive to the binding of phosphate ion, whereas that of histidine–12 is not. This suggests that histidine–119 is involved in the binding of this inhibitor, with histidine–12 partaking in catalysis.

However, there is also evidence which shows that the two histidines appear to be indistinguishable. For example, alkylation of either histidine–12 or –119 leads to enzymic inactivation; it also reduces the binding of phosphate ion, and alkylation is prevented if phosphate ion is bound beforehand. Nuclear magnetic resonance studies indicate this as well, the pK's of the two histidines being similarly affected by alkylation[1], and by the binding of the inhibitor 3′-CMP[2].

A considerable amount of emphasis is sometimes placed on the X-ray structures of the enzyme in this respect. For example, the phosphate group lies between histidine–12 and histidine–119, but there appears to be better interaction between this group on the substrate and histidine–119 than histidine–12. In particular, it has been suggested[3], on the basis of the X-ray evidence, that the —OR group binding site is histidine–119, and that histidine–12 interacts with the 2′-OH on the ribose ring. This implicates both histidines in catalysis. Further, an interaction between the pyrimidine ring and the enzyme has been deduced.

Objections to X-ray crystallography may be raised here, namely (1) the observed interactions are for an *unreactive* complex, not a reactive substrate, and (2) the crystal structure may not correspond to the structure in solution. The first point is easily overcome, since the differences between substrate and inhibitor structure are fairly well known, and the lack of reactivity can be readily explained in many cases (as in chymotrypsin). On the second point, there is evidence both ways, depending on the particular enzyme (see Chapter 1). For ribonuclease, Winstead and Wold[4] have pointed out that the crystals were obtained from concentrated solutions of buffer (such as 3M ammonium sulphate). They have found that there is a difference in the properties of the enzyme in dilute and concentrated buffer salts:

(1) hydrolysis of cyclic cytidylate was enhanced in high salt concentrations, but reduced for RNA as substrate;
(2) inhibition by RNA and phosphate of the enzyme is eliminated at high concentrations of salt;
(3) RNAse S and RNAse A were found to have similar physical and catalytic properties at low salt concentrations, but different properties at high.

[1] Meadows, *et al.*, *Proc. natn. Acad. Sci. U.S.A.*, **60**, 766 (1968), *loc. cit.*
[2] Meadows, *et al.*, *ibid.*, **58**, 1307 (1967) and *loc. cit.*, and D. H. Meadows and O. Jardetzky, *ibid.*, **61**, 406 (1968).
[3] G. C. K. Roberts, *et al.*, *Proc. natn. Acad. Sci. U.S.A.*, **62**, 1151 (1969).
[4] J. A. Winstead and F. Wold, *J. biol. Chem.*, **240**, 3694 (1965).

While this objection may be overcome by the fact that n.m.r. studies have on the whole agreed quite well with X-ray data, it should be noted that there are some inconsistencies in the n.m.r. data, as has been pointed out above.

In conclusion, it can be seen that the precise assignment of the roles of the histidines–12 and –119 is not really possible at the present time. It is not unreasonable, however, that one or both of them could affect binding to a certain extent as well as being involved in catalysis. This would reconcile some of the ambiguities in the above evidence.

Similar contradictions are found in the evidence concerning the role of lysine–41 in catalysis. Modification of this residue renders the enzyme inactive, but the inhibitor 2′-CMP has been observed to bind to this inactive enzyme[1]. This appears to show that the residue is involved in catalysis rather than binding. However, a different conclusion was reached by other workers[2] in what is an extension of previous work by Klee and Richards[3]. Ribonuclease was guanadinated using two different reagents. The one (1-guanadino-3,5-dimethylpyrazole nitrate, GDMP) alkylated only nine of the ten residues present in the enzyme and the resulting derivative was still active. The other (*o*-methyl isourea) alkylated all ten lysines, producing an inactive enzyme. The tenth lysine was shown to be lysine–41. Even in the fully guanidinated enzyme, histidines–12 and –119 in the active site could still be alkylated in the usual way. This indicated that removal of the basicity of these histidine residues was not the factor responsible for the loss in enzyme activity. These authors concluded that the two histidines are involved in catalysis, and the lysine–41 in binding.

Despite the certain amount of confusion about the exact roles of the various groups, a considerable amount of information is known about the mechanisms, and a number of such mechanisms have been proposed. One of these, due to Hammes[4], uses one histidine only in the catalytic step, to act as a general acid in the first step of the reaction (cyclization), and then as a general base in step 2. However, stereochemical studies show that more than one group must be involved in the catalysis. These studies have centred on whether or not pseudorotation can occur at the phosphorus atom in the phosphordiester group on the substrate[5]. Phosphates have a trigonal bipyramid structure, the oxygen atoms being either apical (two) or equatorial (three) as indicated in Fig. 10.5. A number of semi-empirical rules have been developed on the basis of numerous organic experiments with phosphates and other compounds[6]: (1) ring strain in cyclic phosphates is minimized in a 5-membered ring if the ring spans one apical and one equatorial position; (2) more electronegative (especially negatively-charged) groups occupy apical

[1] E. A. Barnard and H. Ramel, *Nature*, **195**, 243 (1962).
[2] D. M. Glick and E. A. Barnard, *Biochim. biophys. Acta*, **214**, 326 (1970).
[3] W. R. Klee and F. M. Richards, *J. biol. Chem.*, **229**, 489 (1957).
[4] G. G. Hammes, *Acct chem. Res.*, **1**, 321 (1968).
[5] For a review on this topic see F. H. Westheimer, *Acct chem. Res.*, **1**, 70 (1968).
[6] F. H. Westheimer, *loc. cit.*

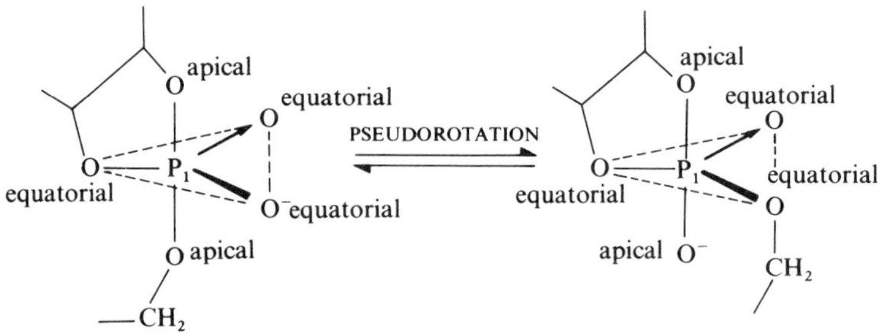

FIG. 10.5.   Pseudorotation about the phosphorus atom during ribonuclease action.

positions; (3) pseudorotation may occur; and (4) groups must enter and leave such a cyclic phosphate (on formation or decomposition) from apical positions.

Usher and coworkers[1] have pointed out that, in the second step of the reaction, a mechanism involving pseudorotation must have the leaving group in an adjacent position, i.e. apical if the attacking group is equatorial. The alternative would be the so-called in-line mechanism, with both attacking and leaving groups apical. The former mechanism must apply for a mechanism such as that suggested by Hammes. (The alternative is a considerable movement of the catalytic group or the substrate, for which there is not much evidence.) The isolation of two enantiomers of uridine-2',3'-cyclic phosphorothioate by Eckstein[2] led to this study using H_2O^{18} to determine the geometry of the ring-opening and ring-closing step. For an in-line mechanism the one isomer is expected to contain no excess of O^{18}, while the other isomer should have one equivalent of the label. This was found to occur for both the chemical and enzymic hydrolysis, indicating that the second step in ribonuclease catalysis occurred by an in-line mechanism. This does not involve pseudorotation. Since the first step is almost the exact reversal of the second it is likely to have the same mechanism. This indicates that two separate groups must be involved in catalysis. For further discussion of these points the reader is referred to the review of Richards and Wyckoff[3].

Mechanism of Witzel

On the basis of considerations such as those outlined above a number of workers have proposed specific detailed mechanisms for the ribonuclease reaction. These will now be considered.

The mechanism of Witzel[4] incorporates the effect of substitution in the pyrimidine ring on the catalytic rate constant at high substrate concentrations by utilizing the 2'-carbonyl group of the ring to polarize the 2'-hydroxyl

[1] D. A. Usher, D. I. Richardson, and F. Eckstein, *Nature*, **228**, 665 (1970): see also D. A. Usher, *Proc. natn. Acad. Sci. U.S.A.*, **62**, 661 (1969).
[2] F. Eckstein, *J. Am. chem. Soc.*, **92**, 4718 (1970).
[3] F. M. Richards and H. W. Wyckoff, in *The Enzymes* (3rd edn.), ed. P. D. Boyer, Vol. 4, p. 647, 1971, Academic Press, New York and London.
[4] H. G. Gassen and H. Witzel, *Eur. J. Biochem.*, **1**, 36 (1967).

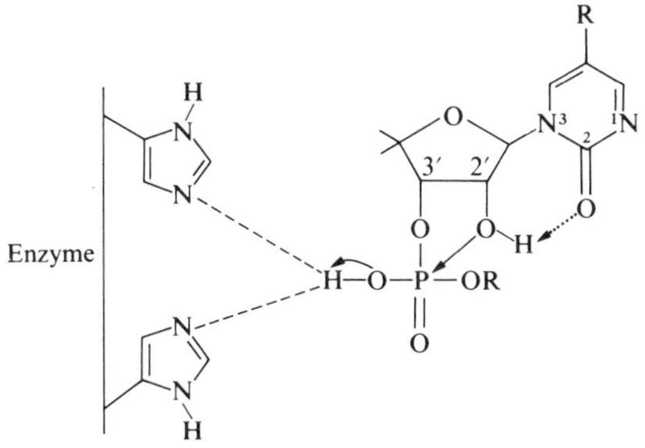

FIG. 10.6. Essential feature of Witzel's mechanism for ribonuclease action.

on the ribose ring (see Fig. 10.6). The two histidine residues, –12, and –119, are not used in the catalytic step at all, but rather for binding the substrate at the phosphate group. It has been shown, however, that there are binding sites on the enzyme for parts of the substrate in addition to the phosphate moiety. Further, at least one of the histidines is involved in the catalysis, and the pH dependence of the reaction would be expected, on the basis of this mechanism, to reflect the ionization of the substrate, and not the ionization of the groups on the enzyme, as is the case.

Mechanism of Rabin

In Rabin's mechanism[1] the two histidine residues are invoked in the catalytic step as an acid–base pair, the one accepting the proton from the 2'-hydroxyl on the ribose ring in the cyclization step, and replenishing it in the second step. The second histidine provides the proton for the formation of the alcohol. The pyrimidine 2'-carbonyl group is also utilized as shown in Fig. 10.7. (In an earlier version of this mechanism[2], the pyrimidine ring was not used at all, contrary to experimental evidence already presented). A similar mechanism has been proposed by Roberts[3] and coworkers, who identify histidine–12 as interacting with the 2'-hydroxyl on the sugar ring, and histidine–119 with the P—O bond and with water. This latter histidine residue is also implicated in binding in their mechanism, consistent with the ambiguous nature of some of the evidence discussed above.

The mechanism of Wang

In Wang's mechanism[4] (see Fig. 10.8) the acidic catalytic group is considered as being either a lysine or a histidine. The basic one is a histidine

[1] A. Deavin, A. P. Mathias, and B. R. Rabin, *Nature*, **211**, 252 (1966).
[2] D. Findlay, A. G. Herries, A. P. Mathias, B. R. Rabin, and C. A. Ross, *Biochem. J.*, **85**, 652 (1962).
[3] G. C. K. Roberts, *et al.*, *Proc. natn. Acad. Sci. U.S.A.*, **62**, 1151 (1969).
[4] J. H. Wang, *Science*, **161**, 328 (1968).

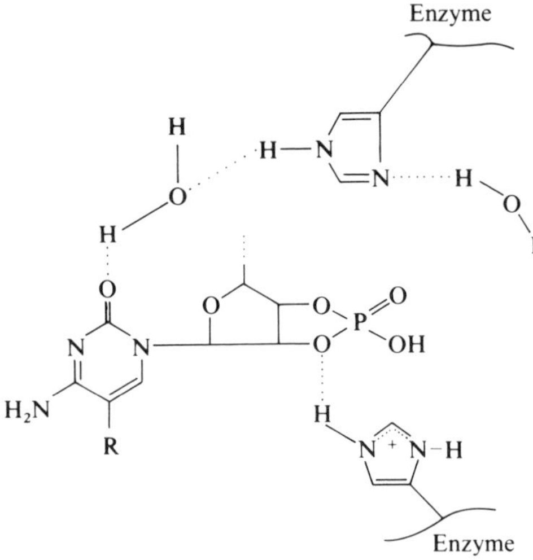

Fig. 10.7.   Essential feature of Rabin's mechanism for ribonuclease action.

residue, forming a hydrogen bond with the 2′-hydroxyl. This histidine also protonates the leaving alkoxyl group, as in the Rabin mechanism. The acidic group is used mainly for neutralizing the negative charge on the phosphate group, i.e. in binding. Account is taken in this mechanism of pseudorotation possibilities, and the acidic group assists here by supplying and then retrieving its proton in the cyclization process. Supplying the proton keeps the group in an apical position; reprotonation returns it to an equatorial one, and the leaving alcohol group to an apical position. This mechanism involves pseudorotation, which has been shown as unlikely to occur.

The mechanism of Ramsden and Laidler

In this mechanism[1], aspartate–121† and lysine–41 are specifically included, both as binding groups (see Fig. 9.7, p. 285). The former hydrogen bonds with the N—H group of the pyrimidine ring, accounting for the inhibition requirements of Ukita and coworkers (see p. 299). The latter hydrogen bonds with an oxygen atom, on the phosphate group. In catalysis, the histidines act again as an acid–base pair, with a type of charge-relay system, similar to that invoked by Wang. The pyrimidine ring holds a water molecule by hydrogen bonding, in a manner analogous with the later version of Rabin's mechanism; but instead of the other end of the water molecule hydrogen bonding with a histidine residue, it does so with the P—O bond to be broken. The imidazoles

---

† There is no strong evidence that aspartate–121 is involved in the reaction, but suggestions that it is have been made by W. H. Stein, *Brookhaven Symp. Biol.*, **13**, 104 (1960), and by C. B. Anfinsen, *J. biol. Chem.*, **221**, 405 (1956).

---

[1] E. N. Ramsden and K. J. Laidler, *Can. J. Chem.*, **44**, 2597 (1966).

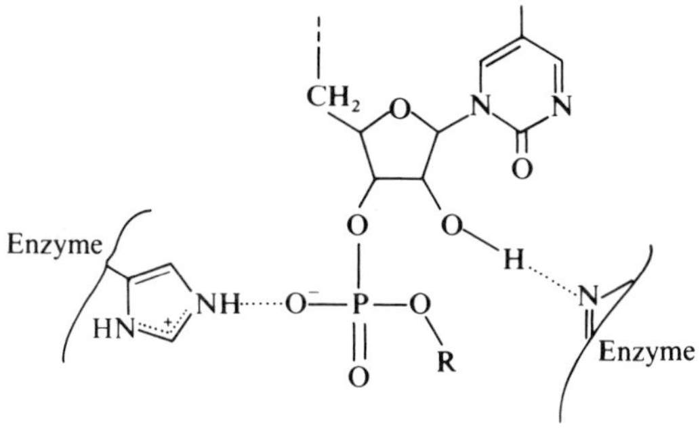

FIG. 10.8.   Essential feature of Wang's mechanism for ribonuclease action.

do not need to re-ionize during catalysis, however, as is the case in the mechanism of Rabin.

Temperature-jump relaxation studies

Hammes and coworkers[1] have used the temperature-jump relaxation technique, together with the stopped-flow technique, to investigate the number of various enzyme species that are involved in the overall catalysis. An indicator was used; its ionization was temperature dependent, so that the

TABLE 10.1

Relaxation times observed in the ribonuclease system

Enzyme species	Number of relaxation times	Type of reaction involved	Magnitude
E	$1(\tau_1)$	Isomerization	10^{-3}–10^{-4} s
E-C3'P	$2(\tau_2 ; \tau_3)$	Bimolecular; and isomerization	
E-C2',3'P	$2(\tau_5 ; \tau_6)$	Bimolecular; isomerization	10^7–10^8 mol^{-1} for bimolecular;
E-C3',5'P	$2(\tau_7 ; \tau_8)$	Biomolecular; isomerization	10^{-3}–10^{-4} for isomerization

perturbation of the system could readily be followed. It was proved that the indicator itself did not bind to the enzyme to produce any possible artefacts. A number of relaxation times (τ) were observed; these are summarized in Table 10.1.

The relaxation time τ_1 is observed in the enzyme even in the absence of substrate[2]. It involves the catalytic site of the enzyme, since it is not observed when 3'-cytidine monophosphate or 2'-cytidine monophosphate are bound

[1] G. G. Hammes, *Adv. Prot. Chem.*, **23**, 1 (1968); *Acct chem. Res.*, **1**, 321 (1968); G. G. Hammes and F. G. Walz, *J. Am. chem. Soc.*, **91**, 7179 (1969).
[2] T. C. French and G. G. Hammes, *J. Am. chem. Soc.*, **87**, 4669 (1965).

to the enzyme. Only a certain portion of the active site appears to be involved, however, since sulphate (a known inhibitor of the enzyme) does not affect this relaxation time, though it binds to the active site[1]. This relaxation time is independent of substrate concentration, and involves a conformational change. pH studies indicate that a group of pK about 6 (probably a histidine residue) is involved. Since carboxymethylation of histidine–119 prevents this isomerization, it would appear that it is histidine–119 that is concerned. Since the rate corresponding to τ_1 is lower in D_2O than in H_2O[2] a hydrogen bond to this histidine may be present. Hammes suggests that it may be a hydrogen bond to a negative carboxyl residue in the enzyme.

In the presence of substrate (3′-CMP, and uridine-3′-phosphate) a new relaxation time (τ_2) is observed, becoming shorter as the substrate concentration is increased. This is a bimolecular process, with one of the enzyme isomers binding the substrate more strongly than the other. When the enzyme was saturated with a substrate, a new relaxation time (τ_3) was observed. This is associated with the isomerization of the enzyme–substrate complex. It is also independent of substrate concentration, but pH dependent.

Similar results have been observed for the other (cyclic) phosphates, cytidine-2′,3′-phosphate and cytidylyl-3′,5′-cytidine[3].

Independent evidence for conformational changes in ribonuclease catalysis has been obtained. For example, the rate of digestion by subtilisin of the enzyme is considerably reduced in the presence of inhibitors[4]. Since the point of cleavage by subtilisin is considerably removed from the binding site of the inhibitors used, a conformational change is indicated[5].

The simplest mechanism which must be invoked to explain the above results is indicated in Fig. 10.9.

By following the effect of pH on the various relaxation processes, Hammes and coworkers found that three groups were involved, of pK's 5, 6 and 6·7. This would seem to implicate three histidine residues in these processes. Hammes suggests that two of these are histidines–12 and –119; and the third, on the basis of his work and the X-ray and n.m.r. data already discussed, is postulated as histidine–48.

Chymotrypsin (E.C.3.4.4.5)

Chymotrypsin is a pancreatic enzyme which is concerned with the hydrolysis of proteins during the digestive process. It belongs to the class of proteolytic enzymes, or proteases; it catalyses the hydrolysis of peptide bonds which are on the carboxyl terminal side of aromatic and other non-polar residues.

[1] H. A. Saroff and W. A. Garroll, *J. biol. Chem.*, **237**, 3384 (1962).
[2] R. A. Cathou and G. G. Hammes, *J. Am. chem. Soc.*, **86**, 3240 (1964).
[3] J. E. Erman and G. G. Hammes, *ibid.*, **88**, 5607 and 5614 (1966).
[4] G. Markus, E. A. Barnard, B. A. Castellani, and D. Saunders, *J. biol. Chem.*, **243**, 4070 (1968).
[5] Further evidence for this is given by R. E. Cathou, G. G. Hammes, and P. R. Schimmel, *Biochemistry*, **4**, 2687 (1965).

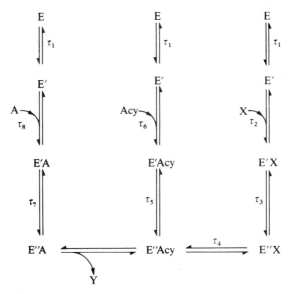

FIG. 10.9. The simplest mechanism required to explain the temperature-jump relaxation studies of Hammes and coworkers. The substrate A is cytidylyl-3',5'-cytidine while Acy is cytidine-2',3'-cyclic phosphate. The enzyme exists in two isomeric forms E and E'. The products are X and Y.

Activation

All the pancreatic enzymes are formed from inactive precursors, the zymogens, in an activation process that involves the removal of a few amino acid residues. The manner in which chymotrypsin is formed from its precursor chymotrypsinogen, which consists of a single polypeptide chain, is shown in Fig. 10.10[1]. Another of the pancreatic enzymes, trypsin, catalyses this activation process, as does chymotrypsin itself. The usual form of active chymotrypsin is α-chymotrypsin. There also exists δ-chymotrypsin which, as seen from Fig. 10.10, has been formed from chymotrypsinogen by the splitting out of a dipeptide; it therefore contains two polypeptide chains. In the formation of α-chymotrypsin an additional dipeptide has been split out, so that α-chymotrypsin consists of three separate polypeptide chains, which are designated as A, B and C.

Structure

α-Chymotrypsin has been prepared in pure crystalline form, its amino-acid sequence determined[2], and its X-ray structure established in some detail.

[1] See P. Desneuelles, in *The Enzymes*, **4**, 93 (1960); C. F. Jacobsen, *C.r. Trav. Lab. Carlsberg, Ser. Chim.*, **25**, 325 (1947).
[2] For details of the amino-acid sequence see B. S. Hartley, *Nature*, **201**, 1281 (1964).

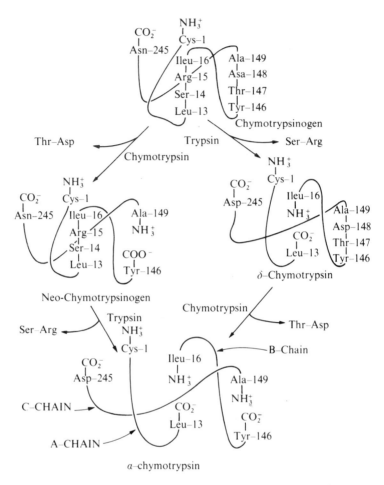

FIG. 10.10.   The conversion of chymotrypsinogen into chymotrypsin, catalysed by the enzymes trypsin and chymotrypsin. The diagram gives an approximate idea of the relative positions of some of the groups, but for simplicity the chains are shown purely schematically (see also Fig. 1.2, p. 11).

The X-ray work has been described briefly in Chapter 1, and Fig. 1.2 (p. 11) gives a schematic representation of the structure of α-chymotrypsin as determined by D. M. Blow and his coworkers. Figure 10.11 gives a more detailed view of the groups that are in the neighbourhood of the active centre of the enzyme.

A vast number of kinetic and mechanistic studies have been carried out on this enzyme. The present account covers only a selection of the relevant publications, and is concerned with what we believe to be the main aspects of the problem of arriving at an understanding of the detailed molecular mechanism.

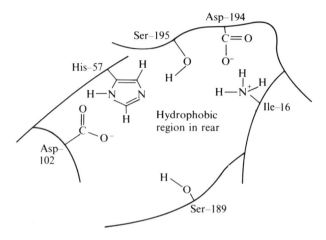

FIG. 10.11. A schematic representation of the essential groups at the active centre of chymo-trypsin (adapted from H. P. Kasserra and K. J. Laidler, *Can. J. Chem.*, **47**, 4031 (1969)).

Active groups

The evidence that chymotrypsin has one active centre per molecule has been discussed in Chapter 1. An understanding of the mechanism of action of the enzyme involves a knowledge of what amino acid residues are essential for activity, and of what role each of these groups plays in catalysis. The first matter, the identity of the groups, is well understood, and will now be considered; the second is still a source of controversy, and is discussed later.

Some indication of the nature of the active groups is provided by studies of the effect of pH on the k_c and K_A values, with a variety of different substrates. The type of curve obtained[1], compared with trypsin, is shown in Fig. 5.8 (p. 159). The same type of behaviour is shown by both amide and ester substrates; for the former acylation, and for the latter deacylation, is rate limiting (see later). The usual type of analysis of the pH behaviour (see Chapter 5) leads to the conclusion that k_c/K_A is affected by two ionizing groups of pK values ~ 7 and ~ 8.5 while k_c (whether corresponding to acylation or deacylation) is affected by the group of pK ~ 7 only. The only group in proteins having a pK of about 7 is histidine. Furthermore, studies of the effect of solvent on the pK values indicate that the group of pK ~ 7 is neutral, while that of pK ~ 8.5 is cationic[2], when the enzyme is in its active form.

Further evidence for the participation of histidine comes from several other types of study. Chemical modification, using the photooxidation

---

[1] See, for example, M. L. Bender, G. E. Clement, F. J. Kézdy, and H. d'A. Heck, *J. Am. chem. Soc.*, **86**, 3680 (1964); A. Himoe, P. C. Parks, and G. P. Hess, *J. biol. Chem.*, **242**, 919 (1967); H. Kaplan and K. J. Laidler, *Can. J. Chem.*, **45**, 547 (1967).
[2] H. Kaplan and K. J. Laidler, *ibid.*, **45**, 539, 547 (1967).

technique[1], indicated that oxidation of histidine–57 led to complete loss of catalytic activity. Inhibition of the enzyme by substrate analogues such as tosylphenylalanylchloromethylketone (TPCK)[2] again caused loss of activity, and the inhibitor was located as covalently attached to histidine–57. A variety of other substrate analogue inhibitors have subsequently been developed; some of them are used to determine the 'operational molarity' of the enzyme (i.e. the operative number of active sites in the solution)[3]. Model compounds, possessing an imidazole group, are found to be able to carry out similar non-enzymic hydrolyses, indicating that the histidine is involved in catalysis and that it is involved as a base[4]. Confirming proof came from the X-ray analysis of the inhibited enzyme[5], where histidine–57 was observed to be in the active site, close to the bound inhibitor molecule.

That serine–195 is essential to the action of the enzyme has also been conclusively demonstrated. The inhibitor diisopropylfluorophosphate (DFP) completely inactivates chymotrypsin (and the other pancreatic enzymes)[6]. Amino acid analysis of the resulting derivative indicated that the inhibitor becomes attached to the serine hydroxyl residue. A certain amount of controversy resulted over these experiments, since it was possible that the modification of the serine might cause steric interference in the active site, leading to indirect loss of activity; or that the inhibitor was actually bound to histidine–57, and was subsequently transferred, by a non-catalytic process, to the serine residue, perhaps during the amino-acid hydrolysis procedure. The first objection has been conclusively overcome by experiments on the subtilopeptidases, A, B and C (the subtilisins); these are bacterial proteases which are known to react by a mechanism similar to that of chymotrypsin[7]. In the subtilisins the serine —OH was converted to —SH without interfering with the rest of the enzyme[8], since no other sulphur groups are present. Steric hindrance by a bulky group is not possible here, and considerable inactivation occurred; there was not total loss because the —SH group is still catalytically active in the same way, having similar properties to the —OH. The second objection was overcome by comparing the spectra of DFP-chymotrypsin with that of model serine–phosphate compounds and model histidine–phosphate compounds[9], and also by comparing their

[1] D. E. Koshland, D. H. Strumeyer, and W. J. Ray, *Brookhaven Symp. Biol.*, **15**, 101 (1962).

[2] B. Ong, E. Shaw, and G. Schoellmann, *J. biol. Chem.*, **240**, 694 (1965).

[3] G. R. Schonbaum, B. Zerner, and M. L. Bender, *J. biol. Chem.*, **236**, 2930 (1965); B. F. Erlanger, S. N. Boxbaum, R. A. Sack, and A. G. Cooper, *Archs Biochem. Biophys.*, **19**, 542 (1967); B. F. Erlanger, *et al.*, *Biochem. J.*, **118**, 421 (1970); D. V. Roberts, *et al.*, *ibid.* (1971).

[4] See e.g. W. P. Jencks and J. Carriuolo, *J. Am. chem. Soc.*, **83**, 1743 (1961).

[5] D. M. Blow, J. J. Birktoft, and B. S. Hartley, *Nature*, **221**, 337 (1969); T. A. Steitz, R. Henderson, and D. M. Blow, *J. molec. Biol.*, **46**, 337 (1969); R. Henderson, *J. molec. Biol.*, **54**, 341 (1970).

[6] B. S. Hartley, *A. Rev. Biochem.*, **29**, 45 (1960).

[7] A. N. Glazer, *Proc. natn. Acad. Sci.*, **59**, 996 (1968), and references quoted therein.

[8] K. E. Neet and D. E. Koshland, *Proc. natn. Acad. Sci.*, **56**, 1606 (1966); K. E. Neet, A. Nanci, and D. E. Koshland, *J. biol. Chem.*, **243**, 6392 (1968); L. Polgar and M. L. Bender, *Biochemistry*, **6**, 610 (1967); *Proc. natn. Acad. Sci. U.S.A.*, **64**, 1335 (1969).

[9] J. Mercouroff and G. P. Hess, *Biochem. biophys. Res. Commun.*, **11**, 283 (1963); M. L. Bender, *J. Am. chem. Soc.*, **84**, 2582 (1964); S. A. Bernhard, S. J. Lau, and H. Noller, *Biochemistry*, **4**, 1118 (1965).

respective rates of hydrolysis[1]. Both experiments showed a greater similarity between enzyme and model serine compounds than between enzyme and model histidine compounds, indicating that the acylenzyme (see below) does form at serine–195.

The photo-oxidation processes discussed above, and others, also modified methionine–192. However, it was shown[2] that this led to only partial inactivation of the enzyme, and not to total loss of activity as with the modification of histidine–57, indicating that methionine–192 was involved not as a catalytic group, but in maintaining the proper enzyme structure. The X-ray studies quoted above indicate that this is so, methionine being at the active site.

Aspartic acid–102 has been shown by X-ray studies to be at the active centre. It is thought that there is a hydrogen bond between this residue and the N-terminal residue of one of the chains, isoleucine–16. This ion pair has been suggested as maintaining the conformation of the active site under certain conditions, but not under others, causing activity or loss of activity depending on the circumstances. This particular point will be discussed in some detail below.

Nature of the catalytic reaction

The nature of the groups involved either directly or indirectly in chymotrypsin catalysis having been established, it remains to determine the exact nature of the catalytic reaction, and then to assign functional roles to the various groups, relating them to the detailed mechanism. This latter controversial point will be dealt with later. At this stage, the overall nature of the catalytic process and the roles of aspartic acid–102 and isoleucine–16 will be discussed.

It has now been conclusively established, by a variety of different experimental approaches, that the minimal mechanism of hydrolysis by chymotrypsin involves two intermediates (see Chapter 3):

$$E + A \underset{k_{-1}}{\overset{k_1}{\rightleftharpoons}} EA \overset{k_2}{\searrow} EA' \overset{k_3}{\rightarrow} E + Y.$$
$$X$$

The first stage involves the formation of a Michaelis–Menten complex between enzyme and substrate; this is followed by an acylation process in which the substrate is attached to serine–195 and the first product X is released. There is finally a deacylation process, yielding the second product Y and the free enzyme. Some of the experimental evidence for this mechanism will now be reviewed.

[1] B. M. Anderson, E. H. Cordes, and W. P. Jencks, *J. biol. Chem.*, **236**, 455 (1961).
[2] W. B. Lawson and H. J. Schramm, *J. Am. chem. Soc.*, **84**, 2017 (1962); H. Weiner, C. W. Batt, and D. E. Koshland, *J. biol. Chem.*, **241**, 2687 (1966); W. J. Ray and D. E. Koshland, *J. Am. chem. Soc.*, **85**, 1977 (1963).

(1) The earliest indication of the existence of the second intermediate, or acyl-enzyme, came from the observation, in a stopped-flow experiment†, of a burst of *p*-nitrophenol in the pre-steady-state reaction of chymotrypsin with the non-specific substrate, *p*-nitrophenylacetate[1]. After the burst, the alcohol was released at a constant rate, equal to the steady-state rate of hydrolysis. This was interpreted as implying that there was a rapid formation of an intermediate followed by a rate-determining breakdown of the intermediate to yield the first product. The rate of formation of this intermediate did not correspond to the equilibrium normally observed for the formation of a Michaelis-type complex between enzyme and substrate or inhibitor, indicating that it was a distinct intermediate, and not the Michaelis complex. Such studies have subsequently been carried out on numerous other non-specific substrates[2]. When acylation is the slow step (as with amides) such a burst is not seen and would not be expected on the basis of the above mechanism. Unfortunately, similar studies with specific substrates have not usually been possible, since such substrates are hydrolysed too rapidly for the stopped-flow technique to be used. This problem has, however, been overcome in a number of ways, as follows.

(2) At low pH, the reaction is very much slower than at neutral pH. Bender and coworkers[3] observed spectral changes on combining the enzyme with specific substrates such as *N*-acetyl-L-tryptophan, which were accompanied by a reduction in the number of active sites, indicating the formation of a relatively stable intermediate. Using a pH-jump stopped-flow technique to convert the inactive low-pH acyl enzyme to the active form at the required pH, it was found that the observed rate of reaction with *N*-(2-furoyl)-acryloyl tryptophan methyl ester corresponded with k_c determined from steady-state experiments, indicating that k_3 (deacylation) is the rate-determining step for this substrate[4]. In some cases[5] it has been possible to correlate changes in absorbance of substrate, product and acyl enzyme with such a mechanism; this provides more rigorous proof.

(3) Yet another method involves causing chymotrypsin to react with a competitive inhibitor, proflavin, to form an enzyme–inhibitor complex with a distinctive spectrum[6]. This inhibitor is then displaced by substrate when the

---

† In certain special cases, such as the hydrolysis of trimethylacetate, stopped-flow techniques are not needed as the pre-steady-state lasts several minutes. [M. L. Bender, F. J. Kézdy, and F. C. Wedler, *J. chem. Educ.*, **44**, 84 (1967).]

[1] B. S. Hartley and B. A. Kilby, *Biochem. J.*, **56**, 288 (1954); H. Gutfreund and J. M. Sturtevant, *Proc. natn. Acad. Sci.*, **42**, 719 (1956); F. J. Kézdy and M. L. Bender, *Biochemistry*, **1**, 1097 (1962).

[2] M. L. Bender and B. Zerner, *J. Am. chem. Soc.*, **84**, 2550 (1962); G. R. Schonbaum, B. Zerner, and M. L. Bender, *J. biol. Chem.*, **236**, 2930 (1961); S. A. Bernhard, S. J. Lau, and H. Noller, *Biochemistry*, **4**, 1108 (1965).

[3] F. J. Kézdy, G. E. Clement, and M. L. Bender, *J. Am. chem. Soc.*, **86**, 3690 (1964).

[4] C. G. Miller and M. L. Bender, *J. Am. chem. Soc.*, **90**, 6850 (1968).

[5] E. Charney and S. A. Bernhard, *J. Am. chem. Soc.*, **89**, 2726 (1967).

[6] S. A. Bernhard and H. Gutfreund, *Proc. natn. Acad. Sci. U.S.A.*, **53**, 1238 (1965); K. G. Brandt, A. Himoe, and G. P. Hess, *J. biol. Chem.*, **242**, 3973 (1967); A. Himoe, K. G. Brandt, R. J. DeSa, and G. P. Hess, *ibid.*, **244**, 3483 (1969).

latter is added, but, since the proflavin competes effectively with substrate, the rate is considerably lower than normal. This has allowed many specific substrates to be examined using stopped-flow techniques. It has also allowed rates to be studied over a wide range of pH, which the method (2) above did not allow. This removed the doubt as to whether results obtained at very low pH were applicable at other pH's or not.

(4) A method that does not appear to have been used with chymotrypsin, but has had much success with trypsin, involves the monitoring of a single turnover of enzyme reacting with substrate[1]. Instead of having the enzyme concentration very much lower than that of substrate, as in most experiments, the concentrations are similar so that each enzyme molecule on the average reacts with only one substrate molecule. The rate of liberation of the product corresponding to acylation (or deacylation) may then be measured, and the rate compared with the steady-state rate. This determines which step is rate-determining, and provides further evidence for the existence of the acyl-enzyme intermediates.

(5) Steady-state experiments have been conducted on a variety of different kinds of substrates[2]. Most of the esters react with chymotrypsin with the same value of k_c, despite the fact that their basicities and structures have varied over a wide range. This has been taken as an indication that all of these substrates react via a common intermediate. Some support for this has been obtained with studies in which nucleophiles have been allowed to compete with water for the acyl enzyme (see p. 202). Here, the rate has depended on the nature of the nucleophile, but has been independent of the nature of the acyl-enzyme. Perhaps the most convincing steady-state evidence is that the constants K_A, K_{eq} and k_c do, in fact, correspond with the various microscopic constants (determined by pre-steady-state kinetics). Since the values of these macroscopic constants depend on the mechanism chosen, agreement between prediction and experiment is fairly good evidence of the correctness of the mechanism.

(6) Different types of inhibition behaviour have been observed[3] with α-chymotrypsin depending on the substrate used and on the rate-determining step for that substrate. For nicotinyl-L-tryptophanamide (with $k_2 \ll k_3$) inhibition with indole is simple competitive; for N-acetyl-L-tryosine ethyl ester (with $k_3 \ll k_2$), inhibition is simple non-competitive; and for methyl hippurate (with $k_2 \simeq k_3$), it is mixed. The results indicate that indole is not bound to the EA complex for nicotinyl-L-tryptophanamide; that with ATEE it binds to free and acyl enzyme, blocking deacylation, and that for

[1] T. E. Barman and H. Gutfreund, *Proc. natn. Acad. Sci. U.S.A.*, **53**, 1243 (1965); *Biochem. J.*, **101**, 411 (1966).
[2] For a review of this aspect, see M. L. Bender and coworkers, *J. Am. chem. Soc.*, **86**, 3674, 3697, 3704 (1964).
[3] H. Kaplan and K. J. Laidler, *Can. J. Chem.*, **45**, 559 (1967), and references therein; R. J. Foster and C. Niemann, *J. Am. chem. Soc.*, **77**, 3365 (1955); T. H. Applewhite, R. B. Martin, and C. Niemann, *ibid.*, **80**, 1457 (1958).

methyl hippurate a mixture of binding patterns occurs. The observed inhibition behaviour is compatible with acyl enzyme formation.

It would thus seem that the existence of a second intermediate has been well established. That this intermediate is, indeed, an acyl enzyme, and not some other intermediate, has had its most direct evidence from comparison of the spectra of acyl enzymes with those of model compounds. This has been made possible by the preparation of several stable acyl-enzymes. Other evidence tends to be more indirect.

pH Dependence

One problem to be resolved was: what is responsible for the drop in activity at high pH as registered in plots of k_c/K_A as a function of pH? Application of the steady-state treatment to the two-intermediate mechanism (see p. 78) leads to the following values for the various constants in terms of the microscopic constants:

$$K_A = \frac{k_{-1} + k_2}{k_1} \cdot \frac{k_3}{k_2 + k_3} \qquad k_c = \frac{k_2 k_3}{k_2 + k_3} \qquad \frac{k_c}{K_A} = \frac{k_2}{(k_{-1} + k_2)/k_1}. \quad (1)$$

Amides:

$$k_3 \gg k_2; \qquad K_A = \frac{k_{-1} + k_2}{k_1} \qquad k_c = k_2. \quad (2)$$

Esters:

$$k_2 \gg k_3; \qquad K_A = \frac{k_{-1} + k_2}{k_1} \cdot \frac{k_3}{k_2} \qquad k_c = k_3. \quad (3)$$

The alkaline limb of the k_c/K_A versus pH profile is the result of the dependence of one or more of the microscopic constants in K_A on pH. It was assumed by Bender and coworkers[1], that k_{-1}/k_1 (the binding parameter) would probably not be pH-dependent above pH 8. Since the same type of pH curve is observed for esters and amides, and since the common feature in K_A is k_2, it seemed reasonable to conclude that the drop in activity at high pH was due to the pH dependence of k_2, and to a change in the rate-determining step from step 3 to step 2 at higher pH, for substrates for which at neutral pH stage 3 was rate determining. This received apparent support from studies on p-nitrophenylacetate, where both k_c and K_A were observed to be pH-dependent between pH's 8 and 10. Since this substrate is an ester, with $k_3 = k_c$, the pH dependence at high pH seemed best explained as mentioned above. However, this substrate is not typical of most chymotrypsin substrates.

It soon became evident that the above explanation was not correct. This arose as a result of studies of the dependence on pH of (1) inhibitor binding,

[1] For a review, see M. L. Bender and F. J. Kézdy, *A. Rev. Biochem.*, **34**, 49 (1965).

(2) spectral characteristics, (3) titrations, (4) substrate binding and (5) ionic strength-dependent conformational changes for several species of chymotrypsin, notably its zymogen, and α and δ enzymes†. These aspects will now be considered.

(1) *Inhibitor binding.* The first indications that the explanation of Bender and coworkers was not correct came from studies of the binding of the inhibitor diisopropylfluorophosphate (DFP) to chymotrypsin[1]. The binding process was observed to involve a group of pK around 8·5, corresponding to the pK responsible for loss of enzymic activity (k_c/K_A) at high pH. This implied that binding (K_s) is the process affected by pH, and not catalysis (k_2).

(2) *Spectral studies.* These included monitoring the specific rotation[2], circular dichroism and optical rotatory dispersion and UV absorption at 290 nm[3], fluorescence[4] and Moffit parameters[5] (see Chapter 1, p. 13) as as function of pH. The optical rotations of both the zymogen and the acetylated DFP-chymotrypsin are pH-independent, while that of the active acetylated chymotrypsin (obtained by activating the acetylated zymogen with trypsin as described above) is pH-dependent, the group involved again having a pK of around 8·5. Since the only major difference between δ-chymotrypsin and chymotrypsinogen is the existence of isoleucine–16 as a free amino group in the former, it was suggested that this residue was responsible for both the structural and catalytic pH-dependence at high pH, and that the two were related. Furthermore, since the specific rotation of acetylated δ-chymotrypsin at neutral pH resembles that of the DFP enzyme, while that at high pH resembles that of the zymogen, it was suggested that the change in conformation proposed as causing the loss of catalytic activity at high pH was the same conformational change as between chymotrypsinogen and the DFP-enzyme. A scheme based on two pH-dependent conformational forms of chymotrypsin has been proposed[6], to account for the observed behaviour.

(3) *Titration studies.* The pH-dependent conformational change gained additional support from studies[6] in which the enzymes were titrated over a range of pH values. The procedure involved titration of acetylated α-chymotrypsin, and of normal δ-chymotrypsin, and comparison with acetylated and normal zymogen in order to determine whether there was any difference in titrable functional groups. (Not all titrable groups in an enzyme are titrated

† For a review of this work, see G. P. Hess, J. McConn, E. Ku, and G. McConky, *Phil. Trans. R. Soc.*, **B257**, 89 (1970).

[1] A. Himoe and G. P. Hess, *Biochem. biophys. Res. Commun.*, **23**, 234 (1966); A.-Y. Moon, J. M. Sturtevant, and G. P. Hess, *J. biol. Chem.*, **240**, 4204 (1965).

[2] H. L. Oppenheimer, B. Labouesse, and G. P. Hess, *J. biol. Chem.*, **241**, 2720 (1960).

[3] J. McConn, G. D. Fasmann, and G. P. Hess, *J. molec. Biol.*, **39**, 551 (1969).

[4] Y. D. Kim and R. Lumry, *J. Am. chem. Soc.*, **93**, 1003 (1971).

[5] H. Parker and R. Lumry, *ibid.*, **85**, 483 (1963).

[6] B. Labouesse, R. H. Havsteen, and G. P. Hess, *Proc. natn. Acad. Sci. U.S.A.*, **48**, 2137 (1962); A.-Y. Moon, J. M. Sturtevant, and G. P. Hess, *loc. cit.*; B. Labouesse, H. Oppenheimer, and G. P. Hess, *Biochim. biophys. Res. Commun.*, **14**, 313, 318 (1964).

under these conditions, since some groups may be 'buried' and unavailable to solvent.) The results indicated that δ-chymotrypsin had one more amino group than the zymogen, and that it also had one more amino group than the DFP–enzyme. This group had a pK of around 8·5, and was shown by amino acid analysis to be the α-amino group of isoleucine–16. The fact that the DEP–enzyme. This group had a pK of around 8·5, and was shown by suggestion that isoleucine–16 becomes 'buried' on binding of inhibitors; this provided additional evidence that the pH-dependence at high pH is the result of a conformational change associated with binding.

(4) *Substrate binding.* The combination of chymotrypsin with substrate during catalysis has been shown by several workers to involve the uptake of one mole of protons per mole of substrate (and enzyme) at high pH, in the same way as was found for the binding of inhibitors[1]. This has been confirmed for the hydrolysis of specific amide substrates. It has also been demonstrated that substrate binding is associated with a conformational change[2]. Again, the group involved had a pK of around 8·5. Since the substrates were amides, K_s was conclusively implicated in the pH dependence for the first time. Other kinetic studies led to a similar conclusion[3]. k_c was found to have a pH dependence involving a group of pK 7 (see Fig. 5.8). Results for α- and δ-chymotrypsin, while being quantitatively different, appeared to follow a similar pattern (but see below)[4]. Other kinetic studies have led to a similar conclusion about the pH-dependence of binding.

(5) *Conformational changes.* There is evidence for a pH-dependent conformational change from a variety of different workers[5]. The evidence also suggests that the conformations of chymotrypsinogen and chymotrypsin are different at neutral pH[6] (as suggested above) and similar at high pH[7]. X-ray evidence also suggests this[8].

Lumry and coworkers[9] have identified, by thermodynamic studies, the existence of several different states of the enzyme, depending on conditions of temperature and pH.

Studies of the effect of ionic strength on the physical[10] and kinetic[11] properties of chymotrypsin have also indicated a pH-dependent conforma-

[1] J. Keizer and S. A. Bernhard, *Biochemistry*, **5**, 4127 (1966); D. M. Glick, *ibid.*, **7**, 3391 (1968); F. C. Wedler and M. L. Bender, *J. Am. chem. Soc.*, **91**, 3894 (1969).

[2] P. Valenzuela and M. L. Bender, *Biochim. biophys. Acta*, **235**, 411 (1971); also page 319.

[3] A. Platt and C. Niemann, *Proc. natn. Acad. Sci. U.S.A.*, **50**, 817 (1963); W. A. Mutakis and C. Niemann, *ibid.*, **51**, 397 (1964); M. L. Bender, M. J. Gibian, and D. J. Whelan, *ibid.*, **56**, 833 (1966).

[4] A. Himoe, P. C. Parks, and G. P. Hess, *J. biol. Chem.*, **242**, 919 (1967); A. Himoe, K. G. Brandt, and G. P. Hess, *ibid.*, 3963 (1967).

[5] M. L. Bender and F. C. Wedler, *J. Am. chem. Soc.*, **84**, 3052 (1967).

[6] H. Neurath, J. A. Rupley, and W. J. Dreyer, *Archs Biochem. Biophys.*, **65**, 243 (1956); G. D. Fasmann and R. J. Foster, *Fedn Proc.*, **24**, 472 (1965).

[7] R. Biltonen, R. Lumry, V. Madison, and H. Parker, *Proc. natn. Acad. Sci. U.S.A.*, **54**, 1412 (1965).

[8] J. Kraut, J. D. Roberts, H. T. Wright, and Ng H. Xuong, *Biochemistry*, **9**, 1997 (1970).

[9] Y. D. Kim and R. Lumry, *J. Am. chem. Soc.*, **93**, 1003 (1971).

[10] A. Kurosky, J. E. S. Graham, J. W. Dixon, and T. Hofmann, *Can. J. Biochem.*, **49**, 529 (1971).

[11] P. Valenzuela and M. L. Bender, *Proc. natn. Acad. Sci. U.S.A.*, **63**, 1214 (1969); W. J. Canady, *et al.*, *J. biol. Chem.*, **246**, 1129, 1135 (1971); *Archs Biochem. Biophys.*, **134**, 253 (1969).

tional change. These studies have included ionic strength effects on the binding of substrate, and on the availability of isoleucine–16 to various reagents such as nitrous acid. In the work of Bender *et al.*, isoleucine–16 was found to be accessible to reagent (i.e. exposed) at low ionic strength (as suggested for high pH), and inaccessible (i.e. buried within the enzyme) at high ionic strength (as at low or neutral pH). The critical value at which this change occurs is $I = 0.5$M. A group of pK 9 was also implicated. In the work of Canady *et al.*, the same pK group was involved, as revealed by ultraviolet and circular dichroism studies.

All of these various approaches have led to the same conclusion: there is a group of pK 8·5 involved in pH-dependent conformational changes, which depend on ionic strength; a group of similar pK is involved in the pH-dependence at high pH of k_c/K_A; this group is isoleucine–16, since chymotrypsinogen and chymotrypsin differ by this group only. This conclusion is supported by X-ray evidence (at low pH) which reveals the existence of an ion pair between aspartic acid–102 and isoleucine–16. It is suggested that, at high pH, isoleucine–16 becomes deprotonated, the ion pair being broken and the integrity of the active site destroyed (see Fig. 10.12). Hess and coworkers

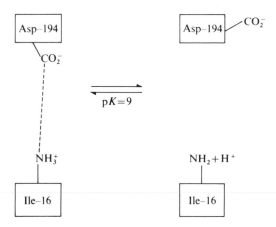

FIG. 10.12. Schematic representation of the conformational change occurring when the isoleucine-16 group of chymotrypsin changes its state of ionization (see H. P. Kasserra and K. J. Laidler, *Can. J. Chem.*, **47**, 4031 (1969)).

have taken this conformational change a step further[1], by applying the steady state treatment to the model shown in Fig. 10.13. For the expressions obtained for k_c and K_A, the original paper should be referred to, as many new terms are defined. The model is based on certain assumptions: (1) there is a pH-dependent equilibrium between two major conformations of the enzyme at alkaline pH, controlled by a single ionizing group on the enzyme; (2) this

[1] Himoe, Parks, and Hess, *loc. cit.*

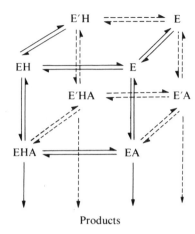

Products

FIG.13. Scheme of Himoe, Parks, and Hess for the formation of chymotrypsin–substrate complexes, showing ionization equilibria, and equilibria between two main conformations, E and E′, of the enzyme.

equilibrium accounts for the pH behaviour of the enzyme above pH 8; (3) neither substrate nor inhibitor binds to the inactive form of the enzyme at high pH (dotted lines in Fig. 10.13).

The mechanism leads to certain predictions. For instance, the pH-dependence of K_A for specific amide substrates should be accompanied by proton uptake. Also, conversion of the active to the inactive form should occur at higher pH in the presence of substrate than in the absence. This can be seen in the light of the conformational changes discussed above for binding. These predictions have been borne out by experiment[1].

However, certain evidence has been produced which appears to contradict the above explanation. Most of this evidence concerns the specific assignment of isoleucine–16 as the group of pK 8·5 which is involved in the alkaline limb of the curve k_c/K_A against pH.

(1) *Titration experiments*. Marini and Martin[2] performed titration experiments, similar to those of Hess *et al.* discussed above, on α-chymotrypsin, and observed that the titration curves for the enzyme and the DFP-enzyme were identical between pH 5 and 10, i.e., no group of pK 8·5 was present in the one and not in the other, in contrast to the previous results. Marini and Martin and coworkers suggest that Hess *et al.* did not take into account the autolysis of chymotrypsin, and that, rather than DFP-chymotrypsin having one less titrable amino group than the enzyme, it is the enzyme which has an additional group as a result of autolysis. They found[3] that their titration data implicated

---

[1] Himoe, Brandt, and Hess, *loc. cit.*
[2] M. A. Marini and C. J. Martin, *Biochim. biophys. Acta*, **168**, 73 (1968).
[3] M. A. Marini and C. J. Martin, *Eur. J. Biochem.*, **19**, 153, 162 (1971).

two groups of pK 7·75, and not one of 7·25 and one of 8·5, as had previously been thought.

(2) *Alkylation experiments.* Martin and coworkers[1] have also alkylated chymotrypsin, both α and δ forms, with reagents such as formaldehyde and two amidinating agents. In the case of formaldehyde, the pK of the amino group present in chymotrypsin and not in the zymogen (isoleucine–16) was found to be 5·0 in the modified enzyme, and 7·8 in the normal enzyme. Amidination raised the pK of isoleucine–16 to around 12. In both cases, however, the enzyme was active (in the latter case, there was no change in specific activity; in the former there was some loss of activity, ascribed to alkylation of a histidine residue as well); and the pK of the alkaline limb of the k_c/K_A vs pH profile remained unchanged. Similar results were obtained using the succinylated enzyme[2], again with no change in specific activity; here, a negative charge is introduced.

These results indicate that isoleucine–16 is not needed for catalytic activity in either α or δ chymotrypsin; and that it is not the group involved in loss of activity in the former.

(3) *Kinetic experiments.* Valenzuela and Bender[3] have studied the kinetics of δ chymotrypsin and of the succinylated enzyme obtained in the usual way by activation of the succinylated zymogen. The modified enzyme reacted with acetic anhydride and nitrous acid, becoming inactivated with accompanying loss of isoleucine–16. The enzyme–substrate complex formed with indoleacryloyl ester, however, did not react with these reagents at all, indicating a difference in conformation for the free and complexed enzyme. Similar changes were observed by Dixon and Hofmann[4] for the rates of deamination of DFP- and tosyl-chymotrypsins.

Kinetic experiments[5] show that, contrary to the assumed kinetic identity of α and δ chymotrypsins, they are probably quite different. The former has a K_A which is independent of pH above pH 10; the latter, however, does not appear to level off. It would appear unwise to assume that the pH profiles of these two enzymes can be explained in the same way. Valenzuela and Bender[6] suggest that in the case of δ-chymotrypsin the conformational change may well be governed by isoleucine–16, but in α-chymotrypsin the activity loss might be governed by the other α-amino group, alanine–149.

Action of histidine–57

A question of considerable importance in the mechanism of catalysis of chymotrypsin is whether the histidine–57 is a general base catalyst or acts as

[1] S. J. Agarival, C. J. Martin, T. T. Blair, and M. A. Marini, *Biochem. biophys. Res. Commun.*, **43**, 510 (1971).

[2] T. T. Blair, M. A. Martin, S. P. Agarival, and C. J. Martin, *FEBS Lett.*, **14**, 86 (1971).

[3] P. Valenzuela and M. L. Bender, *Biochem. biophys. Acta*, **235**, 411 (1971).

[4] J. W. Dixon and T. Hofmann, *Can. J. Biochem.*, **48**, 671 (1970); A. Kurosky, *et al.*, *loc. cit.*

[5] P. Valenzuela and M. L. Bender, *loc. cit.*

[6] P. Valenzuela and M. L. Bender, *Biochem. biophys. Acta*, **235**, 411 (1971); *J. Am. chem. Soc.*, **93**, 3783 (1971).

a nucleophile. Three lines of evidence have a bearing on this problem: (1) D_2O isotope effects, (2) evidence concerning the existence of a tetrahedral intermediate (necessary for nucleophilic catalysis), and (3) substituent effects on the reaction rate.

(1) The rate at which chymotrypsin catalyses the hydrolysis of substrates in D_2O is 2 to 3 times slower than the rate in H_2O[1], for both acylation and deacylation. This is characteristic of the magnitude of the primary isotope effect expected if proton transfer is involved in the rate-determining step of a reaction (see Chapter 7)†. With nucleophilic catalysis little isotope effect is expected. An isotope effect of 2 to 3 has been observed for imidazole catalysis of model compounds[2].

(2) A considerable amount of evidence has been accumulated, largely on the basis of substituent effects (see below), which is consistent with, but does not prove, the existence of a tetrahedral intermediate in the acylation of chymotrypsin[3]. Such an intermediate is necessary for nucleophilic catalysis. There is no evidence to prove the existence or otherwise of such an intermediate; but there is evidence to suggest that the formation or breakdown of the intermediate is not the rate-determining step in chymotrypsin[4] for deacylation. This evidence rests on the assumption that the rates of formation and breakdown of a tetrahedral intermediate via aminolysis will be different from that for hydrolysis. Several acylenzymes were tested in this manner[4] and no significant change in the pK for deacylation (around 7) was observed using amines and water as the nucleophiles.

(3) Numerous studies have been undertaken in which the electron density at the reactive bond in the substrate has been varied by means of different substituents in the aromatic ring. It was hoped to determine what kind of electron movement takes place during catalysis, and in this way to throw some light on the mechanism. In several cases[5] a positive Hammett constant has been observed for both acylation and deacylation, indicating the involvement of an electron-deficient centre during catalysis. This, in turn, suggests the involvement of a base in catalysis, a role filled by histidine–57. Support for this conclusion has been obtained from nucleophile studies[6].

The situation appears to be rather more complicated than this, as indicated by Williams[7]. Anilide substrates have the opposite requirement—electron-

† On the other hand N^{14}/N^{15} studies [M. H. O'Leary and M. D. Kluetz, *J. Am. chem. Soc.*, **92**, 6089 (1970)] lead to the different conclusion that the slow step for amide acylation is the breaking of the amide bond.

[1] M. L. Bender and G. A. Hamilton, *J. Am. chem. Soc.*, **84**, 2570 (1964); M. Caplow and W. P. Jencks, *Biochemistry*, **1**, 833 (1962).
[2] M. L. Bender, E. J. Pollock, and M. C. Neveu, *J. Am. chem. Soc.*, **84**, 595 (1962).
[3] For a review, see M. Caplow, *ibid.*, **91**, 3639 (1969).
[4] Vishnu and M. Caplow, *ibid.*, **91**, 6754 (1969).
[5] M. Caplow and W. P. Jencks, *Biochemistry*, **1**, 883 (1962); M. L. Bender and K. Nakamura, *J. Am. chem. Soc.*, **84**, 2577 (1962); M. Caplow and W. P. Jencks (1962), *loc. cit.*
[6] P. W. Inward and W. P. Jencks, *J. biol. Chem.*, **240**, 1986 (1965).
[7] A. Williams, *Biochemistry*, **9**, 3387 (1970).

donating substituents increase the rate here (ρ is negative)[1]. Williams suggests that the special nature of anilides produces this result, and that an electron-deficient site is required, but also suggests that electrophilic catalysis is possible.

An objection to the general base–general acid mechanism has been raised by Menger and Brock[2]. They observed that a substrate well suited to anchimeric assistance for amide catalysis was hydrolysed many orders of magnitude more slowly by α-chymotrypsin than were other amides. The conclusion is that general base catalysis is not a completely satisfactory mechanism for α-chymotrypsin.

General mechanisms of α-chymotrypsin action

Although the evidence for general-base catalysis is perhaps not as strong as one might wish, it is certainly stronger than the case for nucleophilic catalysis. Most mechanisms of chymotrypsin catalysis assume general-base catalysis. They also have as the rate-determining step a proton transfer, in agreement with the evidence. They differ in the detailed explanations as to how the proton transfer occurs. These differences have a bearing on the interpretations as to why serine–195 is so reactive in the chymotrypsin active site, and why the overall catalytic process is so much more efficient than nonenzymic hydrolysis (excluding the binding processes, which will be discussed later). Some of these mechanisms will be briefly discussed here. They may be divided into two categories: (1) those in which a hydrogen bond between histidine–57 and serine–195 is postulated, and (2) those in which no such bond is used in the mechanism.

(1a) *Wang's mechanism*[3]. This mechanism (see Fig. 10.14) is based primarily on the studies with anilide substrates mentioned above. The hydrolysis of these substrates by chymotrypsin was studied in H_2O and D_2O. Wang reasoned that if proton transfer occurred before the activated complex is formed, it should be possible to observe a correlation between rate of hydrolysis and basicity of the substrate. This is, in fact, what was observed: there is an increase in rate with increasing basicity of substrate, in both solvents, the kinetic isotope effect decreasing with increasing basicity, as predicted by the proposed mechanism. For esters, however, a post-activated-complex protonation is proposed, esters being weaker bases than amides, peptides and anilides.

(1b) *Blow's mechanism*[4]. This mechanism (Fig. 10.15) is based on the X-ray structure of an inactive derivative of chymotrypsin, at low pH. From

[1] W. F. Sager and P. C. Parks, *J. Am. chem. Soc.*, **85**, 2678 (1964); J. H. Wang and L. Parker, *J. biol. Chem.*, **243**, 3729 (1968); T. Inagami, S. S. York, and A. Patchornik, *J. Am. chem. Soc.*, **87**, 126 (1965).
[2] F. M. Menger and H. T. Brock, *Tetrahedron*, **24**, 3453 (1968).
[3] J. H. Wang, *Science*, **161**, 328 (1968); *Proc. natn. Acad. Sci. U.S.A.*, **66**, 874 (1970).
[4] J. J. Birktoft, D. M. Blow, R. Henderson, and T. A. Steitz, *Phil. Trans. R. Soc.*, B**257**, 67 (1970) and previous publications by Blow and coworkers.

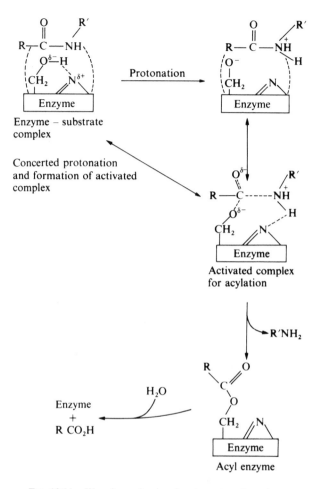

FIG. 10.14. Wang's mechanism for chymotrypsin action.

the positions of the various residues in the active site (see Fig. 10.11), it appears that aspartic acid–102 is involved in an ion pair with isoleucine–16. Serine–195 and histidine–57 also appears to be connected by a hydrogen bond. A similar arrangement has been observed by Henderson for indole-acryloyl-chymotrypsin[1]. This hydrogen-bonded network (or charge relay system) has also been observed for subtilisin[2], which has a similar mechanism as already mentioned.

There are several weaknesses in the above evidence. First, the structure of chymotrypsin at acid pH is not the same as at neutral and alkaline pH, and the same orientation of catalytic residues may not apply. Second, an

[1] R. Henderson, *loc. cit.*
[2] C. S. Wright, R. A. Alden, and J. Kraut, *Nature*, **221**, 235 (1969); D. M. Shotton and H. C. Watson, *Nature*, **225**, 811 (1970).

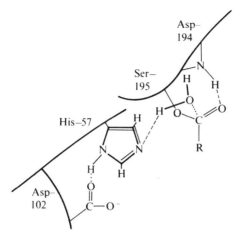

FIG. 10.15. The essential features of the mechanism of Blow and coworkers for the hydrolysis of acyl-chymotrypsin.

inhibited enzyme is not active, so that some feature of the observed crystal does not apply to the active enzyme. Nevertheless, such negative evidence by no means disproves the mechanism. It has the advantage over Wang's mechanism above, in that the special reactivity of serine–195 is well accounted for, as is the role of histidine as a general base. The observation of various water molecules in the active site, some of which are removed on binding of the inhibitor, suggests how water is involved in catalysis, also via a hydrogen bond network, the water coming between the histidine and serine.

(2a) *The Bender and Polgar mechanism*[1]. This mechanism (see Fig. 10.16) suggests that the mode of binding of the substrate, aligning the groups on it with those on the enzyme, is far more important for catalysis than the relationship between the groups on the enzyme itself (contrast Blow's mechanism). Specific substrates fit into the active site properly, and non-specific ones less well. The active site is regarded as being rigid rather than flexible, in order to emphasize the requirement for a proper fit. A water molecule lies between serine–195 and histidine–57, acting as a bridge in the charge-relay system†. Histidine–57 abstracts a proton from serine–195 hydroxyl, and this same proton is given back to the leaving group of the substrate. The carboxyl of aspartic acid–102 is used to stabilize the release of the proton of histidine as it changes from general base to general acid catalysis in going from acylation to deacylation. This binding pattern is supported by results showing that the activation entropies for deacylation of non-specific acyl enzymes vary widely, while the enthalpies remain the

† This bridge system is actually preformed in chymotrypsinogen [S. T. Freer, J. Kraut, J. D. Roberts, H. T. Wright, and Ng. H. Xuong, *Biochemistry*, 9, 1997 (1970).] The main difference in conformation between zymogen and enzyme is the isoleucine–16–aspartate–102 ion pair.

[1] M. L. Bender and L. Polgar, *Proc. natn. Acad. Sci. U.S.A.*, 64, 1335 (1969).

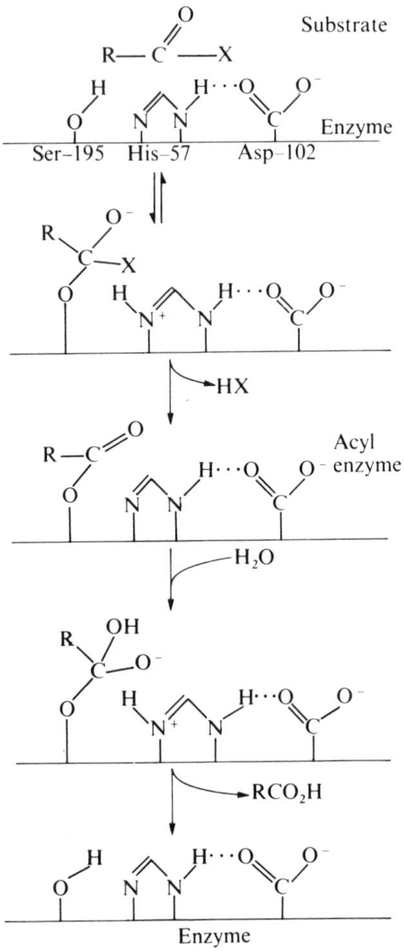

FIG. 10.16.   The mechanism of Bender and Polgar for chymotrypsin action.

same[1]. Further, the entropy change is more negative for specific than non-specific substrates.

(2b) *Kasserra and Laidler's mechanism*[2]. Kasserra and Laidler have developed a composite mechanism which combines what appear to be the best features of previous proposals, and have given explicit structures for the activated complexes.

Binding and specificity: nature of the active site

The mechanism whereby the actual bond is broken in chymotrypsin hydrolysis is only one part of the explanation for the catalytic efficiency of

[1] M. L. Bender, F. J. Kézdy, and C. R. Gunter, *J. Am. chem. Soc.*, **86**, 3714 (1964).
[2] H. P. Kasserra and K. J. Laidler, *Can. J. Chem.*, **47**, 4031 (1969).

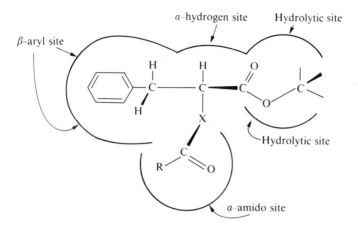

FIG. 10.17.   The binding sites for chymotrypsin, as revealed by the studies of Cohen and coworkers.

this enzyme. As has been mentioned briefly above, the mode of binding of the substrate to the enzyme is at least of equal importance. This binding process is also responsible for the specificity of the enzyme. Some aspects of binding and the chemical nature of the active site will now be briefly discussed.

For some time it has been known that chymotrypsin hydrolyses peptide bonds on the c-terminal side of several amino acid residues, predominantly the aromatic ones, tyrosine, tryptophan and phenylalanine[1]. It also attacks bonds adjacent to other non-polar, aliphatic residues such as leucine and methionine. While aromaticity does lead to greater reactivity, it is not essential; in fact, hexahydro derivatives of aromatic residues are also hydrolysed[2]. The general method for testing such specificity is to allow the enzyme to attack either natural proteins, or synthetic peptides and other substrates, and to examine the new terminal amino acid residues produced after hydrolysis. Some residues are even known to prevent hydrolysis, if placed at certain positions with respect to the bond being hydrolysed.

More recently, experiments along the lines of Schechter and Berger for papain[3] have been conducted with chymotrypsin in order to map the active site[4]. Other experiments, utilizing synthetic substrates of varying sizes, have given some idea of the size of the active site for chymotrypsin.

Another aspect of the specificity of chymotrypsin, apart from the nature and effect of adjacent and other amino acid residues in the substrate, concerns the nature of the other groups in the substrate. For instance, the amido group adjacent to the bond being broken (see Fig. 10.17) is of considerable

[1] M. Bergmann, *Adv. Enzymol.*, **2**, 49 (1942); G. E. Hein and C. Niemann, *Nature*, **210**, 903 (1966) and references therein.
[2] R. R. Jennings and C. Niemann, *J. Am. chem. Soc.*, **75**, 4687 (1953).
[3] I. Schechter and I. Berger, *Phil. Trans. R. Soc.*, B**257**, 249 (1970).
[4] K. Morihara, T. Oka, and H. Tsuzuki, *Biochem. biophys. Res. Commun.*, **35**, 210 (1969).

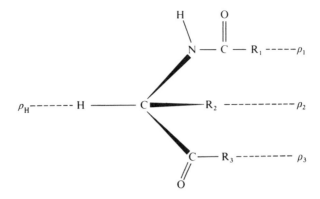

FIG. 10.18.   Binding sites for chymotrypsin.

importance for efficient hydrolysis[1]. If the amide group is lacking, or replaced by oxygen or by methylene, hydrolysis is several orders of magnitude slower than with the 'normal' substrate. N-methylation of the amido group also decreases reactivity considerably[2]. Substrates with a free amino group may, however, be hydrolysed[3].

The requirement for an aromatic or non-polar group adjacent to the bond being hydrolysed is readily understood in the light of various experiments which have indicated that the overall nature of the active site is hydrophobic. These experiments have involved a number of different approaches. Measurement of the K_A's of various inhibitors of the enzyme have indicated that the binding free-energy change is similar to the free-energy change involved in transferring the inhibitors from aqueous to non-polar solvent[4]. Other experiments utilizing this approach include the measurement of the partitioning of competitive inhibitors of chymotrypsin between water and the enzyme, and comparing them with model compounds partitioned between water and various organic solvents[5].

Ingles and Knowles[6] have measured the rates of deacylation of several acyl-chymotrypsins. They varied both the hydrophobicity and the H-bonding capacity of the acyl amino group in order to obtain an estimate of the relative importance of these two forces in binding and catalysis. They found that reducing the hydrophobicity altered k_c markedly, but K_A only

[1] D. W. Ingles and J. R. Knowles, *Biochem. J.*, **108**, 561 (1968); K. J. Laidler and M. L. Barnard, *Trans. Faraday Soc.*, **52**, 497 (1956); S. G. Cohen and S. Y. Weinstein, *J. Am. chem. Soc.*, **86**, 5326 (1964); G. E. Hein and C. Niemann, *Proc. natn. Acad. Sci., U.S.A.*, **47**, 1340 (1961); D. W. Ingles and J. R. Knowles, *Biochem. J.*, **99**, 275 (1966).
[2] R. L. Peterson, K. W. Hubela, and C. Niemann, *Biochemistry*, **2**, 962 (1963); S. G. Cohen, J. Crossby and E. Khedouri, *ibid.*, 820 (1963).
[3] J. E. Purdie and N. L. Benoiton, *Can. J. Biochem.*, **48**, 1058 (1970).
[4] See, e.g., I. V. Berezin, A. V. Leveskov, and K. Martinek, *FEBS Lett.*, **7**, 20 (1970).
[5] J. R. Knowles, *J. theor. Biol.*, **9**, 213 (1965); D. W. Ingles and J. R. Knowles, *Biochem. J.*, **104**, 369 (1967); R. Wildnauer and W. J. Canady, *Biochemistry*, **5**, 2885 (1966).
[6] D. W. Ingles and J. R. Knowles, *loc. cit.*

slightly. Hein and Niemann[1], on the other hand, found in similar experiments that K_A was greatly affected by a change in hydrophobicity in the substrate. Cuppett and Canady[2] suggest that hydrophobicity is essential for both binding and catalysis. The binding, they suggest, is related to the aromatic or non-polar amino acid residue. For catalysis they propose that aromaticity is important for the formation of the ion pair between aspartate–102 and isoleucine–16, these residues being in a non-polar part of the active site. This has received support from several experiments, such as those of Berezin and Martinek[3]. These workers found for one or two substrates that the free energy of activation of chymotrypsin catalysis varied linearly with the free energy of binding. This indicates that the same forces important for binding the substrate (hydrophobic) are important in the activated complex.

Recent experiments have led to an extremely detailed knowledge of the nature of binding and the active site. An indication that more than one aromatic binding site was present in the active site came from the observation[4] that DFP-chymotrypsin could still bind substrates and inhibitors

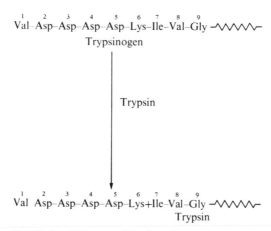

FIG. 10.19.   The conversion of trypsinogen into trypsin.

containing such aromatic groups. Confirmation of this, and further details about other binding sites, including that for the acylamido group, has been obtained in a series of experiments by Cohen and coworkers[5]. This involved synthesis of a variety of substrates with changes in the substituents at various positions in the substrate, and observing the effects on catalysis. The overall picture of the active site resulting from this work is illustrated in Fig. 10.17.

[1] G. E. Hein and C. Niemann, *Proc. natn. Acad. Sci. U.S.A.*, **47**, 341 (1961).
[2] C. C. Cuppett and W. J. Canady, *J. biol. Chem.*, **245**, 1069 (1970).
[3] I. V. Berezin and K. Martinek, *FEBS Lett.*, **8**, 261 (1970).
[4] W. Cohen and B. F. Erlanger, *J. Am. chem. Soc.*, **82**, 3928 (1960).
[5] S. G. Cohen, A. Milovanovic, R. M. Schultz, and S. Y. Weinstein, *J. biol. Chem.*, **244**, 2664 (1969) and references therein; S. G. Cohen and R. M. Schultz, *ibid.*, **243**, 2667 (1968) and references therein.

The experiments of Cohen and coworkers have also thrown light on the stereospecificity of chymotrypsin hydrolysis. Generally, only the L-enantiomers of the substrates are hydrolysed by this enzyme. However, in some cases, with rigid, synthesized substrates, the D-enantiomers were hydrolysed, sometimes even preferentially over the L-isomers[1]. It turns out that certain of the sites in Fig. 10.18 for binding have less stringent restrictions on the type of group that can be accommodated. This can lead to non-specific binding, of the same sort as discussed by Polgar and Bender above.

Several attempts have been made to determine quantitatively the relative importance of hydrophobicity and other factors in the binding process; among them are the investigations of Ingles and Knowles, discussed above[2]. A useful approach to this has been varying the substituents on the substrate to determine the effects on k_c and K_A. Correlations[3] have been made with parameters which give an estimate of the contributions of hydrophobicity[4], electronic[5] and steric effects[6] and polarizability[7]. Some of the general conclusions[8] that have been arrived at are as follows (see Fig. 10.18):

(1) Site ρ_2: the inductive and steric effects have the same hydrophobic parameters (π), for several series of substrates. It would appear that these effects are both involved in this site, which has been shown to be the hydrophobic (aromatic) binding site.

(2) Site ρ_3: hydrophobicity is important here again, but only to about 25 per cent of the extent as it is for site ρ_2. Steric and inductive effects here also have the same dependence on hydrophobicity. This site, and site ρ_2, would appear to be fairly stringent in their requirements, having two parameters associated with them.

(3) Site ρ_1: at this site binding correlated not with hydrophobicity but more with polarizability (P_s). Here again, both inductive and steric effects are dependent on polarizability.

The overall conclusion is that both polarizability and hydrophobicity are important for binding (as indicated in certain qualitative experiments above). In fact, polarizability may be even more important than hydrophobicity, since there is a better correlation, from the regression analysis, of binding with P_s than with π for sites ρ_1 and ρ_3.

On the one hand, groups have been identified as present in the active site; some of these are not involved in the chemical reaction, but in binding. On

[1] S. G. Cohen and R. M. Schultz, *loc. cit.*

[2] D. W. Ingles and J. R. Knowles, *loc. cit.*, for a review of this type of work see G. Hansch and E. Coates, *J. pharm. Sci.*, **59**, 731 (1970).

[3] G. Hansch and E. Coates, *loc. cit.*

[4] T. Fujita, J. Iwasa, and C. Hansch, *J. Am. chem. Soc.*, **86**, 5175 (1964).

[5] J. E. Leffler and E. Grunwald, in *Rates and Equilibria of Organic Reactions*, John Wiley and Sons, New York (1963).

[6] R. W. Taft, in *Steric Effects in Organic Chemistry* (ed. M. S. Newman), John Wiley and Sons, New York (1956), p. 644.

[7] D. Agin, L. Hersh, and D. Holtzman, *Proc. natn. Acad. Sci., U.S.A.*, **53**, 952 (1965).

[8] G. E. Hein and C. Niemann, *J. Am. chem. Soc.*, **84**, 4482, 4495 (1962); W. B. Lawson and H. J. Schramm, *ibid.*, **84**, 2017 (1962); G. Hansch and E. Coates, *loc. cit.*

the other hand, specific sites on the enzyme have been identified as responsible for binding. It remains to determine whether these specific sites on the enzyme can be linked with any specific groups in the active site.

Evidence along these lines stems from two types of investigation, chemical modification and X-ray analysis. The ρ_1 area is associated with methionine–192. Chemical modifications of the residue after the bifunctional 'substrate' had acylated the serine–195, demonstrated this point[1]. X-ray studies by Blow and Henderson have confirmed that methionine is probably involved in binding. The residue moves a significant amount when an inhibitor binds to the enzyme from its position in the native molecule. The aromatic binding site, ρ_2, has been clearly demonstrated by X-ray work (the 'tosyl hole'). It appears that tryptophan–215 might be involved in a charge transfer reaction here[2]. Site ρ_3 has been identified as histidine and serine by a variety of experimental methods, already discussed above, as well as X-ray work. Other binding groups, corresponding to the binding sites of Cohen *et al.*, are also evident from the X-ray studies.

It would appear that, while certain details are yet to be fully investigated, the evidence obtained so far has allowed a reasonable correlation of the mode of chymotrypsin catalysis with the known physical structure of the enzyme, and the groups in the active site.

Trypsin (E.C.3.4.4.4)

Trypsin is another member of the family of pancreatic proteolytic enzymes; it catalyses the hydrolysis of peptide bonds involving arginine and lysine amino acid residues, as well as certain other groups (see below). It has many structural and mechanistic features in common with chymotrypsin.

Activation

Like chymotrypsin, trypsin is present in the digestive system as an inactive zymogen precursor, which is protected against autodigestion. Conversion of this zymogen, trypsinogen, to the active enzyme is catalysed by trypsin itself and also by enterokinase. The activation process occurs as in Fig. 10.19, and involves the removal of a hexapeptide from the N-terminal end of trypsinogen[3]. The presence of calcium ions facilitates the rate of activation, and also reduces the amount of inert protein which results from the splitting of additional bonds in trypsinogen, in addition to the primary one between lysine–6 and isoleucine–7[4]. There are two binding sites for calcium to trypsinogen[5]. One of these sites has a high affinity for a calcium ion, the binding of which causes a conformational change which protects the molecule

[1] Lawson and Schramm, *loc. cit.*
[2] D. S. Sigman and E. R. Blout, *J. Am. chem. Soc.*, **89**, 1767 (1967).
[3] P. Desnuelle and C. Fabre, *Biochim. biophys. Acta*, **18**, 29 (1955).
[4] M. R. MacDonald and M. Kunitz, *J. gen. Physiol.*, **25**, 53 (1941).
[5] M. Delaage and M. Lazdunski, *Biochem. biophys. Res. Commun.*, **28**, 590 (1967); P. Desnuelle and M. Lazdunski, *Biochem. biophys. Res. Commun.*, **29**, 235 (1967).

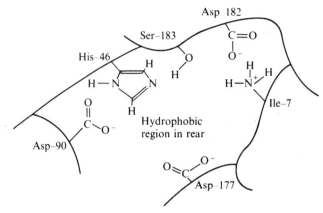

F<small>IG</small>. 10.20.   A schematic representation of the essential groups at the active centre of trypsin (adapted from H. P. Kasserra and K. J. Laidler, *Can. J. Chem.*, **47**, 4031 (1969)). Note the similarity to chymotrypsin (Fig. 10.11); the main difference is the presence of the negatively-charged ASP–177.

from forming inert protein[1]. The second site has a lower affinity for calcium and does not involve a conformational change, but is responsible for acceleration of the activation process and an increase in the specificity of splitting: it is thought to involve the binding of calcium ions to the four aspartic acid residues 2–5. This would be necessary in view of the specificity of trypsin for residues containing positive, not negative, charges. Presumably the presence of the calcium ion offsets the unfavourable nature of these acidic residues, allowing trypsin to activate the zymogen.

As with chymotrypsin, the mechanism of action will be discussed in terms of the groups present in the active centre, the overall nature of the reaction, and finally an assignment of the roles of the groups involved. The groups believed to be present are considered first.

Serine–183

Trypsin, like chymotrypsin and other serine proteases, is alkylated by diisopropylfluorophosphate (DFP) at its reactive serine residue[2]. The resulting enzyme is completely inactive, although it can be reactivated with nucleophiles such as hydroxylamine; this indicates that the serine has been alkylated at its hydroxyl group[3]. These experiments indicate that this serine residue, number 183, is essential for catalysis. Other alkylations include the use of bromoacetone[4], guanidine derivatives[5] and various active site reagents[6].

[1] See also F. Seydoux and J. Yon, *Eur. J. Biochem.*, **3**, 42 (1967).
[2] See A. K. Balls and E. F. Jansen, *Adv. Enzymol.*, **13**, 32 (1952) for a detailed discussion.
[3] W. Cohen, M. Lache, and B. F. Erlanger, *Biochemistry*, **1**, 486 (1962).
[4] J. G. Beeley and H. Neurath, *Biochemistry*, **7**, 1239 (1968).
[5] W. B. Lawson, M. D. Leafer, A. Tewes, and G. J. S. Rao, *Hoppe-Seyler's Z. phys. Chem.*, **349**, 251 (1968).
[6] D. D. Schroeder and E. Shaw, *Archs biochem. Biophys.*, **142**, 340 (1971).

Histidine–46

pH studies, which will be discussed in more detail later, have indicated that a group of pK 7 is involved in catalysis. Such a pK is usually indicative of a histidine residue, and active-site reagents have confirmed this. Trypsin is, for example, completely inactivated by reaction with the substrate analogue tosyl-lysine chloromethylketone (TLCK). The loss of activity was correlated with alkylation of a histidine residue[1]. Subsequent work showed that the N-3 of the imidazole ring of histidine–46 was involved[2].

Aspartic acid–177

Indications of the presence of an essential carboxyl group have been obtained largely as a result of chemical modification studies. Two carboxyls can be protected from modifications by competitive inhibitors of trypsin, and this implies that they are near to the active site[3]. Eyl and Inagami found that aspartic acid–177 was the major residue modified; since the modified enzyme could still bind DFP, they concluded that the loss in activity was a result of loss of specificity, i.e. that asp–177 is involved in binding (see below). Shaw and coworkers[4] modified three carboxyl residues, and found that different active site 'titrants' had their reactivity with trypsin modified in different ways as a result of the modification. They suggested that one of the carboxyl groups is the specificity site, one is involved in an ion pair with isoleucine–7, and the third is involved in the charge relay system, by analogy with chymotrypsin.

Other evidence for carboxyl involvement comes from studies of the inhibition of trypsin by alkylammonium ions[5]. This work will be referred to again later.

Finally, kinetic studies at low pH[6] implicate a group of pK 4 in k_2/K_A; this implies that the group is free to ionize in the free enzyme, but not in the Michaelis complex, indicating that the group is involved in binding. Further identification of the carboxyl group as aspartic acid–177 comes from a study by Smith and Shaw[7] and by analogy with the primary structure of chymotrypsin, which will be discussed in more detail later.

Isoleucine–7

The evidence for the involvement of this residue will be dealt with in the section on pH studies. Suffice to say that it is thought not to be involved in

[1] M. Mares-Guia and E. Shaw, *Fedn Proc.*, **22**, 528 (1963).
[2] P. H. Petra, W. Cohen, and E. Shaw, *Biochem. biophys. Res. Commun.*, **21**, 612 (1965); E. Shaw and S. S. Springhorn, *ibid.*, **27**, 391 (1967).
[3] A. Eyl and T. Inagami, *Biochem. biophys. Res. Commun.*, **38**, 149 (1970); *J. biol. Chem.*, **246**, 738 (1971); H. Nakayama, K. Tanizawa, and Y. Kanaoka, *Biochem. biophys. Res. Commun.*, **40**, 537 (1970).
[4] G. Feinstein, P. Bodlaender, and E. Shaw, *Biochemistry*, **8**, 4949 (1969).
[5] T. Inagami, *J. biol. Chem.*, **239**, 787 (1964); M. Mares-Guia and E. Shaw, *ibid.*, **240**, 1578 (1965); E. J. East and C. G. Trowbridge, *Archs Biochem. Biophys.*, **125**, 834 (1968).
[6] J. A. Stewart, H. S. Lee, and J. E. Dobson, *J. Am. chem. Soc.*, **85**, 1537 (1963).
[7] R. Smith and E. Shaw, *J. biol. Chem.*, **244**, 4709 (1969).

either binding or catalysis directly, but only indirectly, as is isoleucine–16 in chymotrypsin, in maintaining the conformation of the active site.

pH Studies

The results generally observed with trypsin for a variety of substrates are shown schematically in Fig. 5.8 (p. 159) where they are compared with those for chymotrypsin. They reveal that a group of pK around 7 is involved in the rate at high substrate concentration and also in k_c/K_A, the rate at low substrate concentration. This applies to specific[1] and non-specific ester substrates[2], and also to specific amide substrates[3]. These results imply that the histidine–46 group is free to ionize in the free enzyme and is also used in acylation for esters and in deacylation for amides. In contrast with chymotrypsin (see Fig. 4.8), k_c falls off at high pH, depending on a group of pK around 9·5. k_c/K_A, however, remains constant at high pH, which implies that the group involved in k_c (deacylation for esters, see below) is not involved in complex formation or in acylation.

The group of pK 7 is histidine–46. The group of pK 9·5 is more difficult to assign. It appears that the fall-off in rate of hydrolysis is the result of a conformational change, observed by several workers to occur as a function of pH[4]. Results of Kasserra and Laidler[5] on the variation of the specific rotation of the enzyme with pH have indicated that the pH dependence of the conformational change follows the deprotonation of a group of pK around 9–10, the same pK as for the k_c dependence on pH discussed above. It appears that, as in the case of chymotrypsin, the two phenomena are causally related. The group involved has been suggested as isoleucine–7 at the N-terminus of trypsin, by analogy with chymotrypsin. Evidence for this[6] comes from deamination studies using nitrous acid; loss of activity following deamination was found to correlate with loss of isoleucine–7. Evidence that this group is not a lysine residue (which might have the same pK for its ε-amino group as the α-amino group in isoleucine–7) is provided by the observation that acetylated trypsin has the same activity as trypsin itself[7]. Since all tyrosines and ε-amino groups are acetylated during this procedure, it seems unlikely that these groups are involved in the conformational change.

Figure 10.20 gives a schematic representation of the groups that are believed to be close to the active centre of trypsin, as revealed by X-ray, pH and other studies.

[1] H. Gutfreund, *Trans. Faraday Soc.*, **51**, 441 (1955); J. J. Bechet, *J. chim. Phys.*, **61**, 584 (1964); **62**, 1095 (1965); T. Inagami and J. M. Sturtevant, *Biochim. biophys. Acta*, **38**, 64 (1960).
[2] H. P. Kasserra and K. J. Laidler, *Can. J. Chem.*, **47**, 4021 (1969).
[3] J. Chevalier and J. Yon, *Biochim. biophys. Acta*, **112**, 116 (1966); R. Shukuya and K. Watanabe, *J. Biochem.*, **44**, 481 (1957).
[4] A. D'Albis and J. J. Bechet, *Biochim. biophys. Acta*, **140**, 435 (1967); M. Lazdunski and M. Delaage, *ibid.*, **140**, 417 (1967).
[5] H. P. Kasserra and K. J. Laidler, *Can. J. Chem.*, **47**, 4021 (1969).
[6] S. T. Scrimger and T. Hofmann, *J. biol. Chem.*, **242**, 2528 (1967).
[7] J. Labouesse and M. Gervais, *Biochemistry*, **2**, 215 (1967); see also J. Chevalier, J. Yon, G. Spotorno, and J. Labouesse, *ibid.*, **5**, 480 (1968).

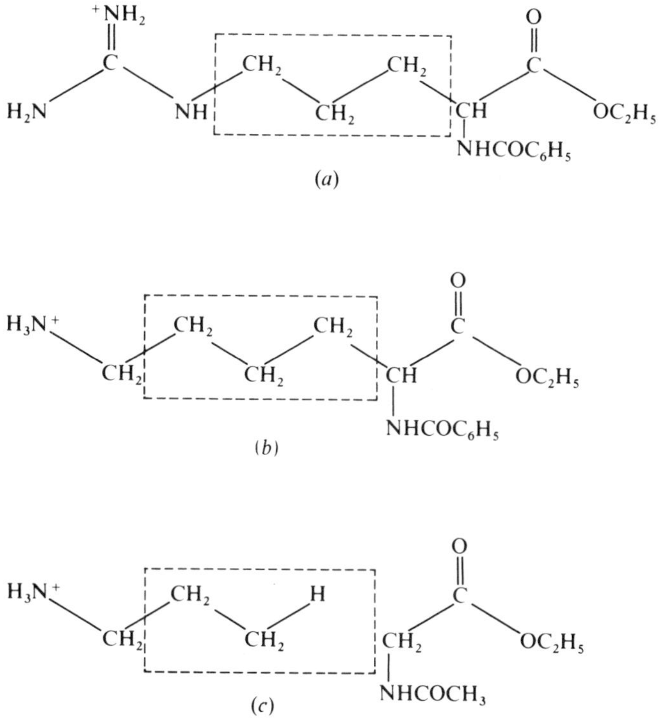

$$^+NH_2$$

$$H_2N \quad C \quad NH \quad CH_2 \quad CH_2 \quad CH_2 \quad CH \quad \overset{O}{\overset{\|}{C}} \quad OC_2H_5$$

NHCOC$_6$H$_5$

(a)

$$H_3N^+ \quad CH_2 \quad CH_2 \quad CH_2 \quad CH_2 \quad CH \quad \overset{O}{\overset{\|}{C}} \quad OC_2H_5$$

NHCOC$_6$H$_5$

(b)

$$H_3N^+ \quad CH_2 \quad CH_2 \quad CH_2 \quad H \quad CH_2 \quad \overset{O}{\overset{\|}{C}} \quad OC_2H_5$$

NHCOCH$_3$

(c)

FIG. 10.21. Baker's proposal for the modifying effects of alkylammonium ions on the trypsin system. (*a*) and (*b*) are two specific substrates; (*c*) shows a non-specific substrate to the right, and a modifier; the two together occupy the same positon on the enzyme as the specific substrates. The dotted regions are held to a hydrophobic site.

Evidence for the acyl-enzyme intermediate

Several workers have carried out pre-steady state studies on ester substrates of both specific and non-specific substrates of trypsin. In order to be able to do this for specific esters, it is necessary to slow down the reaction somewhat, in the same way as was done for chymotrypsin. Proflavin, a competitive inhibitor of trypsin, provides one way of doing this[1]. For N-benzoyl arginine ethyl ester (BAEE) k_2 (about 2×10^4 s^{-1}) is much greater than k_3 (about 15·5 s^{-1}), so that the rate-determining step is deacylation, an intermediate (the acyl enzyme) being involved. Similar results were obtained for tosyl arginine methyl ester (TAME) with thionine used as the dye binding to trypsin in order to follow the reaction[2]. Kasserra and Laidler[3] followed the pre-steady state kinetics of the non-specific substrate N-carbobenzoxyl-L-alanine p-nitrophenyl ester (CANE) using conditions such that the enzyme

[1] S. A. Bernhard and H. Gutfreund, *Proc. natn. Acad. Sci. U.S.A.*, **53**, 1238 (1965).
[2] A. Himoe, *J. biol. Chem.*, **245**, 1836 (1970).
[3] H. P. Kasserra and K. J. Laidler, *Can. J. Chem.*, **47**, 1793 (1969).

concentration was much greater than the substrate concentration. The theory (see pp. 185–192) indicates that these conditions reveal the value of k_2, whatever the relative magnitudes of k_2 and k_3. This constant was much larger than the steady-state constant at high substrate concentrations, so that deacylation is also rate-limiting in this case[†].

Direct spectrophotometric observation of the change in concentration of enzyme–substrate complexes has also been possible for non-specific ester substrates of trypsin, such as *N-trans*-cinnamoyl imidazole[1]. This was made possible by lowering the pH to about 5, where the rate is much lower than at neutral pH. Combination of the substrate with enzyme produced spectral changes which can be distinguished from the spectra of free enzyme or substrate. Deacylation was found to be the slow step here also.

Steady-state kinetics, while being somewhat more ambiguous than pre-steady-state analyses, have provided further evidence for an acyl-enzyme intermediate. Several benzoyl-arginine esters have been observed to be hydrolysed by trypsin at the same rate[2] even though their non-enzymic rates of hydrolysis were expected to vary by a factor of ten. Similar experiments were carried out by Bender and coworkers[3]. These studies indicate that de-acylation is the rate-determining step for esters[‡].

Nucleophile studies provide further evidence for an intermediate in trypsin hydrolysis[4]. The results indicate an acceleration by nucleophiles such as methanol of the rate (k_c) at high substrate concentrations. This result implies that acylation is the slow step, since water is normally involved in this step, and it is a poorer nucleophile than methanol. While other solvent effects of such nucleophiles have been observed[5], there is evidence[6] that the dielectric effects cannot explain the increase in k_c. Seydoux and Yon[7] found that isopropanol, at lower concentrations than methanol, lowered the dielectric constant of water to the same extent from 78·5 to 75·7; however, only methanol produced the increase in k_c, isopropanol having no effect. Since both solvents are expected to have similar dielectric effects under these conditions, the nucleophile effect appears to be genuine.

---

† This contradicts the conclusion arrived at by F. Seydoux and J. Yon (*Biochem. biophys. Res. Commun.*, **44**, 745 (1971), based on several unproven assumptions, that non-specific substrates have acylation as the rate-limiting step.

‡ See also E. T. Elmore, D. V. Roberts, and J. J. Smith, *Biochem. J.*, **102**, 728 (1967); M. J. Baines, J. B. Bair, and D. T. Elmore, *ibid.*, **90**, 470 (1967); M. L. Bender and F. J. Kézdy, *J. Am. chem. Soc.*, **87**, 4954 (1965) for similar studies.

[1] M. L. Bender and E. T. Kaiser, *J. Am. chem. Soc.*, **84**, 2556 (1962).

[2] G. W. Schwert and M. A. Eisenberg, *J. biol. Chem.*, **179**, 665 (1949).

[3] M. L. Bender, J. V. Killheffer, and F. J. Kézdy, *J. Am. chem. Soc.*, **86**, 5330 (1964).

[4] F. Seydoux and J. Yon, *Eur. J. Biochem.*, **3**, 42 (1967).

[5] T. Inagami and J. M. Sturtevant, *Biochim. biophys. Acta*, **38**, 64 (1960); L. M. del Castillo, *et al.*, *ibid.*, **128**, 578 (1966); M. Castanede-Agullo, *et al.*, *ibid.*, **139**, 398 (1967); H. P. Kasserra and K. J. Laidler, *Can. J. Chem.*, **48**, 1793 (1970).

[6] F. Seydoux and J. Yon, *Biochem. biophys. Res. Commun.*, **44**, 745 (1971).

[7] F. Seydoux and J. Yon, *Eur. J. Biochem.*, **3**, 42 (1967).

In fact, Seydoux and Yon find that their nucleophile results can best be explained by assuming a mechanism involving separate binding sites for water and nucleophile:

$$E + A \underset{}{\overset{K_s}{\rightleftharpoons}} EA \overset{k_2}{\searrow} EA' \underset{N}{\overset{W}{\rightleftharpoons}} EA'W \overset{k_3^*}{\rightarrow} E + Y$$
$$\searrow_X \qquad\qquad \searrow EA'N \overset{k_4^*}{\rightarrow} E + Z.$$

Unfortunately, no pre-steady state studies with specific amide substrates of trypsin appear to have been carried out. It is assumed by analogy with chymotrypsin (and from steady-state studies) that acylation is the rate-determining step[1]. It is also assumed that $(k_{-1} + k_2)/k_1 = K_A$ for amides; it is known, however, that for some esters at least, $(k_{-1} + k_2)/k_1 \gg K_A$. Hence, for esters, K_A cannot be interpreted in terms of binding, whereas for amides it can.

While trypsin certainly appears to involve a two-intermediate mechanism, there is considerable evidence that, at least in some cases, other types of chemical intermediate may be involved in addition to, or apart from, the acyl-enzyme. Bernhard and Gutfreund[2] were able to trap the intermediate in hydrolysis by both chymotrypsin and trypsin, by dropping the pH very rapidly using a stopped-flow instrument. Raising the pH subsequently enabled them to follow the reaction of this intermediate in the stopped-flow system. When this was done, they observed release of ethanol, as expected during acylation. They interpreted this as indicating that the actual rate-determining step occurred before acylation; this might possibly involve a tetrahedral intermediate. Other evidence of this type of intermediate comes from pre-steady-state analysis by Johannin and Yon[3]. They observed formation of an intermediate using a spectrophotometer with a rate of formation different from the rate of formation and decomposition of EA, and different from the rate of liberation of alcohol.

Mechanism of trypsin catalysis

A comparison of the mechanisms of trypsin and chymotrypsin action has been given by Kasserra and Laidler[4]. There is a considerable amount of evidence indicating that the structures of the two enzymes are basically the same, even though the X-ray structure of trypsin is not available at the time of writing. The spectra of the two enzymes are very similar[5] and the difference spectra between enzyme combined with substrate and free enzyme are also very similar[6]. The primary sequences of amino acids show a considerable

[1] H. Gutfreund and J. M. Sturtevant, *Proc. natn. Acad. Sci. U.S.A.*, **42**, 719 (1956).
[2] S. A. Bernhard and H. Gutfreund, *Proc. natn. Acad. Sci. U.S.A.*, **53**, 243 (1965).
[3] G. Johannin and J. Yon, *Biochem. biophys. Res. Commun.*, **25**, 320 (1966).
[4] H. P. Kasserra and K. J. Laidler, *Can. J. Chem.*, **47**, 4031 (1969).
[5] B. Jirgensons, *J. biol. Chem.*, **241**, 147 (1966).
[6] M. L. Bender and E. T. Kaiser, *J. Am. chem. Soc.*, **84**, 2556 (1962).

degree of homology[1]. The four disulphide bonds that are common to the two enzymes occur in similar positions in the sequence; the two additional ones in trypsin can be omitted (as evidenced by model building) without much alteration in the structure of trypsin[2]. Topographical considerations also indicate similarity[3]. In addition, the X-ray structures of the two chymotrypsinogens A and B and trypsinogen are very similar[4].

The kinetics of hydrolysis of several non-specific substrates are very similar for the two enzymes[5]. These and other kinetic and mechanistic similarities between the two enzymes are summarized in Table 10.1.

From the evidence already discussed for these two enzymes, several common mechanistic features emerge. The histidine is very probably used to remove a proton from the serine $-OH$ group during the acylation process; hence the requirement for the unprotonated form. A proton must be transferred to the leaving alcohol group as well. This could be achieved in a concerted system as in the mechanism of Kasserra and Laidler (see p. 324). The histidine is used in the deacylation step in the unprotonated form as well; it presumably aids in splitting the water molecule which acts as a nucleophile in this step, producing the second product, and also in reprotonation of the serine $-OH$. Deacylation is looked on as the exact reverse of

TABLE 10.1a

Evidence relating to complex formation in chymotrypsin and trypsin

Kind of evidence	Chymotrypsin	Trypsin
pH studies at low substrate concentrates	Bell shaped curve; pK of ~ 7 and ~ 9	Sigmoid curve; pK of ~ 7; no falling-off up to pH ~ 10
pH studies in mixed solvents	Group of pK ~ 7 is a cationic acid; group of pK ~ 9 is a neutral or ion-pair acid	—
pH studies at low pH	Carboxyl group (pK ~ 4) present	Carboxyl group (pK ~ 4) present
Chemical and kinetic studies	Single histidine residue involved in reaction, but not in complex formation	Single histidine residue involved in reaction, but not in complex formation
Chemical and kinetic studies	N-terminal isoleucine–16 (pK ~ 9) indirectly participates in the binding process	—

[1] See, for example, B. S. Hartley, J. R. Brown, D. L. Kauffman, and L. B. Smillie, *Nature*, **207**, 1157 (1967).
[2] B. S. Hartley, *et al.*, *loc. cit.*, and B. Keil and F. Sorn, *FEBS Symp.*, No. 1, p. 37 (1964).
[3] K. Shibata, *et al.*, *Biochim. biophys. Acta*, **81**, 323 and **86**, 477 and **93**, 346 (1964).
[4] L. E. Smillie, *et al.*, *Nature*, **218**, 343 (1968).
[5] See, for example, J. A. Stewart and L. E. Ouellet, *loc. cit.*, M. L. Bender and E. T. Kaiser, *loc. cit.*; T. Inagami and J. M. Sturtevant, *loc. cit.*

TABLE 10.1b

Evidence relating to acylation of chymotrypsin and trypsin

Kind of evidence	Chymotrypsin	Trypsin
Chemical and kinetic studies	Acylation involves the hydroxyl group of serine–195	Acylation involves the hydroxyl group of serine–183
pH studies with amides, for which acylation is rate-limiting	Rate falls off at pH <7.5; no fall-off in alkaline solution up to pH 9.5	Rate falls off at pH <7.5; no fall-off in alkaline solution up to pH 9.5
Substituent effects	Proton transfer is involved in acylation	—
Deuterium isotope effects	Proton transfer is involved in acylation	—
pH studies at low concentrations of esters, for which deacylation is rate-limiting	Both ionizing groups are apparently involved in acylation	Single ionizing group ($pK \approx 7$) participates in acylation

TABLE 10.1c

Evidence relating to deacylation of chymotrypsin and trypsin

Kind of evidence	Chymotrypsin	Trypsin
pH studies with N-α substituted esters, for which deacylation is rate limiting	Sigmoid curve; pK of ~ 7; no falling-off up to pH ~ 10	Bell-shaped curve; pK of ~ 7 and ~ 9.5. Apparently both groups involved in deacylation
pH studies with esters on which the α-amino group is not substituted	Bell-shaped curve; pH optimum shifted from pH ~ 8 to pH ~ 6	Bell-shaped curve; pH optimum shifted from pH ~ 8 to pH ~ 6
Effect of added nucleophiles	Deacylation is first-order in nucleophile (e.g. methanol)	—
Kinetic studies	No intermediate detected in deacylation	—
Deuterium isotope effects	Proton transfer is involved in deacylation	Proton transfer is involved in deacylation

For references to the evidence see H. P. Kasserra and K. J. Laidler, *Can. J. Chem.*, **47**, 4031 (1969).

acylation, with the $-OR$ group of the alcohol replaced by an $-OH$ group in water.

The differences in the pH profiles of trypsin and chymotrypsin at high pH can be explained as follows. The $-NH_3^+$ of the terminal isoleucine group in both enzymes loses a proton, leading to a conformational change, as discussed above, which destroys the integrity of the active site. For chymotrypsin, the conformational change occurs during complex formation. This affects the acyl-enzyme, and slows down the deacylation process (for esters). For trypsin, deacylation is reduced by the conformational change affecting the acyl-enzyme, but not the Michaelis complex.

Specificity and binding

As has been mentioned, specific substrates for trypsin contain a positive charge on the amino acid adjacent to the bond being broken[1]; examples of such amino acids are lysine and arginine. The enzyme will also hydrolyse S-2-aminoethyl cysteine residues, which also contain a positive charge, and are approximately the same length as the other two residues[2]. Neutral substrates are also hydrolysed, although at a rate of about one thousandth of that for the specific substrates[3]. Hydrolysis is stopped by di-*N*-methylation of the substrate, although it is merely slowed down by mono-alkylation[4].

The question of specific (with positive charge) and non-specific (without positive charge) substrates has received considerable attention, especially with reference to whether the different substrates bind at different sites. Most of the evidence points to overlapping sites for binding of the two kinds of substrate[5]. For instance, certain alkylammonium ions are known to inhibit specific substrate hydrolysis[6]. For non-specific substrates, such as *N*-acetyl glycine ethyl ester, small alkyl ammonium ions (methyl or ethyl) did not inhibit, although larger ones (benzoyl or *n*-butyl) did[7]. In fact, the smaller alkylammonium ions actually accelerated the hydrolysis of the non-specific substrate. It would appear that the positive charge on the alkyl ammonium group provides a better alignment of the catalytic groups, in the same way as for a specific substrate. Larger modifiers presumably have no accelerative effect because they cannot fit in to the active site. Baker pictures the modifier effect as in Fig. 10.21. Similar acceleration by alkylammonium ions has also been observed[8].

Sanborn and Hein[9] found that removal of the positive charge affected binding more than catalysis, and concluded that the positive charge is related largely to the specificity site (i.e. the anionic site, aspartic acid–177) on

[1] H. Neurath and G. W. Schwert, *Chem. Rev.*, **46**, 69 (1950); M. Bergmann, J. S. Fruton, and H. Pollok, *J. biol. Chem.*, **127**, 643 (1939).
[2] H. Lindley, *Nature*, **178**, 647 (1956); S-S. Wang and F. H. Carpenter, *Biochemistry*, **6**, 215 (1967).
[3] T. Inagami and H. Mitsuda, *J. biol. Chem.*, **239**, 1388, 1395 (1964); T. Inagami and J. M. Sturtevant, *Biochim. biophys. Acta*, **38**, 64 (1960) (for ATEE); M. L. Bender and E. T. Kaiser, *J. Am. chem. Soc.*, **84**, 2556 (1962) (*N-trans*-cinnamoyl imidazole); and J. A. Stewart and L. Ouellet, *Can. J. Chem.*, **37**, 751 (1959) (*p*-nitrophenyl acetate).
[4] L. Benoiton and J. Deneault, *Biochim. biophys. Acta*, **113**, 613 (1968).
[5] See B. R. Baker, *Design of Active Site Directed Irreversible Inhibitors*, pp. 5ff, John Wiley and Sons, New York (1967) for a review.
[6] T. Inagami, *J. biol. Chem.*, **239**, 787 (1964).
[7] T. Inagami and T. Murachi, *J. biol. Chem.*, **239**, 1388, 1395 (1964).
[8] D. Ponzi and G. E. Hein, *Biochem. biophys. Res. Commun.*, **25**, 60 (1966); F. Seydoux and J. Yon, *Biochem.*, **10**, 2284 (1971); V. Kallen-Trummer, *et al.*, *Biochemistry*, **9**, 3580 (1970).
[9] B. M. Sanborn and G. E. Hein, *Biochemistry*, **7**, 3616 (1968).

the enzyme. Their results and conclusions may be summarized as follows:
(1) the rates of positive and neutral substrates are affected differently by
modifiers Q of different charge types, at the same $[Q]/K_Q$ values; (2) a given
modifier affects the hydrolysis of positive and neutral substrates differently;
(3) the reaction of TLCK with trypsin is unaffected by neutral modifiers but
is affected by positively-charged modifiers; and (4) equilibrium dialysis
experiments by Sanborn and Bryan[1] indicate that the inhibition is not
simple competitive for the modifiers used. It would appear that, while neutral

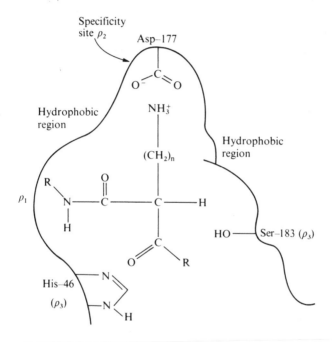

FIG. 10.22. Schematic representation of the binding of a specific substrate to trypsin.

molecules generally inhibit the hydrolysis of charged substrates, they can
inhibit, enhance or have no effect on the *binding* of such substrates. Also,
positive modifiers generally enhance the binding of neutral substrates, as
discussed above. In other words, the two sites do appear to overlap. Further
evidence supporting this conclusion has been obtained[2].

A second aspect relating to specificity and binding is that better hydrolysis
is obtained by trypsin on substrates having the amino end acylated. As with

[1] B. M. Sanborn and W. P. Bryan, *Biochemistry*, **7**, 3624 (1968).
[2] C. G. Trowbridge, A. Krehbiel, and M. Laskowski, *Biochemistry*, **2**, 843 (1963); and S.
Howard and J. Mehl, *Biochim. biophys. Acta*, **105**, 594 (1965).

chymotrypsin, however, substrates having a free amino group are hydrolysed. The kinetics of hydrolysis are somewhat different, however, as indicated by differences in pH-activity profiles[1]. Stewart and coworkers have interpreted the kinetic differences in terms of the inclusion in the profile of the ionization of the free amino group on the substrate.

Other differences between the two kinds of substrate include the effect of alcohol in the reaction medium on the rate. The rate of liberation of titrable acid in alcoholic medium compared to the rate in pure aqueous medium is higher for blocked substrates and lower for substrates with a free amino group. Furthermore, the rate is pH-independent for the acylated substrates, and pH-dependent for the free ones[2]. The difference was again attributed to the presence of the free amino group. Proof of this was obtained by the observation that there was a linear correlation between the ratio of trans-esterification rate to hydrolysis rate and the ratio of the amount of ionized to uncharged forms of the α-amino groups in the substrate[3]. It was suggested[4] that the alkaline limb of the pH-rate profile might in this case be due to the ionization of the amino group on the substrate rather than to a conformational change. The correspondence of pKs might, perhaps, be fortuitous. The effect of nucleophiles on k_c and K_A also differs for the two substrate types[5]; no reason for this has been given.

As was briefly mentioned above, the length of the substrate chain is also fairly critical for binding. If a methylene group is added to or subtracted from the lysine or arginine chain, a much lower activity is obtained[6]. On the whole, trypsin, like chymotrypsin, is specific for L-amino acids[7], although some neutral short-chain molecules can be hydrolysed as the D-antipodes[8]. Studies similar to those on chymotrypsin with substrates of fixed geometry would presumably reveal more about the stereospecificity of trypsin.

As with chymotrypsin, esters are hydrolysed about 300 times faster than amides[9]. Another important aspect of trypsin binding is that, as for chymotrypsin, there is a hydrophobic contribution. Inhibitor binding studies provide evidence for this contribution in addition to the charge-charge interaction, not found for chymotrypsin[10]; a tyrosine residue is believed to be involved in this interaction. Further studies indicate that both binding and

[1] L. M. del Castillo, *et al.*, *Biochim. biophys. Acta*, **191**, 354 (1969); J. H. Seely and N. L. Benoiton, *Can. J. Biochem.*, **48**, 1122 (1970); J. A. Stewart, *et al.*, *Biochim. biophys. Res. Commun.*, **42**, 1220 (1971).

[2] L. M. del Castillo, Y. Bustamante, and M. Castanede-Agullo, *Biochim. biophys. Acta*, **128**, 578 (1966).

[3] M. del Castillo, *et al.*, *Biochim. biophys. Acta*, **191**, 354 (1969).

[4] *ibid.*, **191**, 362 (1969).

[5] F. Seydoux and J. Yon, *Eur. J. Biochem.*, **3**, 42 (1967).

[6] K. Kitagawa and H. Izumiya, *J. Biochem.*, **46**, 1159 (1959).

[7] B. F. Erlanger, N. Kokowski, and W. Cohen, *Archs Biochem. Biophys.*, **95**, 27 (1961).

[8] H. P. Kasserra and K. J. Laidler, unpublished results.

[9] S. A. Bernhard, *Biochem. J.*, **59**, 506 (1955).

[10] M. Mares-Guia and E. Shaw, *J. biol. Chem.*, **240**, 1579 (1965).

catalysis are improved by this hydrophobic interaction; this is again analogous to the situation for chymotrypsin[1]. The evidence suggests that the
hydrophobicity produces a favourable entropy change on binding, the
enthalpy changes for numerous compounds being close to zero[2].

That the binding site is hydrophobic in nature can be seen from the
similarity between the free-energy change on binding and the free-energy
change on transferring benzene from an aqueous solution to pure liquid
benzene.

The binding of trypsin substrates has been discussed in terms of four
major groupings on the substrate, and four 'sub-sites' on the enzyme. The
situation is illustrated schematically in Fig. 10.22. It is to be noted that the
hydrophobic site lies between the specificity sites, as indicated by the results
referred to above. The binding parameters of the D and L isomers of α-*N*-
benzoylarginine *p*-nitroanilide are the same, but the D-isomer is not
hydrolysed. This indicates a change in the orientation of the catalytic groups
with respect to the bond to be broken. It appears that the carboxamide
portion binds in the same way for substrate and inhibitor. The H and
RCONH groups (corresponding to the ρ_1 site on the enzyme) presumably
interchange binding sites in a similar way to the analogous situation for
chymotrypsin. This would prevent proper attack on the carbonyl by the
reactive serine residue on the enzyme.

The ρ_2 site is the specificity site, and has already been discussed in some
detail. The ρ_3 site relates to catalysis; since the X-ray structure is not available,
and comparable studies to those of Cohen *et al.* for chymotrypsin have not
been carried out with trypsin, little can be said about this site, except that it
contains the catalytic residues histidine and serine.

Difference spectra studies for binding of the inhibitor benzamidine to
trypsin indicate that binding produces a change in the substrate, and no
significant change in the enzyme[3]. This is of some interest from the point of
view of theories of enzyme action (see Chapter 8).

Lactate dehydrogenase (E.C.1.1.1.27)†

A considerable amount of work has been carried out on lactate dehydrogenase; both kinetic and structural investigations have been made, including
a preliminary study[4] of the X-ray crystallographic structure of this enzyme.
Many of the features of the mechanism of action of this enzyme appear to
be common to other dehydrogenases.

† For a review, see G. W. Schwert and A. D. Winer, *The Enzymes*, 7, 127 (1963) (ed. P. D. Boyer, H. Lardy, and K. Myrbäck, Academic Press, New York).

[1] M. Mares-Guia, E. Shaw, and W. Cohen, *ibid.*, **242**, 5777 (1967).
[2] T. Inagami, *J. biol. Chem.*, **239**, 787 (1964); M. Mares-Guia and A. F. S. Figueiredo, *Biochemistry*, **9**, 3223 (1970).
[3] E. J. East and C. G. Trowbridge, *Archs Biochem. Biophys.*, **125**, 334 (1968).
[4] M. G. Rossman, *et al.*, *Cold Spring Harb. Symp. quant. Biol.*, **36**, 179 (1972).

Lactate dehydrogenase catalyses the following process:

Nicotinamide adenine dinucleotide (NAD^+) (Coenzyme 1)

Reduced nicotinamide adenine dinucleotide (NADH) (Reduced Coenzyme 1)

The enzyme has a molecular weight of 130 000 to 145 000[1]. It has four subunits, each of which binds one molecule of substrate and coenzyme[2]. That each of these binding sites is enzymically active is indicated by the correlation between reduction in binding of coenzyme and reduction in enzyme activity with time, on heat denaturation of the enzyme[2].

[1] D. B. S. Millar, *J. biol. Chem.*, **237**, 2135 (1962); H. J. Fromm, *J. biol. Chem.*, **238**, 2938 (1963).
[2] H. J. Fromm, *loc. cit.*; S. F. Velick, *J. biol. Chem.*, **233**, 1455 (1958); D. B. S. Millar, *ibid.*, **234**, 1149 (1959); R. S. Criddle, C. H. McMurray, and H. Gutfreund, *Nature*, **220**, 1091 (1968).

Studies on this enzyme have been complicated by the fact that it comprises subunits which exist in two structurally and kinetically similar (but not identical) forms. These are named H and M, since the former predominates in heart tissue, and the latter in muscle. These subunits combine together to form a total of five tetrameric isoenzymes, H_4, H_3M, H_2M_2, HM_3 and M_4. These isozymes, and the H and M monomers themselves, differ from one another in their electrophoretic mobility[1]; they differ by one electrical charge, which suggests that the monomers differ in one amino acid residue. In order to simplify the interpretation of kinetic and structural studies most investigations have been made with either the H_4 or the M_4 enzyme.

Compulsory order kinetics

Several lines of evidence have pointed to an ordered, compulsory sequence of addition and release of substrates and products for lactate dehydrogenase. One investigation[2] involved the use of equilibrium isotope exchange (see p. 266) on the bovine heart and rabbit muscle enzymes. At pH 7·9 large amounts of the substrate L-lactate almost completely prevent isotopic exchange between NAD^+ and NADH, while exchange between pyruvate and lactate continues. At low concentrations of lactate, isotopic exchange between these coenzymes does occur. No exchange takes place in the absence of lactate.

For a ping pong mechanism (see p. 123) NAD^+/NADH exchange would continue irrespective of the concentration of the other substrate, since each substrate reacts with the enzyme to give product directly. Such a mechanism is, therefore, ruled out by the above results. A random-order reaction (see p. 119) would require both exchanges to reach a maximum and remain at the maximum value with increase in substrate concentration. Since this is contrary to the results, such a mechanism does not apply.

A mechanism involving compulsory binding of coenzyme first, followed by substrate, is consistent with the observed results. The binary complex, $E–NAD^+$ is first formed. In the presence of only small amounts of lactate, this binary complex may react to give products, or dissociate back to enzyme plus free coenzyme. In this way, NAD^+/NADH isotopic exchange will result once equilibrium is established, as is observed. At high lactate concentrations, however, the enzyme exists almost entirely as the ternary complex $E–NAD^+$-lactate. The compulsory order for dissociation (to yield isotopic exchange) requires lactate to dissociate first, followed by NAD^+. Since this does not occur to any significant extent at high lactate concentrations, no exchange can take place, as is observed experimentally. The compulsory binding order mechanism is strongly indicated by these equilibrium isotope exchange experiments.

[1] C. L. Markert and K. Moller, *Proc. natn. Acad. Sci. U.S.A.*, **45**, 753 (1959).
[2] E. Silverstein and P. D. Boyer, *J. biol. Chem.*, **239**, 3901 (1964).

At pH 9·7, however, the reaction appears to be only partially ordered, since $NAD^+/NADH$ isotopic exchange does occur, although to a lesser extent than that for lactate/pyruvate.

A second indication for a compulsory ordered sequence was obtained by Thomson, Darling and Bordner[1] from kinetic isotope effects for rabbit muscle lactate dehydrogenase. They synthesized NAD^+ with deuterium at the fourth carbon atom in the nicotinamide ring, both in the transferable and non-transferable positions (see p. 259). As expected, there was found to be a primary kinetic isotope effect on V for NADH when the deuterium was substituted in the transferable position, but only a secondary effect for the other position. More important, the kinetic isotope effects on the constants K_A, K_B and K'_A in the general rate equation (see p. 115) were found to be consistent with what would be predicted for compulsory binding (p. 121), and inconsistent with the expectations from the other mechanisms.

Detailed kinetic studies on this enzyme provided further evidence for a compulsory order of binding, with coenzyme binding first (in both directions of reaction). On the assumption that both substrates react with the enzyme before dissociation of either product, the rate obeys the following equation:

$$v = \frac{V[A].[B]}{K'_A K_B + K_B[A] + K_A[B] + [A].[B]}.$$

This applies, however, to both compulsory and random order reactions (see p. 126). Early work by Winer, Schwert, and coworkers[2] established that binding of reduced coenzyme to lactate dehydrogenase results in a change in fluorescence of the former, so that the reaction of the binary complex, E–NADH, with substrate (pyruvate) can be followed spectrophotometrically. The fluorescence pattern was changed further on binding of the substrate. These experiments confirmed the existence of binary complexes, and strongly suggested the formation of ternary complexes (see below). They also demonstrated that coenzyme had to bind first in order for reaction to occur. Coenzyme analogues such as triphosphopyridine nucleotide (TPNH, NADPH) did not form such reactive binary complexes with the enzyme. This indicates that the complexes observed were real and specific to the enzyme reaction, and not artifacts. Similar results were obtained by Fromm[3], who observed spectral changes related to the enzyme when coenzyme and substrate were bound. This work also proved the existence of the binary complex in the reverse direction.

[1] J. F. Thomson, J. J. Darling, and L. F. Bordner, *Biochim. biophys. Acta*, **85**, 177 (1964); see R. H. Baker and H. R. Mahler, *Biochemistry*, **1**, 35 (1962).

[2] A. D. Winer, W. B. Novoa, and G. W. Schwert, *J. Am. chem. Soc.*, **79**, 6571 (1957); see also B. Chance and J. B. Nielands, *J. biol. Chem.*, **199**, 383 (1952); and R. A. Alberty, *J. Am. chem. Soc.*, **75**, 1925 (1953).

[3] H. J. Fromm, *J. biol. Chem.*, **238**, 2938 (1963).

A thorough kinetic study of lactate dehydrogenase was undertaken by Zewe and Fromm[1], using the technique of product inhibition (see p. 127). The usual primary and secondary plots were carried out, with the results pointing to an ordered addition.

A similar approach to this last one was employed by Gutfreund and coworkers[2], who used absorption of visible light by coenzyme as a means of monitoring the reaction. Binary complexes between pig heart lactate dehydrogenase and NAD^+ and also NADH were readily detected, and carefully studied. Oxamate, a competitive inhibitor for pyruvate, prevented the dissociation of NADH from the enzyme. Without oxamate, NADH could dissociate normally. This implies that NADH must bind first, followed by pyruvate. In another study[3] it was shown that NAD^+ must bind before lactate.

Ternary complexes

As early as 1953 it was shown that direct, stereo-specific transfer of hydrogen takes place between NADH and the alpha carbon of pyruvate[4]. It is highly unlikely that such a specific transfer would occur via an intermediary group on the surface of the enzyme. That ternary complexes exist in the reaction catalysed by lactate dehydrogenase has also been shown by many of the results discussed above. However, it remains to determine their significance in the kinetics and mechanism of catalysis. The experiments with oxamate led to the conclusion than an unreactive complex, E–NADH–oxamate, is formed[5]. It seemed reasonable to suggest that pyruvate probably forms a similar reactive complex in the actual reaction.

Furthermore, the calculated Haldane relationship (p. 83) for a Theorell–Chance mechanism did not agree with the observed one, but that calculated on the basis of an ordered ternary-complex mechanism did agree[6].

Abortive ternary complexes

The existence of abortive or dead-end ternary complexes (i.e. ternary complexes which do not lead to reaction, see p. 126) has a profound effect on the interpretation of results obtained from kinetic studies with two-substrate enzymes. For instance, product inhibition studies normally allow a definite distinction to be made among the four basic mechanisms for such enzymes. The presence of abortive ternary complexes, however, alters the kinetic

[1] V. Zewe and H. J. Fromm, *J. biol. Chem.*, **237**, 1668 (1962).

[2] H. d'A. Heck, C. H. McMurray, and H. Gutfreund, *Biochem. J.*, **108**, 793 (1968).

[3] H. Gutfreund, R. Cantwell, C. H. McMurray, R. S. Criddle, and G. Hathaway, *ibid.*, **106**, 683 (1968).

[4] F. A. Loewus, P. Ofner, H. F. Fisher, F. H. Westheimer, and B. Vennesland, *J. biol. Chem.*, **202**, 699 (1953).

[5] H. d'A. Heck, *et al.*, *loc. cit.*: F. A. Loewus, *et al.*, *loc cit.*

[6] V. Zewe and H. J. Fromm, *Biochemistry*, **4**, 785 (1965); M. T. Hakala, A. J. Glaid, and G. W. Schwert, *J. biol. Chem.*, **221**, 191 (1956).

behaviour, and it is often not possible to make the usual distinction (see p. 126). It turns out that the lactate dehydrogenase system has at least two abortive ternary complexes (see scheme on p. 351), namely $E-NAD^+$-pyruvate and E–NADH-lactate. Several lines of evidence have led to this conclusion.

Kinetic results obtained by Zewe and Fromm[1], when interpreted on the assumption of the existence of the two abortive ternary complexes, agreed well with this mechanism. These workers were able to measure the dissociation constants for the two complexes. Earlier work had already demonstrated the existence of the abortive complex, E–NADH-lactate[2]. The work with oxamate discussed above supported the existence of the complex $E-NAD^+$-pyruvate.

It has been possible, in some instances, to isolate abortive ternary complexes, using substrate analogues[3]. While such isolation by no means establishes the kinetic significance of such abortive complexes, it is certainly a strong indication that they probably exist in the normal catalytic process. It has even been possible to isolate the complex $E-NAD^+$-pyruvate[4]. This was shown, using isotopically labelled materials, to be present in a $1:1:1$ ratio of enzyme : NAD^+ :-pyruvate. Its formation could also be followed spectrophotometrically. It was not possible to form a similar stable complex of E–NADH-lactate. These workers suggested that such a complex probably is not important in the lactate dehydrogenase catalysis. The work discussed above, however, suggests that this is not the case. Care must therefore be taken when interpreting such complex isolation studies. The abortive complex, $E-NAD^+$-pyruvate may well be the cause of substrate inhibition by pyruvate, which has been frequently observed with this enzyme. It may also correspond to the complex whose X-ray structure has been determined (see p. 350). It may even be the same as that detected kinetically by Gutfreund and his coworkers.

The rate-determining step

A further matter that has been investigated in some detail with this enzyme relates to the rate-determining step. This problem has turned out to be closely connected with conformational changes in the enzyme. For muscle lactate dehydrogenase (predominantly M_4), conformational changes have been invoked by several groups of workers. A marked change in the order of binding of NADH to the enzyme was observed in fluorometric titration experiments[5] from 0·43 to 1·5 with increasing concentration of NADH. The most marked change was for the first molecule of NADH.

[1] V. Zewe and H. J. Fromm, *J. biol. Chem.*, **237**, 1668 (1962).
[2] A. D. Winer, W. B. Novoa, and G. W. Schwert, *J. Am. chem. Soc.*, **79**, 6571 (1957).
[3] H. A. Lee, R. H. Cox, S. L. Smith, and A. D. Winer, *Fedn Proc.*, **25**, 711 (1966).
[4] G. DiSabato, *Biochem. biophys. Res. Commun.*, **33**, 688 (1968).
[5] S. R. Anderson and G. Weber, *Biochemistry*, **4**, 1948 (1965).

Optical rotatory dispersion measurements[1] showed that the change with pH in alpha-helix structure correlated with the dependence on pH of the rate of pyruvate reduction, indicating a conformational change on binding (or reaction) of pyruvate. Oxalate, a competitive inhibitor for pyruvate, was also found to alter the secondary structure, but only in the presence of coenzyme. Czerlinski and Schreck[2] showed that NADH binds to rabbit muscle M_4 lactate dehydrogenase in two distinct steps, one of which corresponds to a structural change[2]. (This contrasts with the data for H_4, see below.) They suggested three possibilities for this behaviour: (1) a structural change in NADH itself (see below); (2) removal of a water molecule from the zinc atom on binding of coenzyme, or (3) a helix-type conformational change in the enzyme.

Rapid-reaction techniques

In their investigations of pig H_4 lactate dehydrogenase, Gutfreund and coworkers[3] determined which was the rate-determining step in the reaction. They found that the rate of dissociation of the E–NADH complex for pig's heart (H_4) lactate dehydrogenase was faster than the turnover rate at pH 6, but about the same as the turnover rate at pH 7 and above. The first mole of NADH was produced very much more rapidly ($k = 70\,s^{-1}$) than the rest, which was produced at the steady-state rate of about $k = 4\,s^{-1}$. This result suggests that the dissociation of E–NADH is the rate-determining step[4] (i.e. at pH 6). This, however, was also observed at pH 7, where E–NADH dissociation occurs at the same rate as the turnover rate; this result localized the rate-determining step to a point after the reduction of pyruvate, and before the dissociation of the E–NADH complex (since interconversion of the two ternary complexes, E–NAD-lac and E–NADH-pyr was much faster than the turnover rate; i.e. interconversion could not be the slow step). The authors suggested that the rate-determining step might be the dissociation of pyruvate from the ternary complex, or a rearrangement of the E–NADH complex. The same investigation showed that the rate-determining step for the reverse reaction (oxidation of lactate) was the dissociation of the E–NADH binary complex.

Heck[5] has confirmed, by temperature-jump relaxation studies on pig's heart (H_4) lactate dehydrogenase, that the rate-determining step for pyruvate reduction is indeed the dissociation of NADH from the binary complex. He found that the rate of dissociation of NADH from the binary complex was

[1] I. A. Bolotina, D. S. Markovitch, M. V. Volkenstein, and P. Zavodsky, *Biochim. biophys. Acta*, **132**, 271 (1967).

[2] G. H. Czerlinski and G. Schreck, *J. biol. Chem.*, **239**, 913 (1964).

[3] Gutfreund, *et al.*, *Biochem. J.*, **106**, 683 (1968); Gutfreund, Cantwell, McMurray, Criddle, and Hathaway, *loc. cit.*

[4] H. d'A. Heck, *et al.*, *ibid.*, **793** (1968); Heck, McMurray, and Gutfreund, *loc. cit.*

[5] H. d'A. Heck, *J. biol. Chem.*, **244**, 4375 (1969).

the same as the rate measured in the stopped-flow studies of Gutfreund *et al.* This was interpreted as implying the lack of a slow conformational change at this point, i.e. that NADH dissociation, and not a conformational change, was the slow step.

Criddle *et al.*[1] however, arrived at a different conclusion. For the reduction of pyruvate at alkaline pH values, the formation of the E–NADH-pyr complex from pyruvate + E–NADH was found to be the rate-determining step. At pH 6, however, the interconversion of two E–NADH binary complexes appeared to be the slow step. This interconversion was detected by relaxation studies, which revealed a relaxation time slow enough to be consistent with an isomerization of the E–NADH complex. The two binary complexes were distinguishable, one being observed only if the enzyme had just participated in a lactate to pyruvate conversion. Since pyruvate is known to dissociate rapidly from its ternary complex with E–NADH, the conversion of one binary complex into another appeared to be the rate-determining step.

The existence of a conformational change (or isomerization) at this E–NADH binary complex was supported by other work by Heck[2], despite his reaching a different conclusion as to the slow step for pyruvate reduction. He observed that the binding of oxamate (a competitive inhibitor for pyruvate binding) was pH-dependent for bovine heart lactate dehydrogenase; however, it was pH-independent for pig's heart lactate dehydrogenase, calling for caution when comparing data from different species or tissues, even for the same isozyme. No detectable optical rotatory dispersion change was observed for binding of oxamate. Entropies of binding were obtained, and were observed to become more positive at high pH values. On the basis of an electrostatic model for binding of the negatively charged oxamate to the enzyme, these values would be expected to decrease. The entropy of dissociation at higher pH values was found to be higher (more positive), in compensation. Heck suggests that the increase in pH may, in fact, produce a looser conformation of the enzyme, even though this change was not detectable by optical rotatory dispersion; this effect seems to be consistent with Gutfreund's results.

The nature of the conformational change

Other experiments have been able to shed still more light on this particular point. Nuclear magnetic resonance studies have revealed that NAD^+ and NADH actually exist in two conformational forms in solution[3], the so-called elongated and collapsed forms. Possibly the isomerization of the E–NADH complex is, in reality, an isomerization of the coenzyme. Nuclear magnetic

[1] R. S. Criddle, C. H. McMurray, and H. Gutfreund, *Nature*, **220**, 1091 (1968).
[2] H. d'A. Heck, *loc. cit.*
[3] O. Jardetzky and N. G. W. Wade-Jardetzky, *J. biol. Chem.*, **241**, 85 (1966).

resonance experiments with chicken muscle (M_4) enzyme indicate that this may be the case[1]. It was found that the energy barrier between the folded and open forms of NADH was not really significant, ΔH being $+ 5$ kcal and ΔS being $+ 19$ e.u. The coenzyme remained folded on interacting with the enzyme[2]. However, because of the different environments of the protons in the various parts of the NADH molecule, it was possible to show that both the adenine and pyridine moieties were immobilized on binding, with the purine moiety facilitating binding of the pyridine ring; without the purine ring, binding was not complete.

Another interesting study on binding was carried out on dogfish muscle (M_4) lactate dehydrogenase[3]. The free energies of binding of several inhibitors (NAD^+ analogues) of this enzyme were measured, and were correlated with structural changes observed by X-ray crystallography when a crystal of enzyme was penetrated by inhibitors. The K_Q values of the various inhibitors corresponded well with the concentrations of inhibitor required to produce space group changes in the crystal. It was found that adenosine monophosphate alone could cause such a change, but not adenosine. Further, no adenosine analogue could produce the space group changes unless it had a $5'$–phosphate group; this indicates the importance of these parts of the coenzyme. Although the nicotinamide mononucleotide portion of the coenzyme could not bind on its own to the enzyme, without this portion, the binding energy was reduced by 50 per cent. It appeared that the adenosine monophosphate part of the molecule was bound first and that it generated, by means of conformational change, a binding site for the rest of the coenzyme. This provides further evidence for the suggestion that the rate-determining step is dependent on a conformational change.

Further evidence for a conformational change as the rate-determining step for reaction of lactate dehydrogenase was obtained by deuterium kinetic-isotope effects[4]. L-lactate and d-C-2-L-lactate were used as substrates, and measurements made of the initial velocities of the substituted and un-substituted substrate at pH 8.6. For Michaelis–Menten kinetics the isotope ratio is given by

$$\frac{v_H}{v_D} = \frac{V^H[A]^H(K_A^D + [A]^D)}{V^D[A]^D(K_A^H + [A]^H)}$$

where $[A]$ is the concentration of lactate and V and K_A the kinetic parameters. The concentrations $[A]^H$ and $[A]^D$ were kept the same in the separate

[1] R. H. Sarma and N. O. Kaplan, *Proc. natn. Acad. Sci. U.S.A.*, **67**, 1375 (1970).
[2] X-ray studies have confirmed this; see M. G. Rossman, *et al.*, *Cold Spring Harb. Symp. quant. Biol.*, **36**, 179 (1972).
[3] A. McPherson, *J. molec. Biol.*, **51**, 39 (1970).
[4] A. M. Cantwell and D. Dennis, *Biochem. biophys. Res. Commun.*, **41**, 1166 (1970).

experiments, and since V^H was found to be equal to V^D, the equation reduces to:

$$\frac{v_H}{v_D} = \frac{K_A^D + [A]}{K_A^H + [A]}.$$

The experimental results were found to fit this equation well. The fact that the same value of V was found for both isotopes indicates that at high substrate concentrations there is no kinetic isotope effect; in other words, transfer of hydrogen is not involved in the rate-determining step.

At low concentrations of substrate, on the other hand, a normal primary kinetic-isotope effect (v_H/v_D greater than unity) was observed. This indicates an isotope effect on K_A, and that hydrogen-transfer may be the rate-determining step at low substrate concentrations. It also indicates that release of bound NADH (shown to have an inverse kinetic-isotope effect by Thomson and Nance[1]) may not be significant in determining the overall rate, since there can be only one rate-determining step, and this is associated with lactate at low concentrations.

These experiments illustrate the importance of determining kinetic-isotope effects over a range of substrate concentrations, as was emphasized in Chapter 8.

Some preliminary X-ray crystallographic studies have been made on the M_4 enzyme[2]. Unfortunately, only one subunit could be induced to crystallize as an abortive ternary complex with NAD^+ and pyruvate. Furthermore, the sequence of this enzyme is not yet known in full. The resolution of the ternary complex was not sufficiently good to reveal much more than a general confirmation of some of the binding data discussed above. The highly resolved native enzyme has nothing bound, so that deductions are difficult to make. For both H_4 and M_4 isozymes, an essential thiol peptide has been isolated[3]; this is seen to be near to a cleft in the molecule, where the binding sites for NAD^+ and pyruvate appear to be.

Whereas the kinetics of lactate dehydrogenase have been worked out in considerable detail, including the rate-determining step and conformational changes, the groups involved in the activity are still not known with any certainty. A specific tyrosine and a histidine residue have been shown to be important for H_4 activity, since modification of them partially inactivates the enzyme[4]. Apart from the necessity for a thiol group, no further details appear to be known.

[1] J. F. Thomson and L. S. Nance, *Biochim. biophys. Acta*, **99**, 369 (1965).

[2] M. J. Adams, *et al.*, *Nature*, **227**, 1098 (1970); see also M. G. Rossman, *et al.*, *loc. cit.*

[3] T. P. Fondy, J. Everse, G. A. Driscoll, F. Castillo, F. E. Stolzenbach, and N. O. Kaplan, *J. biol. Chem.*, **240**, 4219 (1965); J. J. Holbrook, G. Pfleiderer, K. Mella, M. Volz, M. Leskowac, and R. Jeckel, *Eur. J. Biochem.*, **1**, 476 (1967).

[4] G. Pfleiderer, *et al.*, *Z. phys. Chem.*, **350**, 473 (1969).

In the light of the foregoing discussion the following overall mechanism can be suggested for lactate dehydrogenase:

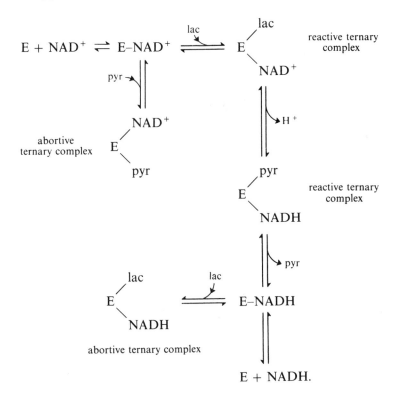

11. Sigmoid Kinetics and Allostery

WHEREAS many enzyme-catalysed reactions exhibit a hyperbolic relationship between the rate and the substrate concentration, as represented by the Michaelis–Menten equation, there are a number of exceptional types of behaviour. Sometimes, as illustrated schematically in Fig. 11.1, there is a sigmoid relationship; the slope of the plot, instead of decreasing steadily as in the Michaelis–Menten case, first *increases* and then decreases. Some of the

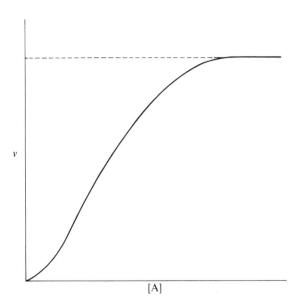

FIG. 11.1.   Schematic representation of sigmoid kinetics.

various explanations that have been put forward to explain sigmoid behaviour are considered in the present chapter. One of the many explanations is related to the phenomenon of *allostery* (p. 370), and a a result sigmoid kinetics is by some workers referred to as allosteric kinetics. This usage, however, will be avoided in the present book, since it is misleading for two main reasons:

(1) It is not allostery itself, but behaviour that is often, but not invariably, related to allostery that provides an explanation for sigmoid kinetics.
(2) There are many other explanations for sigmoid kinetics besides the behaviour that is sometimes related to allostery.

Equations of sigmoid kinetics

Before discussing the various mechanisms for sigmoid kinetics we consider in a formal way the simplest types of equations, relating the velocity v and the substrate concentration [A], that are consistent with the behaviour. The simplest such equation is

$$v = \frac{j[A]^2}{k + m[A]^2} \tag{1}$$

where j, k and m are constants. At low substrate concentration $v \approx j[A]^2/k$ and this quadratic equation accounts at low [A] values for the increase in slope with increasing [A] ($dv/dA \approx 2j[A]/k$). At higher [A] values the $m[A]^2$ term begins to become important, and the curve finally levels off at a v value of j/m.

A more general, and more useful, equation that can be fitted to most cases of sigmoid kinetics is[1]

$$v = \frac{i[A] + j[A]^2}{k + l[A] + m[A]^2}. \tag{2}$$

Sigmoid kinetics are, however, only obtained provided that certain conditions are satisfied as far as the four constants are concerned. This may best be seen with reference to double logarithmic plots of the numerator N, and the denominator D, of expression (2); such plots are shown in Fig. 11.2. A plot of log N against log A consists of two straight lines, of slope 1 below $A = i/j$ and of 2 above $A = i/j$ (there is, of course, a rounding off and not a sharp inflection point). Similarly, a plot of log D against log A will be three straight lines, of slope zero below $A = k/l$ and of 2 above $A = l/m$; the slope is unity in between. In order for sigmoid kinetics to be obtained, with an initial increase in slope followed by a decrease, it is necessary for the curves to be superimposed in the manner shown in Fig. 11.2. The condition for the constants is thus†:

$$\frac{k}{l} > \frac{i}{j}.$$

It is to be noted that eqn (1), with $i = l = 0$, satisfies this condition.

Another useful way of looking at the situation is with reference to the two diagrams in Fig. 11.3. With reference to the upper diagram there are four

---

† The condition $l/m > k/l$ is not essential; if this is not satisfied the slope of the log D curve changes directly from 0 and 2, and sigmoid behaviour is still obtained.

[1] W. Ferdinand, *Biochem. J.*, **98**, 278 (1966).

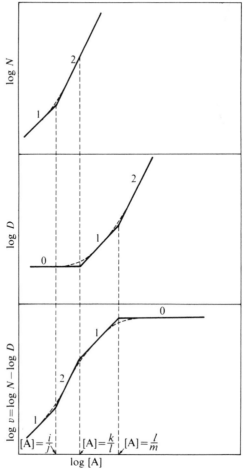

FIG. 11.2.   Double-logarithmic plots illustrating the form of eqn (2).

points a, b, c, and d at which the following conditions apply:

$$a: \quad \frac{dv}{d[A]} = \frac{v}{[A]}; \quad \frac{d \log v}{d \log [A]} = 1$$

$$b: \quad \frac{dv}{d[A]} = \frac{2v}{[A]}; \quad \frac{d \log v}{d \log [A]} = 2$$

$$c: \quad \frac{dv}{d[A]} = \frac{v}{[A]}; \quad \frac{d \log v}{d \log [A]} = 1$$

$$d: \quad \frac{dv}{d[A]} = 0; \quad \frac{d \log v}{d \log [A]} = 0.$$

These conditions are represented in the double-logarithmic plots in Fig. 11.3.

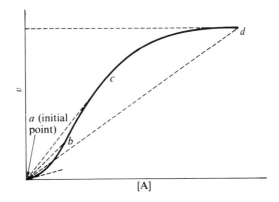

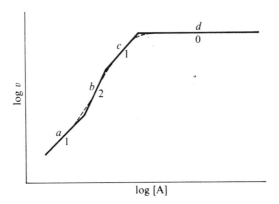

FIG. 11.3   Plots illustrating the form of eqn (3).

A still more general equation for sigmoid kinetics is

$$v = \frac{f + j[A]^n}{g + m[A]^n} \tag{3}$$

where the index n is two or greater, and f and g are functions of $[A]$ involving powers not higher than $n - 1$. The limiting rate V at high substrate concentrations is

$$V([A] \to \infty) = \frac{j}{m} \tag{4}$$

and

$$V - v = \frac{jg - mf}{m(g + m[A]^n)} \tag{5}$$

whence

$$\frac{v}{V-v} = \frac{m(f + j[A]^n)}{jg - mf}.$$ (6)

A plot of $\log[v/(V-v)]$ against $\log[A]$ thus has a slope of n at sufficiently high $[A]$ values. Such a plot is commonly known as a *Hill plot*, having first been used by A. V. Hill[1] in his studies of the binding of O_2 to haemoglobin.

Lineweaver–Burk plots of $1/v$ against $1/[A]$ are most easily considered with reference to eqn (1):

$$\frac{1}{v} = \frac{k + m[A]^2}{j[A]^2}$$ (7)

$$= \frac{k}{j[A]^2} + \frac{m}{j}.$$ (8)

The variation of $1/v$ against $1/[A]$ will therefore be as shown in Fig. 11.4. If this equation applies a straight line will be obtained if $1/v$ is plotted against $1/[A]^2$ (Fig. 11.4).

Mechanisms that do not lead to sigmoid kinetics

It may be useful at the outset to indicate a few mechanisms that may lead to non-hyperbolic behaviour, but which by themselves cannot lead to sigmoid kinetics.

Michaelis-type kinetics involving a series of intermediates

The scheme

$$E + A \rightleftharpoons EA \searrow EA' \rightarrow EA'' \rightarrow \ldots E + Y$$
$$\downarrow$$
$$X$$

leads to an equation of the form

$$v = \frac{V[A]}{K_A + [A]}$$ (9)

and therefore to hyperbolic behaviour; the special case of two intermediates, EA and EA', was considered on p. 78.

Catalysis by two enzymes, or by an enzyme having two or more catalytic sites per molecule

It has sometimes been suggested that sigmoid kinetics may arise if the catalysis is brought about by more than one enzyme, or if the enzyme molecule contains more than one active site, of differing catalytic activities.

[1] A. V. Hill, *J. Physiol.*, **40**, 190 (1910).

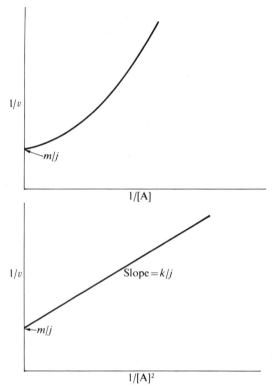

FIG. 11.4. (*a*) A schematic Lineweaver–Burk plot, for the case in which eqn (1) applies; (*b*) a plot of $1/v$ against $1/[A]^2$.

The following analysis, however, shows that non-hyperbolic kinetics can then arise, but that the behaviour is not sigmoid.

The form of the equation that will apply to two enzymes, or two active sites per molecule, is

$$v = \frac{V[A]}{K_A + [A]} + \frac{V'[A]}{K'_A + [A]} \tag{10}$$

$$= \frac{(K'_A V + K_A[V'])[A] + (V + V')[A]^2}{K_A K'_A + (K_A + K'_A)[A] + [A]^2}. \tag{11}$$

This is of the form of eqn (2). It was shown above that sigmoid kinetics can only arise if $kj > il$ and this with reference to eqn (11) requires that

$$K_A K'_A (V + V') > (K'_A V + K_A V')(K_A + K'_A).$$

This is obviously not true, since the right-hand side contains the terms of the left-hand side plus two additional terms.

The substrate contains an impurity, which forms a complex with it

Suppose that the substrate A contains an impurity Q that forms with it a complex that is not acted upon by the enzyme:

$$A + Q \rightleftharpoons AQ$$

$$K_Q^A = \frac{[A].[Q]}{[AQ]}. \tag{12}$$

The total concentration of Q, $[Q]_0$, will be related to the total concentration of A, $[A]_0$, by

$$[Q]_0 = a[A]_0 \tag{13}$$

where a is a fractional constant. It will be assumed that $[A]_0 \gg [Q]_0$, so that the difference between $[A]$ and $[A]_0$ can be ignored; however

$$[Q]_0 = [Q] + [AQ]. \tag{14}$$

Equation (12) gives

$$K_Q^A = \frac{[A]_0([Q]_0 - [AQ])}{[AQ]}$$

or

$$[AQ] = \frac{[A]_0[Q]_0}{K_Q^A + [A]_0} = \frac{a[A]_0^2}{K_Q^A + [A]_0}. \tag{16}$$

The concentration of free A is slightly less than that of $[A]_0$, being given by

$$[A] = \frac{[A]_0 K_Q^A + [A]_0^2 - a[A]_0^2}{K_Q^A + [A]_0}. \tag{17}$$

The rate of the enzyme-catalysed reaction is therefore

$$v = \frac{V[A]}{K_A + [A]}$$

$$= \frac{V([A]_0 K_Q^A + [A]_0^2 - a[A]_0^2)/(K_Q^A + [A]_0)}{K_A + ([A]_0 K_Q^A + [A]_0^2 - a[A]_0^2)/(K_Q^A + [A]_0)} \tag{19}$$

$$= \frac{V K_Q^A[A]_0 + V(1 - a)[A]_0^2}{K_A K_Q^A + (K_A + K_Q^A)[A]_0 + (1 - a)[A]_0^2}. \tag{20}$$

The condition that kj must be greater that il requires that

$$V K_A K_Q^A(1 - a) > V K_Q^A(K_A + K_Q^A)$$

and since this is impossible, there cannot be sigmoid kinetics.

It is important to note that when the *enzyme* contains an impurity that combines with the substrate there *can* be sigmoid kinetics (p. 359). The difference is that when the substrate contains the impurity the concentration of the latter is very low at low substrate concentrations; it has therefore a negligible effect on reducing the rate under these conditions, and the initial dependence of rate on $[A]_0^2$ is not obtained.

The substrate contains an impurity which inhibits by combining with the enzyme

The general equation for an inhibited system (eqn (89) on p. 94) is

$$v = \frac{V[A]}{K_A\left(1 + \frac{[Q]}{K_Q^s}\right) + \left(1 + \frac{[Q]}{K_Q^i}\right)[A]}. \tag{21}$$

If Q is an impurity in A, eqn (13) applies, and the rate is thus

$$v = \frac{V[A]_0}{K_A + \{(K_A a/K_Q^s) + 1\}[A]_0 + (a/K_Q^i)[A]_0^2} \tag{22}$$

which, lacking a term in $[A]_0^2$ in the numerator, is not of the right form to give sigmoid kinetics (see eqn (2)).

The enzyme contains an impurity which inhibits by combining with the enzyme

The treatment of this case is similar to the above; the only difference is that Q is now a constant (at constant $[E]_0$) instead of being proportional to A. Equation (21), with Q constant, is obviously not of the correct form to give sigmoid kinetics.

Mechanisms leading to sigmoid kinetics

The enzyme contains an impurity which combines with the substrate[1]

An extreme case of this is that the inhibiting impurity combines irreversibly with the substrate, rendering it inactive; the effective substrate concentration will therefore be less than $[A]_0$ by a constant amount. If the behaviour is otherwise simple, the variation of rate with substrate concentration will therefore be as shown in Fig. 11.5; the curve is hyperbolic and is displaced to the right by an amount equal to the concentration of substrate that is attached to the inhibitor.

If the binding of the impurity to the substrate is not complete there will be a rounding-off of the curve, as shown in Fig. 11.5, corresponding to sigmoid kinetics. Unfortunately, the steady-state treatment of this case involves the solution of quadratic equations and yields equations that are not helpful in leading to an understanding of the problem. The following treatment of limiting cases is, however, useful.

Consider the variation of the concentration $[A]$ of free substrate A with the total concentration $[A]_0$. If $[A]_0$ is large compared with the concentration

[1] J. Westley, *Enzymic catalysis*, Harper and Row, New York, 1969.

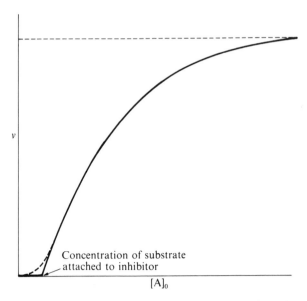

Fig. 11.5.   The firm line shows the behaviour when the enzyme contains an impurity which is irreversibly bound to the substrate, rendering it inactive. The dotted line shows the rounding off expected if the binding is reversible.

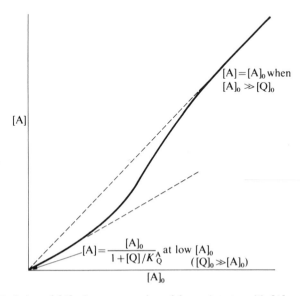

Fig. 11.6.   Variation of [A], the concentration of free substrate, with $[A]_0$, the total concentration, when there is an impurity present in the enzyme, which becomes attached to substrate.

$[Q]_0$ of impurity, $[A]$ is approximately equal to $[A]_0$, as shown in Fig. 11.6. At lower concentrations of $[A]_0$, when $[Q]_0 \gg [A]_0$, the value of $[A]$ will be substantially less than that of $[A]_0$. Thus

$$[A]_0 = [A] + [AQ] \tag{23}$$

$$= [A]\left(1 + \frac{[Q]_0}{K_Q^A}\right) \tag{24}$$

so that the initial slope of the plot of $[A]$ against $[A]_0$ is $1/(1 + [Q]/K_Q^A)$ (see Fig. 11.6). The variation of $[A]$ with $[A]_0$ is thus sigmoid, and since the rate at low $[A]_0$ is proportional to $[A]$, there will be sigmoid kinetics.

It has been seen in the preceding discussion that in simple enzyme systems an impurity will only lead to sigmoid kinetics if it is an impurity present in the enzyme and if it combines with the substrate. Table 11.1 summarizes the conclusions.

TABLE 11.1

Sigmoid kinetics produced in simple enzyme systems by the presence of an impurity

	Impurity combines with:	
Impurity present in:	Substrate	Enzyme
Substrate	NO	NO
Enzyme	YES	NO

Two forms of the enzyme

Sigmoid behaviour becomes possible if the enzyme can exist in two forms of different activities. A specific mechanism of this type has been suggested by Rabin[1]:

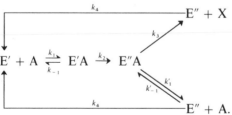

Of the two enzyme forms E′ and E″, the latter is the more active; E″A can give rise to E″ and the product X, but E′A cannot give product directly. As the substrate concentration increases there is an abnormal increase in rate, because A forces the conversion of E′ into E″; at low A the rate thus varies with A to a higher power than the first, so that sigmoid kinetics results.

[1] B. R. Rabin, *Biochem. J.*, **102**, 220 (1967); see C. Cennamo, *J. theor. Biol.*, **21**, 260 (1968); C. Frieden, *J. biol. Chem.*, **245**, 5788 (1970); G. R. Ainsley, J. P. Shill and K. E. Neet, *J. biol. Chem.*, **297**, 7088 (1972).

Application of the steady-state treatment to this reaction scheme gives

$$v = \frac{k_1 k_3 \left\{ \dfrac{k_2}{k_3 + k'_1}[A] + \dfrac{k_2 k'_{-1}}{k_4(k_3 + k'_1)}[A]^2 \right\}[E_0]}{k_{-1} + k_2 + k_1 \left\{ 1 + \dfrac{k_2}{k_4} + \dfrac{k_2}{k_3 + k'_1} \right\}[A] + \dfrac{k_1 k_2 k'_{-1}}{k_4(k_3 + k'_1)}[A]^2} \tag{25}$$

which is of the same form as eqn (2). Thus, provided that the rate constants satisfy the inequality, there can be sigmoid kinetics.

Binding of two molecules of substrate to one molecule of enzyme[1]

Sigmoid kinetics can arise if the enzyme can form the complex EA_2 which undergoes reaction more rapidly than EA; this gives rise to a quadratic dependence of rate on A at low substrate concentrations. A special case is when EA is inactive:

$$E + A \underset{k_{-1}}{\overset{k_1}{\rightleftharpoons}} EA \xrightarrow{k_2[A]} EA_2 \xrightarrow{k_3} E + A + X.$$

The steady-state equation is

$$v = \frac{k_1 k_2 k_3 [E]_0 . [A]^2}{k_{-1}k_3 + (k_1 + k_2)k_3[A] + k_1 k_2[A]^2}. \tag{26}$$

This is of the form of eqn (2); since $i = 0$ the condition $kj > il$ is necessarily satisfied. Worcel, Goldman, and Cleland[2] have obtained evidence that this mechanism applies to the enzyme nicotinamide adenine dinucleotide oxidase.

It is to be noted that the mechanism implies *site interaction*. If the enzyme binds a molecule of substrate at each of two or more identical sites, with no interaction between the sites, normal hyperbolic behaviour is obtained.

The substrate acts as a modifier

Frieden[3] has suggested a mechanism in which the substrate can form an active complex EA and also an inactive one AE by binding at another site. There is also the possibility of binding at both sites, to give AEA. The scheme is

[1] A. Worcel, D. S. Goldman, and W. W. Cleland, *J. biol. Chem.*, **240**, 3399 (1965); a more general mechanism, involving the attachment of *n* molecules of substrate, has been considered by D. E. Atkinson, J. A. Hathaway, and E. C. Smith, *J. biol. Chem.*, **240**, 2682 (1965) and by K. Taketu and B. M. Pogell, *J. biol. Chem.*, **240**, 651 (1965).
[2] *Loc. cit.*
[3] C. Frieden, *J. biol. Chem.*, **239**, 3522 (1964).

The steady-state treatment of this case leads to a very complicated expression. If E, EA, AE and AEA are assumed to be essentially at equilibrium, with the equilibrium constants indicated above, the resulting rate equation is

$$v = \frac{[E]_0\{k[A] + (k'[A]^2/K_A'')\}}{K_A + \{1 + (K_A/K_A')\}[A] + ([A]^2/K_A'')}. \tag{27}$$

This is of the form of eqn (2), and leads to sigmoid kinetics provided that

$$(K_A/K_A'')k' > \{1 + (K_A/K_A')\}k.$$

This requires that reaction of AEA is sufficiently rapid compared with that of EA.

Two-substrate systems

When two substrates A and B are undergoing reaction by a random-ternary complex mechanism (see p. 119) the general form of the rate equation is[1]

$$v = \frac{k[A].[B](a + b[A] + c[B])}{1 + p[A] + q[B] + r[A].[B] + s[A]^2 + t[B]^2 + u[A]^2.[B] + w[A].[B]^2} \tag{28}$$

If [A] is varied at constant [B] the form of the equation is

$$v = \frac{k'(a + c')[A] + k'b[A]^2}{1 + q' + t' + (p + r' + v'')[A] + (s + u')[A]^2}. \tag{29}$$

This is of the form of eqn (2), and therefore can lead to sigmoid kinetics. Atkinson and Walton[2] have obtained experimental results that are consistent with this mechanism.

Alternative pathways

Two-substrate systems will also lead to sigmoid kinetics if they occur by two or more alternative pathways. For example, a reaction might occur by a ping-pong mechanism and a ternary complex mechanism, as shown in Fig. 11.7. Sweeny and Fisher[3], and Griffen and Brand[4] consider numerous combinations of mechanism, and show that they lead to sigmoid kinetics.

Interacting Subunits

Several explanations for sigmoid kinetics are based on the idea that certain enzyme molecules are composed of a number of subunits, which interact with one another. The first theory of this type was proposed by Monod,

[1] K. J. Laidler, *Trans. Faraday Soc.*, **52**, 1374 (1956); see W. Ferdinand, *Biochem. J.*, **98**, 278 (1966); H. D. Engers, W. A. Bridger, and N. B. Madsen, *Biochemistry*, **9**, 3281 (1970).
[2] D. E. Atkinson and G. M. Walton, *J. biol. Chem.*, **240**, 757 (1965).
[3] J. R. Sweeny and J. R. Fisher, *Biochemistry*, **7**, 561 (1968); see J. R. Fisher and V. D. Hoagland, *Adv. biol. med. Phys.*, **12**, 163 (1968).
[4] C. C. Griffin and L. Brand, *Archs Biochem. Biophys.*, **126**, 856 (1968).

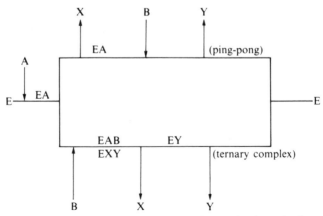

Fɪɢ. 11.7. Alternative ping-pong and ternary-complex mechanisms, leading to sigmoid kinetics.

Wyman, and Changeux[1], in a paper which also introduced the concept of *allostery*. This term (see p. 370) refers to the attachment of a substrate or modifier at a site other than the catalytic site, with resulting modification of the enzyme. It is important to recognize that allostery in itself does not provide an explanation for sigmoid kinetics, and that enzyme molecules can have subunits without there being an allosteric effect. It is the existence of interacting subunits that gives rise, *under certain circumstances*, to sigmoid kinetics.

The treatment of Monod *et al.* is based on the following ideas:

(1) Enzyme molecules are oligomers, consisting of a number of identical subunits known as protomers.

(2) All of the subunits in a given molecule are in the same conformation. This is an extreme case of interaction between subunits; if one subunit changes its conformation, the others in the same molecule must do so also. Other treatments, to be considered later (p. 368), involve a more limited type of interaction.

(3) If each subunit can exist in two conformations there are two forms, E^r and E^t, of the enzyme molecule; in the first, all subunits are in the one conformation, in the second all are in the other conformation.

The way in which these postulates lead to sigmoid kinetics is as follows. Suppose that the equilibrium constant†

$$\frac{[E^t]}{[E^r]} \equiv L \qquad (30)$$

† Monod, *et al.*, refer to L as the 'allosteric constant'. This terminology seems unfortunate, however, since the existence of this equilibrium between the conformers has no necessary connection with allostery.

[1] J. Monod, J. Wyman, and J.-P. Changeux, *J. molec. Biol.*, **12**, 88 (1965).

is very large; [E'] and [E'] denote the concentrations of the respective forms with no substrate attached. The free enzyme thus exists largely in the E' form. Suppose, however, that a molecule of substrate is much more strongly bound to E' than it is to E'; i.e.

$$K_t \equiv \frac{[E'] \cdot [A]}{[E'A]} \gg \frac{[E'] \cdot [A]}{[E'A]} \equiv K_r.$$

As substrate is added to the system the binding is largely to E', and the effect of this is to displace the equilibrium in the direction E' → E'. This facilitates binding of additional substrate molecules. The concentration of enzyme–substrate complexes thus increases with [A] to a higher power than the first; the rate increases in the same way, so that sigmoid kinetics can be observed.

A simple quantitative treatment of the problem is as follows, for an oligomer consisting of four subunits. The various equilibria are represented in Fig. 11.8, in which one conformer of the subunits is shown as a circle, and

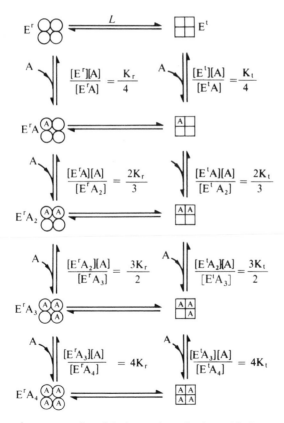

FIG. 11.8.   Schematic representation of the interacting subunit model of Monod, Wyman, and Changeux, with four subunits.

the other as a square. Each subunit is assumed to be capable of binding one substrate molecule, the dissociation constants for binding being the same apart from statistical factors, which are shown in the figure†.

The fraction of sites which have substrate molecules bound to them is given by

$$\bar{Y}_A = \frac{\begin{array}{c}[E^rA] + 2[E^rA_2] + 3[E^rA_3] + 4[E^rA_4] + \\ + [E^tA] + 2[E^tA_2] + 3[E^tA_3] + 4[E^tA_4]\end{array}}{4\{[E^r] + [E^rA] + [E^rA_2] + [E^rA_3] + [E^rA_4] + \\ + [E^t] + [E^tA] + [E^tA_2] + [E^tA_3] + [E^tA_4]\}} \tag{31}$$

$$= \frac{[E^rA]\left\{1 + 3\left(\dfrac{[A]}{K_r}\right) + 3\left(\dfrac{[A]^2}{K_r^2}\right) + \left(\dfrac{[A]^3}{K_r^3}\right)\right\} + \\ + [E^tA]\left\{1 + 3\left(\dfrac{[A]}{K_t}\right) + 3\left(\dfrac{[A]^2}{K_t^2}\right) + \left(\dfrac{[A]^3}{K_t^3}\right)\right\}}{4\left[[E^r]\left\{1 + 4\left(\dfrac{[A]}{K_r}\right) + 6\left(\dfrac{[A]^2}{K_r^2}\right) + 4\left(\dfrac{[A]^3}{K_r^3}\right) + \left(\dfrac{[A]^4}{K_r^4}\right)\right\} + \\ + [E^t]\left\{1 + 4\left(\dfrac{[A]}{K_t}\right) + 6\left(\dfrac{[A]^2}{K_t^2}\right) + 4\left(\dfrac{[A]^3}{K_t^3}\right) + \left(\dfrac{[A]^4}{K_t^4}\right)\right\}\right]} \tag{32}$$

$$= \frac{[E^rA](1 + \alpha)^3 + [E^tA](1 + c\alpha)^3}{4\{[E^r](1 + \alpha)^4 + [E^t](1 + c\alpha)^4\}} \tag{33}$$

where

$$\alpha = \frac{[A]}{K_r} \quad \text{and} \quad c = \frac{K_r}{K_t}. \tag{34}$$

Since

$$\frac{[E^tA]}{[E^rA]} = \frac{K_r}{K_t} \cdot \frac{[E^t]}{[E^r]} = cL \tag{35}$$

eqn (33) reduces as follows:

$$\bar{Y}_A = \frac{[E^rA]\{(1 + \alpha)^3 + cL(1 + c\alpha)^3\}}{4[E^r]\{(1 + \alpha)^4 + L(1 + c\alpha)^4\}} \tag{36}$$

$$= \frac{\alpha(1 + \alpha)^3 + \alpha cL(1 + c\alpha)^3}{(1 + \alpha)^4 + L(1 + c\alpha)^4}. \tag{37}$$

In general, for n subunits, the result is

$$\bar{Y}_A = \frac{\alpha(1 + \alpha)^{n-1} + \alpha cL(1 + c\alpha)^{n-1}}{(1 + \alpha)^n + L(1 + c\alpha)^n}. \tag{38}$$

† For example, there are 3 ways for A to become attached to E^rA, and two ways for E^rA_2 to lose an A; this leads to the statistical factor 2/3 in the expression for $E^rA \cdot A/E^rA_2$.

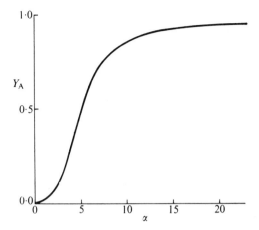

FIG. 11.9.   Theoretical curve for the saturation function $\overline{Y}_A$ against α, for $n = 4$ and $c = 0.10$.

Under certain conditions a plot of $\overline{Y}_A$ against α (which is proportional to A) is sigmoid. For example, Fig. 11.9 shows the plot for $n = 4$, $L = 1000$ (i.e. $[E^t] \gg [E^r]$) and $c = 0(K_t \gg K_r)$. This situation was discussed qualitatively above. Cooperativity is greatest for either of the combinations:

$$[E^t] \gg [E^r]\ (L\ \text{large})\quad \text{and}\quad K_t \gg K_r\ (c\ \text{small})$$

or

$$[E^r] \gg [E^t]\ (L\ \text{small})\quad \text{and}\quad K_r \gg K_t\ (c\ \text{large}).$$

When $c = 1$, and also when L is negligibly small,

$$\overline{Y}_A = \frac{\alpha}{1 + \alpha} = \frac{[A]}{K_r + [A]}. \tag{39}$$

The behaviour will then be hyperbolic and not sigmoid. Also, when L is very large

$$\overline{Y}_A = \frac{c\alpha}{1 + c\alpha} \tag{40}$$

which again corresponds to hyperbolic behaviour.

It is to be noted[1] that the above treatment involves the tacit assumption that the chemical reaction is sufficiently slow that it does not disturb the equilibria in Fig. 11.8. A steady-state treatment of the problem has been given by Wong, Endrenyi, and Chan[2].

[1] See C. Frieden, *J. biol. Chem.*, **242**, 4045 (1967).
[2] J. F. T. Wong and L. Endrenyi, *Can. J. Biochem.*, **49**, 568 (1971); L. Endrenyi, M. S. Chan, and J. F. T. Wong, *ibid.*, **49**, 581 (1971).

The equations of Monod, Wyman, and Changeux have been successfully applied to several systems, notably haemoglobin[1] and aspartate trans-carbamylase[2]. The temperature-jump method (see p. 39) has been applied[3] to glyceraldehyde-3-phosphate dehydrogenase, and values of L, c and K_r have been determined.

The treatment outlined above has involved the hypothesis that the enzyme form having the weaker substrate affinity is the predominant form in the absence of substrate; addition of substrate increases the proportion of the other form and the affinity of the whole population of enzyme molecules for substrate. This increases progressively as more molecules of substrate are bound. Enzymes of this type are called enzymes of type K, since there is a change in the affinity of the enzyme for the substrate, and hence there is no constant Michaelis constant K_A.

For some systems, however, this treatment has not proved satisfactory, and it has been necessary to postulate a second type of oligomeric enzyme, of type V. In these enzymes both conformers have the same affinity for the substrate, and there is therefore no effect of a substrate molecule on the combination of another substrate molecule with the enzyme. However, there is a difference in the k_c values for the two enzyme forms: i.e., when the substrate is attached to one form it undergoes reaction more rapidly than when it is attached to the other. Enzymes of type V show hyperbolic, and not sigmoid kinetics, but they can be detected on the basis of studies with inhibitors.

An alternative model for interacting subunits

The model of Monod, Wyman, and Changeux involved the hypothesis that all of the subunits in a given protein molecule exist in the same conformation at a given time; a change of conformation of one subunit requires a change in all of the others. This is the most extreme type of interaction possible. Koshland and his coworkers[4] have developed mechanisms of the same kind, but based upon a more general hypothesis in which hybrid conformational states can exist; that is, one subunit can change its conformation without producing any change in the others. However, there is some interaction between subunits in a given molecule, in that a change in con-

---

[1] Monod, Wyman, and Changeux, *loc. cit.*

[2] J.-P. Changeux, J. C. Gerhart, and H. K. Schachman, *Biochemistry*, 7, 531 (1968); J. P. Changeux and M. M. Rubin, *ibid.*, 7, 553 (1968).

[3] K. Kirschner, F. E. B. S. Symposium on *Regulation of Enzyme Activity and Allosteric Interactions* (eds. E. Kvamme and A. Pihl), Associated Press, London and New York, 1968, p. 39.

[4] D. E. Koshland, G. Nemethy, and D. Filmer, *Biochemistry*, 5, 365 (1966); J. E. Haber and D. E. Koshland, *Proc. natn. Acad. Sci. U.S.A.*, 58, 2087 (1967); D. E. Koshland, A. Conway, and M. E. Kirtley, F.E.B.S. Symposium on *Regulation of Enzyme Activity and Allosteric Interactions* (eds. E. Kvamme and A. Pihl), Associated Press, London and New York, 1968, p. 131. It is to be noted that the Koshland model permits 'negative cooperativity' whereas the Monod one does not. One can obtain normal hyperbolic kinetics from both models by suitable choice of constants (A. J. Cornish-Bowden and D. E. Koshland, *Biochemistry*, 9, 3337 (1970)).

formation of one subunit changes the relative stability of conformations of neighbouring subunits. The affinity of the two conformations for the substrate is different; addition of substrate thus disturbs the equilibrium between the conformational forms, and therefore can give rise to sigmoid kinetics under suitable conditions.

(a)   Square system

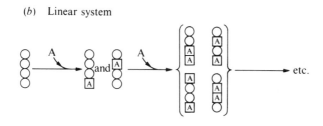

(b)   Linear system

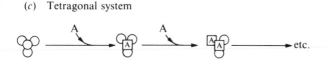

(c)   Tetragonal system

(d)   Concerted mechanism: all subunits
change conformations together

FIG. 11.10.   The four basic interacting-subunit mechanisms of Koshland and coworkers.

Koshland and coworkers put forward four basic mechanisms, which are illustrated in Fig. 11.10. It is assumed that when a substrate molecule becomes attached to a subunit in one conformation, represented by a circle, it induces its conversion into the other conformation, represented by a square. When there is a conformational change in one subunit it favours (but does not compel) a corresponding change in a neighbouring subunit, but not in a non-neighbouring one. In the square system, for example, it is assumed that there are no diagonal interactions, each subunit interacting with two others. In the tetragonal system, on the other hand, each subunit interacts with each of the other three. The concerted mechanism (d) is equivalent to that of Monod, Wyman, and Changeux.

Allostery

During recent years much attention has been devoted to the fact that enzymes are often found to be influenced by substances which are not substrate analogues and which are apparently not attached to the enzyme at the active site to which the substrate becomes attached. These substances are referred to as *modifiers* or as *effectors*, and they may be activators or inhibitors. The fact that they are attached at a different site from the substrate site has led to the phenomenon being referred to as *allostery*. (Greek *allos*, other; *stereos*, solid.)

It is important to note that, according to this definition of allostery, the basic model of Monod, Wyman, and Changeux is not as such an allosteric model, since there is no postulate that there are sites other than those to which the substrate molecules are attached. There is considerable confusion in the literature on this point since these workers, and many following them, have referred to the model as an allosteric one. Only if the treatment is extended to include additional sites for attachment of the modifier, with a resulting effect on the catalytic activity, can the mechanism be regarded as allosteric.

Another point of confusion in the literature is that it is often considered that allostery provides an explanation for sigmoid kinetics. This is not the case; allostery simply implies a special type of inhibition or activation, and the kinetic equations derived on pp. 89–111 apply to it. (The fact that conformational changes may occur does not affect the argument, since these changes are reflected in the inhibition constants.) The kinetic equations for inhibited or activated systems do not lead to sigmoid kinetics unless special postulates are made, such as those considered earlier in this chapter.

It will be clear from what has been said that sigmoid kinetics provides no evidence for allostery. The most direct evidence for allostery is that the enzyme can be made insensitive to the modifier (effector) without any loss of catalytic activity. This clearly indicates that the site to which the modifier is attached is different from the catalytic site. This being so, the modifier must act by inducing a conformational change which alters the activity at the catalytic site.

Some of the principles considered above will now be illustrated for two enzymes, aspartate transcarbamylase and glyceraldehyde-3-phosphate dehydrogenase.

Aspartate transcarbamylase (E.C.2.1.3.2)

Aspartate transcarbamylase catalyses the reaction

$$O=C\begin{matrix} NH_2 \\ \\ O \end{matrix} \quad + \quad \begin{matrix} CO_2^- \\ | \\ CH_2 \\ | \\ CH \\ H_3^+N \quad CO_2^- \end{matrix} \quad \rightarrow$$

carbamyl phosphate   aspartate

$$\rightarrow \quad O=C\begin{matrix} N^+H_3 \quad CO_2^- \\ \\ | \\ CH_2 + HPO_4^{2-} \\ NH-CH \\ CO_2^- \end{matrix}$$

carbamyl aspartate

This reaction is the first in a series of six reactions which occur in the bio-synthesis of uridine triphosphate and cytidine triphosphate. Many aspects of its kinetics have been studied[1]; only those relevant to a discussion of sigmoid kinetics will be considered here. Product inhibition studies[1] reveal that carbamyl phosphate binds first, followed by aspartate. The enzyme is one of the many so-called *regulatory* enzymes, its particular mechanism being *feedback inhibition*. By this is meant that a product of the biosynthetic route (in this case, cytidine triphosphate or uridine triphosphate) inhibits the enzyme near the start of the synthesis, thus providing a means of controlling the synthesis. When sufficient product has built up the initial step is slowed down and therefore the subsequent steps. Many such enzymes have been observed to exhibit sigmoid kinetic behaviour, and also to comprise several subunits. Since the product does not resemble the substrate of the inhibited enzyme, it was suggested that the inhibition took place at an allosteric site, removed from the catalytic active site. It was also suggested that interaction between the subunits was responsible for the sigmoid kinetic behaviour. This has been proved correct for some enzymes, including aspartate transcarbamylase; however, as has been discussed earlier in this chapter, such interaction is by no means essential for explaining sigmoid behaviour.

Ligand interaction

Several interesting phenomena have been observed in studies on aspartate transcarbamylase. So-called *homotropic* effects have been found in the

[1] R. W. Porter, M. O. Modebe, and G. R. Stark, *J. biol. Chem.*, **244**, 1846 (1969), and references therein.

interaction between the enzyme and the substrates aspartate[1] (and its analogue, succinate[2]) and carbamyl phosphate[3]. By this is meant that binding of one molecule of the relevant ligand influences the binding of subsequent ligands of the same sort. (*Heterotropic* effects imply that binding of one type of ligand affects the binding of another type.) For aspartate transcarbamylase the binding of one molecule of substrate facilitates the binding of the subsequent molecules of substrate, in a cooperative fashion (as opposed to an antagonistic homotropic effect), resulting in sigmoid behaviour. Such sigmoidicity is pH-dependent as illustrated in Fig. 11.11.

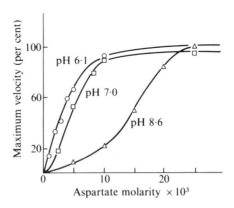

FIG. 11.11.   Sigmoid kinetics in the action of aspartae *trans*-carbamylase; the curves show the effect of pH (from J. C. Gerhart and A. B. Pardee, *Cold Spring Harb. Symp. quant. Biol.*, **28**, 491 (1963)).

The various ligands which affect the catalysis of aspartate *trans*-carbamyl-ase influence the behaviour of one another with respect to the catalytic activity of the enzyme. For instance, the binding of the substrates to the enzyme is greatly reduced in the presence of cytidine triphosphate (CTP), the end product of the biosynthetic pathway, and an inhibitor of aspartate transcarbamylase[4]. Both substrates are affected in a similar manner by CTP.

In addition, CTP inhibition of the enzyme is reduced as the concentration of either substrate is increased. This is an antagonistic heterotropic effect. The inhibitory effect is almost completely removed when both substrates, aspartate and carbamyl phosphate, are present in saturating amounts. (Adenosine triphosphate, an activator of the enzyme, also removes CTP inhibition; it does so in a competitive way, and probably binds at the same site.)

[1] J. C. Gerhart and A. B. Pardee, *J. biol. Chem.*, **237**, 891 (1962); *Cold Spring Harb. Symp. quant. Biol.*, **28**, 491 (1963).
[2] J.-P. Changeux, J. C. Gerhart, and H. K. Schachman, *Biochemistry*, **7**, 531 (1968).
[3] M. R. Bethell, K. E. Smith, J. S. White, and M. E. Jones, *Proc. natn. Acad. Sci. U.S.A.*, **60**, 1442 (1968), and references therein.
[4] Gerhart and Pardee, *loc. cit.*; Changeux, *et al.*, *loc. cit.*

In the presence of small amounts of carbamyl phosphate, succinate (the substrate analogue for aspartate) is able to remove the inhibitory effect of CTP to a lesser extent than in the presence of saturating amounts; the one substrate has a dramatic effect on the behaviour of the other.

Several experiments have shown that the CTP inhibits in a partially competitive manner. This, together with the above and other evidence, is an indication that the substrates and inhibitor all have independent binding sites, in the sense that they are non-overlapping. The binding of each is, however, affected by the presence of the others. As will be discussed below, these interactions probably occur via a conformational change associated with ligand binding, which affects the kinetics of the catalysis.

Subunit studies

Several studies were undertaken in order to clarify the situation. Gerhart and Schachman[1] studied the rate of reaction of thiol groups in aspartate transcarbamylase with p-mercuribenzoate, and also the sedimentation velocity pattern of the enzyme, in the presence of the substrate carbamyl phosphate and the substrate analogue succinate. McClintock and Markus[2] studied thiol group reactivity, and the rate of digestion of the enzyme by trypsin (to monitor conformational changes) and also ligand binding, as a function of the concentrations of substrates. These studies led to the conclusion that the activity of the enzyme, the binding of various ligands to the enzyme, and the changes in physical structure observed, all had the same dependence on the concentrations of the substrates carbamyl phosphate and aspartate (or succinate). The sigmoidicity in activity is, in other words, a direct consequence of binding of the ligands mediated via a change in conformation subsequent to or during ligand binding. This is why sigmoidicity disappears when the subunit structure of the protein is destroyed (see below). As an example, the sedimentation velocity experiments indicated that the enzyme contracts when cytidine triphosphate binds to it, and also when succinate and carbamyl phosphate are absent (when the enzyme is not active); on the other hand, it expands when it is in a catalytically active state, as in the absence of cytidine triphosphate or in the presence of the two substrates. Thiol reactivity depended in a similar way on ligand binding. As will be seen below, relaxation studies by Hammes and coworkers led to similar results.

Subsequent studies proved conclusively that cytidine triphosphate binds at a separate site from the substrates, and also showed that sigmoidicity was directly related to the subunit nature of the enzyme[3]. Treating the enzyme with mercurials, or heating it and then rapidly cooling it, led to desensitization of the enzyme to the inhibitor CTP, but still allowed the

---

[1] J. C. Gerhart and H. K. Schachman, *Biochemistry*, **7**, 538 (1968).
[2] D. K. McClintock and G. Markus, *J. biol. Chem.*, **243**, 3855 (1968); **244**, 36 (1969).
[3] J. C. Gerhart and H. K. Schachman, *Biochemistry*, **4**, 1054 (1965).

enzyme to bind aspartate and to carry out its catalytic function; however, the regulatory function appeared to have been destroyed. It was ultimately possible to separate the enzyme, after such treatment, into two types of subunit[1]. The larger molecular weight subunit was responsible for the catalytic activity (and in fact had a higher specific activity than the native enzyme itself). The smaller subunit was catalytically inactive, but was able to bind cytidine triphosphate, the regulatory ligand. Further support for separate binding sites came from the observation of distinct relaxation times for the binding of CTP and succinate.

Numerous attempts have been made to determine the number of subunits present in native aspartate transcarbamylase, and the number of ligand sites at each subunit. The results to date have been contradictory. Ligand binding studies[1] and end group amino acid analysis[2] indicated that each enzyme molecule had four regulatory and two catalytic subunits; and that each catalytic subunit bound two equivalents of the substrate analogue, succinate, while each regulatory subunit had one binding site for CTP (or its analogue bromo-CTP).

Subsequent studies have indicated that there are six polypeptide *chains* in each enzyme molecule of both regulatory and catalytic subunits. These studies have included molecular weight and amino acid analysis[3], hybridization of native and chemically modified subunits[4], and X-ray crystallography[5]. While the estimations of the number of chains per subunit appear to be reliable, the number of each type of subunit per molecule is as yet uncertain. The X-ray evidence indicates that there are probably six of each kind of subunit, i.e. one polypeptide chain corresponds to one subunit. More experimental evidence is, however, required to clarify this point.

Following the separation of the subunits, experiments were performed on them to determine their kinetic function in the intact enzyme. The catalytic residue on its own was found to have normal Michaelis–Menten kinetics[6] compared with the sigmoid behaviour of the native enzyme. This indicates that the regulatory subunit is involved in CTP binding, and also in the sigmoidicity. The mercurial experiments quoted above support this. Most of the thiol groups are on the regulatory subunit, with only a few on the catalytic one; however, thiol modification corresponded with conformational change and with ligand binding and activity, as a function of ligand concentration. Markus, McClintock and Bussell have produced evidence suggesting that the relevant changes arise via changes in the quaternary

[1] J.-P. Changeux, J. C. Gerhart, and H. K. Schachman, *loc. cit.*; G. G. Hammes and C.-W. Wu, *Biochemistry*, **10**, 1046, 1051 (1971); G. G. Hammes, R. W. Porter, and C.-W. Wu, *Biochemistry*, **9**, 2992 (1970).
[2] J.-P. Changeux, J. C. Gerhart, and H. K. Schachman, *Biochemistry*, **7**, 531 (1968).
[3] K. Weber, *J. biol. Chem.*, **243**, 543 (1968).
[4] K. Weber, *Nature*, **218**, 1116 (1968).
[5] E. A. Meighen, V. Pigiet, and H. K. Schachman, *Proc. natn. Acad. Sci. U.S.A.*, **65**, 234 (1970).
[6] D. C. Wiley, *et al.*, *Cold Spring Harb. Symp. quant. Biol.*, **36**, 285 (1972).

structure of the enzyme[1]. The catalytic subunit also undergoes conformation changes, however, as indicated by transient n.m.r. studies of Sykes and coworkers[2].

Relaxation studies

Relaxation studies have revealed differences in the behaviour of the separate subunits as compared with the native enzyme, including differences in rate-determining steps. Eckfeldt and coworkers[3] found that the rate-determining step in the binding of bromo-CTP for native aspartase transcarbamylase was a conformational change; for the regulatory subunit, however, binding was merely a bimolecular process. Hammes and coworkers[4] found a similar situation with the catalytic subunit, for which binding was a bimolecular process; however, a conformational process associated with binding was the slower step for the native enzyme.

Despite the above differences between native enzyme and separate subunits, there are many similarities in behaviour. Carbamyl phosphate enhances the binding of succinate in both the native enzyme and the catalytic subunit alone; and adenosine triphosphate removes the inhibitory effect of CTP in both the native enzyme and the regulatory subunit alone. It would seem that certain functions of the enzyme require both subunits to be present at once; other functions do not, and involve no interaction between subunits.

Comparison of models

Several attempts have been made to distinguish whether the sigmoid behaviour of aspartate transcarbamylase can be explained by the Monod model, by the Koshland model, or in some other way. The evidence tends to be more qualitative than quantitative, with some exceptions. Gerhart and Schachman[5] worked with succinate and other ligands and were the first to interpret their results in terms of the Monod model. Also, Changeux and Reuben[6] worked with succinate and found that the Monod model provided the best quantitative fit to the data. The relaxation experiments of Sykes and coworkers[7], also with succinate, led to a similar conclusion. Hammes and coworkers[8] suggested that the sequential model of Koshland and his coworkers should allow the detection of a spectrum of relaxation processes, one for each distinct enzyme form. Only one or two have been observed by most workers, a result which tends to favour the Monod model. However, the results may be equally well explained by a mechanism involving

[1] G. Markus, D. K. McClintock, and J. B. Bussell, *J. biol. Chem.*, **246**, 762 (1971).
[2] B. D. Sykes, P. G. Schmidt, and G. R. Stark, *J. biol. Chem.*, **245**, 1180 (1970).
[3] J. Eckfeldt, G. G. Hammes, S. C. Mohr, and C.-W. Wu, *Biochemistry*, **9**, 3353 (1970).
[4] G. G. Hammes and W.-C. Wu, *Biochemistry*, **10**, 1051 (1971).
[5] J. C. Gerhart and H. K. Schachman, *loc. cit.*
[6] J.-P. Changeux and M. M. Reuben, *Biochemistry*, **7**, 553 (1958).
[7] B. D. Sykes and coworkers, *loc. cit.*
[8] G. G. Hammes and C.-W. Wu, *Biochemistry*, **10**, 1051, 2150 (1971); G. G. Hammes, R. W. Porter, and G. R. Stark, *ibid.*, p. 1046.

a conformational change associated with the binding of any one substrate molecule (except the last) as the slow process relative to all other binding steps, without the need to invoke either model.

On the other hand, Bethell and coworkers worked with aspartate and obtained results that they suggest are not in agreement with a simple two-state concerted model for the effect of carbamyl phosphate on the velocity of reaction of the enzyme. They argue that if only two forms of the enzyme exist, sigmoid kinetics would not be expected for one substrate in the presence of saturating amounts of the other. While several forms may exist, however, it is still possible for two forms to predominate kinetically. McClintock and Markus[2] used aspartate as the second substrate (instead of succinate as used by the other groups), and found that the sequential model is indicated. They made a plot of the substrate concentration required to produce a given degree of saturation for ligand binding against the substrate concentration required to produce the same degree of kinetic saturation. This plot was linear, which is consistent with the sequential mechanism, each molecule in turn producing a further change in conformation (and binding); the concerted mechanism would lead to a curved plot.

It would appear that the type of kinetics followed is, in the first place, not an easy matter to resolve experimentally. Second, the kinetic behaviour depends rather critically on the exact conditions used, as indicated by the pH-dependence of sigmoidicity cited above, and by the different kinetics observed for aspartate, the substrate, and its analogue, succinate. With the exception of the results for aspartate, however, most of the experimental work appears to support the Monod model for this enzyme.

Glyceraldehyde-3-phosphate dehydrogenase (E.C.1.2.1.12)

Similar experimental approaches have been used in the study of glyceraldehyde-3-phosphate dehydrogenase. The reaction catalysed is:

glyceraldehyde-3-phosphate   1,3-diphosphoglycerate

[1] M. R. Bethell, *et al.*, *loc. cit.*
[2] D. K. McClintock and G. Markus, *loc. cit.*

The situation is complicated by the fact that several different forms of the enzyme have been used by various workers, the mechanisms of these enzymes being different. The two which have received the most attention to date are the enzymes from yeast and muscle, which will now be discussed. Sigmoid kinetics have been observed as a function of coenzyme (NAD^+) concentration for both enzymes; the substrate, glyceraldehyde-3-phosphate, however, does not exhibit such kinetics, but normal Michaelis–Menten kinetics[1].

Yeast enzyme

Kirschner and coworkers[2] have gathered a considerable amount of evidence indicating that this enzyme acts according to the Monod (concerted) two-state model, discussed earlier in this chapter. The experimental techniques used were both relaxation and stopped-flow. The conditions under which the experiments are performed are fairly critical. As shown in Fig. 11.12,

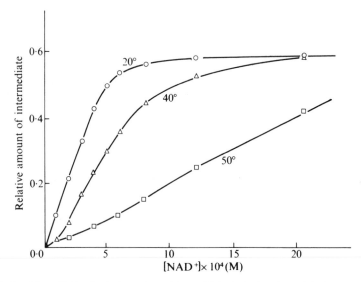

FIG. 11.12. Sigmoid kinetics in the action of glyceraldehyde-3-phosphate dehydrogenase (from K. Kirschner, *FEBS Symposium on 'Regulation of Enzyme Activity and Allosteric Interactions'*, 1968). The ordinate is not the rate, but is the relative concentration of intermediate, as determined spectrophotometrically.

sigmoidicity is temperature-dependent for this enzyme and most of the work was carried out at pH 8·5 and 40°C. Initially, three relaxation times were detected; with improved instrumentation five were subsequently found[3], but the additional relaxation processes did not affect the overall mechanism.

---

[1] J. Teipel and D. E. Koshland, *Biochim. biophys. Acta*, **198**, 183 (1970).
[2] K. Kirschner, M. Eigen, R. Bittman, and B. Voigt, *Proc. natn. Acad. Sci. U.S.A.*, **56**, 1661 (1966); K. Kirschner, *J. molec. Biol.*, **58**, 51 (1971).
[3] K. Kirschner, E. Gallego, I. Shuster, and D. Goodall, *J. molec. Biol.*, **58**, 29 (1971).

The variations of the three important relaxation times with NAD^+ concentration are illustrated in Fig. 11.13. Only τ_3 is related to the sigmoid kinetics, as can be seen from the figure. The rates corresponding to τ_1 and τ_2 are much larger than those related to τ_3, and were assigned to the two different forms of the enzyme; the rate corresponding to τ_3, being very much smaller, is possibly indicative of a conformational change, associated with $T \rightleftharpoons R$.

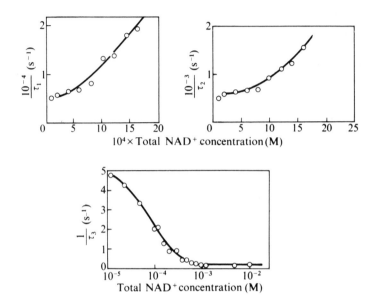

FIG. 11.13. Variations of the three relaxation times τ_1, τ_2, and τ_3 with NAD^+ concentration, for glyceraldehyde-3-phosphate dehydrogenase (from K. Kirschner, E. Gallego, I. Schuster, and D. Goodall, *loc. cit.*).

Analysis of the data indicated a reasonably good fit to the Monod model; in fact, it was shown[1] that the T form was inactive, but able to bind NAD^+ and substrate. Support for this came from a variety of other sources. The sedimentation velocity pattern and the optical rotatory dispersion curves for the enzyme were examined as a function of NAD^+ concentration[2], and the resulting dependence also fitted the Monod model. The degree of volume and shape change of crystals of the enzyme was observed (using small angle X-ray diffraction) to depend on NAD^+ concentration in the same way[3]. The apo-enzyme (obtained by removing all traces of NAD^+) contracted in volume as NAD^+ molecules became bound to it, in a non-linear fashion.

[1] K. Kirschner, *loc. cit.*
[2] R. Jaenicke and W. B. Gratzer, *Eur. J. Biochem.*, **10**, 158 (1969).
[3] H. Durschlag, *et al.*, *FEBS Lett.*, **4**, 75 (1969); *Eur. J. Biochem.*, **19**, 9 (1971).

This is evidence against a simple sequential mechanism (with all binding constants equal), since such a mechanism would predict a linear dependence (as for aspartate transcarbamylase). The monitoring of conformational changes as a function of ligand binding, using the sulfhydryl reactivity method, supported the above conclusions[1].

An investigation of this enzyme by Cook and Koshland[2] led to a slight modification of the above conclusions, since a mixture of positive and negative cooperativity was observed; the four dissociation constants are related to each other in the following manner:

$$K_1 \; < \; K_2 \; > \; K_3 \; > \; K_4$$

negative   positive
cooperativity   cooperativity

The non-equality of the binding constants was indicated by a break in the slope of the line of the Hill plot, as shown in Fig. 11.14. Koshland's examination, being rather more quantitative in nature, was able to detect the deviation

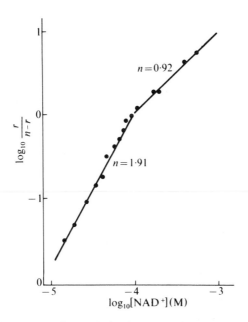

FIG. 11.14. Hill plot (see eqn (6), p. 356) for the results of an equilibrium dialysis study on glyceraldehyde-3-phosphate dehydrogenase (from R. A. Cook and D. E. Koshland, *Biochemistry*, **9**, 3337 (1970); r is the concentration of intermediate and n its maximum concentration.

[1] K. Kirschner, *loc. cit.*
[2] R. A. Cook and D. E. Koshland, *Biochemistry*, **9**, 3337 (1970).

from the normal Monod model, which would have the following relationship between constants:

$$K_1 < K_2 = K_3 = K_4.$$

A possible explanation for the difference in results is that Koshland prepared his enzyme in a slightly different way.

Rabbit muscle enzyme

This is the first enzyme for which it has been conclusively shown that the simple two-state Monod model does not apply. Experiments by several workers[1] showed that negative cooperativity applied here, i.e. binding of one molecule of NAD^+ discourages the binding of the second and subsequent molecules. The Monod model as originally stated has no provision for this, since the binding of substrate is proposed to produce the *active* form of the enzyme, i.e. only positive cooperativity is expected. Here, the first molecule is bound very strongly ($K \approx 10^{-11}$) and the last very weakly ($K \approx 10^{-5}$). On this point there has been a certain amount of controversy, since some workers have been unable to detect conformational changes on the binding of the fourth ligand[2]. Conway and Koshland point out, however, that there need not necessarily be a correlation between conformational change and activity, unless this is specifically demonstrated. On the other hand, extension of the concentration range has recently led to the observation of such a change for the binding of the fourth ligand[3].

An alternative mechanism for the rabbit muscle enzyme has, however, been presented. Malhotra and Bernhard[4] found that only two equivalents of an active site reagent, β-(2-furyl)-acryloyl phosphate, which forms an acyl-enzyme with glyceraldehyde-3-phosphate dehydrogenase, could be introduced per mole of enzyme, unless the NAD^+ concentration was considerably lowered, when it was possible to introduce four equivalents per mole. The second pair of active site reagent molecules reacted about twenty times less rapidly than the first pair. Since the enzyme has four subunits, it was suggested that there might be two different types of subunit, existing in pairs, instead of four equivalent subunits. Such an arrangement, Malhotra and Bernhard suggest, is capable of explaining the observed results. Evidence for two different pairs of subunits has been obtained for lobster muscle enzyme[5].

[1] A. Conway and D. E. Koshland, *Biochemistry*, 7, 4011 (1968); J. J. M. de Vijlder and E. C. Slater, *Biochim. biophys. Acta*, 132, 207 (1967); I. Listowsky, *et al.*, *J. biol. Chem.*, 240, 4253 (1965); J. J. M. de Vijlder and E. C. Slater, *Biochim. biophys. Acta*, 167, 23 (1968).
[2] S. F. Velick, *J. biol. Chem.*, 233, 1455 (1958); J. J. M. de Vijlder and B. J. M. Hansen, *Biochim. biophys. Acta*, 178, 434 (1969).
[3] N. C. Price and G. R. Radda, *Biochim. biophys. Acta*, 235, 27 (1971).
[4] O. P. Malhotra and S. A. Bernhard, *J. biol. Chem.*, 243, 1243 (1962).
[5] J. I. Harris, B. P. Meriweather, and J. H. Park, *Nature*, 198, 154 (1963); R. N. Perham and J. I. Harris, *J. molec. Biol.*, 7, 316 (1963).

Trentham[1] has examined lobster and sturgeon muscle GPDH using the stopped-flow technique, and suggests that the predominant mechanism may involve conformational changes as a consequence of the chemical reaction itself, as suggested by Malhotra and Bernhard.

Once again, it can be seen that a decision as to the exact type of mechanism causing sigmoidicity is a difficult task, and can often not be done unambiguously. It would appear, however, that the rabbit muscle enzyme *probably* behaves according to the sequential mechanism of Koshland and coworkers.

Another enzyme exhibiting negative cooperativity is 3′, 5′-cyclic-adenosine-monophosphate diesterase[2].

---

[1] D. R. Trentham, *Biochem. J.*, **122**, 59, 71 (1971).

[2] T. R. Russell, W. J. Thompson, F. W. Schneider and M. M. Appleman, *Proc. natn. Acad. Sci. U.S.A.*, **69**, 1791 (1972).

12. Supported Enzyme Systems

MOST of the kinetic work on enzymes has been carried out with both the enzyme and the substrate present in free solution. However, in biological systems the enzymes are often not in free solution but are attached to structural material and, therefore, do not have the mobility that they have in free solution. Sometimes they are attached to the membranes of cells, in which case the substrate molecules must diffuse through the membrane material in order to be acted upon by these enzymes. The kinetic behaviour is then different from the behaviour in free solution.

In view of this, it is of importance for kinetic investigations to be made on systems in which the enzyme molecules have become attached to a support. Up to the present time only a few kinetic studies of this kind have been made, but enough has been done for some of the principles to be fairly well understood. The object of this chapter is to discuss some of this work†. Four main topics will be considered: (1) the various ways in which enzymes can become attached to membranes and other supports; (2) general kinetic principles which arise in systems of this kind; (3) these principles as applied to some experimental results; and (4) the rather special aspect of flow systems involving supported enzymes.

An understanding of the kinetics of supported enzyme systems will lead to a deeper knowledge of the behaviour of enzymes in living systems. In addition, investigations of this kind have some technological significance, in view of certain applications of supported enzyme systems; some of these are referred to later.

General aspects of supported enzyme systems

Classification of supported enzymes

The terminology used in this field has been somewhat varied; the expressions immobilized enzymes, solid-supported enzymes, insoluble enzyme derivatives, insolubilized enzymes and matrix-bound enzymes have been used. In the present book we use the term supported enzyme (or solid-supported enzyme if the support is a solid) when we are referring to an enzyme which is either distributed more or less uniformly through a support or is attached to its surface. The term insolubilized enzyme seems best reserved for an enzyme which has been modified and rendered less soluble by a reagent which acts upon certain groups on the enzyme—for example, enzymes treated with glutaraldehyde to yield an insoluble product.

Figure 12.1 shows a possible classification of supported enzyme systems.

† For further details see K. J. Laidler and P. V. Sundaram, in *Chemistry of the Cell Interface* (ed. H. D. Brown), Academic Press, New York, 1971; R. Goldman, L. Goldstein and E. Katchalski, in *Biochemical Aspects of Reactions on Solid Supports* (ed. G. R. Stark), pp. 1–78, Academic Press, New York, 1971.

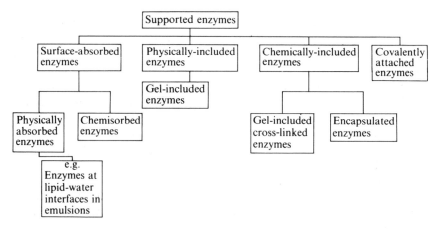

Fig. 12.1. A classification of supported-enzyme systems. This classification applies to both solid and liquid (e.g. emulsion) supports.

Attachment of enzymes to supports

Only in the last ten years or so have efforts been directed toward developing methods for making supported enzymes and studying their properties. Most of the enzymes studied so far are hydrolytic enzymes which in general do not require coenzymes; their molecular weights range from 50 000 to about 500 000.

There are a few cases in which enzymes have been studied when present at an interface, such as a lipid–water interface in an emulsion. This has been done, for example, with catalase by Fraser, Kaplan, and Schulman[1]. Enzymes have been confined in tiny collodion and nylon capsules and studied, particularly by Chang and coworkers[2].

The methods employed for attaching enzymes to solids by covalent bonds will not be discussed here, since they have recently been reviewed fully by Silman and Katchalski[3] and by Kay[4]. For the enzyme to remain active, it must obviously be attached by means of groups that are not part of the active centre or are not even indirectly involved in the catalysis. In most cases it is not known which groups are involved in the attachment to the solid. The whole problem of the attachment of enzymes to solids is a very complex one, and the factors are by no means well understood.

Enzymes at interfaces

Enzyme molecules can be present at various types of interface, such as solid–liquid, air–liquid, and liquid–liquid (emulsion). In nature, many

[1] M. J. Fraser, J. G. Kaplan, and J. H. Schulman, *Discuss. Faraday Soc.*, **20**, 44 (1955).

[2] T. M. S. Chang, *Sci. J.*, July, p. 62 (1967); T. M. S. Chang, F. C. MacIntosh, and S. G. Mason, *Can. J. Physiol. & Pharmacol.*, **44**, 115 (1966).

[3] I. H. Silman and E. Katchalski, *A. Rev. Biochem.*, **35**, 873 (1966).

[4] G. Kay, *Process Biochem.*, August (1968); see also P. V. Sundaram and W. E. Hornby, *FEBS Lett.*, **10**, 325 (1970).

enzyme reactions occur at solid–liquid or liquid–liquid interfaces, the liquid phase being aqueous and the solid phase frequently being of hydrophobic character. The attachment of an enzyme at a hydrophilic–hydrophobic interface is a matter of considerable complexity. Protein molecules at an interface will tend to orient themselves so that the polar groups are in the hydrophilic phase and the non-polar groups in the hydrophobic phase. There may be some unfolding of the protein molecule, and this may result in deactivation of the enzyme. Davies and Rideal[1] have discussed the rigidity of the structures of proteins and other polymers as affected by different interfacial conditions.

The adsorption of catalase at oil–water interfaces was studied by Fraser, Kaplan, and Schulman[2]. By using a variety of stabilizers, they succeeded in forming protein layers of various thicknesses adsorbed on emulsion droplets. They studied how the enzymic activity was affected and correlated it with the degree of unfolding.

Supported substrates

A small number of investigations have been carried out on systems in which the substrate, rather than the enzyme, has been attached to a support. As with supported enzymes, the substrate may be attached, physically or covalently, to a surface, may be physically included in the support, or may be covalently attached throughout the support. Using both substrates and inhibitors attached to supports several workers have succeeded in separating one enzyme, in pure form, from a mixture[3].

Work of this kind has been done by Trurnit[4] using bovine serum albumin fixed on glass slides and by McLaren and Estermann[5] using denatured lysozyme adsorbed on kaolinite; in both cases the effect of chymotrypsin on these supported substrates was studied.

Methods of studying supported enzymes

Brief reference will be made to a few of the ways in which the kinetic aspects of supported enzymes have been investigated.

(1) *Slurry.* In a few investigations[6] the supported enzyme has been present as a slurry, i.e., as a stirred solution of substrate containing tiny particles of the support and enzyme. The kinetic methods employed are much the same as when the enzyme is in homogeneous solution.

[1] J. T. Davies and E. K. Rideal, *Interfacial Phenomena*, p. 246, Academic Press, New York (1963).
[2] M. J. Fraser, J. G. Kaplan, and J. H. Schulman, *Discuss. Faraday Soc.*, **20**, 44 (1955).
[3] P. Quatrecasas, *J. biol. Chem.*, **245**, 3059 (1970); J. Denburg and M. DeLuca, *Proc. natn. Acad. Sci.*, **67**, 1057 (1970); V. Kasche, *Biochem. biophys. Res. Commun.*, **38**, 875 (1970).
[4] H. J. Trurnit, *Archs Biochem. Biophys.*, **47**, 251 (1953).
[5] A. D. McLaren and E. F. Estermann, *Archs Biochem. Biophys.*, **61**, 158 (1956); **68**, 157 (1957).
[6] M. D. Lilly, W. E. Hornby, and E. M. Crook, *Biochem. J.*, **100**, 718 (1966).

A modification of this method involves causing the substrate solution to flow into a reaction vessel containing the slurry, the course of reaction being followed continuously, e.g. using a pH-stat.

(2) *Column catalysis or reaction in a packed bed.* In this technique the solid-supported enzyme particles are packed in a column and the substrate solution is perfused through at the desired rate. The effluent is analysed for the amount of product formed. This method has the advantage over the slurry method in that samples for product estimation do not contain small amounts of the enzyme particles and that the complications of stirring are not encountered. The kinetic analysis of reactions in flow systems is considered later.

(3) *Enzyme sheets or membrane-bound enzymes.* Enzymes attached to sheets or membranes (e.g., 2·5 cm diameter cellulose sheets—see Wilson, Kay, and Lilly[1]) are packed between filter paper of similar size and put in a holder. Substrate is perfused through at a definite rate and the effluent analysed, e.g., spectrophotometrically.

(4) *Enzymes included in or covalently attached to gels.* Work has been done[2] with polyacrylamide gels in which enzymes (glucose oxidase, lactic and alcohol dehydrogenases, and trypsin) have either been mechanically trapped or have become attached by covalent bonds (cross-linking). These supported enzymes can be prepared from the monomer by inducing polymerization in the presence of the enzyme.

(5) *Encapsulated enzymes.* A rather special case of supported enzymes involves trapping the enzymes in microcapsules. Chang[3] in particular has made nylon and collodion microcapsules, of membrane thicknesses varying from 200 to 700 Å, containing various enzymes. The free amino groups of the protein may be cross-linked with groups in the capsule membrane. The membrane thickness and size can be controlled, as can the concentration of enzyme within the microcapsules. These capsules have been used in a slurry and in columns. The latter have been used in extracorporeal shunts for the treatment of patients with inborn errors of metabolism[4]; enzymic reactions can be brought about without actually introducing the enzyme into the body, and in this way immunological complications are avoided.

Basic kinetic laws for solid-supported enzymes

The present section is concerned with the general kinetic principles relating to the action of solid-supported enzymes on substrates that are in

[1] R. J. H. Wilson, G. Kay, and M. D. Lilly, *Biochem. J.*, **108**, 845 (1968); R. J. H. Wilson, G. Kay, and M. D. Lilly, *Biochem. J.*, **109**, 137 (1968).

[2] T. Wieland, H. Teterman, and K. Buening, *Z. Naturf.*, **B21**, 1003 (1966); S. J. Updike and G. P. Hicks, *Nature* (Lond.), **214**, 896 (1967); S. J. Updike and G. P. Hicks, *Science*, **158**, 270 (1967); J. K. Inman and H. M. Dintzis, *Biochemistry*, **8**, 4074 (1969).

[3] T. M. S. Chang, *Sci. J.*, July, p. 62 (1967); T. M. S. Chang, F. C. MacIntosh, and S. G. Mason, *Can. J. Physiol. and Pharmacol.*, **44**, 115 (1966).

[4] T. M. S. Chang, *Nature*, **229**, 117 (1971).

free solution. In particular, equations will be developed showing the variation of the reaction rate with the substrate concentration. The experimental results are later discussed in the light of these basic kinetic equations. The final section of this chapter deals with the special problem of the kinetic equations applicable to flow systems when, for example, the substrate is allowed to flow through a solid bed that contains enzymes.

Factors influencing activity of supported enzymes

There are four main reasons why an enzyme may behave differently when supported than when present in free solution.

(1) The enzyme may be conformationally different in the supported state as compared with free solution. It has been seen that kinetic behaviour depends very critically on enzyme conformation, certain changes leading to a complete loss of activity.

(2) In the support the interaction between the enzyme and the substrate takes place in a different environment from that existing in free solution. Studies of enzyme reactions in different solvents have shown that environmental effects can be very profound (see Chapter 7).

(3) There will be partitioning of the substrate between the support and the free solution so that the substrate concentration in the neighbourhood of the enzyme may be different from what it is in free solution. For example, a relatively non-polar substrate will be more soluble in a support which contains a number of non-polar groups than in aqueous solution; a polar substrate will be less soluble. In addition to this, profound effects may arise if both the substrate and the support are electrically charged; this particular situation has been considered in some detail by Goldstein, Levin, and Katchalski[1].

(4) The reaction in the solid support may be to some extent diffusion-controlled. In homogeneous aqueous solution, even the fastest of enzyme-catalysed reactions appear not to be diffusion-controlled, but this is no longer the case when the substrate has to diffuse toward the enzyme through the solid support. The indications are that only in the case of an exceedingly slow enzyme reaction will the reaction be other than diffusion-controlled.

Little information is at present available about factors 1 and 2 for solid-supported enzymes[2]. The kinetic effects of partitioning of substrate between support and solution have been considered by Goldman, Kedem, and Katchalski[3], but only for the case of electrostatic interactions. These workers have also given a treatment of the diffusion problem in a number of special

[1] L. Goldstein, Y. Levin, and E. Katchalski, *Biochemistry*, **3**, 1913 (1964).
[2] See, however, D. Gabel, I. Z. Steinberg, and E. Katchalski, *Biochemistry*, **10**, 4661 (1971) and P. Wahl, M. Kasai, J.-P. Changeaux, and J.-C. Auchet, *Eur. J. Biochem.*, **18**, 332 (1971).
[3] R. Goldman, O. Kedem, and E. Katchalski, *Biochemistry*, **7**, 4518 (1968).

cases; they did not, however, consider the limiting behaviour at high substrate concentrations. Hornby, Lilly, and Crook[1] have also developed a theoretical treatment of the diffusional effects.

Sundaram, Tweedale, and Laidler[2] have developed a general treatment of the kinetic laws applicable to the situation in which an enzyme-containing membrane is present in a solution of a substrate, such that the solution is in contact with the opposite faces of the membrane. To a good approximation this treatment will apply to discs suspended in the substrate solution, provided that the diameter of the discs is large compared with the thickness. Sundaram *et al.* were especially concerned with the limiting behaviour at low and high substrate concentrations and with how the kinetic parameters are affected by the characteristics (e.g., the thickness) of the membrane. They also gave some computer solutions in order to confirm the general conclusions. The following is based on this treatment.

The variation of substrate concentration through an enzyme-containing support is shown schematically in Fig. 12.2. The substrate concentration

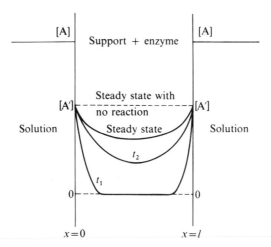

FIG. 12.2. Schematic representation of the variations in substrate concentrations when an enzyme-containing support, of thickness l (cm), is immersed in a substrate solution of concentration [A].

in the free solution is [A] and that just inside the interface is [A']; the ratio [A']/[A] is the partition coefficient P for the substrate between the two phases. After a short period of time t_1, the concentration will fall sharply to zero close to each interface; after a longer period of time the variation may be as shown by the curve marked t_2. Eventually a steady state will be reached.

[1] W. E. Hornby, M. D. Lilly, and E. M. Crook, *Biochem. J.*, **107**, 669 (1968).
[2] P. V. Sundaram, A. Tweedale, and K. J. Laidler, *Can. J. Chem.*, **48**, 1498 (1970); for other treatments see R. Goldman and E. Katchalski, *J. theor. Biol.*, **32**, 243 (1971); A. D. McLaren and L. Packer, *Adv. Enzymol.*, **33**, 245 (1970); V. Kasche, H. Lundquist, R. Bergman, and R. Axen, *Biochem. biophys. Res. Commun.*, **45**, 615 (1971).

If no reaction were occurring, the steady state would correspond to a horizontal line, and this situation is approached if the enzyme is extremely inactive or the membrane very thin.

Attainment of the steady state

It is important to consider first the time it takes for steady state to be essentially established. An exact solution of the presteady-state diffusion equations, for the case in which reaction is occurring as well as diffusion, does not seem to be possible. A reasonably reliable estimate of the time for the establishment of steady state will, however, be given by a treatment in which only diffusion is occurring. The nonsteady-state solution (Barrer[1]) for the system represented in Fig. 12.2 is that concentrations a of substrate within the membrane vary with the distance x and the time t according to

$$a = [A'] \left[1 - \frac{2}{\pi} \sum_{n-1}^{\infty} \frac{1 - \cos n\pi}{n} \sin \frac{n\pi x}{l} \exp\left(-\frac{Dn^2\pi^2 t}{l^2} \right) \right] \qquad (1)$$

where D is the diffusion constant. At the larger t values, at which the steady state is more or less established, the leading term, with $n = 1$, will be most important, so that approximately

$$a = [A'] \left[1 - \frac{4}{\pi} \sin \frac{\pi x}{l} \exp\left(-\frac{D\pi^2 t}{l^2} \right) \right]. \qquad (2)$$

In order to have an arbitrary criterion for the establishment of the steady state, we may ask at what time, at $x = l/2$, does the concentration a differ by 10 per cent from the steady-state value [A']. This time τ is defined by

$$\frac{4}{\pi} \exp\left(-\frac{D\pi^2 \tau}{l^2} \right) = \frac{9}{10} \qquad (3)$$

which reduces to

$$\tau = 0.257 \frac{l^2}{D}. \qquad (4)$$

A typical value of D for a low molecular weight substrate is $3 \times 10^{-6} \text{ cm}^2 \text{ s}^{-1}$ and hence

$$\tau = 8.5 \times 10^4 \, l^2 \text{ s}. \qquad (5)$$

If the membrane is 1 cm thick, the time to come within 10 per cent of steady-state conditions is thus of the order of 10^5 s or 28 hours. For a membrane 1 mm thick the time required is 10^3 s or 17 minutes.

When chemical reaction is occurring in addition, the times will be not as great since the steady-state concentrations that are attained in the membrane

[1] R. M. Barrer, *Diffusion in and through Surfaces*, p. 15, Cambridge University Press, London and New York (1941).

are not so high. It seems safe to conclude that under ordinary circumstances, with membranes less than 1 mm thick, steady state will be essentially established within a few minutes. With membranes much thicker than this, however, it is unsafe to assume that steady-state conditions apply.

Slow enzymic reaction

If the enzyme reaction is low (i.e., the enzyme is very inactive or is present in low concentrations), there will be little depletion of substrate through the support and the reaction will not be diffusion-controlled; the treatment is then very simple. The rate of reaction per unit volume of support is

$$v = \frac{k'_c [E]_m \cdot [A']}{K'_A + [A']} \text{ mol litre}^{-1} \text{ s}^{-1} \tag{6}$$

where $[E]_m$ is the number of moles of enzyme present per litre of support and k'_c and K'_A are the true catalytic constant and Michaelis constant, respectively, for the reaction within the membrane (they may differ from the values k_c and K_A in the free solution because of factors 1 and 2). Suppose that $\{E\}$ is the total number of moles of enzyme present in a support of thickness l (centimetres) and cross-sectional area s (square centimetres); $[E]_m$ is then $10^3 \{E\}/sl$ mol litre^{-1}, and the total rate of reaction in the support of volume sl (millilitres) is given by

$$v' = \frac{k'_c [E]_m \cdot [A'] sl/10^3}{K'_A + [A']} = \frac{k'_c \{E\} \cdot [A']}{K'_A + [A']} \text{ mol s}^{-1}. \tag{7}$$

The concentration $[A']$ is equal to $P[A]$, where P is the partition coefficient; the rate is thus

$$v = \frac{k'_c \{E\} \cdot [A]}{(K'_A/P) + [A]}. \tag{8}$$

The apparent Michaelis constant is K_A/P. The rate at low substrate concentrations is

$$v'_{low} = \frac{P k'_c}{K'_A} \{E\} \cdot [A]. \tag{9}$$

If the same amount of enzyme were distributed in 1 litre of free solution, the rate at low substrate concentrations would be $k_c \{E\}/K_A$; the rate on the support thus differs by the factor P, the partition coefficient, apart from differences between k'_c and k_c and between K'_A and K_A, brought about by factors 1 and 2.

At high substrate concentrations, however, eqn (7) leads to the result that

$$v'_{high} = k'_c \{E\} \tag{10}$$

so that the rate will be the same on the supported enzyme as in free solution, provided that k'_c is the same as k_c.

It is evident that systems of this kind, where the enzyme is of low activity (the enzyme being kept at sufficiently low concentrations), will be valuable for studying the influence of factors 1 and 2 on the activity of the enzyme.

More active enzyme systems

If the rate of the enzymic reaction is greater (enzyme more active or present at higher concentrations), there will be a significant variation in substrate concentration through the support, and the effects of diffusion will then be more important. According to Fick's second law of diffusion, the rate of accumulation of substrate at any cross section, due to diffusion, is given by

$$D\left(\frac{\partial^2 a}{\partial x^2}\right)_t \quad \text{mol litre}^{-1}\,\text{s}^{-1}$$

where D (square centimetres per second) is the diffusion constant, and the concentration a is in moles per litre. In the steady state this must exactly balance the rate of removal of substrate by reaction, which is equal to

$$\frac{k'_c[E]_m \cdot a}{K'_A + a} \quad \text{mol litre}^{-1}\,\text{s}^{-1}.$$

The equation to be solved is, therefore,

$$D\left(\frac{\partial^2 a}{\partial x^2}\right)_t = \frac{k'_c[E]_m \cdot a}{K'_A + a}. \tag{11}$$

An explicit solution of this equation cannot be obtained except in the limiting cases $a \ll K'_A$ and $a \gg K'_A$. These limiting solutions will be obtained first, after which some numerical solutions will be given.

Case I. *Low substrate concentrations.* When $a \ll K'_A$, eqn (11) may be written as

$$\frac{\partial^2 a}{\partial x^2} = \alpha^2 a \tag{12}$$

where α (per centimetre) is equal to $(k'_c[E]_m/DK'_A)^{\frac{1}{2}}$. The solution of this equation, subject to the boundary conditions $x = 0$, $a = [A']$ and $x = l$, $a = [A']$, is

$$\alpha = [A']\frac{\sinh \alpha x + \sinh \alpha(l - x)}{\sinh \alpha l}. \tag{13}$$

The gradients da/dx at $x = 0$ and $x = l$ are

$$\frac{da}{dx} = \mp \alpha[A']\frac{\cosh \alpha l - 1}{\sinh \alpha l}. \tag{14}$$

The rate of entry of substrate at each of the surfaces is, therefore,

$$v = -sD\left(\frac{\mathrm{d}a}{\mathrm{d}x}\right)_{x=0} = sD\left(\frac{\mathrm{d}a}{\mathrm{d}x}\right)_{x=l} \tag{15}$$

$$= \alpha sD[\mathrm{A}']l\frac{\cosh \alpha l - 1}{\sinh \alpha l}. \tag{16}$$

This is equal to the rate of exit of product, since in the steady state there is no accumulation of substrate or product within the membrane; the net rate of formation of product in a membrane of cross-sectional area s is thus

$$v' = 2\alpha sD[\mathrm{A}']\frac{\cosh \alpha l - 1}{\sinh \alpha l}. \tag{17}$$

The rate per unit volume of support is

$$\frac{v'}{sl} = \frac{2\alpha D[\mathrm{A}']}{l} \cdot \frac{\cosh \alpha l - 1}{\sinh \alpha l} \tag{18}$$

$$= \frac{k'_c[\mathrm{E}]_m}{K'_\mathrm{A}}[\mathrm{A}'] \cdot \frac{2}{\alpha l} \cdot \frac{\cosh \alpha l - 1}{\sinh \alpha l}. \tag{19}$$

This result was first obtained by Goldman, Kedem, and Katchalski. The function $F = 2(\cosh \alpha l - 1)/\alpha l \sinh \alpha l$ is plotted in Fig. 12.3; it has its maximum value of unity when αl approaches zero. The maximum rate attainable when a given amount of enzyme $[\mathrm{E}]_m$ is present in 1 litre of support is thus obtained when the thickness l approaches zero, and is equal to

$$v = \frac{k'_c[\mathrm{E}]_m \cdot [\mathrm{A}']}{K'_\mathrm{A}} = \frac{k'_c[\mathrm{E}]_m \cdot P[\mathrm{A}]}{K'_\mathrm{A}}. \tag{20}$$

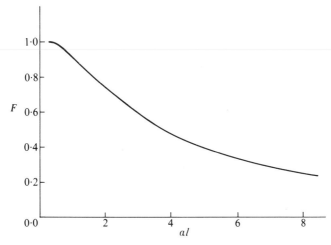

Fig. 12.3. Function $F = 2(\cosh \alpha l - 1)/(\alpha l \sinh \alpha l)$ plotted against αl.

If instead this amount of enzyme were present in 1 litre of free solution, and if the substrate concentration were sufficiently low that the kinetics were first order, the rate would be

$$v = \frac{k_c\{E\} \cdot [A]}{K_A} \text{ mol s}^{-1}.$$

(21)

The limiting rate for the supported enzyme, when the membrane is made thinner, therefore differs from that in homogeneous solution by the factors

$$\frac{k_c'}{k_c}, \frac{K_A}{K_A'}, P = \frac{[A']}{[A]}.$$

The form of the function F is of some interest; the value is very close to unity (within 1 per cent) for αl values less than about 0·3, after which there is a substantial fall. The value of 0·3 for αl, therefore, represents something of a critical value, below which the rates for the supported enzyme will correspond closely to those in free solution (apart from factors 1 and 2), and above which they will have lower values. The quantitative significance of this is discussed below.

Case II. High substrate concentrations. When $a \gg K_A'$, eqn (11) becomes

$$\frac{\partial^2 a}{\partial x^2} = \beta$$

(22)

where β (units of mol cm^{-5}) is equal to $10^{-3} k_c'[E]_m/D$. The solution of this, subject to the boundary conditions $x = 0$, $a = [A']$ and $x = l$, $a = [A']$, is

$$a = [A'] - \tfrac{1}{2}\beta x(l - x).$$

(23)

The concentration gradients at $x = 0$ and $x = l$ are

$$\frac{da}{dx} = \pm\tfrac{1}{2}\beta l.$$

(24)

The rate of entry of substrate through area s at each side of the support is, therefore,

$$v' = \tfrac{1}{2}\beta sD.$$

(25)

The total rate of product formation in a membrane of cross-sectional area s (square centimetres) is thus

$$v' = \beta sDl \text{ mol s}^{-1}$$

(26)

$$= 10^{-3} k_c'[E]_m[A]l \text{ mol s}^{-1}.$$

(27)

The rate per litre of support is thus

$$v' = k_c'[E]_m = k_c'\{E\} \text{ mol litre}^{-1}\text{s}^{-1}.$$

(28)

If the same amount of enzyme is present in 1 litre of solution, the rate is

$$v = k_c\{E\} \text{ mol litre}^{-1}\,s^{-1}. \tag{29}$$

The limiting, high-substrate, concentration rates now differ only by the factor k'_c/k_c. Experimental studies at high substrate concentrations will thus reveal this factor, uncomplicated by the partitioning effect.

Case III. The general case. Solutions of the general equation (11) can only be obtained by numerical means. One can, however, infer the approximate form of the rate vs. substrate concentration dependence on the basis of the solutions for Cases I and II. This is shown schematically in Fig. 12.4.

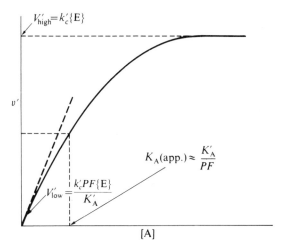

FIG. 12.4.   Plot of v' against substrate concentration in the free solutions. The rate v' is the rate when $\{E\}$ moles of enzyme are present in unit volume of support.

The limiting rate at high substrate concentrations is $k'_c\{E\}$ [eqn (26)] and at low concentrations, $k'_c PF\{E\} . [A]/K'_A$, where P is the partition coefficient and F the function plotted in Fig. 12.3. If Michaelis–Menten kinetics are obtained for the supported enzyme, then the rate expression will be

$$v' = \frac{k_c(\text{app.})\{E\} . [A]}{K_A(\text{app.}) + [A]} \tag{30}$$

where

$$k_c(\text{app.}) = k'_c \tag{31}$$

and

$$K_A(\text{app.}) = \frac{K'_A}{PF}. \tag{32}$$

The value of K_A(app.) (apparent Michaelis constant) extrapolated to low membrane thickness is thus K'_A/P, and if P is measured directly, K'_A can be evaluated and compared with the value in free solution.

Solution of eqn (11) can be carried out numerically by reducing the equation to two first-order equations. The equation can be written as

$$\frac{d^2 a}{dx^2} = \frac{ca}{b+a} \tag{33}$$

where c is $k'_c[E]_m/D$ and b is K'_A. The substitution $z = da/dx$ leads to

$$\tfrac{1}{2}z^2 = \int \frac{ca}{b+a} \, da + B \tag{34}$$

where B is a constant of integration. This equation can be integrated directly and gives

$$z = \frac{da}{dx} = \pm 2^{\frac{1}{2}} \left[ca - cb \ln \left(1 + \frac{a}{b} \right) + B \right]^{\frac{1}{2}}. \tag{35}$$

This can be integrated numerically over the required interval of x, subject to the boundary conditions

$$x = 0, a = [A'] \quad \text{and} \quad x = l, a = [A'].$$

Since these do not permit the evaluation of c in eqn (35), a rough estimate of da/dx at $x = 0$ is made, and eqn (35) is integrated until $x = l$. The initial value da/dx can then be adjusted, by trial and error, until $[A]'_{x=l} = [A]'_{x=0}$ to the desired degree of accuracy.

A number of computer calculations have been carried out, and they confirm the general conclusions drawn above. Examples are shown in Figs. 12.5 and 12.6; the left-hand curves show concentration-distance profiles, and the right-hand curve is a plot of rate against the logarithm of $[A']$, the substrate concentration just inside the surface. Figure 12.5 relates to the following conditions:

$$l = 1 \text{ mm}; \quad K'_A = 10^{-3}\text{M}; \quad c = 0.85 \text{ mol cc}^{-5} \text{ cm}^{-5}.$$

The latter condition corresponds, for example, to the following values:

$$k'_c = 8.5 \text{ s}^{-1}; \quad [E]_m = 10^{-7} \text{ M}; \quad D = 10^{-6} \text{ cm}^2 \text{ s}^{-1}.$$

It is seen from Fig. 12.5 that if the substrate is equally soluble in the support and in the solution, the K_A(app.) value would be 1.26×10^{-3}M, i.e. slightly greater than K'_A. The k_c(app.) value, however, is 8.5 s^{-1}, exactly the same as k'_c. The rate constant at low substrate concentrations, k_c(app.)/K_A(app.), is $6.75 \times 10^{-3}\text{M}^{-1} \text{ s}^{-1}$, slightly less than the value of 8.5×10^3 for the enzyme in free solution. If the partition coefficient is not unity but is P, the K_A(app.) values will be reduced by the factor P, and the k_c(app.)/K_A(app.) values raised by the factor P.

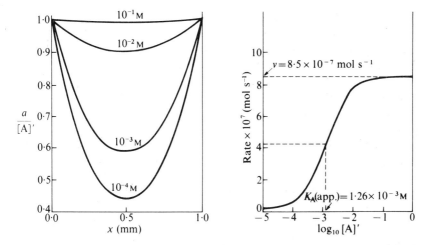

FIG. 12.5. Computer solutions for the kinetics of a solid-supported enzyme, under the following conditions: enzyme concentration in support, 10^{-7} M; $k_c = 8.5\text{ s}^{-1}$; $D = 10^{-6}\text{ cm}^2\text{ s}^{-1}$; $K_A = 10^{-3}$; thickness of support = 1 mm. The left-hand curve shows the variations in substrate concentration through the membrane, for the four surface concentrations indicated. The rates in the right-hand figure correspond to one litre of support containing 10^{-7} mol enzyme.

Figure 12.6 shows the results of similar calculations for a membrane of thickness 0·1 mm, other conditions being the same. In line with the earlier discussion, the $K_A(\text{app.})$ value (for $P = 1$) is now much closer to the K'_A value; it is indeed, almost exactly 10^{-3} M. Similarly, the $k_c(\text{app.})$ and $k_c(\text{app.})/K_A(\text{app.})$ values are indistinguishable from the values for the enzyme in free solution. Again, if P is other than unity, the $k_c(\text{app.})$ value remains unchanged, but $K_A(\text{app.})$ and $k_c(\text{app.})/K_A(\text{app.})$ are altered.

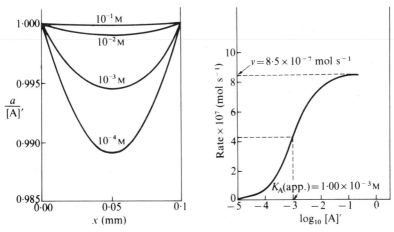

FIG. 12.6. Similar plots to those shown in Fig. 12.5 for a membrane 0·1 mm thick.

General conclusions

The main points that have been revealed by the treatment are as follows:

(1) If k_c and K_A are unchanged by the support, and $P = 1$, the rates at high substrate concentrations will be the same for a given amount of supported enzyme as for the same amount of enzyme per litre of free solution. At lower substrate concentrations the rates will tend to be less for the supported enzyme, and the Michaelis constants higher because of diffusional effects.

(2) However, as the membrane is made thinner (at constant membrane volume), the lower-concentration rates for the support will tend to approach those for the enzyme in free solution. The computer calculations have shown that, under typical conditions, identical rates have been obtained for a membrane 0.1 mm thick.

(3) The higher the enzymic activity (i.e., the higher k_c' or $\{E\}_m$) the lower will be the membrane thickness at which the rates become the same.

(4) If $k_c' \neq k_c$, the high substrate concentration rates will differ by the ratio k_c'/k_c but will not be affected by the partition coefficient P.

(5) If $K_A' \neq K_A$ and $P \neq 1$, the low-concentration rates will differ by the factor $Pk_c'K_A/K_A'k_c$ and the apparent Michaelis constants by the factor K_A'/PK_A.

Points 2 and 3 can be put into a more quantitative form with reference to Fig. 12.3. The value of F is very close to unity for values of αl less than about 0.3 and falls fairly strongly at higher values. The critical thickness below which the rates on the supported enzyme correspond to those in free solution, therefore, corresponds to an αl value of about 0.3. The value of α for the parameters used in the computer calculations is about 30 cm^{-1}, so that the critical value should be about $0.3/30 = 0.01$ cm. The computer calculations did in fact show that at $b = 0.01$ cm the limit had been reached.

Significance of the partition coefficient P

In the foregoing treatments an important role has been played by the partition coefficient P, which is the ratio of the concentration of substrate just inside the support to that in the bulk solution. It is necessary to consider this coefficient in further detail. Two situations are to be distinguished—one in which the support can be regarded as homogeneous and the other in which it is heterogeneous, the enzyme action taking place in only part of the support (e.g., in pores).

In the homogeneous case no particular problem arises. The enzyme normally will be distributed uniformly throughout the support. The substrate will partition itself between the bulk solution and the surface of the support, and the distribution within the support will be as discussed. The magnitude of the partition coefficient P could be determined directly by separate experiments with the support to which no enzyme has been attached, since the enzyme will normally have little effect on the coefficient. The magnitude of P may be very different from unity in certain cases. For example, a substrate having a large number of non-polar groups will be more soluble in a support containing many non-polar groups than in pure water; P will then be much greater than unity. A highly polar substrate, however, may be much more soluble in water than in the support and P will be less than unity. Another important factor, to which Katchalski and coworkers have called attention in particular (see, especially, Goldstein, Levin, and Katchalski[1]), relates to the electrostatic charges on the substrate and support. Thus, if the substrate and support bear opposite charges, P will tend to be large; if they bear like charges, P will tend to be small. Some examples of this are to be found later.

When the support is heterogeneous the above principles still apply, but certain complications arise. Suppose that a fraction θ of the support consists of pores and the remainder of inert material. If $\{E\}$ mol of enzyme are attached to Al millilitres of support the effective concentration will not be $10^3 \{E\}/Al$ since the enzyme will be present only in the pores; instead it will be $10^3 \{E\}/\theta Al$. This factor alone will tend to produce an enhancement in rate, as compared with the homogeneous case, by a factor of $1/\theta$. However, the fact that the substrate can only dissolve in the pores leads to a reduction in rate, the rate being θ times what it would have been if the pore phase constituted the entire support. These two effects exactly cancel. The conclusion is, therefore, that the rate in a porous support will be the same as in a homogeneous support of the same dimensions, if the composition of the material in the pores is the same as that in the homogeneous support.

However, the partition coefficient to be used in the theoretical treatment must relate to the distribution of substrate between the bulk solution and the volume within the porous phase, and not between the bulk solution and the support as a whole. The effective partition coefficient P_{eff}, equal to the ratio of the concentrations of substrate in the support and in the bulk solution, will be smaller than the true coefficient P by the factor θ (i.e., $P_{eff} = \theta P$). If θ can be determined by separate experiments, then P can be calculated and used in the theoretical treatment.

Electrostatic effects

One factor that has an important effect on the magnitude of the partition coefficient P is the electrostatic one. If both the support and the substrate are electrically charged and have opposite signs, P will be abnormally large because of the electrostatic forces between them; if the signs are the same, the value of P will tend to be small.

A quantitative treatment of this effect has been developed by Goldstein, Levin, and Katchalski[1] in terms of the electrical potential ψ of the support

[1] L. Goldstein, Y. Levin, and E. Katchalski, *Biochemistry*, **3**, 1913 (1964).

relative to the bulk solution. This potential is positive if the support bears an excess of positive charges, and negative if it bears an excess of negative charges. Suppose that the substrate molecules bear a charge ze, where e is the charge on the H^+ ion and z is a positive or negative integer. The energy of a substrate molecule on the support is $ze\psi$ with respect to the bulk solution, and, according to the Boltzmann principle, the distribution of the substrate between the support and the bulk solution is given by

$$P_{el} = \frac{[A']}{[A]} = e^{-ze\psi/kT}. \tag{36}$$

Here $[A']$ is the concentration of substrate in the support and $[A]$ that in the bulk solution; e is the base of the natural logarithms.

The P_{el} is the magnitude of the partition coefficient arising from electrostatic effects alone. If this were the only factor, eqns (30) to (32) will apply to a good approximation with expression (36) used for P. At sufficiently low membrane thicknesses $F = 1$, so that the rate equation becomes

$$v = \frac{k'_c\{E\}.[A]}{K'_A\,e^{ze\psi} + [A]}. \tag{37}$$

It is to be noted that, as discussed, the electrostatic interactions have no effect in the limit of high substrate concentrations.

The apparent Michaelis constant for the supported system is

$$K_A(\text{app.}) = K'_A\,e^{ze\psi/kT}. \tag{38}$$

Thus when z and ψ are of the same sign (e.g., a positively charged substrate and a positively charged support) the apparent K_A is greater than the true value; when they are of opposite signs, $K_A(\text{app.})$ is less than the true K'_A.

Unfortunately, it is difficult to make a reliable estimate of the magnitude of ψ for a particular support. The procedure used by Goldstein, Levin, and Katchalski[1] was to calculate ψ values from the experimental results; some examples are considered later.

Theory of pH effects

The influence of pH on the rates of enzyme reactions will, in general, be different for a supported enzyme compared with an enzyme in free solution.

Goldstein, Levin, and Katchalski have treated the problem in terms of the distribution of hydrogen ions between the bulk solution and the support. They have developed a quantitative treatment for the situation in which the support contains ionizing groups and bears an electrostatic charge. This treatment follows the same lines as that for the distribution of substrate between support and free solution. The relative concentration of hydrogen

[1] L. Goldstein, Y. Levin, and E. Katchalski, *Biochemistry*, **3**, 1913 (1964); an alternative procedure, of calculating the zeta potential, has been carried out by A. D. McLaren, *Enzymologia*, **21**, 356 (1960).

ions in the near neighbourhood of a charged polyelectrolyte gel is different from that of the solution with which it is in equilibrium; an enzyme embedded in a gel will thus be exposed to a pH which is different from that in the bulk of the solution. If the polyelectrolyte is positively charged, the hydrogen-ion concentration will be lower than in the bulk of the solution, whereas, if the polyelectrolyte is negative, the hydrogen-ion concentration will be higher. The pH-activity curve of an enzyme attached to a support should, therefore, be displaced toward lower pH values when the support is positively charged and toward higher pH values when the support is negatively charged.

A quantitative treatment of these effects proceeds as follows. Suppose that the support bears a positive charge which creates a positive potential ψ. The ratio of the concentration of hydrogen ions in the support, $[H^+]'$, is then related to that in the bulk solution, $[H^+]$, by the Boltzmann factor:

$$\frac{[H^+]'}{[H^+]} = e^{-e\psi/kT} \tag{39}$$

where e is the charge on the H^+ ion, and e the base of the natural logarithms. Since for the free solution

$$pH = -\log_{10}[H^+] \tag{40}$$

and for the support,

$$pH' = -\log_{10}[H^+]' \tag{41}$$

it follows that

$$\Delta pH \equiv pH' - pH = \frac{e\psi}{2\cdot303kT}. \tag{42}$$

The factor 2·303 is for conversion from natural to common logarithms. These pH shifts cause corresponding shifts in the pK values, as shown schematically in Fig. 12.7. Some applications of this treatment are considered later (p. 404).

The electrostatic effects on the distribution of hydrogen ions and charged substrate molecules can be looked upon in a slightly different and more realistic way than that envisioned by Goldstein, Levin, and Katchalski[1]. These workers regard the potential ψ as extending uniformly throughout the bulk of the support, whereas actually on a microscopic scale there will be profound variations in potential through the support. The shifts in pK values are better interpreted in terms of local electrostatic effects than in terms of an average potential. Consider, for example, the ionization

$$-B^+-H \overset{K}{\rightleftharpoons} -B + H^+.$$

[1] L. Goldstein, Y. Levin, and E. Katchalski, *Biochemistry*, 3, 1913 (1964).

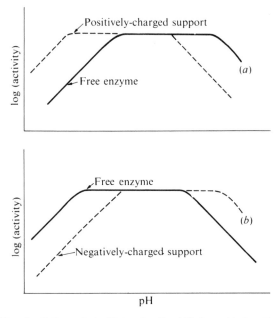

FIG. 12.7. Profiles of activity against pH showing the shifts brought about by (a) a positively-charged support and (b) a negatively-charged support.

If a CO_2^- group is brought into proximity to this ionizing group, the ionization

$$\begin{matrix} \}B^+ - H \\ \}CO_2^- \end{matrix} \;\overset{K'}{\rightleftharpoons}\; \begin{matrix} \}B \\ \}CO_2^- \end{matrix} + H^+$$

will occur with greater difficulty because of the attraction of the $-CO_2^-$ group for the proton. The dissociation constant K' will thus be less than K; i.e., $pK' > pK$. This way of looking at the problem leads to the same qualitative conclusion that an excess of negatively charged groups on the support will increase the pK values, whereas positive groups will reduce the pK values. These shifts will not, however, be related to bulk hydrogen-ion concentrations in the manner envisioned by Goldstein, Levin, and Katchalski[1].

A similar point of view is to be preferred with reference to the K_A shifts.

Effects of temperature and of inhibitors

The theory could readily be extended to cover the effects of temperature and of inhibitors on the kinetics of supported enzyme systems. This will not be considered here, however, since hardly any experimental work has been done in these areas.

[1] L. Goldstein, Y. Levin, and E. Katchalski, *Biochemistry*, **3**, 1913 (1964).

Application to experimental results

Systems for which electrostatic effects are unimportant

Several kinetic aspects of solid-supported enzyme systems have been investigated experimentally, but few systematic studies have been made, and the whole subject is still in its infancy. Table 12.1 gives a very brief summary of some of the kinetic investigations. It would appear that in most of the studies αl was below the critical value of 0.3 so that diffusional effects were unimportant. Partition coefficients between the support and the free solution have apparently never been measured, but one would expect them to be not too far different from unity in a number of cases where there are no electrostatic interactions between the enzyme and the support.

For example, Katchalski and coworkers[1] have studied a number of systems in which an enzyme has been attached to an uncharged support, and they have found the k_c(app.) and K_A(app.) values to be little changed. Similar evidence has been obtained by Hornby, Lilly, and Crook[2], who found that the neutral support *p*-aminobenzyl cellulose caused little change in the K_A(app.) value for the action of adenosinetriphosphate (ATP)-creatine phosphotransferase on ATP.

Similarly, when the substrate is neutral there is generally little effect on K_A values, even when the support is charged. For example, Money and Crook (quoted by Hornby, Lilly, and Crook[2]) found that with the uncharged substrate, *N*-acetyltyrosine ethyl ester, the attachment of chymotrypsin to CM-cellulose–70, a negatively charged support, brought about only a small change in the K_A value. Table 12.2 gives some K_A values for systems such as these where there are no electrostatic interactions between the substrate and the support.

A study over a range of pH values has been made by Sundaram[3] for urease adsorbed on kaolinite. For this system, in which the substrate has no net charge, the K_A values are always somewhat larger for the bound enzyme than for the attached enzyme. Figure 12.8 shows a plot of $V_{max}\,(=k_c(\text{app.})\{E\})$ values against pH. The pH profiles are similar in shape, with the adsorbed enzyme having a somewhat higher activity throughout the pH range investigated.

These results indicate that when either the support or the substrate is uncharged, so that electrostatic interactions are unimportant, the support does not bring about any important change in the kinetic parameters. However, there is often a small but significant increase in the K_A(app.) value when the enzyme is attached to a support in such systems. This effect

[1] E. Katchalski, *Polyamino Acids, Polypeptides, Proteins*, Proc. Int. Symp., p. 283 (1962); A. Bar-Eli and E. Katchalski, *J. biol. Chem.*, **238**, 1690 (1963); E. Riesel and E. Katchalski, *J. biol. Chem.*, **239**, 1521 (1964); I. H. Silman, M. Albu-Weissenberg, and E. Katchalski, *Biopolymers*, **4**, 441 (1966). A recent experimental test of the theory has been carried out by P. S. Bunting and K. J. Laidler, *Biochemistry*, **11**, 4477 (1972).
[2] W. E. Hornby, M. D. Lilly, and E. M. Crook, *Biochem. J.*, **107**, 669 (1968)
[3] P. V. Sundaram, Doctoral Thesis, University of London (1967).

TABLE 12.1

Activity studies on bound enzymes†

Enzyme	Support	Remarks	Ref.
Trypsin	p-Amino-DL-phenylalanine-L-leucine copolymer	Neutral support; pH optimum unchanged	[1]
Chymotrypsin	p-Amino-DL-phenylalanine-L-leucine copolymer	Neutral support; pH optimum unchanged	[2]
Papain	p-Amino-DL-phenylalanine-L-leucine copolymer	Neutral support; pH optimum unchanged	[2]
	p-Amino-DL-phenylalanine-L-leucine copolymer	K_A for BAEE unchanged	[3]
Trypsin	Maleic anhydride-ethylene copolymer	Negatively charged support; pH optimum raised by 2·5 units	[4]
Chymotrypsin	Maleic anhydride-ethylene copolymer	pH Optimum raised	[4]
	DEAE cellulose	Positively charged support; pH optimum lowered	[5]
Aminocylase	DEAE cellulose	pH Optimum lowered	[6]
Chymotrypsin	Kaolinite	Negatively charged support; pH optimum raised by 2 units	[7]
Urease	Kaolinite	Marked activation, k_c raised; pH optimum unchanged	[8]
Catalase	Lipid-water interface	Reduced activity	[9]

† Abbreviations; DEAE-diethylaminoethyl; BAEE-*N*-benzoylarginine ethyl ester.

[1] A. Bar-Eli and E. Katchalski, *J. biol. Chem.*, **238**, 1690 (1963).
[2] E. Katchalski, *Polyamino Acids, Polypeptides, Proteins, Proc. Int. Symp.*, 1st, 1961, p. 283 (1962).
[3] I. H. Silman, Doctoral Thesis, Hebrew University, Jerusalem, Israel (1964).
[4] L. Goldstein, Y. Levin, and E. Katchalski, *Biochemistry*, **3**, 1913 (1964).
[5] M. Pecht, Master's Thesis, Hebrew University, Jerusalem, Israel (1966).
[6] T. Tosa, T. Mori, N. Fuse, and I. Chibata, *Enzymologia*, **32**, 153 (1967).
[7] A. D. McLaren and E. F. Estermann, *Archs Biochem. Biophys.*, **68**, 157 (1957).
[8] P. V. Sundaram, Doctoral Thesis, University of London, London (1967).
[9] M. J. Fraser, J. G. Kaplan, and J. H. Schulman, *Discuss. Faraday Soc.*, **20**, 44 (1955).

TABLE 12.2

Apparent Michaelis constants for enzymes attached to uncharged solid supports†

Enzyme	Support	Substrate	Apparent K_A (M)	Ref.
ATP-creatine	None	ATP	6.5×10^{-4}	[1]
phosphotransferase	*p*-Aminobenzylcellulose	ATP	8.0×10^{-4}	
Chymotrypsin	None	ATEE	2.7×10^{-4}	[2]
	CM cellulose –70	ATEE	5.6×10^{-4}	

† Abbreviations: ATP—adenosine triphosphate; ATEE—*N*-acetyltyrosine ethyl ester; CM-O-carboxymethyl cellulose.

[1] W. E. Hornby, M. D. Lilly, and E. M. Crook, *Biochem. J.*, **107**, 669 (1968).
[2] Money and Crook, quoted by W. E. Hornby, M. D. Lilly, and E. M. Crook, *Biochem. J.*, **107**, 669 (1968).

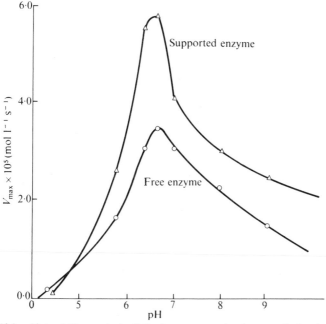

FIG. 12.8. Plots of V_{max} against pH for free urease and urease attached to kaolinite.

is understandable in terms of the treatment of diffusion given earlier; because of the necessity for diffusion, the concentration of the substrate close to the enzyme is somewhat less than that in the bulk phase, and this will lead to less binding and, therefore, to a higher K_A(app.) value.

The finding of Sundaram[1] that the V_{max} values are all higher for the bound enzyme than for the unbound requires some effect in addition to diffusion.

[1] P. V. Sundaram, Doctoral Thesis, University of London (1967).

The V_{max} values relate to enzyme saturation, so that the depletion of substrate near the enzyme has no effect on these values. Several suggestions can be offered for the increased V_{max} values, but it appears most likely that there is a true environmental effect, the k_c(app.) value being greater when the enzyme is supported. This might be due to a change in the effective dielectric constant of the surroundings, which will have an effect on the electrostatic interactions between the enzyme and the substrate.

Charged supports

When there are charges on both the substrate and the support, the apparent Michaelis constant for the bound enzyme is significantly different from its value for the free enzyme. This is seen in Table 12.3. For the ATP-creatine phosphotransferase system, for example, there is a tenfold increase in K_A(app.) when the enzyme is attached to the polyanionic support CM-cellulose–90, and little change when the enzyme is attached to the neutral *p*-aminobenzyl cellulose. This increase in K_A(app.) when the negatively-charged support is used can be correlated with the electrostatic repulsion between the support and the negatively-charged substrate, which will have the effect of decreasing the concentration of substrate in the neighbourhood of the enzyme, and of decreasing the binding between enzyme and substrate. In the ficin system, however, where the substrate is positively charged, there is a tenfold decrease in K_A(app.) when the enzyme in attached to the negatively charged support. These observations can be readily explained in terms of electrostatic interactions between substrate and support. Unlike charges on substrate and support will lead to enhancement of the substrate concentration close to the surface (P is increased), and this will lead to additional binding, i.e., to a lower K_A(app.). Conversely, like charges will decrease the local substrate concentration in comparison with that in the bulk solution (P is lowered) and will lead to a decrease in binding and to a higher K_A(app.).

Support for the conclusion that these changes in K_A(app.) are due to electrostatic effects is provided by the observation of A. Katchalsky[1] that the effects are almost entirely eliminated by increasing the ionic strength of the solution. Increasing the ionic strength, like increasing the dielectric constant, has the effect of decreasing the electrostatic forces between charges and will thus reduce the attractive or repulsive forces between the support and the substrate.

Another effect that can be clearly related to electrostatic interactions is the shift in the pH optimum when, for a charged substrate, the enzyme is attached to a charged support. An example of this is seen in Fig. 12.9; the enzyme is trypsin and the substrate is *N*-benzoylargininamide, which is positively charged. When the enzyme is attached to a negatively charged support, the pH optimum is displaced by approximately 2·6 pH units

[1] A. Katchalsky, *Biophys. J.*, **4**, 9 (1964).

TABLE 12.3

Apparent Michaelis constants for enzymes attached to solid supports†

Enzyme	Support	Charge on support	Substrate	Charge on substrate	Apparent K_m (M)	Ref.
ATP-creatine transphosphorase	None		ATP	—	6.5×10^{-4}	[1]
	p-Aminobenzylcellulose	0	ATP	—	8.0×10^{-4}	
	CM cellulose-90	—	ATP	—	7.0×10^{-3}	
Ficin	None		BAEE	+	2.0×10^{-2}	[2]
	CM cellulose-70	—	BAEE	+	2.0×10^{-3}	
Chymotrypsin	None		ATEE	0	2.7×10^{-4}	[1]
	CM cellulose-70	—	ATEE	0	5.6×10^{-4}	
Trypsin	None		BAA	+	6.8×10^{-3}	[3]
	Maleic acid-ethylene copolymer	—	BAA	+	2.0×10^{-4}	
Papain	None	0	BAEE	+	1.9×10^{-2}	[4]
	p-Aminophenylalanine-L-leucine copolymer	—	BAEE	+	1.9×10^{-2}	
Bromelain	None		BAEE	+	0.11‡	[5]
	CM cellulose	—	BAEE	+	0.049‡	

† Abbreviations: ATP-adenosine triphosphate; BAEE-*N*-benzoylarginine ethyl ester; ATEE-*N*-acetyltyrosine ethyl ester; BAA-*N*-benzoylarginineamide; CM, *O*-carboxymethyl cellulose.
‡ At and ionic strength of 0.2M.

¹ W. E. Hornby, M. D. Lilly, and E. M. Crook, *Biochem. J.*, **107**, 669 (1968).
² W. E. Hornby, M. D. Lilly, and E. M. Crook, *Biochem. J.*, **98**, 420 (1966).
³ L. Goldstein, Y. Levin, and E. Katchalski, *Biochemistry*, **3**, 1913 (1964).
⁴ E. Katchalsky, *Polyamino Acids, Polypeptides, Proteins, Proc. Int. Symp.*, 1st, 1961, p. 283 (1962).
⁵ C. W. Wharton, E. M. Crook, and K. Brocklehurst, *Eur. J. Biochem.*, **6**, 572 (1968b).

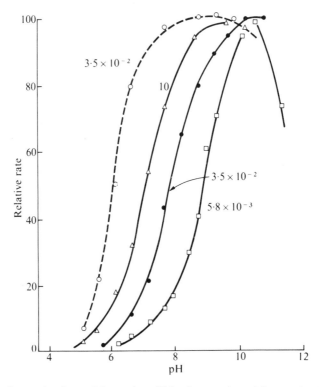

FIG. 12.9. Curves showing activity against pH for free trypsin and for trypsin attached to a negatively-charged support (copolymer of maleic acid and ethylene). The curve for the free enzyme is dashed; the other curves are for the supported enzyme. Ionic strength (in mol l^{-1}) is shown beside each curve.

toward more alkaline values. As the ionic strength is increased, the pH-activity curve shifts back toward more acid pH values, approaching the curve for pure trypsin.

As mentioned earlier, Goldstein, Levin, and Katchalski[1] developed a theory of these pH shifts in terms of an electrostatic potential ψ. They did not develop an explicit treatment of ψ in terms of the properties of the polyelectrolyte support, but calculated ψ values from the ΔpH values observed experimentally for supported and unsupported enzymes. Some values for trypsin supported on a copolymer of maleic acid and ethylene are shown in Table 12.4 for various ionic strengths. The pH values [see eqn (48)] are those corresponding to 50 per cent of maximal activity for the free enzyme, whereas the pH' values are equal to the pH at which the native enzyme shows a catalytic activity equal to that of the bound enzyme. The ψ values obtained in this way are consistent with values obtained in other

[1] L. Goldstein, Y. Levin, and E. Katchalski, *Biochemistry*, **3**, 1913 (1964).

TABLE 12.4

Calculated electrostatic potentials
for trypsin attached to a support†

Ionic strength (mol per litre)	ΔpH	$\psi \times 10^3$ (volts)
0.6×10^{-2}	-2.4 ± 0.2	-145 ± 15
1.0×10^{-2}	-2.0	-124 ± 13
3.5×10^{-2}	-1.6	-96 ± 12
0.2	-1.3	-81 ± 10
1.0	-0.4	-27 ± 8

† Support—copolymer of maleic acid and ethylene.

ways for this polyelectrolyte. The ψ values are seen to decrease with ionic strength.

Another estimate of the value of ψ can be made by seeing how the apparent Michaelis constant is changed when the enzyme is attached to the charged support. The equation that applies is eqn (38). Table 12.5 shows some results, at two ionic strengths, for the trypsin–maleic acid–ethylene system. The ψ values calculated from eqn (38) are consistent with those given in Table 12.4 calculated from the pK shifts.

TABLE 12.5

Calculated electrostatic potentials
for trypsin attached to a support†

Ionic strength (mol litre^{-1})	K_m (app.) $\times 10^3$ (mol litre^{-1})	$K'_m \times 10^3$ (mol litre^{-1})	$\psi \times 10^3$ (volts)
0.04	0.2 ± 0.05	6.9 ± 1.0	-92 ± 12
0.50	5.2 ± 0.5	6.9 ± 1.0	-7 ± 3

† Support—copolymer of maleic acid and ethylene; substrate-benzoyl-L-arginine amide ($z = 1$).

Kinetics of flow systems

A few investigations have been made of the kinetics of enzyme reactions in which the enzyme is attached to a water-insoluble polymer, and the substrate solution is allowed to flow through it. Besides providing useful fundamental information which may be valuable in connection with a study of enzymes in situ, these studies are also relevant to processes in which enzymes are allowed to catalyse reactions of industrial interest; the latter aspects have been discussed by Lilly and Sharp[1].

Two situations are of particular interest. In the first, the substrate solution moves steadily through a packed bed of catalyst in such a way that a cross

[1] M. D. Lilly and A. K. Sharp, *Chem. Engr, Lond.*, **215**, CE12 (1968).

section of the fluid moves as does an imaginary piston; in this case there is said to be pellet or plug flow. In the second type of system the catalyst system is continuously stirred, and the reactor is then known as a continuous feed stirred-tank (CFST) reactor.

Pellet flow through a packed bed

The kinetic equations for pellet flow of substrate solution through a packed bed of catalysts attached to a water-insoluble polymer were first worked out by Lilly, Hornby, and Crook[1], whose treatment is as follows. The general rate equation for an enzyme-catalysed reaction may be written as

$$\frac{-d[A]}{dt} = \frac{k_c[E]_0 \cdot [A]}{K_A + [A]} \tag{43}$$

and the integration of this equation subject to the boundary condition $[A] = [A]_0$ when $t = 0$ gives rise to

$$K_A \ln \frac{[A]_0}{[A]} + [A]_0 - [A] = k_c[E]_0 t. \tag{44}$$

In a flow system where there is pellet flow the time t that the system spends in contact with the catalyst is

$$t = \frac{V_f}{Q} \tag{45}$$

where V_f is the 'void' volume or 'free' volume and Q the rate of flow. The total enzyme concentration $[E]_0$ is equal to the total number of moles of enzyme, $[E]_t$, in the reaction system divided by the total volume V_t of the system:

$$[E]_0 = \frac{[E]_t}{V_t} \tag{46}$$

the fraction f of substrate converted is

$$f = \frac{[A]_0 - [A]}{[A]_0}. \tag{47}$$

Insertion of eqns (45), (46), and (47) into eqn (44) gives

$$f[A]_0 - K_A \ln (1 - f) = \frac{k_c[E]_t V_f}{Q V_t}. \tag{48}$$

This may be written as

$$f[A]_0 - K_A \ln (1 - f) = \frac{k_c[E]_t \beta}{Q} \tag{49}$$

[1] M. D. Lilly, W. E. Hornby, and E. M. Crook, *Biochem. J.*, **100**, 718 (1966).

and

$$f[A]_0 - K_A \ln(1 - f) = \frac{C}{Q} \tag{50}$$

where β, equal to V_f/V_t, is the *voidage* of the column, and C, equal to $k_c[E]_t\beta$ is referred to as the *reaction capacity* of the packed bed reactor.

These equations relate the fraction f of substrate converted during the passage through the column to the flow rate Q, in terms of the initial substrate concentration $[A]_0$, the Michaelis constant K_A, and the reaction capacity of the column. The way in which f varies with Q/C is shown in Fig. 12.10 for various values of K_A and $[A]_0$. For zero flow rate (infinite

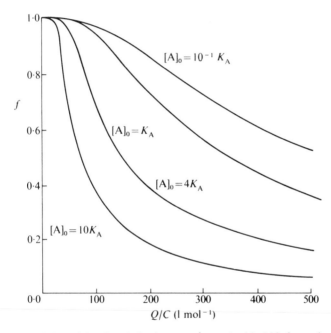

FIG. 12.10. Variation of fraction f of substrate converted with Q/C, for various values for $[A]_0/K_A$ {see eqn (54)}.

residence time), there is complete conversion; for infinite flow rate, there is no conversion.

In order to determine K_A and C/Q from experimental data, $f[A]_0$ may be plotted against $-\ln(1 - f)$; a schematic plot is shown in Fig. 12.11. The intercept on the $f[A]_0$ axis is C/Q and the slope, $-K_A$. From C/Q, if Q, B, and E are known, the kinetic constant k_c can be calculated. Figure 12.12 shows a plot given by Lilly, Hornby, and Crook[1] for the hydrolysis of

---

[1] M. D. Lilly, W. E. Hornby, and E. M. Crook, *Biochem. J.*, **100**, 718 (1966); S. P. O'Neill, M. D. Lilly, and P. N. Rowe, *Chem. Engng Sci.*, **26**, 173 (1971).

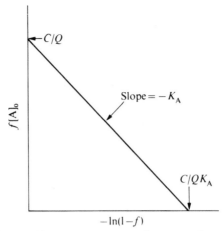

FIG. 12.11.   Schematic plot of $[A]_0$ against $\ln(1-f)$ {see eqn (50)}. The slopes and intercepts are shown.

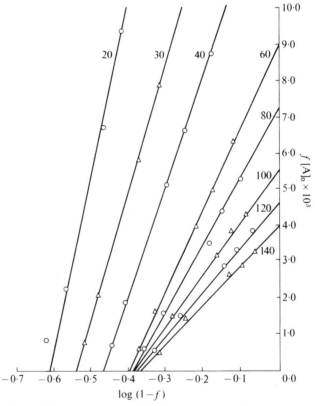

FIG. 12.12.   Plot of $f[A]_0$ against $\ln(1-f)$ for the hydrolysis of N-benzoylarginine ethyl ester by a column of cellulose ficin. The numbers on the curves indicate the flow-rates Q (in ml h^{-1}).

N-benzoylarginine ethyl ester by a column of CM-cellulose-ficin. If the theory were strictly applicable, the slopes of these plots, equal to $-K_A$, would be independent of flow rate, but some variation is seen. This may be due to the fact that the flow is more complex than envisaged; each cellulose-ficin particle is surrounded by a diffusion layer[1], and the reaction rate may be controlled to some extent by the rate of diffusion of the substrate through this layer.

Continuous feed stirred-tank (CFST) reactors

Kinetic equations for flow of substrate solution through a continuous feed stirred-tank reactor have been given by Lilly and Sharp[2]. It is assumed that there is perfect micromixing throughout the vessel. The substrate balance equation is

rate of entrance of substrate = rate of exit of substrate +
+ rate of removal of substrate
by the reaction.

If Q is the rate of flow of the substrate (in litres s^{-1}), the rates of entrance and exit are

$$\text{rate of entrance} = Q[A]_0 \tag{51}$$

$$\text{rate of exit} = Q[A]_f \tag{52}$$

when $[A]_f$ is the final concentration of substrate. The rate of removal of the substrate in the stirred tank of liquid volume V_1 is

$$v = \frac{k_c[E]_0 \cdot [A]_i V_1}{K_A + [A]_i} \tag{53}$$

the substrate concentration within the tank being $[A]_i$. Then

$$Q([A]_0 - [A]_f) = \frac{k_c[E]_0 \cdot [A]_i V_1}{K_A + [A]_i}. \tag{54}$$

If the total amount of enzyme present is $\{E\}$ and the volume of the liquid and the stirred tank is V_t,

$$[E]_0 = \frac{\{E\}}{V_t}. \tag{55}$$

With the definitions

$$f = \frac{[A]_0 - [A]_i}{[A]_0} \tag{56}$$

[1] F. Helfferich, *Ion Exchange*, Chap. 8, McGraw-Hill, New York (1962).
[2] M. D. Lilly and A. K. Sharp, *Chem. Engr, Lond.*, **215**, CE12 (1968).

and

$$\varepsilon = \frac{V_1}{V_t},$$ (57)

eqn (48) becomes

$$f[A]_0 Q = \frac{k_c[E]\varepsilon_r}{K_A/[A]_0(1 - f) + 1}$$ (58)

or

$$f[A]_0 + \frac{f}{1 - f}K_A = \frac{k_c\{E\}\varepsilon_r}{Q} = \frac{C_t}{Q}$$ (59)

where C_t, equal to $k_c\{E\}\varepsilon$, is known as the *reaction capacity* for the stirred tank reactor. Equation (59) somewhat resembles eqn (50) and gives a similar variation of P with Q/C_T. Lilly and Sharp[1] have made comparisons of the behaviour of the two types of reactors. It is evident that if the stirring is not complete in the CFST reactor, there will be deviations from eqn (59). These workers have investigated the effect of stirrer speed on the kinetic behaviour, using carboxymethylcellulose-chymotrypsin with acetyltyrosine ethyl ester as substrate, and found that the reaction rate increases with the rate of stirring.

[1] M. D. Lilly and A. K. Sharp, *Chem. Engr, Lond.*, **215**, CE12 (1968).

13. The Denaturation of Proteins

WHEN proteins, either in the dry state or in solution, are heated or treated in various other ways they generally undergo some characteristic changes which result from alterations in their three-dimensional structures. Proteins that have undergone such changes are known as *denatured* proteins in contrast to fresh or *native* proteins, and the process that occurs is known as *denaturation*. The main features of the denaturation process and the properties of the denatured proteins have been discussed in various reviews[1]. The present treatment is confined largely to the kinetic aspects of the denaturation process, and especially to the denaturation of enzymes, which leads to a loss of their catalytic activity and is then specifically referred to as *inactivation* or *deactivation*.

The terms denaturation and inactivation have been applied rather loosely to various types of changes occurring in proteins. It is actually rather difficult to define denaturation in a precise manner, since different types of treatment (for example, application of heat and addition of urea) may being about different changes in a given protein. According to an early definition[2], denaturation is a change in the natural protein whereby it becomes insoluble in solvents in which it was previously soluble. Another definition is that given by Neurath and coworkers[3]: denaturation is any non-proteolytic modification of the unique structure of a native protein, giving rise to definite changes in chemical, physical or biological properties. This definition, which might perhaps be considered to be too broad, excludes the mere hydrolysis of the protein, but does not exclude the possibility that there is dissociation of the protein into smaller fragments, or that there is aggregation into larger molecules; such changes do, in fact, commonly occur under the action of denaturing agents.

The main agents that bring about protein denaturation or enzyme inactivation are:

(1) Heat.
(2) Very high hydrostatic pressures (5000–10 000 atm).
(3) Irradiation by ultraviolet light or the various kinds of ionizing radiation, such as cathode rays and α-particles.

---

[1] See especially, H. Neurath, J. P. Greenstein, F. W. Putnam, and J. O. Erickson, *Chem. Rev.*, **34**, 157 (1944); F. W. Putnam, in *The Proteins* (ed. H. Neurath and K. Bailey), Vol. 1B, Chapter 9, Academic Press, New York, 1954; C. Tanford, *Adv. Protein Chem.*, **23**, 121, 223 (1968); C. Tanford, *Physical Chemistry of Macromolecules*, John Wiley and Sons, New York, 1961, pp. 624–639; C. Tanford, *Adv. Protein Chem.*, **24**, 1 (1970).
[2] H. Wu, *Chin. J. Physiol.*, **5**, 321 (1931).
[3] Neurath, Greenstein, Putnam, and Erickson, *loc. cit.*

(4) Ultrasonic waves (the effect here is probably simply due to local temperature increases[1]).

(5) Acids and bases.

(6) Other chemical agents such as organic solvents (e.g. alcohol), urea, guanidinium salts, etc.

(7) Enzymes[2].

Of the above, the first few may be classed as physical agencies, the last three as chemical ones. However, it is important to realize that the various agents act in collaboration with one another. Any protein solution can be denatured by the addition of a certain amount of heat, and also by the addition of acid; very often, denaturation can also be brought about by the addition of less heat together with the addition of a smaller amount of acid. The exact relationship between these two particular factors, heat and pH, is indeed a very complicated one, and is considered in some detail later. In a similar way, urea can act in collaboration with temperature and pH to bring about the process of denaturation.

Basically what happens when a protein is denatured is that there are changes in the secondary and tertiary bonding existing in the protein molecules; this results in a change in the general shape of the molecule. It has been seen that the catalytic effectiveness of an enzyme depends very critically on the relative positions of groups that may be on different portions of the polypeptide chains; a change in the overall structure of the protein may therefore lead to a complete loss in enzymic activity.

Apart from this loss of catalytic activity found with enzymes, a number of other property changes are frequently found to occur when a protein is denatured; not all of them, however, will necessarily occur with a given protein. The most important of these changes are:

(1) Decrease in solubility.

(2) Loss of crystallizability.

(3) Change in overall molecular shape.

(4) Increase in chemical reactivity.

(5) Increase in susceptibility to attack by proteolytic enzymes.

(6) Loss of biological activity (for enzymes, loss of catalytic activity; if a protein is a hormone it loses its power to regulate biological functions).

In some cases the process of denaturation is reversible, the protein returning to its original form when the denaturing agent is removed. Sometimes, for example, a protein that has been denatured by mild heating in solution regains its original characteristics on cooling. In other cases the

[1] V. F. Griffing, *J. chem. Phys.*, **20**, 939 (1952).

[2] It has been suggested, in particular by Linderstrøm-Lang [K. Linderstrøm-Lang, *Proteins and Enzymes*, Stanford University Press (1952); K. Linderstrøm-Lang, R. D. Hotchkiss, and G. Johnsen, *Nature*, **142**, 966 (1938); see also H. P. Lundgren, *J. biol. Chem.*, **138**, 293 (1941)] that the initial effect of proteolytic enzymes on protein substrates is to bring about denaturation, after which the proteins are more sensitive to the hydrolytic action of the enzyme.

process of denaturation produces a state from which the protein cannot return. On the whole, denaturation under gentle conditions (mild heating or small pH changes) is more likely to produce reversible denaturation, while more drastic conditions produce an irreversible effect. Sometimes denaturation is partly reversible and partly irreversible.

Order of reaction

Since processes of protein denaturation involve the transformation of a single reactant species, the protein itself, it might be thought that the reactions would be unimolecular, and hence be kinetically of the first order. This presumption is, however, by no means justified when the experimental data are examined, and many examples of orders higher than unity have been discovered. It is unfortunate that many of the earlier investigators proceeded on the basis of the assumption of first-order kinetics, making no proper determinations of the order; in some cases the data have been calculated and quoted on the mere assumption that the kinetics are first-order. In the present section a review will first be given of the experimental evidence; afterwards the significance of the results will be discussed.

In investigating the question of the order of protein denaturations it is of the greatest importance to distinguish clearly between the two orders of reaction, the order with respect to concentration (or the *true* order), and the order with respect to time. The distinction between these two orders has been discussed in Chapter 2. Since it is only at the beginning of any reaction that the conditions are known with certainty it is the order with respect to concentration, or true order, that is the more fundamental quantity. However, for a full understanding of the mechanism it is necessary for both orders to be known.

The main experimental data† on reaction orders for protein denaturation are summarized in Table 13.1. It will be seen that the majority of investigators have merely determined the order with respect to time, by following the course of disappearance of native protein. In the few cases in which both orders have been determined they are by no means always the same.

The data will not be discussed in detail; only the more significant points will be considered. The case of the heat inactivation of pepsin is an interesting one in view of the number of times it has been investigated and the apparently divergent results that have been obtained. The first detailed study of the reaction was made by Tammann[1] who calculated first-order rate coefficients but found them to decrease markedly as the reaction proceeded; this indicates the order with respect to time to be greater than unity. Arrhenius[2] quoted data, obtained by various colleagues, which suggested that the order

---

† It is to be noted that most of these results were obtained a good many years ago; remarkably little work has been done in this field during the last twenty years or so.

[1] G. Tammann, *Z. phys. Chem.*, **18**, 426 (1892).
[2] S. Arrhenius, *Immunochemie*, Leipzig, p. 57 (1907).

TABLE 13.1

Orders of reaction for protein denaturations

Protein	n_c	n_t	Remarks	Investigators	Ref.
Heat denaturation					
Pepsin	—	>1	—	Tammann	[1]
Pepsin	—	1	—	Arrhenius	[2]
Pepsin	—	3/2	—	Michaelis and Rothstein	[3]
Pepsin	1	1	Crystalline pepsin used; impurities cause the rate coefficients to fall as reaction proceeds	Northrop	[4]
Pepsin	1	1		Steinhardt	[5]
Pepsin	1–5	1–5	Crystalline pepsin used n_c and n_t increase with decreasing pepsin concentration	Casey and Laidler	[6]
Trypsin	—	1	—	Arrhenius	[2]
Trypsin	1	>1	Impurities cause the rate coefficients to fall as reaction proceeds	Northrop	[7]
Trypsin	—	1	—	Pace	[8]
Trypsin	—	>2	Between pH 2 and 9	Kunitz and Northrop	[9]
Chymotrypsinogen	1	1	Reversible	Eisenberg and Schwert	[10]
Chymotrypsinogen	>1	>1	Irreversible	Eisenberg and Schwert	[10]
Egg albumin	—	~1	—	Chick and Martin	[11]
Egg albumin	—	~1	—	Lewis	[12]
Egg albumin	—	~1	—	Cubin	[13]
Egg albumin	1	1	—	Gibbs, Bier, and Nord	[14]
Ovalbumin	1	3–6	—	Haurowitz, DiMois, and Tekman	[15]

Protein			Notes	Authors	Ref.
Tobacco mosaic virus	~0	<1	—	Lauffer and Price	[16]
Bovine plasma albumin	—	1	From pH 0·8 to 4·0; deviations at higher pH	Levy and Warner	[17]
β-Lactoglobulin	—	1	—	Groves, Hipp, and McMeekin	[18]
β-Lactoglobulin	—	1	—	Christensen	[19]
Haemoglobin	—	1	—	Chick and Martin	[20]
Haemoglobin	—	1	—	Haurowitz, Hardin, and Dicks	[21]
Oxyhaemoglobin	—	1	—	Hartridge	[22]
Oxyhaemoglobin	—	1	—	Lewis	[23]
Horse carbonylhaemoglobin	—	1	—	Zaiser and Steinhardt	[24]
Ferrihaemoglobin	1	1	—	Steinhardt and Zaiser	[25]
Edestin	—	>1	—	Bailey	[26]
Emulsin	—	>1 at 40° and 50°C; 1 above 60°C	—	Tammann	[1]
Malt amylase	—	>1	—	Lüers and Wasmund	[27]
Invertase	—	>1	—	von Euler and Laurin	[28]
Pancreatic lipase	—	1	—	McGillivray	[29]
Peroxidase	—	1	—	Zilva	[30]
Trypsinogen	—	1	—	Pace	[31]
Enterokinase	—	1	—	Pace	[32]
Ricin	—	1	—	Levy and Benaglia	[33]
Influenza A virus	Negative	3/2	—	Lauffer and Carnelly	[34]
Luciferase	1	>1	—	Chase	[35]
ATP-ase	0·6	1	—	Pelletier and Ouellet	[36]

TABLE 13.1—Continued

Protein	n_c	n_t	Remarks	Investigators	Ref.
Heat denaturation					
β-Amylase	<1	1	—	Nakayama and Kono	[37]
Urea denaturation					
Tobacco mosaic virus	1	<1	—	Lauffer	[38]
Influenza A virus	<1	<1	—	Lauffer, Wheatley, and Robinson	[39]
Diphtheria antitoxin	1	>1	—	Wright and Schomaker	[40]
Ovalbumin	1	>1	—	Simpson and Kauzmann	[41]
Alcohol denaturation					
Haemoglobin	—	1	—	Booth	[42]

[1] G. Tammann, *Z. phys. Chem.*, **18**, 426 (1892).
[2] S. Arrhenius, *Immunochemie*, Leipzig (1907), p. 57.
[3] L. Michaelis and M. Rothstein, *Biochem. Z.*, **105**, 60 (1920).
[4] J. H. Northrop, *J. gen. Physiol.*, **13**, 739 (1930); *Ergebn. Enzymforsch.*, **1**, 302 (1932).
[5] J. Steinhardt, *Nature*, **138**, 74 (1936); *K. danske vidensk. Selsk. Math-fys. Medd.*, **14**, No. 11 (1937).
[6] E. J. Casey and K. J. Laidler, *Science*, **111**, 110 (1950); *J. Am. chem. Soc.*, **73**, 1455 (1951).
[7] J. H. Northrop, *J. gen. Physiol.*, **4**, 261 (1922).
[8] J. Pace, *Biochem. J.*, **24**, 606 (1930).
[9] M. Kunitz and J. H. Northrop, *J. gen. Physiol.*, **17**, 591 (1934).
[10] M. A. Eisenberg and G. W. Schwert, *J. gen. Physiol.*, **34**, 583 (1951).
[11] H. Chick and C. J. Martin, *J. Physiol.*, **40**, 404 (1910); **43**, 1 (1911); **45**, 61, 261 (1912).
[12] P. S. Lewis, *Biochem. J.*, **20**, 978 (1926).
[13] H. K. Cubin, *Biochem. J.*, **23**, 25 (1929).
[14] R. J. Gibbs, M. Bier, and F. F. Nord, *Archs Biochem. Biophys.*, **35**, 216 (1952).
[15] F. Haurowitz, F. DiMoia, and S. Tekman, *J. Am. chem. Soc.*, **74**, 2265 (1952).
[16] M. A. Lauffer and W. C. Price, *J. biol. Chem.*, **133**, 1 (1940).
[17] M. Levy and R. C. Warner, *J. phys. Chem.*, **58**, 106 (1954).
[18] M. C. Groves, N. J. Hipp, and T. L. McMeekin, *J. Am. chem. Soc.*, **73**, 2700 (1951).

[19] L. K. Christensen, *C.R. Trav. Lab. Carlsberg, Ser. chim.*, **28**, 37 (1952).
[20] H. Chick and C. J. Martin, *J. Physiol.*, **40**, 404 (1910).
[21] F. Haurowitz. R. L. Hardin. and M. Dicks, *J. physiol. Chem.*, **58**. 103 (1954).
[22] H. Hartridge, *J. Physiol.*, **44**, 34 (1912).
[23] P. S. Lewis, *Biochem. J.*, **20**, 965, 984 (1926).
[24] E. M. Zaiser and J. Steinhardt, *J. Am. chem. Soc.*, **73**, 5568 (1951).
[25] J. Steinhardt and E. M. Zaiser, *J. Am. chem. Soc.*, **75**, 1599 (1953).
[26] K. Bailey, *Biochem. J.*, **36**, 140 (1942).
[27] H. Lüers and W. Wasmund, *Fermentforsch.*, **5**, 169 (1922).
[28] H. von Euler and I. Laurin, *Hoppe-Seyler's Z. physiol. Chem.*, **108**, 64 (1919).
[29] I. H. McGillivray, *Biochem. J.*, **24**, 891 (1930).
[30] S. S. Zilva, *Biochem. J.*, **8**, 656 (1914).
[31] J. Pace, *Biochem. J.*, **25**, 1485 (1931).
[32] J. Pace, *Biochem. J.*, **25**, 1 (1931).
[33] M. Levy and A. E. Benaglia, *J. biol. Chem.*, **186**. 829 (1950).
[34] M. A. Lauffer and H. C. Carnelly, *Archs Biochem.*, **8**, 265 (1945).
[35] A. M. Chase, *J. gen. Physiol.*, **33**, 535 (1950).
[36] G. E. Pelletier and L. Ouellet, *Can. J. Chem.*, **39**, 265 (1961).
[37] S. Nakayama and Y. Kono, *J. Biochem., Tokyo*, **44**. 25 (1957).
[38] M. A. Lauffer, *J. Am. chem. Soc.*, **65**, 1793 (1943).
[39] M. A. Lauffer, M. Wheatley, and G. Robinson, *Archs Biochem.*, **22**, 467 (1949).
[40] G. G. Wright and V. Schomaker, *J. Am. chem. Soc.*, **70**, 356 (1948).
[41] R. B. Simpson and W. J. Kauzmann, *J. Am. chem. Soc.*, **75**, 5139 (1953).
[42] N. Booth, *Biochem. J.*, **24**, 1699 (1930).

with respect to time is unity. Later, Michaelis and Rothstein[1] obtained results which were consistent with those of Tammann, and they stated that a three-halves order (with respect to time) best fitted their data. In later studies, using crystalline enzyme preparations, Northrop[2] and Steinhardt[3] found that the reaction was accurately a first-order one with respect to time. In addition Northrop noted that when impurities are present the first-order coefficients decrease as the reaction proceeds, indicating that under these conditions the order with respect to time is higher than unity. In none of the above investigations was any attempt made to determine the order with respect to concentration.

In an effort to resolve the apparent discrepancies between the results of these workers, Casey and Laidler[4] carried out a systematic study of this problem, and determined the order with respect to concentration and the order with respect to time over a 200-fold range of pepsin concentrations. In order to determine the order with respect to concentration the initial rates of denaturation were measured at various initial pepsin concentrations, and the order determined from a plot of log (initial rate) against log (initial concentration). The plot is shown in Fig. 13.1 and is seen not to be linear, which means that the reaction does not have a simple order over the entire range of concentrations investigated. However, over a limited range of higher concentrations the slope is unity, so that the reaction is of the first order with respect to concentration within this region. As the pepsin concentration is decreased the order with respect to concentration (equal to the slope) gradually increases until, at the lowest concentrations at which the reaction could conveniently be studied (0·004 per cent pepsin by weight), the reaction was approximately obeying the fifth-order law.

In this investigation the orders with respect to time were approximately equal to the orders with respect to concentration. In the high-concentration region, for example, where the initial rates were directly proportional to the concentrations, the time course of the reactions obeyed the first-order law in a satisfactory manner; the deviations occurred, however, in the later stages of the reaction, and this can be attributed to the increase in order as the protein concentration falls. On the other hand the time course of the reaction using low protein concentrations was not consistent with the first-order law. Figure 13.2 shows a typical plot of enzyme activity against time for a run corresponding to an initial pepsin concentration of about 0·006 per cent by weight. The form of the curve is not at all like that for a first-order reaction, and at first sight suggests that the reaction is approaching an equilibrium corresponding to only partial inactivation; this, however, was shown not to be the case by allowing the solutions to stand for very long

[1] L. Michaelis and M. Rothstein, *Biochem. Z.*, **105**, 60 (1920).

[2] J. H. Northrop, *J. gen. Physiol.*, **13**, 739 (1930).

[3] J. Steinhardt, *Nature*, **138**, 74 (1936); *K. Danske vidensk. Selsk. Math-fys. Medd.*, **14**, No. 11 (1937).

[4] E. J. Casey and K. J. Laidler, *Science*, **111**, 110 (1950); *J. Am. chem. Soc.*, **73**, 1455 (1951).

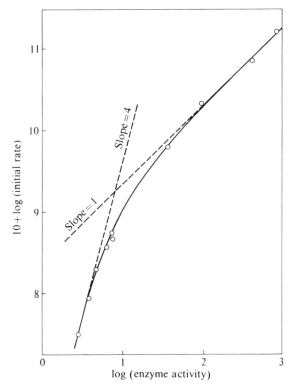

FIG. 13.1.   Plot of $\log_{10} v$ against $\log_{10}$ (enzyme activity) for the inactivation of pepsin (Casey and Laidler, *J. Am. chem. Soc.*, **73**, 1455 (1951)).

periods, after which complete inactivation had occurred. The behaviour was found to be consistent with a mechanism, discussed below, according to which the reaction is intermediate between a first-order and a fifth-order one. The curve shown in Fig. 13.2 is in fact consistent with the kinetic equation (eqn 20 below) which arises from such a mechanism.

In the light of these results that apparent reaction orders vary with the experimental conditions there is no real discrepancy between the results of the previous investigations. Those workers, for example, who obtained good first-order constants were presumably working in the high concentration region where the true order is unity. This is true of Northrop's and Steinhardt's work, the concentrations at which they worked being known to be in the high region. The same apparently applies to the data quoted by Arrhenius; the enzyme concentrations are not known but the value obtained for the activation energy suggests that the concentration was fairly high. On the other hand Tammann, and Michaelis and Rothstein, were presumably working in the region of lower concentration where the first-order behaviour is no longer observed.

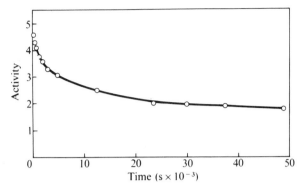

Fig. 13.2.   Plot of activity against time for the inactivation of pepsin (Casey and Laidler, *loc. cit.*).

With other proteins the kinetic orders have not been studied in so detailed a fashion, but there are various indications of orders greater than unity (see Table 13.1). Pace[1] found that the time course of the inactivation of trypsin, trypsinogen, and enterokinase obeyed the first-order law, but Northrop[2], and Kunitz and Northrop[3] found time-orders of higher than unity; in particular it was found that impurities tended to cause the first-order rate coefficients to fall as reaction proceeded. The case of chymotrypsinogen is of interest in that the reversible denaturation is strictly first-order with respect to both concentration and time, whereas the irreversible denaturation has a high order (of about 10) with respect to concentration[4]. The order with respect to time was not determined but the curves suggest that it is greater than unity.

The question of the order of the egg-albumin denaturation has given rise to some confusion; the earlier work has been discussed by Neurath *et al.*[5], who show that the data appear to be consistent with the first-order law. Later work of Gibbs, Bier, and Nord[6] showed that the first-order rate coefficients remained constant as the reaction proceeded and were independent of concentration over a tenfold range; both orders are therefore unity. With ovalbumin, the denaturations by heat[7] and by urea[8] are both of the first order with respect to protein concentration, but are of higher orders with respect to time.

The case of tobacco mosaic virus is of interest in that with both heat[9] and urea[10] denaturations the orders with respect to time are unity. However,

[1] J. Pace, *Biochem. J.*, **24**, 606 (1930).
[2] J. H. Northrop, *J. gen. Physiol.*, **4**, 261 (1922).
[3] M. Kunitz and J. H. Northrop, *J. gen. Physiol.*, **17**, 591 (1934).
[4] M. A. Eisenberg and G. W. Schwert, *J. gen. Physiol.*, **34**, 583 (1951).
[5] H. Neurath, J. P. Greenstein, F. W. Putnam, and J. O. Erickson, *Chem. Revs.*, **34**, 157 (1944).
[6] R. J. Gibbs, M. Bier, and F. F. Nord, *Archs Biochem. Biophys.*, **35**, 216 (1952).
[7] F. Haurowitz, F. DiMoia, and S. Tekman, *J. Am. chem. Soc.*, **74**, 2265 (1952).
[8] R. B. Simpson and W. J. Kauzmann, *J. Am. chem. Soc.*, **75**, 5139 (1953).
[9] M. A. Lauffer and W. C. Price, *J. biol. Chem.*, **133**, 1 (1940).
[10] M. A. Lauffer, *J. Am. chem. Soc.*, **65**, 1793 (1943).

the first-order rate coefficients diminish with increasing protein concentration, a fact that indicates the order with respect to concentration to be less than unity. For the heat denaturation the rate coefficient is roughly inversely proportional to the protein concentration, so that the order with respect to concentration is close to zero.

It is clear from the above survey of the experimental data that widely different types of behaviour are observed even with the limited selection of proteins whose denaturations have been studied. Four main classes of behaviour are to be distinguished, as follows:

(1) Both orders, n_c and n_t, are unity.
(2) The order with respect to concentration is unity, but the order with respect to time is greater than unity.
(3) The order with respect to time is unity, but the order with respect to concentration is less than unity.
(4) Both orders of reaction are greater than unity.

The question of what mechanisms of denaturation are consistent with these four types of behaviour will now be discussed.

Type 1: $n_c = n_t = 1$. When both orders are unity the reactions may be simple unimolecular changes. There is always, however, the possibility that further investigation will disclose complexities requiring more complex mechanisms; for example the low-concentration behaviour of pepsin makes it doubtful that the simple behaviour at higher concentrations can be explained in terms of an elementary unimolecular transformation.

Type 2: $n_c = 1$; $n_t > 1$. Whenever the order with respect to concentration is unity it is possible to formulate a satisfactory mechanism involving the individual protein molecules as reactants in unimolecular processes. However, when the order with respect to time is greater than unity it is necessary to formulate some special feature in the mechanism that will explain the falling-off of the first-order rate coefficients. There are four main possibilities a distinction between which can be made by means of relatively simple experimental tests.

(1) The first possibility is that the denaturation process does not proceed to completion, but reaches an equilibrium state in which a significant proportion of protein remains in the native state. If this is the situation the rate will fall off as reaction proceeds more rapidly than if the process approached a state of complete denaturation, and the calculated first-order constants will fall; this gives an apparent order of the reaction of greater than unity.

This case can be distinguished by allowing denaturation to proceed for a very long period of time, and testing to see whether any native protein remains. If so the kinetic data must be analysed by making use of the appropriate equation for opposing reactions.

(2) A second possibility is that there is some inhibition of the denaturation process by the product of reaction; in other words, the denatured protein has some stabilizing action on the native protein. Such a possibility is by no means unreasonable in view of the known stabilizing effects of impurities. However, no case appears to be known in which this effect has been demonstrated. Experimentally it can be tested by studying the rate of denaturation in the presence of added denatured protein.

(3) The native protein may consist of a number of components N_1, N_2, N_3, etc., and the denaturation process may involve a number of simultaneous unimolecular processes, as indicated by the following scheme.

$$N_1 \xrightarrow{k_1} D$$

$$N_2 \xrightarrow{k_2} D$$

$$N_3 \xrightarrow{k_3} D, \text{ etc.}$$

The different forms of the protein undergo denaturation with the various rate constants k_1, k_2, k_3, etc. If this is the case the time course of the denaturation will follow the kinetic law

$$[D] = [N_1]^0(1 - e^{-k_1 t}) + [N_2]^0(1 - e^{-k_2 t}) + [N_3]^0(1 - e^{-k_3 t}) + \ldots \quad (1)$$

where $[D]$ is the amount of denatured product at time t and $[N_1]^0$, $[N_2]^0$, $[N_3]^0$, etc., are the initial amounts of the various forms N_1, N_2, N_3, etc. Obedience to such a kinetic law will give rise to an apparent order (with respect to time) in excess of unit, but to a true order of unity.

It is possible to test for this mechanism by allowing the protein solution to become partly denatured, recovering the undenatured protein, and again carrying out a kinetic study of the denaturation. If this scheme is the correct one the protein recovered would be different from the original, those components that denature most rapidly being present in relatively smaller proportions than the others; the recovered protein would therefore denature less rapidly than the original. No example of this type of behaviour appears to be known. (See also p. 452 in connection with the urea denaturation of ovalbumin.)

(4) The final possibility is one that probably applies in a number of instances. The protein may undergo denaturation by a number of consecutive steps:

$$N \underset{k_{-1}}{\overset{k_1}{\rightleftharpoons}} D_1 \underset{k_{-2}}{\overset{k_2}{\rightleftharpoons}} D_2 \underset{k_{-3}}{\overset{k_3}{\rightleftharpoons}} \text{ etc.}$$

A number of variants of the scheme are possible, and all of them lead to the result that the order with respect to concentration is unity whereas the order with respect to time is greater than unity. Several simple versions of this general scheme will be considered.

One possibility is that there is an active form N, a completely denatured form D, and a form X which has properties intermediate between N and D, and is formed as a kinetic intermediate:

I $$\text{N} \underset{k_{-1}}{\overset{k_1}{\rightleftarrows}} \text{X} \underset{k_{-2}}{\overset{k_2}{\rightleftarrows}} \text{D.}$$

If a measurement is made of some property y (e.g. optical rotation) as a function of time the kinetic analysis of this system leads to the result that the variation of y with time will obey an equation involving exponential terms:

$$y = y_0 + y_1 e^{-\lambda_1 t} + y_2 e^{-\lambda_2 t} \tag{2}$$

where y_0, y_1 and y_2 are constants, and λ_1 and λ_2 are constants which involve the rate constants k_1, k_{-1}, k_2 and k_{-2}. It is to be noted that eqn (2) is of the same form as eqn (1). A detailed analysis of the kinetic behaviour allows the constants to be determined.

An alternative possibility is that whereas X is intermediate in properties between N and D, it is not formed as a kinetic intermediate; two possibilities exist:

II $$\text{N} \underset{k_{-1}}{\overset{k_1}{\rightleftarrows}} \text{D} \underset{k_{-2}}{\overset{k_2}{\rightleftarrows}} \text{X}$$

and

III $$\text{X} \underset{k_{-1}}{\overset{k_1}{\rightleftarrows}} \text{N} \underset{k_{-2}}{\overset{k_2}{\rightleftarrows}} \text{D.}$$

Both of these schemes lead to the same kind of variation of y with t as given by eqn (2), but the constants λ_1 and λ_2 are related in a different manner to the individual rate constants.

An analysis of these three cases I, II and III has been carried out by Hijazi and Laidler[1] who have shown that in some cases the mechanisms can be distinguished on the basis of the time course of the reactions[1].

In the past, experimental results have usually been interpreted in terms of one or other of these mechanisms, without regard to other possibilities. Simpson and Kauzmann[2], for example, investigated the urea denaturation of ovalbumin, which they followed by measuring the optical rotation. They arrived at the conclusion that consecutive steps are involved by eliminating the other three possibilities, viz. approach to equilibrium, stabilization by products, and simultaneous reactions. They found they could fit their rate curves using three exponential terms (i.e. three rate constants) so that there must be at least two states X_1 and X_2. They assumed that these were kinetic intermediates, i.e., that the mechanism was

$$\text{N} \rightleftarrows \text{X}_1 \rightleftarrows \text{X}_2 \rightleftarrows \text{D.}$$

[1] N. H. Hijazi and K. J. Laidler, *J. chem. Soc. Faraday Trans. I*, **68**, 1235 (1972); see A. Ikai and C. Tanford, *Nature*, **230**, 100 (1971).
[2] R. B. Simpson and W. J. Kauzmann, *J. Am. chem. Soc.*, **75**, 5139 (1953).

However, alternative schemes such as

$$N \rightleftharpoons X_1 \rightleftharpoons D \rightleftharpoons X_2$$

are equally consistent with their results.

Wright and Schomaker[1] studied the urea denaturation of diphtheria anti-toxin, and also found that $n_c = 1$ and $n_t = 1$. They interpreted their results in terms of scheme III above. Similarly, Chase[2] interpreted his results on the heat inactivation of luciferase in terms of a mechanism equivalent to scheme III. Scott and Scheraga[3] have also shown that the ribonuclease inactivation follows an equation of the form of eqn (2).

Further investigation of these systems is required in order for a firm decision to be made as to the mechanisms.

Type 3: $n_c < 1$; $n_t = 1$. When the order with respect to time is unity but the order with respect to concentration is different from unity, the simplest explanation[4] is that the protein contains an impurity. Q, the combination of which gives rise to a complex that denatures at a different rate from the free protein. As long as the denaturations of the free protein N and the complex NQ are simple unimolecular processes, the order with respect to time will be unity. The first-order rate coefficients will, however, vary with the protein concentration owing to shifts in the equilibrium $N + Q \rightleftharpoons NQ$.

If the order with respect to concentration is less than unity it is necessary to suppose that the complex NQ is deactivated more slowly than the native protein N; on concentrating the solution the amount of NQ will increase relative to that of N and the rate will therefore not increase in direct proportion to the concentration of N. The equations corresponding to this explanation are:

$$N + Q \overset{K}{\rightleftharpoons} NQ$$

$$N \xrightarrow{k_1} D$$

$$NQ \xrightarrow{k_2} D + Q.$$

The rate constant k_2 is less than k_1, and K is the equilibrium constant for the complexing.

The kinetic law for this reaction scheme is obtained as follows. The total concentration of native protein is

$$[N]_0 = [N] + [NQ] \tag{3}$$

$$= [N] + K[N].[Q]. \tag{4}$$

[1] G. G. Wright and V. Schomaker, *J. Am. chem. Soc.*, **70**, 356 (1948).
[2] A. M. Chase, *J. gen. Physiol.*, **33**, 535 (1950).
[3] R. A. Scott and H. A. Scheraga, *J. Am. chem. Soc.*, **85**, 3866 (1963).
[4] M. A. Lauffer, *J. Am. chem. Soc.*, **65**, 1793 (1943).

However, the concentration of impurity is proportional to the total concentration of protein,

$$[Q] = k[N]_0 \tag{5}$$

so that

$$[N]_0 = [N] + kK[N] \cdot [N]_0 \tag{6}$$

whence

$$[N] = \frac{[N]_0}{1 + kK[N]_0}. \tag{7}$$

The net rate of denaturation is

$$v = k_1[N] + k_2[NQ] \tag{8}$$

$$= k_1[N] + k_2kK[N] \cdot [N]_0. \tag{9}$$

Together with eqn (7) this gives rise to

$$v = \frac{(k_1 + k_2kK[N]_0)[N]_0}{1 + kK[N]_0}. \tag{10}$$

This equation can account for any order between zero and two. Thus if $kK[N]_0 \gg 1$ and at the same time $k_1 \gg k_2kK[N]_0$, the rate is equal to k_1/kK and the reaction is of zero order. If, on the other hand, $1 \gg kK[N]_0$ and $k_2kK[N]_0 \gg k_1$, the rate is $k_2kK[N]_0^2$, and the order with respect to concentration is two. If k_1 and k_2 are equal the order is unity, the rate being equal to $k_1[N]_0$. These relationships may also be verified in another manner, in terms of the general expression for the order, $d \ln v/d \ln [N]_0$, which is found to be

$$n_c = \frac{d \ln v}{d \ln [N]_0} = \frac{k_1 + 2k_2kK[N]_0 + k_2k^2K^2[N]_0^2}{k_1 + (k_1 + k_2)kK[N]_0 + k_2k^2K^2[N]_0^2}. \tag{11}$$

For the particular case under discussion, $n_c < 1$ and $n_t = 1$, the condition $k_1 \gg k_2$ must arise. Lauffer[1] in fact showed that his data on the urea denaturation of tobacco mosaic virus were consistent with an equation of the form of eqn (10) provided that $kK[N]_0 \gg 1$, but that $k_2kK[N]_0 \gg k_1$ (this means that $k_2 \gg k_1$); under these circumstances

$$v = \frac{k_1}{kK} + k_2[N]_0 \tag{12}$$

which corresponds to an apparent order of less than unity.

It is evident that this type of mechanism, based on the idea of an impurity in the protein, is not capable of accounting for orders in excess of two,

---

[1] M. A. Lauffer, *loc. cit.*

unless it is postulated that more than one molecule of impurity interacts with the protein. If n molecules of impurity react with protein to give NQ_n an analogous treatment to the above gives rise to

$$v = \frac{(k_1 + k_2 kK[N]_0^n)[N]_0}{1 + nkK[N]_0^n}.$$ (13)

This equation can account for orders lying between $n + 1$ and $1 - n$.

It must be emphasized that mechanisms such as the above can account for deviations from first-order kinetics only in so far as concentration is concerned. When the order with respect to time is other than unity some explanation such as those discussed under type 2 above is required.

Type 4: $n_c > 1$; $n_t > 1$. When both orders are in excess of unity a possible explanation may involve a combination of the factors discussed above under types 2 and 3; the high n_c may be due to the interaction of an impurity, the high n_t to incompleteness of reaction, to stabilization by products, or to consecutive or simultaneous reactions. However, a simpler possibility that explains the results satisfactorily if the two orders are equal is that the denaturation is truly polymolecular in character. In some instances there is evidence that the aggregation of protein molecules occurs during denaturation; it is therefore by no means unreasonable that the activated complex might consist of an assembly of a number of protein molecules. Such a mechanism has been proposed specifically for pepsin[1], and has been found to be consistent with the available experimental data.

In the pepsin denaturation the order rises from unity to about 5 as the concentration is diminished, and the mechanism is required to account for this. Suppose that the process of denaturation is a co-operative phenomenon involving n protein molecules, where n is the limiting order at low concentrations. The interaction between the n molecules may be represented by

$$nN \overset{K}{\rightleftharpoons} N_n$$

where N_n is the cluster formed. The cluster is assumed to undergo denaturation more readily than the isolated molecules, perhaps owing to the existence of repulsive forces between the protein molecules. The process of denaturation may therefore be represented as

$$N_n \overset{k}{\rightarrow} D.$$

The equilibrium constant for the clustering is

$$K = \frac{[N_n]}{[N]^n}$$ (14)

[1] E. J. Casey and K. J. Laidler, *J. Am. chem. Soc.*, **73**, 1455 (1951).

and the total initial concentration of protein is

$$[N]_0 = [N] + n[N_n]. \tag{15}$$

Equation (14) therefore becomes

$$K([N]_0 - n[N_n])^n = [N_n] \tag{16}$$

or

$$K[N]_0^n \left(1 - \frac{n[N_n]}{[N]_0}\right)^n = [N_n]. \tag{17}$$

Since $[N_n]$ is always small compared with $[N]_0$ we may expand and accept only the first term; this gives

$$K[N]_0^n \left(1 - \frac{n^2[N_n]}{[N]_0}\right) = [N_n] \tag{18}$$

whence

$$[N_n] = \frac{K[N]_0^n}{1 + n^2 K[N]_0^{n-1}}. \tag{19}$$

The rate of reaction is therefore

$$v = \frac{kK[N]_0^n}{1 + n^2 K[N]_0^{n-1}}. \tag{20}$$

This law corresponds to an nth order reaction at sufficiently low concentrations that $1 \gg n^2 K[N]_0^{n-1}$; the rate is then $kK[N]_0^n$. At high concentrations, however, when $n^2 K[N]_0^{n-1} \gg 1$ the rate is given by

$$v = \frac{k}{n^2}[N]_0 \tag{21}$$

and the reaction is of the first order. The mechanism, of course, provides for the result that the order with respect to time also shows the same change with the protein concentration.

Influence of temperature and pH

The rates of protein denaturations are generally affected very markedly by both temperature and pH, these two factors acting in conjunction with one another. When the temperature is raised the rates are usually increased, although decreases are observed under certain conditions, particularly in urea denaturations. In the majority of instances it is found that the variation of denaturation rate with temperature obeys the Arrhenius law, and in Table 13.2 are quoted some activation energies, calculated in the usual manner from the rate-temperature data. These values are not, however, of great significance except with reference to the effects of other factors, of which pH is one

TABLE 13.2

Energies of activation for protein denaturations

Enzyme	pH	Energy of activation (kcal mol^{-1})	Ref.
Pancreatic lipase	6·0	46·0	[a]
Trypsin	6·5	40·8	[b]
Egg albumin	6·86	129	[c]
Pepsin	4·83	56–147	[d]
ATP-ase	7·0	70	[e]

[a] I. H. McGillivray, *Biochem. J.*, **24**, 891 (1930).

[b] J. Pace, *Biochem. J.*, **24**, 606 (1930).

[c] P. S. Lewis, *Biochem. J.*, **20**, 978 (1926).

[d] E. J. Casey and K. J. Laidler, *J. Am. chem. Soc.*, **73**, 1455 (1951).

[e] L. Ouellet, K. J. Laidler, and M. F. Morales, *Archs Biochem. Biophys.*, **39**, 37 (1952); see G. E. Pelletier and L. Ouellet, *Can. J. Chem.*, **39**, 265 (1961).

of the most important. A detailed analysis of the pH-temperature results is presented later (p. 438–445).

One important feature of the results quoted in Table 13.2 may now be noted however. The activation energies given in the table are considerably higher than those generally obtained for ordinary chemical reactions occurring in the same temperature range as for these protein denaturations. The fact that a reaction will occur at a measurable speed at 30 to 40°C generally requires the activation energy to be no higher than 18 kcal or so; a value of 40 kcal or more usually means that the rate will not become appreciable until much higher temperatures have been reached. The protein denaturations therefore have abnormally high activation energies from the standpoint of their rates, and this means that their entropies of activation are also abnormally high; some values are quoted in Table 13.3, and they are seen to be much higher than usually found in chemical reactions. In order for these matters to be properly interpreted it is necessary for the data to be analysed in terms of a theory of pH dependence. This will now be considered, after which (on pp. 438–445) the theories will be applied to the experimental data for a number of protein systems.

TABLE 13.3

Entropies of activation for protein denaturations

Enzyme	pH	Entropy of activation (e.u. mol^{-1})
Pancreatic lipase	6·0	68·2
Trypsin	6·5	44·7
Egg albumin	6·86	298·4
ATP-ase	7·0	150·0

References as in Table 13.2.

Theory of pH effects

A considerable number of studies have been made on the influence of pH on the rates of protein denaturations, including many enzyme inactivations. In general it is found that the rates pass through a minimum as the pH is varied, and the behaviour is frequently complex. A very accurate study of the pepsin inactivation was made by Steinhardt[1], who over a certain range found that the rate was proportional to the reciprocal of the fifth power of the hydrogen ion concentration. His interpretation of this result was that five successive ionizations are involved, the overall splitting-off of five protons being represented as

$$P \overset{K}{\rightleftharpoons} P' + 5H^+.$$

The dissociation constant K is given by

$$K = \frac{[P'] \cdot [H^+]^5}{[P]} \tag{22}$$

and, if $[P]_0$ is the total pepsin concentration,

$$[P]_0 = [P] + [P']. \tag{23}$$

These equations give rise to

$$[P'] = \frac{K[P]_0}{K + [H^+]^5} \tag{24}$$

and, if it is mainly $[P']$ that undergoes inactivation,

$$v = k[P'] \tag{25}$$

$$= \frac{kK[P]_0}{K + [H^+]^5}. \tag{26}$$

Provided that $[H^+]^5 \gg K$ this corresponds to the inverse fifth power relationship.

A more general formulation of pH effects, which reduces to the Steinhardt treatment as a special case, has been given by Levy and Benaglia[2]. These workers successfully applied their equations to their own data on ricin and to Lewis's data[3] on haemoglobin; the method has also been used to interpret other systems. In the following discussion the notation and methods are slightly different from those used by Levy and Benaglia, but the treatment is equivalent to theirs.

The protein molecule is capable of existing in a number of distinct states of ionization, which may be represented as $P_1, P_2, P_3, \ldots$. It will be supposed that P_1 is the form in which the protein exists in the most highly acid solutions.

[1] J. Steinhardt, *K. danske vidensk. Selsk. Math.-fys. Medd.*, **14**, No. 11 (1937).
[2] M. Levy and A. E. Benaglia, *J. biol. Chem.*, **186**, 829 (1950).
[3] P. S. Lewis, *Biochem. J.*, **20**, 965, 984 (1926).

The molecule P_1 can split off a single proton to give P_2, the charge on which is one less than that on P_1:

$$P_1 \rightleftharpoons P_2 + H^+.$$

The dissociation constant for this process will be written as K_1, so that

$$K_1 = \frac{[P_2] \cdot [H^+]}{[P_1]}. \tag{27}$$

Similarly, P_2 can split off a proton to give P_3, and so on.

If each form of the protein is capable of undergoing denaturation according to a first-order law, the reaction scheme may be represented as follows:

$$P_1 \overset{K_1}{\rightleftharpoons} P_2 \overset{K_2}{\rightleftharpoons} P_3 \overset{K_3}{\rightleftharpoons} P_4 \overset{K_4}{\rightleftharpoons} P_5$$

with downward arrows labelled k_1, k_2, k_3, k_4, k_5 etc.

$$D \quad D \quad D \quad D \quad D$$

Here D represents denatured protein. It will be clear that P_n contains one more positive charge and one more hydrogen atom than P_{n+1}, but for simplicity this is not represented explicitly.

In addition to eqn (27) there is a series of dissociation equations such as the following:

$$K_2 = \frac{[P_3] \cdot [H^+]}{[P_2]}, \qquad K_3 = \frac{[P_4] \cdot [H^+]}{[P_3]}. \tag{28}$$

From these it follows that

$$[P_2] = K_1 \frac{[P_1]}{[H^+]} \tag{29}$$

$$[P_3] = K_2 \frac{[P_2]}{[H^+]} = K_1 K_2 \frac{[P_1]}{[H^+]^2} \tag{30}$$

$$[P_4] = K_1 K_2 K_3 \frac{[P_1]}{[H^+]^3} \tag{31}$$

$$[P_5] = K_1 K_2 K_3 K_4 \frac{[P_1]}{[H^+]^4} \tag{32}$$

and so on. If $[P]_0$ is the total concentration of protein,

$$[P]_0 = [P_1] + [P_2] + [P_3] + [P_4] + \ldots \tag{33}$$

$$= [P_1]\left(1 + \frac{K_1}{[H^+]} + \frac{K_1 K_2}{[H^+]^2} + \frac{K_1 K_2 K_3}{[H^+]^3} + \ldots\right). \tag{34}$$

The net rate of deactivation of protein is given by

$$v = k_1[P_1] + k_2[P_2] + k_3[P_3] + \ldots \tag{35}$$

$$= [P_1]\left(k_1 + \frac{k_2 K_1}{[H^+]} + \frac{k_3 K_1 K_2}{[H^+]^2} + \frac{k_4 K_1 K_2 K_3}{[H^+]^3} \ldots\right). \tag{36}$$

If $\tilde{k}$ is the (pH-dependent) first-order constant for the overall denaturation process,

$$v = \tilde{k}[P]_0 \tag{37}$$

so that

$$\tilde{k} = \frac{k_1 + k_2 K_1/[H^+] + k_3 K_1 K_2/[H^+]^2 + k_4 K_1 K_2 K_3/[H^+]^3 + \ldots}{1 + K_1/[H^+] + K_1 K_2/[H^+]^2 + K_1 K_2 K_3/[H^+]^3 + \ldots}. \tag{38}$$

This relationship has been found to give a satisfactory interpretation of the effects of pH on the rate constants for a number of protein denaturations.

If the individual rate constants k_1, k_2, etc., were all equal, there would be no pH-dependence in the rate of denaturation. The physical significance of this is simply that in this case a change of pH converts protein molecules into forms that denature at the same rates. In order to explain the fact that most denaturation rates pass through a minimum as the pH is varied it is necessary to postulate that the individual k's also pass through a minimum. For example, in the case of ricin the available data suggest that the protein exists in seven different states of ionization, P_1, P_2, P_3, P_4, P_5, P_6, and P_7, and that the individual rate constants are related as follows:

$$k_1 > k_2 > k_3 < k_4 < k_5 < k_6 < k_7.$$

In order to consider the manner in which eqn (38) predicts the influence of pH on the rate of inactivation, it is convenient to discuss how the pH affects the numerator (N) and the denominator (D) of the expression. The variations of $\log N$ and $\log D$ with pH are indicated schematically in Figs. 13.3 and 13.4. At a very low pH the logarithm of N is essentially equal to $\log k_1$, but at a certain pH the second term $k_2 K_1[H^+]$ becomes important; there is therefore an inflexion point where $[H^+] = k_2 K_1/k_1$. Subsequent inflexion points will be found where $[H^+] = k_3 K_2/k_2, k_4 K_3/k_3$. In Figs. 13.3 and 13.4 the curves are shown as segments of straight lines, but the true curves will of course show a rounding-off at the inflexion points.

The logarithm of D has a value of zero at very low pH, and the inflexion points now correspond to $[H^+] = K_1, K_2$, etc.

The exact manner in which Figs. 13.3 and 13.4 are combined to give a curve of $\log v$ against pH depends upon the relative magnitude of the various constants. Figure 13.5 represents the particular reconstruction that is required to account for the data of Levy and Benaglia on ricin. Examination

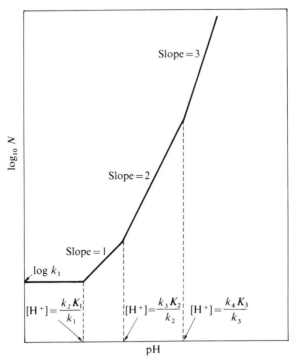

FIG. 13.3.   The variation of $\log_{10} N$ with pH; N is the numerator of the expression in eqn (38). The true curves show a rounding-off at the inflection points.

of the experimental curves, some of which are shown in Fig. 13.6, shows that as the pH is increased the slope takes on successively the following values

$$-2,\ -1,\ 0,\ 1,\ 0,\ 1,\ 2,\ 3.$$

The -2 slope found at the lowest pH's employed may be explained as due to the fact that the dominant terms in the numerator and denominator are k_1 and $K_1 K_2 [H^+]^2$ respectively. The change-over to a slope of -1 must be due to the increase in importance of $k_2 K_1 [H^+]$ with respect to k_1, and this particular inflexion point therefore corresponds to $[H^+] = k_2 K_1 / k_1$. Similarly, the next inflexion point corresponds to $[H^+] = k_3 K_2 / k_2$, and the next to $[H^+] = k_4 K_3 / k_3$. The next inflexion point, however, corresponds to a decrease in slope from 1 to 0, and this can only be explained on the basis of the change in importance of a term in the denominator: $K_1 K_2 K_3 [H^+]^3$ must here become more important than $K_1 K_2 [H^+]$, and this change occurs at $[H^+] = K_3$. The remaining inflexion points, corresponding to increases in slope, occur at $[H^+] = k_5 K_4 / k_4$, $[H^+] = k_6 K_5 / k_5$, and $[H^+] = k_7 K_6 / k_6$.

These various relationships are shown in Fig. 13.5. It is evident that at pH's lower than corresponds to $[H^+] = K_3$, the protein exists mainly as

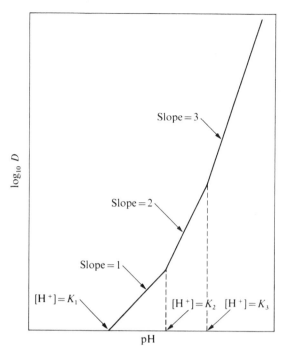

FIG. 13.4. The variation of $\log_{10} D$ with pH; D is the denominator of the expression in eqn (38).

P_3, whereas at higher ones it is mainly P_4. Dissociation constants are represented by bends that are concave to the pH axis ($\frown$), while constants of the type $k_{n+1} K_n / k_n$ are indicated by bends that are convex to this axis ($\smile$). Since there is only one bend of the former type it follows that K_1 and K_2 are too large to be within the range of the experiments, which means that pK_1 and pK_2 are both less than about unity. Similarly K_4, K_5, and K_6 are too small to be revealed by the experiments, which means that pK_4, pK_5, and pK_6 must be greater than about 12. K_4, K_5, and K_6 may correspond to guanidinium groups, but K_1 and K_2 are too large to be associated with normal groups and may be related to hydrogen-bond equilibria such as

$$A - H \cdots B + H^+ \rightleftharpoons A - H + H - B^+.$$

The value of the first-order constant $\tilde{k}$ at the minimum must correspond to k_3, since at this minimum the dominant terms in the numerator and denominator are $k_3 K_1 K_2 [H^+]^2$ and $K_1 K_2 [H^+]^2$ respectively. The quantities that can be obtained from the data are therefore the following:

$$\frac{k_2 K_1}{k_1}, \frac{k_3 K_2}{k_2}, k_3, \frac{k_4 K_3}{k_3}, K_3, \frac{k_5 K_4}{k_4}, \text{etc.}$$

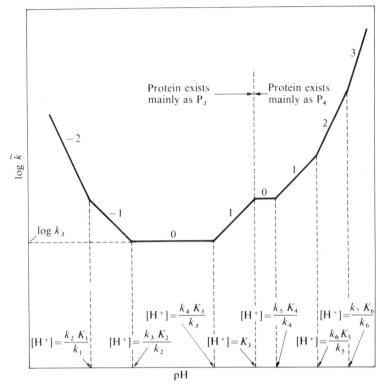

FIG. 13.5. Schematic representation of the pH-dependence of the ricin denaturation.

These can be reduced to the simpler forms:

$$\frac{k_1}{K_1 K_2}, \frac{k_2}{K_2}, k_3, k_4, K_3, k_5 K_4, k_6 K_4 K_5, \text{etc.}$$

The second of these, k_2/K_2, may be shown to correspond to the process

$$P_3 + H^+ \rightarrow P_2^{\neq} \rightarrow \text{denatured products}$$

as follows. The equilibrium between P_2 and P_3, is represented by

$$K_2 = \frac{[P_3] \cdot [H^+]}{[P_2]}, \tag{39}$$

whence

$$k_2[P_2] = \frac{k_2}{K_2}[H^+] \cdot [P_3]. \tag{40}$$

The expression on the left is simply the rate of denaturation of P_2, or the rate with which the system passes over the energy barrier corresponding to

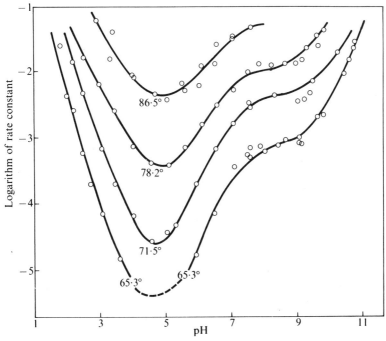

FIG. 13.6.   Logarithm of rate constant plotted against pH, for the denaturation of ricin (data of Levin and Benaglia).

the activated complex $P_2^{\neq}$. The ratio k_2/K_2 is thus simply the rate constant for the reaction between H^+ and P_3 proceeding via the $P_2^{\neq}$ complex. Other relationships of the same type are summarized in Table 13.4, and may easily be derived.

TABLE 13.4
Significance of kinetic constants for denaturation

Constant	Reaction to which it corresponds
$\dfrac{k_1}{K_1 K_2}$	$P_3 + 2H^+ \rightarrow P_1^{\neq} \rightarrow D$
$\dfrac{k_2}{K_2}$	$P_3 + H^+ \rightarrow P_2^{\neq} \rightarrow D$
k_3	$P_3 \rightarrow P_3^{\neq} \rightarrow D$
k_4	$P_4 \rightarrow P_4^{\neq} \rightarrow D$
$k_5 K_4$	$P_4 \rightarrow P_5^{\neq} + H^+ \rightarrow D$
$k_6 K_5$	$P_4 \rightarrow P_6^{\neq} + 2H^+ \rightarrow D$
$k_7 K_6$	$P_4 \rightarrow P_7^{\neq} + 3H^+ \rightarrow D$

$$\text{T ABLE } 13.5$$

The dissociation constant K_3 for ricin

Temperature (°C)	K_3	pK_3
52·9	$3\cdot3 \times 10^{-9}$	8·5
65·3	$3\cdot0 \times 10^{-8}$	7·5
71·5	$4\cdot6 \times 10^{-8}$	7·3
78·2	9×10^{-8}	7·05
86·5	1×10^{-7}	7·0

$\Delta G\,(65°C) = 11\cdot7\,\text{kcal mol}^{-1}$	$\Delta G = 0\cdot00654T^3 - 4\cdot685T^2 + 8706T$
$\Delta H\,(65°C) = 30\cdot0\,\text{kcal mol}^{-1}$	$\Delta H = -0\cdot01309T^3 + 4\cdot685T^2$
$\Delta S\,(65°C) = 54\,\text{e.u.}$	$\Delta S = -0\cdot01963T^2 + 9\cdot370T - 870\cdot6$

Ricin

Levy and Benaglia applied the theory outlined above to their data on the inactivation of ricin, and arrived at values for the various kinetic constants listed in Table 13.4, and for K_3. The values they obtained for K_3 at various temperatures are given in Table 13.5, which also gives the resulting thermodynamical quantities.

The plot of log K_3 against $1/T$ shows a very marked curvature, indicating that the thermodynamical quantities are quite strongly temperature dependent; empirical expressions for this temperature dependence are included in Table 13.5. The pK_3 values were found empirically to fit the equation

$$pK_3 = 6\cdot99 - 0\cdot00143\,(T - 358)^2, \tag{41}$$

where the temperature T is in Kelvins. Equations of this form are satisfactory for describing the dissociations of many acids[1], but the coefficients of the second term are normally very much less (by a factor of 30) than that given above. In other words, the temperature variation of K_3 is very much greater than is normally found, a fact that suggests that the dissociation $P_3 \rightleftharpoons P_4 + + H^+$ may be of an unusual type.

The kinetic constants listed in Table 13.5, when evaluated from the data at various temperatures, were found to give a straight line when their logarithms were plotted against $1/T$. The resulting heats and entropies of activation, together with the free energies at 65°C, are shown in Table 13.6. These values show some rather interesting features. In the first place, they fall into two groups, those for the first three reactions (P_3 as reactant) and those for the remainder (P_4 is the reactant). The $\Delta S^{\neq}$ values are much larger for the first group, and this is partly, but by no means completely, explained by the very large entropy increase (54 e.u.) in going from P_3 to P_4. There is still a difference of about another 50 e.u. to be accounted for, and it is there-

[1] D. H. Everett and W. F. K. Wynne-Jones, *Trans. Faraday Soc.*, **35**, 1380 (1939); H. S. Harned and B. B. Owen, *Chem. Rev.*, **25**, 31 (1939); H. S. Harned and R. A. Robinson, *Trans. Faraday Soc.*, **33**, 973 (1940).

TABLE 13.6

Heats and entropies of activation for the inactivation of ricin

Reaction	$\Delta G^{\neq}$	$\Delta H^{\neq}$	$\Delta S^{\neq}$
$P_3 + 2H^+ \rightarrow P_1^{\neq} \rightarrow D$	20·5	79	173
$P_3 + H^+ \rightarrow P_2^{\neq} \rightarrow D$	26·4	81	161
$P_3 \rightarrow P_3^{\neq} \rightarrow D$	32·3	89	169
$P_4 \rightarrow P_4^{\neq} \rightarrow D$	38·9	53	42
$P_4 \rightarrow P_5^{\neq} + H^+ \rightarrow D$	42·8	63	58
$P_4 \rightarrow P_6^{\neq} + 2H^+ \rightarrow D$	—	—	—
$P_4 \rightarrow P_7^{\neq} + 3H^+ \rightarrow D$	75·4	86	33

fore to be concluded that the three activated complexes $P_1^{\neq}$, $P_2^{\neq}$, and $P_3^{\neq}$ are richer in entropy by about 50 e.u. than are the complexes $P_4^{\neq}$, $P_5^{\neq}$, and $P_7^{\neq}$. This could hardly be due to electrostatic effects, so that presumably the first three complexes have much looser structures than the others. The heats of activation do not show any well-defined regularities.

Another feature of interest about the entropy values in Table 13.6 is that there are no marked changes within the two groups. It would thus appear that although they differ in number of protons (and therefore in total charge), $P_1^{\neq}$, $P_2^{\neq}$, and $P_3^{\neq}$ do not differ much in entropy. Such differences as there are might be due to electrostatic effects, and it would seem that structurally these three complexes are much the same. Rather greater differences are to be found between $P_4^{\neq}$, $P_5^{\neq}$, and $P_7^{\neq}$, but again no striking differences of structure are indicated.

Pepsin

As mentioned previously, Steinhardt's data on pepsin[1] showed that the rate of inactivation was, over a certain pH range, inversely proportional to the fifth power of the hydrogen ion concentration. In terms of eqn (38) this can be interpreted if the dominant term in the numerator is the sixth, and the dominant term in the denominator the first; the rate constant would then be

$$\tilde{k} = \frac{k_6 K_1 K_2 K_3 K_4 K_5}{[H^+]^5}. \tag{42}$$

Steinhardt actually found some deviations from this strict fifth-power dependence, and attributed them to the part played by additional terms in the denominator; in particular the slope of $\log \tilde{k}$ against pH became less than 5 at higher pH values (6·5–7), and this may be explained by the inclusion of the second term in the denominator of eqn (38);

$$\tilde{k} = \frac{k_6 K_1 K_2 K_3 K_4 K_5 / [H^+]^5}{1 + K_1 / [H^+]}. \tag{43}$$

[1] J. Steinhardt, *K. danske vidensk. Selsk. Math.-fys. Medd.*, **14**, No. 11 (1937).

By an analysis of his data along these lines Steinhardt was able to determine a value for K_1 and hence, from the values at different temperatures, the corresponding enthalpy and entropy for the ionization process. The results were:

$$K_1 (25°C) = 1.74 \times 10^{-7}; \qquad pK_1 (25°C) = 6.76$$

$$\Delta H = 9.04 \, \text{kcal mol}^{-1}$$

$$\Delta S = 25.4 \, \text{e.u.}$$

Apart from the entropy value these figures are consistent with the hypothesis that the ionization is that of an imidazole group in histidine. The entropy increase however is much too large to correspond to any simple ionization, and suggests that significant structural changes are involved. The importance of such structural effects has been emphasized by Eyring and Stearn[1].

Steinhardt made the assumption that all the K's, K_1 to K_5, are identical in value and in temperature dependence; from the rates in the region in which the fifth-power law is accurately obeyed it was therefore possible to determine k_6 and the related heat and entropy of activation. The heat of activation in this region is seen from eqn (42) to be $\Delta H_6^{\neq} + 5\Delta H$, and its value was found to be 63.5 kcal; the value of $\Delta H_6^{\neq}$ is therefore

$$63.5 - 5 \times 9.04 = 18.3 \, \text{kcal}.$$

Similarly, the overall entropy of activation is 135.9 e.u., so that $\Delta S_6^{\neq}$ is equal to $135.9 - 5 \times 25.4 = 8.8$ e.u. This value is a reasonable one for a unimolecular process, and there is no need to suppose that profound structural changes occur during this process. It must be emphasized, however, that such changes appear to be occurring during the ionization processes. It should be noted that the above treatment is subject to considerable uncertainty on account of the assumption that all the K's are equal.

Haemoglobin

The kinetics of the denaturation of haemoglobin have been studied by P. S. Lewis[2], who worked at high temperatures (60–70°C), and by Cubin[3], who worked at lower temperatures (18–45°C) in acid solution (pH 4.1 to 4.6). It was shown by Neurath, Greestein, Putnam, and Erickson[4] that Cubin's data indicated the rate to be proportional to the square of the hydrogen ion concentration in the pH region from 4.1 to 4.6. In terms of eqn (38) this would imply that the dominant terms are the first in the numerator and the third

[1] H. Eyring and A. E. Stearn, *Chem. Rev.*, **24**, 253 (1939).
[2] P. S. Lewis, *Biochem. J.*, **20**, 965 (1926).
[3] H. K. Cubin, *Biochem. J.*, **23**, 25 (1929).
[4] H. Neurath, J. P. Greenstein, F. W. Putnam, and J. O. Erickson, *Chem. Rev.*, **34**, 157 (1944).

in the denominator, so that the rate is

$$v = \frac{k_1[\text{H}^+]^2}{K_1 K_2}.$$ (44)

The activation energy in this pH range is $11.8\,\text{kcal mol}^{-1}$, and this corresponds to a heat of activation of $11.8 - 0.6 = 11.2\,\text{kcal mol}^{-1}$.

Equation (38) would predict that as the pH is lowered the dependence on the hydrogen ion concentration will become a linear one; the bend in the curve of $\log k$ against pH will be at $[\text{H}^+] = K_2$. In their analysis of Cubin's data, Neurath *et al.* found evidence for such a change-over, and obtained a value of 2.7 for $\text{p}K_2$ at $25°\text{C}$. From the variation of $\text{p}K_2$ with temperature they concluded that the heat of dissociation of the proton $(-\Delta H_2)$ is $5.6\,\text{kcal}$. In order to obtain the heat of activation corresponding to k_1 it is necessary to add $\Delta H_1 + \Delta H_2$ to the overall activation energy, which equals

$$\Delta H_1^{\ddagger} - \Delta H_1 - \Delta H_2.$$

If ΔH_1 and ΔH_2 are assumed to be equal the resulting $\Delta H_1^{\ddagger}$ is thus

$$11.2 - 2 \times 5.6 = 0\,\text{kcal mol}^{-1}.$$

This zero activation energy implies that the twofold ionization results directly in denaturation without any further reaction.

According to this interpretation of the reaction the slow step in the pH range 4.1–4.6 is the addition of two protons,

$$\text{P}_3 + 2\text{H}^+ \rightleftharpoons \text{P}_1 \rightarrow \text{P}_1^{\ddagger} \rightarrow \text{D}.$$

The analysis has shown that the enthalpies of P_1 and $\text{P}_1^{\ddagger}$ are the same, but that there is an increase of $11.2\,\text{kcal}$ in going from $\text{P}_3 + 2\text{H}^+$ to P_1. The result that $5.6\,\text{kcal}$ are required for each proton is of interest in view of the fact that the breaking of a hydrogen bond requires about $5\,\text{kcal}$. The addition of a proton normally is attended by only a very small heat change, so that the addition of a proton with the breaking of a hydrogen bond, viz.

$$-\text{A}-\text{H}\cdots\text{B}- + \text{H}_3\text{O}^+ \rightleftharpoons -\text{A}-\text{H} + \text{H}-\text{B}^+- + \text{H}_2\text{O}$$

might well be endothermic by about $5\,\text{kcal}$.

The data of P. S. Lewis[1] were interpreted by Levy and Benaglia as fitting the equation

$$\tilde{k} = \frac{k_1 + (k_2 K_1/[\text{H}^+]) + (k_3 K_1 K_2/[\text{H}^+]^2)}{(K_1/[\text{H}^+]) + (K_1 K_2/[\text{H}^+]^2)}$$ (45)

which is a special case of eqn (38). The type of variation of $\log \tilde{k}$ with pH predicted by this equation is shown in Fig. 13.7. The data fit such a curve satisfactorily, and the constants k_2, k_3, K_2, and k_1/K_1 can be determined.

[1] *Loc. cit.*

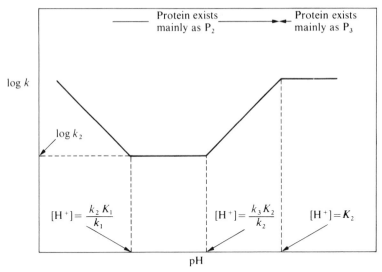

FIG. 13.7.   Variation of $\log_{10} k$ with pH for the haemoglobin denaturation.

TABLE 13.7

Results for haemoglobin denaturation

Process	Constant	Numerical value		Units
		60·5°C	68·5°C	
$P_2 \rightleftharpoons P_3 + H^+$	K_2	$4·0 \times 10^{-8}$	$3·0 \times 10^{-8}$	M
$P_2 \rightleftharpoons H^+ + P_1^{\neq} \rightarrow D$	k_1/K_1	210	2820	$M^{-1} s^{-1}$
$P_2 \rightarrow P_2^{\neq} \rightarrow D$	k_2	0	0	s^{-1}
$P_3 \rightarrow P_3^{\neq} \rightarrow D$	k_3	$4·0 \times 10^{-4}$	$5·9 \times 10^{-8}$	s^{-1}

Levy and Benaglia's values, for the two temperatures employed by Lewis, are shown in Table 13.7. The temperature coefficient associated with k_1/K_1 is extremely large, which means that the corresponding $\Delta H^{\neq}$ and $\Delta S^{\neq}$ values are large; this suggests a profound structural change either during the ionization $P_2 \rightarrow P_1$ or during the subsequent activation process ($P_1 \rightarrow P_1^{\neq}$). No such structural changes appear to be involved in the ionization $P_2 \rightarrow P_3^{\neq}$. The fact that k_2 is essentially zero is significant in that it indicates that at least one ionized form of the protein has considerable stability; a similar result has been obtained with egg albumin, as will be discussed below.

Egg albumin

The influence of pH on the denaturation of egg albumin has been studied by P. S. Lewis[1], by Cubin[2], and by Gibbs, Bier, and Nord[3]. Gibbs, Bier, and

[1] P. S. Lewis, *Biochem. J.*, **20**, 978 (1926).
[2] H. K. Cubin, *Biochem. J.*, **23**, 25 (1929).
[3] R. J. Gibbs, M. Bier, and F. F. Nord, *Archs Biochem. Biophys.*, **35**, 216 (1952).

Nord have analysed their data from the standpoint of the Levy–Benaglia treatment, and a further discussion along similar lines has been presented by Gibbs[1]. The present treatment is consistent with these two publications.

Owing to the presence of salts and to other factors Cubin's results are not comparable with those of Gibbs *et al.*, and will not be considered further. As mentioned earlier, Gibbs *et al.* found the reaction to be strictly of the first order. The data of Gibbs *et al.*, obtained in the pH range of 0·9 to 3·4 can be fitted to an equation of the form

$$k = \frac{k_1[H^+]^2}{K_1 K_2} + \frac{k_2[H^+]}{K_2}. \tag{46}$$

If this is written as

$$k = \frac{k_1 + k_2 K/[H^+]}{(K_1 K_2/[H^+]^2)} \tag{47}$$

it is seen to be a special case of eqn (38) and to correspond to the scheme

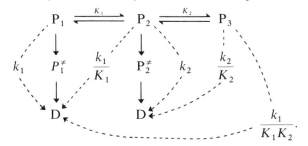

From the data it is possible to determine k_2/K_2 and $k_1/K_1 K_2$; the first of these is the rate constant for the process

$$P_3 + H^+ \longrightarrow P_2^{\neq} \longrightarrow D,$$

while the latter is for

$$P_3 + 2H^+ \longrightarrow P_1^{\neq} \longrightarrow D.$$

Lewis's work was carried out in the more alkaline region of pH 5·0 to 8·0. Towards the acid side of this region eqn (47) was obeyed, but in the alkaline region the equation followed was

$$k = \frac{k_4 K_3}{K_3 + [H^+]}, \tag{48}$$

which may be written as

$$k = \frac{k_4 K_1 K_2 K_3/[H^+]^3}{K_1 K_2/[H^+]^2 + K_1 K_2 K_3/[H^+]^3} \tag{49}$$

[1] R. J. Gibbs, *Archs Biochem. Biophys.*, **35**, 229 (1952).

and is again a special case of eqn (38). The scheme of reactions to which it corresponds is

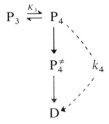

The data indicate that $k_3 = 0$, which means that the particular form P_3 of the protein is completely stable.

By an analysis of the data in this pH region, with the use of eqn (49), it is possible to determine k_4 and K_3 and, since the work was done at two temperatures, to calculate the corresponding heats and entropies.

The results of Gibbs's analysis[1] of the data of Gibbs *et al.* and those of Lewis are summarized in Table 13.8. It is at once apparent that there are

TABLE 13.8

Results for egg albumin denaturation

Process	Constant	Numerical value	ΔH or $\Delta H^{\neq}$	ΔS or $\Delta S^{\neq}$	Worker
$P_3 \rightleftharpoons P_4 + H^+$	K_3	1.97×10^{-9} (65.0°C)	71.6	113	Lewis
$P_3 + 2H^+ \rightarrow P_1^{\neq} \rightarrow D$	k_1/K_1K_2	9.23×10^5 (65.0°C)	183.2	510	Lewis
$P_3 + 2H^+ \rightarrow P_1^{\neq} \rightarrow D$	k_1/K_1K_2	1.46 (44.4°C)	50.0	99.6	Gibbs *et al.*
$P_3 + H^+ \rightarrow P_2^{\neq} \rightarrow D$	k_2/K_2	1.57×10 (65.0°C)	130.1	332	Lewis
$P_3 + H^+ \rightarrow P_2^{\neq} \rightarrow D$	k_2/K_2	1.46×10^{-2} (44.4°C)	36.7	48.5	Gibbs *et al.*
$P_4 \rightarrow P_4^{\neq} \rightarrow D$	k_4	1.90×10^{-4} (65.0°C)	71.6	113	Lewis

serious discrepancies between the results of the two investigations. The reason for this has been discussed by Gibbs. Extrapolation of the results obtained by Gibbs *et al.* to the temperature region employed by Lewis gave rates which were much lower than those obtained by Lewis; conversely, on extrapolation to the lower temperature region, Lewis's rates became much lower than those of Gibbs *et al.* The discrepancy cannot be due to salt effects, and it cannot be explained in terms of successive reactions, since when these are involved the heat of activation must decrease with an increase in temperature. Gibbs therefore concludes that there are two independent paths for the denaturation of egg albumin, paths that occur in addition to the ionization scheme already discussed. It is known that egg albumin

[1] R. J. Gibbs, *loc. cit.*

consists of three different components[1], of which A_1 is converted on ageing into A_2. If this process were accelerated by temperature the rate of denaturation measured at the higher temperatures (by Lewis) might be that of A_2, whereas the rate measured at low temperatures (by Gibbs *et al.*) is that of A_1.

Bovine plasma albumin

Results of a similar type have been obtained by Levy and Warner[2] using bovine plasma albumin. They found that the pH dependence of their rates could be explained by means of the equation

$$v = \frac{k_1[H^+]}{K_1} + k_2 + \frac{k_3 K_2}{[H^+]} + \frac{k_4 K_2 K_3}{[H^+]^2} + \frac{k_5 K_2 K_3 K_4}{[H^+]^3} \qquad (50)$$

which may be seen to arise from eqn (38) if $K_1[H^+]$ is the predominant term in the denominator. Table 13.9 shows the values of the constants obtained

TABLE 13.9

Results for bovine albumin denaturation

Process	Constant	Numerical value (46·2°C)	$\Delta H^{\neq}$	$\Delta S^{\neq}$
$P_2 + H^+ \rightarrow P_1^{\neq} \rightarrow D$	k_1/K_1	5×10^{-5}	23·2	−6·2
$P_2 \rightarrow P_2^{\neq} \rightarrow D$	k_2	5×10^{-6}	24·3	−8·6
$P_2 \rightarrow H^+ + P_3^{\neq} \rightarrow D$	$k_3 K_2$	5×10^{-9}	17·1	−42·4
$P_2 \rightarrow 2H^+ + P_4^{\neq} \rightarrow D$	$k_4 K_2 K_3$	0	—	—
$P_2 \rightarrow 3H^+ + P_5^{\neq} \rightarrow D$	$k_5 K_2 K_3 K_4$	4×10^{-14}	0·6	−124

at 46·2°C, and the corresponding entropies and heats of activation calculated from the temperature dependence of these constants. The zero value of $k_4 K_2 K_3$ implies that $k_4 = 0$, i.e. that the form P_4 is quite stable with respect to denaturation.

Since there is no sign of any levelling off of the rate even when the pH is reduced to as low as 0·8 it can be concluded that the value of pK_1 is less than 0·8. This is too low a value to correspond to any ordinary ionization, and suggests, as did the results with ricin (p. 438), that hydrogen-bond equilibria are involved.

Influence of hydrostatic pressure[3]

Relatively few investigations have been concerned with the effects of high hydrostatic pressures on proteins, and very little is known of the kinetic aspects of denaturations under such conditions. The work that has been

[1] C. F. C. MacPherson, D. H. Moore, and C. G. Longsworth, *J. biol. Chem.*, **156**, 381 (1944).
[2] M. Levy and R. C. Warner, *J. phys. Chem.*, **58**, 106 (1954).
[3] For a detailed review see F. H. Johnson, H. Eyring, and M. J. Polissar, *Kinetic Basis of Molecular Biology*, Chap. 9, John Wiley & Sons, New York (1954).

done may conveniently be divided into two main classes. In investigations of the first type extremely high pressures, of over 50 000 lb in^{-2}, are employed, and the main purpose is to see what permanent effects are induced in the protein molecules; in the case of enzymes, for example, the procedure is frequently to subject the enzyme solution to a high pressure for a definite period of time, to release the pressure, and then to test the enzymic activity. Proteins are generally denatured under such conditions, the effect being a permanent one.

As examples of investigations of this type, in all of which the protein was caused to undergo some denaturation by the action of the pressure, may be mentioned the work of Basset, Macheboeuf *et al.*[1] on lipase amylase, trypsin, and various other substances; of Matthews, Dow, and Anderson[2] on rennin and crystalline pepsin; and of Curl and Jansen[3] on pepsin, trypsin, chymotrypsin, and chymotrypsinogen. Curl and Jansen investigated the effects of pH, enzyme concentration, and other factors on the pressure inactivation, and found some evidence that the process occurs according to the scheme

$$N \rightleftharpoons R$$
$$\downarrow$$
$$D$$

where N is the native enzyme, R a reversibly inactive form, and D an irreversibly inactive form. Pressure favours the conversion of N into both R and D.

Very few kinetic studies have been made of denaturations in the high-pressure range. However, Lauffer and Dow[4] have shown that the coagulation of tobacco mosaic virus at 30°C and under pressures of between 75 000 and 150 000 lb in^{-2} proceeds as a first-order reaction. From the variation of rate with pressure a value for $\Delta V^{\neq}$ of $-100\,000$ c.c. per mole has been estimated, but this very large value would not appear to have any simple significance[5]; probably the mechanism of the reaction is entirely different under the very high pressures.

Investigations of the second type are carried out using more moderate pressures (up to 20 000 lb in^{-2}), and are more concerned with the influence of hydrostatic pressure on the denaturation process as brought about by heat or other agencies. Some of the investigations of this type have been concerned with the effects of moderate pressures on the equilibria between

[1] J. Basset and M. A. Macheboeuf, *C.r. Lebd. Séanc., Acad. Sci., Paris*, **195**, 1431 (1932); J. Basset, M. Lisbonne, and M. A. Macheboeuf, *C.r. Lebd. Séanc., Acad. Sci., Paris*, **196**, 1540 (1933); M. A. Macheboeuf, J. Basset, and G. Levy, *Annls Physiol. Physicochim. biol.*, **9**, 713 (1933); M. A. Macheboeuf and J. Basset, *Ergebn. Enzymforsch.*, **3**, 303 (1934).

[2] J. E. Matthews, R. B. Dow, and A. K. Anderson, *J. biol. Chem.*, **135**, 697 (1940).

[3] A. L. Curl and E. F. Jansen, *J. biol. Chem.*, **184**, 45 (1950); **185**, 713 (1950).

[4] M. A. Lauffer and R. B. Dow, *J. biol. Chem.*, **140**, 509 (1941).

[5] F. H. Johnson, H. Eyring, and M. J. Polissar, *The Kinetic Basis of Molecular Biology*, John Wiley & Sons, New York (1954), p. 297.

native and denatured proteins, and therefore provide information about total volume changes occurring during the denaturation process. Studies have also been made of the effect of hydrostatic pressure on the kinetics of denaturation processes.

The work of this second type has been carried out mainly by Johnson, Eyring, and their collaborators. Thus Brown, Johnson, and Marsland[1] studied the luciferase–luciferin reaction, measuring the intensity of luminescence of *P. phosphoreum* over a range of temperatures and pressures, and found that pressure brings about a reversal of the inactivation process. Eyring and Magee[2] analysed their results and arrived at the conclusion that the effect of pressure on the inactivation equilibrium is consistent with the following relationship for the volume change:

$$\Delta V = -922 \cdot 8 + 3 \cdot 206 T \, (\text{K}). \tag{51}$$

This volume change amounts to 64·6 c.c. at 35°C. There is, however, some uncertainty with regard to the significance of this result, since it is based on an assumed scheme that may not necessarily be valid. The formulation of Magee and Eyring is equivalent to the following one, which more clearly reveals the assumptions involved. The enzyme is considered to be in equilibrium between an active form N and an inactive form D,

$$\text{N} \rightleftharpoons \text{D} \quad (K_i).$$

The active enzyme N and substrate A are assumed to be in equilibrium with a complex,

$$\text{N} + \text{A} \underset{k_{-1}}{\overset{k_1}{\rightleftharpoons}} \text{NA} \quad (K_A)$$

which means that it is assumed that $k_2 \ll k_{-1}$. The complex is also in equilibrium with an inactive form DA, the equilibrium constant being assumed to be the same as for the free enzyme,

$$\text{NA} \rightleftharpoons \text{DA} \quad (K_i).$$

The total enzyme concentration is given by

$$[\text{E}]_0 = [\text{N}] + [\text{NA}] + [\text{D}] + [\text{DA}] \tag{52}$$

$$= [\text{NA}]\{1 + K_A[\text{A}] + K_i(1 + K_A[\text{A}])\} \tag{53}$$

$$= [\text{NA}](1 + K_i)(1 + K_A[\text{A}]). \tag{54}$$

The rate is therefore

$$v = k_2[\text{NA}] = \frac{k_2[\text{E}]_0 \cdot [\text{A}]}{(1 + K_i)(K_A + [\text{A}])}. \tag{55}$$

[1] D. E. S. Brown, F. H. Johnson, and D. A. Marsland, *J. cell. comp. Physiol.*, **20**, 151 (1942).
[2] H. Eyring and J. L. Magee, *J. cell. comp. Physiol.*, **20**, 169 (1942).

In the analysis of the results the effect of pressure on k_2 and K_i was considered, but not that on K_A.

It is clear that the above type of formulation implies that the heat inactivation of the enzyme is 'non-competitive' with respect to the interaction between the enzyme and substrate. Such non-competitive behaviour can only be observed provided that both of the following conditions are satisfied: (i) the heat inactivation of the enzyme is unaffected by the binding of the enzyme with the substrate, and (ii) the enzyme, substrate, and complex are at equilibrium. Unfortunately it is impossible to know whether these conditions are satisfied in this particular enzyme system.

The investigation is, however, of considerable interest since it provided the first evidence of the fact that the denaturation process can be reversed by pressure. Later it was shown, using the same system, that pressure reduces the rate of inactivation[1], so that the $\Delta V^{\neq}$ can also be positive. This has also been demonstrated in direct kinetic studies of the denaturation process using purified proteins. Thus Johnson and Campbell[2] showed that the rate of denaturation of highly purified human serum globulin is reduced by high pressure, but the results were too complex to be interpreted in terms of a $\Delta V^{\neq}$ value. The effect of pressure on the denaturation of tobacco mosaic virus has also been studied[3] and shown to be consistent with a value of 100 c.c. per mole for $\Delta V^{\neq}$.

On the whole the results indicate that very high pressures ($> 50\,000$ lb in^{-2}) increase both the speed and extent of denaturation, and that moderate pressures tend to reduce both the speed and extent of denaturation. These differences show that the denatured states of proteins are very different in the two different pressure ranges. The relationship between the structural changes in proteins and their volume changes is unfortunately a complex one, and not much can be learnt of structure from a study of the volumes. There are several factors involving volume changes through which high pressures might affect the kinetics and equilibria of protein denaturations. In the first place, ionization changes have a marked effect on volume on account of the change in electrostriction of water molecules. Secondly, the shape changes of protein molecules (e.g. from a more compact to a less compact form) may give rise to a volume change. There may, moreover, be volume changes arising from changes in structure of the solvent in the neighbourhood of the protein, or by changes in degree of ionization of the buffer system. In the case of many types of reactions Linderstrøm-Lang and Jacobsen[4] have had outstanding success in predicting volume changes in

[1] F. H. Johnson, H. Eyring, R. Steblay, H. Chaplin, C. Huber, and G. Gherardi, *J. gen. Physiol.*, **28**, 463 (1945).

[2] F. H. Johnson and D. H. Campbell, *J. cell. comp. Physiol.*, **26**, 43 (1945); *J. biol. Chem.*, **163**, 689 (1946).

[3] F. H. Johnson, M. B. Baylor, and D. Frazer, *Archs Biochem.*, **19**, 237 (1948); see M. A. Lauffer and W. C. Price, *J. biol. Chem.*, **133**, 1 (1940).

[4] K. Linderstrøm-Lang and C. F. Jacobsen, *C.r. Lab. Trav. Carlsberg, Ser. chim.*, **24**, 1 (1941).

terms of changes in degree of ionizations. With protein denaturations, however, the matter is obviously much more complex.

Influence of electrolytes

A wide variety of observations on various proteins has shown that electrolytes sometimes increase and sometimes decrease the rate of denaturation. Chick and Martin[1], for example, showed that both ammonium chloride and sodium chloride greatly reduce the rate of denaturation of egg albumin; at the same time they found that these salts reduced the temperature coefficient of the process. Lewis[2] studied the denaturation of oxyhaemoglobin and found that ammonium sulphate increased the rate of denaturation up to a concentration of about 1 M, and diminished it at higher concentrations. He attributed the depressing action of salts at the higher concentrations to their dehydrating power, showing that the order of effectiveness of different salts was the order of their ability of dehydrate. Zilva[3] found that the rate of inactivation of peroxidase was greatly retarded by sodium chloride and various other salts.

Studies of ionic strength effects have also been carried out by Steinhardt and his coworkers. In the case of pepsin, working in a pH range of about 6 to 7 Steinhardt[4] found salts to increase the rate of denaturation to a considerable degree; between ionic strengths of 0·012 and 0·10, for example, the velocity at a given pH increased by a factor of about 40. Steinhardt interpreted his results in terms of the primary and secondary salt effects of Brønsted (p. 62). By an approximate calculation from the data he was able to show that the secondary salt effect was negligible in comparison with the effect observed.

The greater part of the salt effect must therefore be a primary one. For reactions between ions of valences z_A and z_B the Brønsted theory gives rise to the limiting law

$$\log k = \log k_0 + Q z_A z_B \sqrt{u}, \tag{56}$$

where k_0 is the rate at zero ionic strength, u is the ionic strength, and Q is a constant which is numerically very close to unity in water at ordinary temperatures. Protein denaturations are not, however, of the simple ion–ion type, but are probably basically to be regarded as reactions between water molecules and groups on the protein (perhaps charged groups); in the pepsin case, for example, the elementary process has been seen to be, in this pH range,

$$PH_5 \rightleftharpoons P + 5H^+$$

[1] H. Chick and C. J. Martin, *J. Physiol.*, **43**, 1 (1911).
[2] P. S. Lewis, *Biochem. J.*, **20**, 978 (1926).
[3] S. S. Zilva, *Biochem. J.*, **8**, 656 (1914).
[4] J. Steinhardt, *K. danske vidensk. Selsk. Math.-fys. Medd.*, **14**, 1 (1937).

the reaction involving the addition of water molecules. The theory of the influence of ionic strength on reactions of this type is complicated and has not been worked out satisfactorily. The indications are, however, that there will still be a linear dependence of the logarithm of the rate constant on the square root of the ionic strength, but the exact significance of the proportionality factor is not clear.

For the pepsin system Steinhardt found a linear dependence of the logarithm of the rate constant on the square root of the ionic strength at the lower salt concentrations; at higher ones there is a falling off from linearity. The slope, however, is very great, being approximately $+13$. This large slope would appear to imply a considerable separation of opposite charges during the formation of the activated complex.

In the case of carbonyl haemoglobin Zaiser and Steinhardt[1] found that there was again a linear dependence of log k on the square of the ionic strength. The slope was 9 in this case.

Most of the results on the effects of salts on protein denaturation cannot, however, be interpreted in terms of these general salt effects, but require some explanation based on a specific interaction of the ion and the protein. Thus Crewther[2] has made a comprehensive study of the effects of a variety of ions on the thermal denaturation of trypsin, and obtained results that are too complex to be interpreted on the basis of the ionic strengths of the solutions employed. The effects of the various ions depend on a complicated manner on the pH of the solution, suggesting that the results must eventually be explained in terms of the ionizing groups concerned in the denaturation processes and the manner in which they interact with added ions. Detailed interpretation along these lines must await further knowledge of the ionizing group involved and of ion–protein interactions.

Ruegamer[3] has studied the effect of pH and of sodium and calcium ions on the reversible inactivation of coliphage, and has obtained results that are in part explicable in terms of a simple theory of specific ion–protein interactions. The results indicate that the rate of the inactivation process is proportional to the square of the hydrogen ion concentration and inversely proportional to the concentration of calcium ions. Ruegamer explains these results in terms of a competitive interaction between two hydrogen ions and single calcium ions for a critical site on the protein molecule, as follows:

$$PCa + 2H^+ \overset{K}{\rightleftharpoons} PH_2 + Ca^{++}.$$

If it is mainly the PH_2 form that undergoes inactivation, the rate is

$$v = k[PH_2] \tag{57}$$

$$= kK \frac{[PCa] \cdot [H^+]^2}{[Ca^{++}]}. \tag{58}$$

[1] E. M. Zaiser and J. Steinhardt, *J. Am. chem. Soc.*, **73**, 5568 (1951).
[2] W. G. Crewther, *Aust. J. biol. Sci.*, **6**, 597 (1953).
[3] W. R. Ruegamer, *Archs Biochem. Biophys.*, **50**, 269 (1954).

This mechanism satisfactorily explains the experimental results for calcium. Unfortunately the equivalent theory for sodium ions failed to interpret the results, according to which the rate was inversely proportional to the square root of the concentration of these ions.

Kinetics of urea denaturation

The fact that urea brings about the denaturation of proteins, including the inactivation of certain enzymes, has been known at least since the beginning of the century. In 1930 Hopkins[1] studied the urea denaturation of egg albumin and noted several interesting features, some of which are discussed later.

A number of kinetic investigations have been carried out on urea denaturation, particularly with tobacco mosaic virus[2], β-lactoglobulin[3], and ovalbumin[4]. Certain other proteins have been shown to be unaffected by urea.

Several different experimental methods have been employed in the investigations of the kinetics of denaturation by urea. Perhaps the commonest procedure has been to employ a precipitation procedure, the denatured protein usually being less soluble than the native. In other cases the reaction is followed by mixing the solution under investigation with a buffer solution that is so prepared that it induces immediate precipitation. A second general method of following denaturation involves measuring the optical rotation of the solution[5], which frequently changes very considerably on denaturation. In other investigations the change in viscosity has been followed[6].

Order of reaction with respect to protein

The question of the kinetic orders of denaturation processes has been discussed earlier, and a summary of results was given in Table 13.1 (p. 416), which included some information for urea denaturations. Only a very brief account of the main investigations will be included here.

Lauffer[7] studied the urea denaturation of tobacco mosaic virus and found that, in a given run, the first-order law was obeyed satisfactorily; the reaction is therefore of the first order with respect to time. The first-order constants obtained in individual runs, however, were found to vary with the initial protein concentration; they decreased as the protein concentration was increased. The order with respect to concentration is therefore less than unity. Lauffer found empirically that the rate is related to the initial protein concentration $[N]_0$ according to an equation of the same form as eqn (10)

[1] F. G. Hopkins, *Nature*, **126**, 328, 383 (1930).
[2] M. A. Lauffer, *J. Am. chem. Soc.*, **65**, 1793 (1943).
[3] L. K. Christensen, *C.r. Lab. Trav. Carlsberg, Ser. chim.*, **28**, 39 (1952).
[4] R. B. Simpson and W. J. Kauzmann, *J. Am. chem. Soc.*, **75**, 5139 (1953).
[5] R. B. Simpson and W. J. Kauzmann, *loc. cit.*
[6] H. K. Frensdorff, M. T. Watson, and W. J. Kauzmann, *J. Am. chem. Soc.*, **75**, 5167 (1953).
[7] M. A. Lauffer, *J. Am. chem. Soc.*, **65**, 1793 (1943).

on p. 427. He explained this result according to a mechanism, discussed earlier, which involves the interaction between the protein and an inhibitor.

Working with ovalbumin, Simpson and Kauzmann[1] found that the kinetics of the urea denaturation showed different features from those found by Lauffer with tobacco mosaic virus. In a given run the native protein did not decay according to a simple first-order law, but as reaction proceeded the rate slowed down to a greater extent than this law required. From the curves of concentration against time it was possible to determine half-lives, and these were found to be independent of the initial protein concentration. This result is an indication (although not an entirely certain one) of the fact that the denaturation is of the first order with respect to concentration. There is therefore a sharp contrast between these results and those obtained with tobacco mosaic virus.

The fact that the kinetics are first order with respect to concentration excludes the possibility that (as in Lauffer's hypothesis) the rate is affected by an impurity, since such an explanation requires that the reaction will be of a higher order than the first. Simpson and Kauzmann investigated the possibility that the reaction is a simple first-order one and that, in order to account for the falling-off of the rate constants with time, there is some inhibition by products of the denaturation process. They tested this by adding denatured ovalbumin to the reaction system, but found no effect; the possibility is therefore ruled out. The only remaining explanation seems to be that the denaturation is not a simple one but is a complex of simultaneous or consecutive reactions, each one of which must be of the first order in order to account for the independence of the half-life on the concentration.

Two alternative reaction schemes of this type were considered by Simpson and Kauzmann. According to the first, the native protein consists of a number of components N_1, N_2, etc., which undergo denaturation simultaneously:

$$N_1 \xrightarrow{k_1} D$$

$$N_2 \xrightarrow{k_2} D$$

$$N_3 \xrightarrow{k_3} D.$$

According to the second scheme, there is only one type of native protein, but this undergoes denaturation by a series of consecutive steps:

$$N \xrightarrow{k_1} D_1 \xrightarrow{k_2} D_2 \xrightarrow{k_3} D_3 \rightarrow \text{etc.}$$

Experimentally it is possible to distinguish between these alternatives by allowing a protein–urea solution to become partly denatured, recovering the undenatured protein, and again carrying out a study of its denaturation. If the first scheme were the correct one the recovered protein would be

[1] R. B. Simpson and W. J. Kauzmann, *loc. cit.*

different from the original, since those components that denatured most rapidly would be present in smaller amounts than the others; the recovered protein would therefore denature less rapidly than the original. According to the second scheme, however, the recovered protein would have the same characteristics as the original, and would denature at the same rate.

The results of the test were that the recovered protein denatured at essentially the same rate as the original (although some complications were found), so that the second scheme was supported. In order to account for the deviations from first-order kinetics (with respect to time) it is necessary to suppose that the optical rotations of the various proteins N, D_1, D_2, etc., are different; it can then be shown that the rotation at time t is given by an equation of the form of eqn (2) on p. 425. Simpson and Kauzmann actually found that they could fit their rate curves using three exponential terms. It must, however, be emphasized that, as discussed on pp. 423–426, various schemes will give an equation of the form of eqn (2).

Influence of urea concentration

A number of investigations, with a variety of proteins, have shown that the rate of denaturation varies with a very high power of the urea concentration. This has been shown with tobacco mosaic virus[1], fibrinogen[2], erythrocyte catalase[3], β-lactoglobulin[4], and ovalbumin[5]. With the last protein, for example, the rate was found to vary under some conditions with the fifteenth power of the urea concentration. In all cases the order with respect to urea varies somewhat with the conditions, and in particular is found to decrease as the temperature is raised. This may be due to the fact that normal heat denaturation becomes more important, in comparison with urea denaturation, at the higher temperatures.

Various explanations have been offered for the high dependence of rates on urea concentration, but each can be shown to be equivalent to one or the other of two distinctly different theories. The basic difference between the two types of explanation is related to the question of whether the urea interacts with the protein by long-range forces or short-range forces. In the latter case one can speak of specific complexes as being formed between protein and urea, the urea becoming attached to the protein at one or more sites, the total number of which is a definite quantity: there is therefore a definite upper limit to the number of urea molecules that can become attached to a protein molecule. When long-range forces between molecules are involved, on the other hand, it is no longer permissible to think in terms of stoichiometric protein–urea complexes; instead the interaction should be regarded as a solvent effect.

[1] M. A. Lauffer, *J. Am. chem. Soc.*, **65**, 1793 (1943).
[2] E. Mihalyi, *Acta chem., scand.*, **4**, 317 (1950).
[3] H. F. Deutsch, *Acta chem., scand.*, **5**, 1074 (1951).
[4] L. K. Christensen, *C.r. Lab. Trav. Carlsberg, Ser. chim.*, **28**, 39 (1952).
[5] R. B. Simpson and W. J. Kauzmann, *J. Am. chem. Soc.*, **75**, 5139 (1953).

In the case of the urea denaturation of ovalbumin Simpson and Kauzmann[1] discuss these two possibilities in considerable detail, and show that the solvent (long-range) theory can be excluded. Briefly, their main argument is as follows. The only forces in systems such as the ones under consideration that can act over large distances are the electrostatic forces of ions, dipoles and multipoles, and the interaction of a non-conducting solvent such as urea–water mixtures with the fields of such species is best described in terms of the dielectric constant of the solvent. Effects due to long-range forces can therefore be limited to those associated with changes in the dielectric constant of the solvent. If the effect of urea were due to its effect on long-range interactions it would follow that other substances that change the dielectric constant in the same way would have similar effects. This, however, is not the case. Glycine, for example, which increases the dielectric constant of water to an even greater extent than does urea would, on this theory, be expected to be a better denaturant than urea, but it is not. It therefore appears that urea exerts its denaturing action through short-range forces and that it forms specific complexes at definite site on the protein molecule[2].

This being the case, the influence of urea on protein denaturations may be represented by the following general mechanism:

$$N \xrightarrow{k_0} D$$

$$N + U \underset{}{\overset{K_1}{\rightleftharpoons}} NU \xrightarrow{k_1} D$$

$$N + 2U \underset{}{\overset{K_2}{\rightleftharpoons}} NU_2 \xrightarrow{k_2} D$$

$$\cdots\cdots\cdots\cdots\cdots\cdots\cdots\cdots\cdots\cdots\cdots$$

$$N + nU \underset{}{\overset{K_n}{\rightleftharpoons}} NU_n \xrightarrow{k_n} D$$

$$\cdots\cdots\cdots\cdots\cdots\cdots\cdots\cdots\cdots\cdots\cdots$$

$$N + sU \underset{}{\overset{K_s}{\rightleftharpoons}} NU_s \xrightarrow{k_s} D.$$

Here N stands for native protein, D for denatured protein, and NU_n for the complex formed between N and n urea molecules. The constant K_n is for the equilibrium $N + nU \rightleftharpoons NU_n$, and k_n is the first-order rate constant for the denaturation of the complex NU_n. The maximum number of urea molecules that can be attached to a simple protein molecule is written as s. If the protein becomes denatured in a series of steps, as on p. 452, the above mechanism must be regarded as applying to each step.

[1] R. B. Simpson and W. J. Kauzmann, *J. Am. chem. Soc.*, **75**, 5139 (1953).
[2] C. Tanford (see especially *Adv. Protein Chem.*, **24**, 1 (1970)) has emphasized that it is not necessary to conclude that the urea molecules become attached to groups that are exposed in the surface of the native protein; it is more realistic to suppose that the binding sites for urea, etc., appear after unfolding of the native structure has led to the exposure of previously buried parts of the molecule.

The total concentration of native protein is given by

$$[N]_0 = [N] + [NU] + [NU_2] + \ldots + [NU_n] + \ldots + [NU_s] \tag{59}$$

$$= [N](1 + K_1[U] + K_2[U]^2 + \ldots + K_n[U]^n + \ldots + K_s[U]^s). \tag{60}$$

The net rate of denaturation is given by

$$v = k_0[N] + k_1[NU] + k_2[NU_2] + \ldots + k_n[NU_n] + \ldots + k_s[NU_s] \tag{61}$$

$$= [N](k_0 + k_1 K_1[U] + k_2 K_2[U]^2 + \ldots + k_n K_n[U]^n + \ldots +$$
$$+ k_s K_s[U]^s). \tag{62}$$

The first-order rate constant, equal to $v/[N]_0$, is therefore

$$\bar{k} = \frac{k_0 + k_1 K_1[U] + k_2 K_2[U]^2 + \ldots + k_n K_n[U]^n + \ldots + k_s K_s[U]^s}{1 + K_1[U] + K_2[U]^2 + \ldots + K_n[U]^n + \ldots + K_s[U]^s}. \tag{63}$$

This may be written as

$$\bar{k} = \frac{\sum_{n=0}^{s} k_n K_n[U]^n}{\sum_{n=0}^{s} K_n[U]^n}. \tag{64}$$

This equation is considerably more complex than the simple power law

$$\bar{k} = \bar{k}_0[U]^{\bar{n}} \tag{65}$$

which is usually used in the analysis of the experimental data; here $\bar{n}$ has been written for the power obtained empirically from the data. If in eqn (63) one term in the numerator, and one term in the denominator, predominate, the simple law (65) will of course be obeyed; for example, a fifteenth-power law would be obtained if the sixteenth term ($k_{15} K_{15}[U]^{15}$) in the numerator of eqn (63) were predominant, and the first term in the denominator (unity) were much larger than the others. If single terms do not predominate in this way eqn (65) must be regarded as an approximation and the power $\bar{n}$ as a mean value. The exact significance of $\bar{n}$ under these circumstances is revealed by the following treatment.

Equation (65) gives rise to

$$\bar{n} = \frac{\mathrm{d} \ln \bar{k}}{\mathrm{d} \ln [U]} \tag{66}$$

and eqn (64) to

$$\ln \bar{k} = \ln \sum_{n=0}^{s} k_n K_n[U]^n - \ln \sum_{n=0}^{s} K_n[U]^n. \tag{67}$$

Differentiation with respect to [U] gives

$$\frac{\mathrm{d}\ln \bar{k}}{\mathrm{d}[U]} = \frac{\sum\limits_{n=0}^{s} nk_nK_n[U]^{n-1}}{\sum\limits_{n=0}^{s} k_nK_n[U]^n} - \frac{\sum\limits_{n=0}^{s} nK_n[U]^{n-1}}{\sum\limits_{n=0}^{s} K_n[U]^n} \tag{68}$$

so that

$$\bar{n} = \frac{\mathrm{d}\ln \bar{k}}{\mathrm{d}\ln [U]} = \frac{\sum\limits_{n=0}^{s} nk_nK_n[U]^n}{\sum\limits_{n=0}^{s} k_nK_n[U]^n} - \frac{\sum\limits_{n=0}^{s} nK_n[U]^n}{\sum\limits_{n=0}^{s} K_n[U]^n}. \tag{69}$$

The second term on the right-hand side of this expression is simply the average value of n, i.e. the average number of urea molecules attached to a native protein molecule; this quantity may be written as $\bar{n}_p$. The first term on the right-hand side of eqn (69), since it involved k_n, is simply the average number, $\overline{n^{\neq}}$, of urea molecules that are attached to protein molecules that are in the activated state between the native state and the denatured one. This may be readily seen from the fact that k_n is proportional to $K_n^{\neq}$, the equilibrium constant between native protein molecules and those in the activated state.

Equation (69) may therefore be written as

$$\bar{n} = \overline{n^{\neq}} - \bar{n}_p. \tag{70}$$

Expressed in words, the mean order of the reaction with respect to urea is the difference between the mean number of urea molecules bound in the activated state and the mean number bound in the initial native state.

Influence of pH

Very little work has been done on the influence of pH on the rates of urea denaturations. Simpson and Kauzmann[1] investigated the matter over a limited pH range, for the urea denaturation of ovalbumin, and found that the rate was independent of pH in the range from 7 to 9. This is to be contrasted with the behaviour in the case of heat denaturation, for which Lewis[2] found that the pH range over which the rate is constant is much narrower.

Quantitative theoretical treatments of pH effects in urea denaturations must await further experimental studies in this field.

Influence of temperature

Whereas the rates of heat denaturations generally show very large temperature coefficients and correspondingly large energies of activation (see

[1] R. B. Simpson and W. J. Kauzmann, *J. Am. chem. Soc.*, **75**, 5139 (1953).
[2] P. S. Lewis, *Biochem. J.*, **20**, 978 (1926).

Table 13.2), the rates of urea denaturations are usually associated with very small or even negative temperature coefficients. The first clear-cut indication of this was obtained by Hopkins[1], who worked with ovalbumin in 7M urea at pH 6 and followed the denaturation by a solubility method; he found that the rate diminished as the temperature was raised from 0° to 37°C. Working under similar conditions, but following the denaturation by the change in optical rotation, Simpson and Kauzmann found that the rate passes through a minimum at about 20°C; their plot of the results is shown in Fig. 13.8.

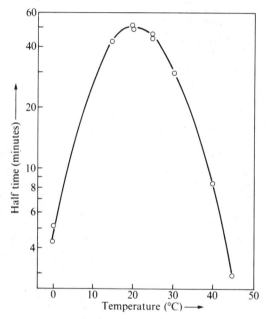

FIG. 13.8. Plot of half-life against temperature, for the urea denaturation of ovalbumin (Simpson and Kauzmann, *J. Am. chem. Soc.*, **75**, 5139 (1953)).

A minimum in the rate against temperature curve was also obtained by Lauffer[2] for the urea denaturation of tobacco mosaic virus.

Other studies that have revealed low or negative temperature coefficients have been carried out with diphtheria anti-toxin[3], erythrocyte catalase[4], and β-lactoglobulin[5].

Various theoretical treatments of temperature coefficients in urea denaturations have been presented: most of them were based on the original postulate

[1] F. G. Hopkins, *Nature*, **126**, 328, 383 (1930).
[2] M. A. Lauffer, *J. Am. chem. Soc.*, **65**, 1793 (1943).
[3] G. G. Wright and V. Schomaker, *J. Am. chem. Soc.*, **70**, 356 (1948).
[4] H. F. Deutsch, *Acta chem., scand.*, **4**, 317 (1950).
[5] C. F. Jacobsen and L. K. Christensen, *Nature*, **161**, 30 (1948); L. K. Christensen, *C.r. Lab. Trav. Carlsberg, Ser. chim.*, **28**, 37 (1952).

of Hopkins[1] that a complex was formed between the protein and urea, this complex being dissociated when the temperature was raised. Lauffer suggested the existence of a series of protein–urea complexes, NU, NU_2, etc., and proposed that the rate of denaturation steadily increases by a constant increment as each urea molecule is added, and that the heat of activation decreases by a constant increment; this, of course, is equivalent to saying that the rate increases by a constant factor as each urea molecule is added. An equivalent theory has been developed by Simpson and Kauzmann, who applied it to give a quantitative interpretation of their results for the ovalbumin denaturation.

The ideas of Hopkins, Lauffer, Simpson, and Kauzmann may be formulated as follows with reference to the scheme of reactions on p. 454. The total number of urea molecules that may be adsorbed and have an effect on the rate of denaturation is s. It is assumed that the association constant for a urea molecule is the same for each of the s sites; if the value of this association constant for a single site is K, the association constant K_n for the adsorption of n urea molecules is given by

$$K_n = {}_sC_n K^n \tag{71}$$

where ${}_sC_n$, equal to $s!/n!(s-n)!$, is the number of ways of choosing n sites from the total number of s. The assumption that the rate constant is increased by a constant factor for each addition of a urea molecule is expressed by the equation

$$k_n = a^n k_0 \tag{72}$$

where a is the constant factor.

The net rate constant for the denaturation is given by eqn (64), and with eqns (71) and (72) this becomes

$$\bar{k} = \frac{k_0 \sum\limits_{n=0}^{s} {}_sC_n a^n K^n [\mathrm{U}]^n}{\sum\limits_{n=0}^{s} {}_sC_n K^n [\mathrm{U}]^n}. \tag{73}$$

Both summations are simply binomial expansions, and the equation becomes

$$\bar{k} = \frac{k_0(1 + aK[\mathrm{U}])^s}{(1 + K[\mathrm{U}])^s}. \tag{74}$$

Taking logarithms,

$$\ln \bar{k} = \ln k_0 + s \ln (1 + aK[\mathrm{U}]) - s \ln (1 + K[\mathrm{U}]) \tag{75}$$

[1] F. G. Hopkins, *Nature*, 126, 328, 383 (1930).

whence

$$\frac{d \ln \bar{k}}{d\,[U]} = \frac{sK(a-1)}{(1 + aK[U])(1 + K[U])} \tag{76}$$

and

$$\bar{n} = \frac{d \ln \bar{k}}{d \ln [U]} = \frac{sK(a-1)[U]}{(1 + aK[U])(1 + K[U])}. \tag{77}$$

The apparent heat of activation of the reaction is given by

$$\Delta H^{\neq} = RT^2 \left(\frac{\partial \ln \bar{k}}{\partial T}\right)_u \tag{78}$$

and application of this to eqn (77) gives

$$\frac{\Delta H^{\neq}}{RT^2} = \frac{\partial \ln k_0}{\partial T} + \frac{s[U]\left(a\dfrac{\partial K}{\partial T} + K\dfrac{\partial a}{\partial T}\right)}{1 + aK[U]} - \frac{s[U]\dfrac{\partial K}{\partial T}}{1 + k[U]} \tag{79}$$

$$= \frac{\partial \ln k_0}{\partial T} + \frac{saK[U]}{1 + aK[U]}\left(\frac{\partial \ln K}{\partial T} + \frac{\partial \ln a}{\partial T}\right) - \frac{sK[U]}{1 + K[U]}\frac{\partial \ln K}{\partial T}. \tag{80}$$

If k_0, K, and a obey the laws

$$\frac{\partial \ln k_0}{\partial T} = \frac{H_0^{\neq}}{RT^2} \tag{81}$$

$$\frac{\partial \ln K}{\partial T} = \frac{H}{RT^2} \tag{82}$$

and

$$\frac{\partial \ln a}{\partial T} = \frac{h}{RT^2} \tag{83}$$

where $\Delta H^{\neq}$, H, and h are temperature-independent, eqn (80) gives rise to

$$\Delta \bar{H} = \Delta H_0^{\neq} - \frac{saK[U]}{1 + aK[U]}(H + h) + \frac{sK[U]}{1 + K[U]}H. \tag{84}$$

The significance of $\Delta H_0^{\neq}$ is that it is the heat of activation for denaturation in the absence of urea. H is the heat of adsorption of a urea molecule on the protein, and h is the reduction in the heat of activation for denaturation when a single additional urea molecule is adsorbed.

Equation (84) may be put into a simpler form by relating the functions containing [U] to the quantities $\bar{n}^{\neq}$ and $\bar{n}_p$, where $\bar{n}_p$ is the mean number of urea molecules attached to the protein in its initial state, and $\overline{n^{\neq}}$ the mean number attached in the activated state. The number $\bar{n}_p$, is given by

$$\bar{n}_p = \frac{\sum\limits_{n=0}^{s} n K_n [\text{U}]^n}{\sum\limits_{n=0}^{s} K_n [\text{U}]^n} \tag{85}$$

$$= \frac{\sum\limits_{n=0}^{s} n_s C_n K^n [\text{U}]^n}{\sum\limits_{n=0}^{s} {}_s C_n K^n [\text{U}]^n} \tag{86}$$

$$= \frac{s K [\text{U}] \sum\limits_{n=0}^{s} {}_{s-1} C_n K^n [\text{U}]^n}{\sum\limits_{n=0}^{s} {}_s C_n K^n [\text{U}]^n} \tag{87}$$

$$= \frac{s K [\text{U}] (1 + K[\text{U}])^{s-1}}{(1 + K[\text{U}])^s} \tag{88}$$

$$= \frac{s K [\text{U}]}{1 + K[\text{U}]}. \tag{89}$$

Similarly, the mean number of urea molecules attached to each protein molecule in the activated state, $\overline{n^{\neq}}$, is given by

$$\overline{n^{\neq}} = \frac{\sum\limits_{n=0}^{s} n k_n K_n [\text{U}]^n}{\sum\limits_{n=0}^{s} k_n K_n [\text{U}]^n} \tag{90}$$

$$= \frac{s a K [\text{U}]}{1 + a K[\text{U}]}. \tag{91}$$

Equation (84) may therefore be written as

$$\Delta H^{\neq} = \Delta H_0^{\neq} - \overline{n^{\neq}}(H + h) + \bar{n}_p H \tag{92}$$

$$= \Delta H_0^{\neq} - \bar{n}(H + h) - \bar{n}_i h \tag{93}$$

using eqn (70).

Author Index

Abeles, R. H., 250
Adams, M. J., 10, 34, 350
Agarival, S. J., 319
Agin, D., 328
Ainsley, G. R., 361
Alberty, R. A., 83–4, 117, 121, 134, 137–8, 149, 183, 344
Albu-Weissenberg, M., 401
Alden, R. A., 10, 11, 322
Aldridge, W. N., 203
Allen, S. H. G., 141
Allewell, N. M., 298
Altman, C., 80, 119, 122–3, 134
Ambrose, J. F., 31
Ammon, R., 174–5
Anderson, A. K., 227, 446
Anderson, B. M., 311
Anderson, D. G., 299
Anderson, L., 141
Anderson, S. R., 346
Anfinsen, C. B., 6, 19, 21, 34, 297–8, 304
Anthony, R. S., 71
Appleman, M. M., 381
Applewhite, T. H., 111, 171, 313
Armstrong, H. E., 180
Arnone, A., 10
Arrhenius, S., 46, 180–1, 198, 199(fig.), 201, 209, 211–3, 243, 243(fig.), 244, 415, 418, 421, 429
Ashmore, P. G., 59
Atassi, M. Z., 21
Atkinson, D. E., 362–3
Atlas, D., 261
Auchet, J.-C., 386
Axen, R., 387

Bach, A., 177
Bader, R. F. W., 246
Bailey, K., 413, 419
Baines, M. J., 334
Bair, J. B., 334
Baker, B. R., 333(fig.), 338
Baker, R. H., 139, 344
Baldescheieler, J. D., 67
Balinsky, D., 141
Balls, A. K., 25–6, 31, 204, 330
Bamann, E., 169, 174–5
Bamford, C. H., 47
Barel, O. A., 262
Bar-Eli, A., 401–402
Barker, H. A., 133, 266, 268
Barman, T. E., 67, 203, 313
Barnard, E. A., 33, 298, 301, 306
Barnard, M. L., 33, 153, 165(fig.), 216, 218, 326
Barrer, R. M., 388
Barton, D. H. R., 6

Basset, J., 446
Batt, C. W., 311
Baumann, J. B., 31
Baumann, W. K., 261
Baxendale, J. H., 256, 210
Bay, V. W., 10
Baylor, M. B., 448
Beardell, A. J., 209, 230
Bechet, J. J., 153, 332
Beeley, J. G., 330
Begg, C., 7
Bell, R. P., 237, 243
Belleau, B., 246
Bello, J., 10
Bellow, J., 298
Benaglia, A. E., 419, 431, 433, 437–8, 441–3
Bender, M. L., 27, 33–4, 64, 67, 84, 86, 153, 204–208, 246, 251, 251(fig.), 270, 286, 309–10, 312–17, 319–20, 323–4 (fig.), 328, 334–6, 338
Benoiton, N. L., 326, 338, 340
Bentham, J., 229–30, 230
Bentley, R., 264, 268
Berezin, I. V., 326–7
Berger, A., 261, 261(fig.)
Berger, I., 325
Berghauser, J., 34
Bergman, R., 387
Bergmann, F., 84
Bergmann, M., 325, 338
Bernhard, S. A., 8, 33, 206, 288, 310, 312, 316, 333, 335, 340, 380–81
Bethell, M. R., 372, 376
Bethge, P. H., 10, 262
Beytia, E., 141
Bier, C. J., 10
Biellmann, J.-F., 279
Bier, M., 418, 422, 442
Bigeleisen, J. W., 237
Biltonen, R., 316
Birktoft, J. J., 10, 20, 33, 284, 310, 321
Bittman, R., 377
Bizzazaro, S. A., 261
Bjerrum, N., 50
Blair, T. T., 319
Blake, C. C. F., 8, 10, 288
Bliss, C. I., 76
Bloomfield, V., 117, 134, 138, 149
Blout, E. R., 13, 329
Blow, D. M., 10, 20, 33, 272, 284, 308, 310, 321, 323(fig.), 329
Blum, J. J., 71, 216
Bock, R. M., 83
Bodlaender, P., 331
Bolotina, I. A., 347

Bonnischen, R. K., 32, 209, 210, 256
Booth, N., 419
Bordner, L. F., 140, 344
Botts, J., 119, 209
Bourquelot, E., 83
Boxbaum, S. N., 310
Boyer, P. D., 134, 139–40, 294, 302, 341, 343
Bradshaw, R. A., 32–4
Brand, L., 363
Brandt, K. G., 312, 316, 318
Branson, H. R., 14
Brandts, J. F., 63
Bray, R. C., 67
Bredig, G., 210
Breecher, A. S., 204
Bridger, W. A., 139, 141, 363
Briggs, D. R., 26, 31
Briggs, G. E., 70, 72–3
Brock, H. T., 321
Brocklehurst, K., 27, 206, 405
Brønsted, J. N., 50, 61, 62, 282, 449
Brown, A., 291
Brown, A. J., 68–70, 72, 74
Brown, C. G., 282–3
Brown, D. E. S., 228, 447
Brown, D. M., 295
Brown, J. F., 282
Brown, J. R., 336
Brubacher, L. J., 34, 206
Bruice, T. C., 291
Bryan, W. P., 339
Buchanan, J., 225
Buchert, A. R., 33
Buckman, T., 12
Budenstein, Z., 269
Buening, K., 385
Bunting, P. S., 181, 401
Bunton, C. A., 237
Burris, C. T., 223, 225
Burk, D., 75, 76, 76(fig.), 88, 89(fig.), 93–4, 96–7, 116, 121, 210, 356, 357(fig.)
Bussell, J. B., 374–5
Bustamante, Y., 340
Butler, J. A. V., 207, 215

Cabib, E., 200, 201, 201(fig.), 205
Caldin, E. F., 244
Campbell, D. H., 448
Canady, W. J., 64, 153, 316–17, 326–7
Canfield, R. E., 6, 21
Cantwell, A. M., 349
Cantwell, R., 345, 347
Caplow, M., 320
Cardinale, G. J., 250

Carleson, K., 33
Carnelly, H. C., 419
Carpenter, F. H., 338
Carriøuolo, J., 310
Carter, H. E., 260
Carver, I. P., 13
Cascales, M., 141
Casey, E. J., 215, 256, 418, 420, 421–2(fig.), 428, 430
Castanede-Agullo, M., 334, 340
Castellani, B. A., 306
Castillo, F., 34, 350
Cathou, R. A., 306
Cennamo, C., 361
Chan, M. S., 367
Chance, B., 38, 71, 114, 123, 209, 210, 256, 344–5
Chang, T., 27
Chang, T. M. S., 383, 385
Changeux, J.-P., 364, 365(fig.), 368, 372, 386, 374–5
Chaplin, H., 448
Chappelet, D., 31
Charney, E., 312
Chase, A. M., 419, 426
Chen, D. T. Y., 209, 223
Chevalier, J., 332
Chiba, H., 32, 34
Chibata, I., 402
Chick, H., 418–19, 449
Chipman, D. M., 293
Chodat, R., 177
Chow, R. B., 29
Christensen, L. K., 419, 451, 453, 457
Clarke, E., 264
Clarke, L. C., 65
Cleland, W. W., 76, 96, 114, 118–19, 123, 130, 134–8, 139–41, 362
Clement, G. E., 153, 251, 309, 312
Coates, E., 328
Cohen, J. A., 29
Cohen, J. S., 12, 297, 299
Cohen, S. G., 325–8
Cohen, W., 28, 33, 327, 330–1, 340–1
Cohn, M., 67, 264, 269
Colowick, S. P., 268–9, 279, 287
Conn, E. E., 279
Conway, A., 368, 380
Cook, R. A., 140, 379, 379(fig.)
Cooke, J. P., 298
Cookson, R. C., 6
Cooper, A. G., 310
Coppola, J. C., 10
Cordes, E. H., 1, 311
Corey, R. B., 14, 17
Cori, C. F., 32
Cori, G. T., 32, 269
Cornish-Bowden, A. J., 368
Cotton, F. A., 10
Cox, R. H., 346
Crestfield, A. M., 33, 297
Crewther, W. G., 450

Criddle, R. S., 342, 345, 347–8
Crook, E. M., 206, 384, 387, 401, 403, 405, 408–409
Crossby, J., 326
Cubin, H. K., 418, 440–2
Cullen, G. E., 73
Cuppett, C. C., 64, 327
Curl, A. L., 227, 446
Czerlinski, G. H., 347
Czopak, H., 34

D'Albis, A., 332
Dahlquish, F. W., 294
Dalziel, K., 117, 136–8, 139–40
Darling, J. J., 140, 344
Darvey, I. G., 182
Davidsohn, H., 143–4, 146
Davies, D. R., 8
Davies, J. T., 384
Davis, B. J., 5
Davis, R. P., 65
Deavin, A., 299, 303
Debye, P., 50, 52–3, 56–8, 62
Decroly, P., 162
Degani, Y., 6
Delaage, M., 329, 332
del Castillo, L. M., 334, 340
DeLuca, M., 384
Denburg, J., 384
Deneault, J., 338
Dennis, A. W., 141
Dennis, D., 349
Dennis, E. A., 299
DeSa, R. J., 312
de Smedty, J., 50
Desneuelles, P., 307
Desnuelle, P., 329
Determan, H., 2
Deutsch, H. F., 453, 457
de Vijlder, J. J. M., 380
Dickerson, R. E., 1, 10, 18, 262
Dickinson, F. M., 140
Dicks, M., 419
DiMoia, F., 418, 422
Dintzis, H. M., 385
Di Sabato, G., 13, 34, 346
Dixon, M., 97
Dixon, G. H., 33
Dixon, J. W., 316, 319
Dixon, M., 1, 157
Dobson, J. E., 331
Doherty, D. G., 26, 31, 268
Dolin, M. I., 141
Domagk, G. E., 34
Domschke, W., 34
Dorsey, J. K., 141
Doudoroff, M., 133, 266–9
Douglas, J., 139
Douzou, P., 67
Dow, R. B., 227, 446
Drenth, J., 10
Dreyer, W. J., 316
Driscoll, G. A., 34, 350
Duggan, E. L., 162
Durschlag, H., 378
Dutler, H., 261

Eadie, G. S., 76, 77(fig.), 89
East, E. J., 331, 341
Eckfeldt, J., 375
Eckstein, F., 302
Edman, P., 7
Endrenyi, L., 367
Egan, R., 25
Eigen, M., 39, 283–4, 377
Eisele, B., 34
Eisenberg, D., 10
Eisenberg, M. A., 200, 206, 334, 418, 422
Ellison, J. S., 34
Elmore, E. T., 334
Endrenyi, L., 367
Engel, H. J., 34
Engers, H. D., 139, 363
Engle, R. R., 26, 33
Epand, R. M., 12, 202, 297
Epstein, C. J., 19, 21
Erickson, J. O., 413, 422, 440
Erlanger, B. F., 310, 327, 330, 340
Erman, J. E., 306
Estermann, E. F., 384, 402
Ethier, M., 209, 216, 219(fig.), 220
Evans, M. G., 210, 224, 256
Everett, D. H., 438
Everse, J., 34, 350
Eyl, A., 331
Eyring, H., 45, 53, 227–9, 440, 445–8

Fabre, C., 329
Farnham, S. B., 20
Fasman, G. D., 315–16
Feder, J., 206
Feinstein, G., 331
Fellows, R. E., 32
Fendley, J. A., 243
Ferdinand, W., 353, 363
Fernley, H. N., 203, 206
Figueiredo, A. F. S., 341
Filmer, D., 368
Findlay, D., 34, 161, 299, 303
Fischer, E. H., 32, 69, 255
Fisher, H. F., 250, 279, 345
Fisher, J. R., 363
Fishman, W. H., 34
Fitting, C., 269
Fleischer, G. A., 166
Folsch, G., 34
Fondy, T. P., 34, 350
Ford, G. C., 10
Foster, R. J., 86, 113, 166, 171–2, 172–3(fig.), 313, 316
Fraenkel, G., 67
Franzen, J. S., 20
Fraser, M. J., 383–4, 402
Frazer, D., 448
Freis, F. L., 38
Freer, S. T., 10, 323
Freiss, S. L., 256
French, T. C., 305
Frenkel, J., 209
Frensdorff, H. K., 451
Fridborg, K., 10

Frieden, C., 83–4, 361–2, 367
Friedkin, M., 269
Friess, S. L., 256
Fromm, H. J., 80, 136, 139–41, 342, 344–6
Frost, A. A., 35, 47
Fruchter, R. G., 297
Fruton, J. S., 268, 338
Fuhlbrigge, A. R., 83–4
Fujita, T., 328
Fuse, N., 402

Gabel, D., 386
Gallego, E., 377, 378(fig.)
Garces, E., 141
Garroll, W. A., 306
Gassen, H. G., 302
Gates, V., 294
Geis, I., 1, 18, 262
Gensler, R. L., 229
Gerhart, J. C., 368, 372, 372(fig.), 373–5
Gervais, M., 33, 332
Gherardi, G., 448
Ghosh, N. K., 34
Gibbs, R. J., 418, 422, 442–5
Gibian, M. J., 316
Glaid, A. J., 345
Glantz, R. R., 84, 84(fig.), 215, 256
Glasstone, S., 45
Glazer, A. N., 19, 33, 262, 310
Glick, D. M., 301, 316
Gold, A. M., 139
Goldberger, R. F., 19
Goldfinger, R. F., 21
Goldman, D. S., 362
Goldman, R., 382, 386–7, 391
Goldstein, L., 382, 386, 397–400, 402, 405–6
Goodall, D., 377, 378(fig.)
Graham, J. E. S., 316
Gratzer, W. B., 378
Green, D. W., 8
Green, J. R., 278
Greenstein, J. P., 413, 422, 440
Griffin, C. C., 363
Griffing, V. F., 414
Grisolia, S., 141
Groves, M. C., 418
Grunberg-Manago, M., 269
Grunwald, E., 328
Gunter, C. R., 208, 286, 324
Gutfreund, H., 33, 65, 67, 181, 203, 206, 254, 288, 312–13, 332–3, 335, 342, 345–8

Haber, F., 278
Haber, J. E., 368
Hagihawa, B., 262
Hakala, M. T., 121, 345
Haldane, J. B. S., 70, 72–3, 81, 83–6, 121, 164, 174, 180, 229, 287, 345
Hamann, S. D., 220, 225
Hamilton, C. L., 288–9

Hamilton, E. A., 253
Hamilton, G. A., 320
Hamilton, P. B., 5
Hammes, G. G., 39, 284, 299, 301, 302, 305–6, 307(fig.), 373, 374–5
Hammond, B. R., 33
Hammond, G. S., 288–9
Hanes, C. S., 117–18, 134
Hansch, G., 328
Hansen, B. J. M., 380
Hanson, K. R., 76
Hanson, T. L., 141
Hanss, M., 65
Harbron, E., 244
Hardin, R. L., 419
Hardman, K. D., 298
Harker, D., 10, 298
Harned, H. S., 438
Harris, D. O., 291
Harris, J. I., 32, 380
Harshman, S., 26, 33
Harting, J., 268
Hartley, B. S., 10, 20, 26, 31, 33, 189, 206, 284, 307, 310, 312, 336
Hartridge, H., 419
Hartsuck, J. A., 10, 262
Hass, G. M., 32, 34
Hassid, W. Z., 133, 266, 268
Hasting, J., 71
Hathaway, G., 345, 347
Hathaway, J. A., 362
Haurowitz, F., 418–19, 422
Havir, E., 76
Havsteen, R. H., 315
Hayashi, A., 67
Hayes, J. E., 26, 71
Hazen, E. E., 10
Heck, H. d'A., 153, 251, 309, 345, 347, 348
Hedrich, J. L., 32
Heiderer, G. P., 32
Hein, G. E., 205, 206, 261, 288–9, 325–328, 338
Heinricksen, R. L., 298
Helfferich, F., 411
Helmreich, E., 32
Henderson, R., 10, 20, 33, 310, 321–2, 329
Hendrickson, D. W., 252
Henri, V., 68, 166–68, 170–71, 174, 176
Hermans, J., 63, 250
Herries, D. G., 296, 299, 303
Hersh, L., 328
Herskovitz, T. T., 11
Hess, G. P., 33, 205, 309–10, 312, 315–18, 318(fig.)
Hiatt, R. R., 139
Hicks, G. P., 385
High, D. F., 10
Hijazi, N. H., 195, 425
Hill, A. V., 356, 379(fig.)
Hill, R. B., 29
Hill, R. L., 32, 34

Himoe, A., 309, 312, 315–8, 318(fig.), 333
Hinberg, I., 202
Hinkle, P. M., 206, 208
Hinshelwood, C. N., 35, 256
Hipp, N. J., 418
Hiromi, K., 57
Hirs, C. H. W., 6, 297–8
Hirschmann, M., 261
Hitchcock, D. J., 171
Hoagland, V. D., 363
Hoare, J. P., 84, 86, 256
Hofmann, T., 316, 319, 332
Hofstee, B. H. J., 76
Holbrook, J. J., 34, 350
Holtzman, D., 328
Hommes, F. A., 73, 182
Hopkins, F. G., 451, 457–8
Horecker, B. L., 27
Hornby, W. E., 383–4, 387, 401, 403, 405, 408–9
Horwitz, A., 12
Hoshino, O., 299
Hotchkiss, R. D., 414
Howard, S., 339
Hsu, R. Y., 141
Hubela, K. W., 326
Huber, C., 448
Hückel, E., 50, 52–3, 56, 62
Hulett, J. R., 243
Hunt, L., 63
Hussain, S. S., 34

Ikai, A., 425
Inagami, T., 298, 321, 331–2, 334, 336, 338, 341
Ingles, D. W., 290, 326, 328
Ingold, C. K., 256, 280
Ingraham, L. I., 119
Ingram, V. M., 8
Inman, J. K., 385
Inward, P. W., 320
Irie, M., 299
Ishii, S., 86
Iwasa, J., 328
Izumiya, H., 340

Jackson, M. R., 27
Jacobsen, C. F., 307, 448, 457
Jaenicke, R., 378
James, A. T., 76
James, E., 139
James, S., 33
Jandorf, B. J., 25
Jang, R., 25–6, 31
Jansen, E. F., 25–6, 31, 227, 330, 446
Jansonius, J. N., 10
Jardetsky, O., 12, 66, 297, 299, 300, 348
Jaureguí-Adell, J., 6
Jeckel, D., 32, 34
Jeckel, R., 34, 350
Jencks, W. P., 22, 59, 237, 246, 254, 272, 275, 280, 287–8, 291, 310–11, 320

Jennings, R. R., 171, 325
Jirgensons, B., 335
Johannin, G., 335
Johansen, G., 76
Johnsen, G., 414
Johnson, F. H., 228–9, 445–8
Johnson, L. N., 298
Johnson, R. M., 139
Johnston, R. B., 268–9
Jolles, J., 6
Jones, M. E., 372
Jornvall, H., 32
Josephson, K., 83, 146, 147(fig.)
Jung, M. J., 279

Kaiser, E. T., 206, 334–6, 338
Kallen-Trummer, V., 338
Kanaoka, Y., 331
Kanarek, L., 32
Kandt, K. G., 312
Kannan, K. K., 10
Kaplan, H., 33, 102, 106, 111,
    149–50, 153, 158, 161, 309, 312
Kaplan, J. G., 383–4, 402
Kaplan, N. O., 32, 34, 268–9, 279,
    349–50
Kartha, G., 8, 10, 298
Kasai, M., 386
Kasche, V., 384, 387
Kasparian, M., 244
Kassell, B., 29
Kasserra, H. P., 158–9, 162, 181,
    192, 206, 284, 309(fig.), 317(fig.)
    324, 330(fig.), 332–7, 340
Kastenschmidt, J., 32
Kastenschmidt, L. L., 32
Katchalsky, A., 404
Katchalski, E., 382–3, 386–7,
    391, 397–402, 405–6
Kauffman, D. L., 6, 33, 336
Kaufman, S., 207–8, 215, 219,
    256
Kauzmann, W. J., 20–1, 419,
    422, 425, 451–4, 456, 457,
    457(fig.), 458
Kay, G., 383, 385
Kedem, O., 386, 391
Keesom, W. H., 50
Keil, B., 336
Keilin, D., 71
Keizer, J., 33, 316
Keller, P. J., 32
Kézdy, F. J., 33, 153, 206, 208,
    251, 286, 309, 312–14, 324, 334
Khedoun, E., 326
Kilby, B. A., 26, 31, 189, 206, 312
Killheffer, J. V., 334
Kim, Y. D., 315–16
King, E. L., 80, 119, 122–3, 134
Kirkpatrick, D. S., 139
Kirkwood, J. G., 57
Kirsch, J. F., 206, 208
Kirshner, K., 368, 377–8, 378
    (fig.), 379
Kirtley, M. E., 368

Kistiakowsky, G. B., 31
Kitagawa, K., 340
Klee, W. A., 298, 301
Klotz, I. M., 20–1, 26
Kluetz, M. D., 320
Knowles, J. R., 254, 290, 326, 328
Ko, S. H. G., 206
Koekoek, R., 10
Koenig, D. F., 10, 288
Kokowski, N., 340
Kono, Y., 419
Kopka, M. L., 10
Korey, S. R., 269
Kornberg, A., 269
Koshland, D. E., 29, 33, 114, 135,
    254, 257, 264–6, 268–71, 275,
    285, 286(fig.), 290–1, 294,
    310–11, 368–9, 375, 377, 379,
    379(fig.), 380–1
Kotani, M., 67
Kowalski, A., 269
Kraut, J., 8, 10, 11, 316, 322–3
Krebs, E. G., 32
Krehbiel, A., 339
Kridl, A. G., 31
Krupta, R. M., 86–8, 102, 105,
    106, 111, 149, 153, 157, 205,
    206
Ku, E., 205, 315
Kuhn, R., 146, 174–5
Kumar, D. S. V. S., 6
Kunitz, M., 329, 418, 422
Kurosky, A., 316, 319
Kustin, K., 39
Kustind, K. J., 284
Kvamme, E., 141, 368

Labouesse, B., 33, 315
Labouesse, J., 33, 332
Lache, M., 330
Laidler, K. J., 33–5, 43, 45, 53, 57,
    73, 84, 86–8, 102, 105, 106, 111,
    119, 121, 149–50, 153, 157–9,
    161–2, 165(fig.), 176, 181–2,
    185, 192, 195, 198, 199(fig.),
    202, 204–206, 215–16,
    218, 219(fig.), 220, 223, 225,
    225(fig.), 230, 232–3, 237, 242,
    256, 280, 284–5, 285(fig.), 296,
    298–9, 304, 309, 309(fig.), 313,
    317(fig.), 324, 326, 330(fig.),
    332–7, 340, 363, 382, 387, 401,
    418, 420, 421–2(fig.), 425, 428,
    430
Landskroener, P. A., 57, 209, 280
Langmuir, I., 14, 181
Lardy, H., 341
Lardy, H. A., 141
Laskowsky, M., 11, 339
Lau, S. J., 206, 310, 312
Lauffer, M. A., 418–19, 422, 426–
    7, 446, 448, 451–53, 457–8
Laurin, I., 419
Laursen, R. A., 31

Lawrence, A. J., 65
Lawrence, D. J. R., 220
Lawrence, L., 256
Lawson, W. B., 33, 311, 328–30
Lazdunski, C., 31, 34, 208
Lazdunski, M., 31, 34, 208, 329,
    332
Leaback, D. H., 5
Leafer, M. D., 33, 330
Lee, H. A., 346
Lee, H. S., 149, 331
Leffler, J. E., 328
Leggett-Bailey, J., 1
Lench, S. J., 278
Lentz, P. J., 10
Lenz, D. E., 205, 206
Leskowac, M., 34, 350
Letort, M., 36–7
Leveskov, A. V., 326
Levin, Y., 386, 397–400, 402,
    405–406, 437
Levine, D., 28, 203, 206
Levit, S., 261
Levy, G., 446
Levy, H. R., 279
Levy, M., 418–19, 431, 433, 438,
    441–3, 445
Lewis, E. S., 38
Lewis, P. S., 418–19, 430–31,
    440–45, 449, 456
Lewis, W. C., McC., 180
Liljas, A., 10
Lilly, M. D., 384–5, 387, 401, 403,
    405, 407–409, 411–12
Lin, T. Y., 294
Linderstrøm-Lang, K., 414, 448
Lindley, H., 338
Lineweaver, H., 75, 76, 76(fig.),
    88, 89(fig.), 93–4, 96–7, 116,
    121, 211, 356, 357(fig.)
Ling, R., 76
Linsley, K. B., 31
Lipmann, F., 34
Lipscomb, W. N., 10, 17, 262
Lisbonne, M., 446
Listowsky, I., 380
Little, G., 27
Loewus, F. A., 258, 279, 345
Loewus, M. W., 26, 31
Longsworth, C. G., 445
Lowe, G., 34, 206
Lowry, 61
Lucas, E. C., 206
Ludwig, M. L., 10
Lüers, H., 419
Lumry, R., 76, 84, 84(fig.), 215,
    256, 315–16
Lundgren, H. P., 414
Lundin, J., 10
Lindquist, H., 387

MacDonald, M. R., 329
Macheboeuf, M. A., 446
MacIntosh, F. C., 383, 385
MacPherson, C. F. C., 445

McAulay, J., 57–8
McClintock, D. K., 373–6
McConkey, G., 205, 315
McConn, J., 205, 315
McConnell. H. M., 12
McDonald, C. E., 204
McFadden, B. A., 141
McGillivray, I. H., 419, 430
McLafferty, F. W., 7
McLaren, A. D., 230, 232, 384, 387, 399, 402
McMeekin, T. L., 418
McMurray, C. H., 342, 345, 347–8
McPherson, A., 10, 349
Madison, V., 316
Madsen, H. B., 32
Madsen, N. B., 139, 363
Magee, J. L., 227–8, 447
Mahler, A. H., 134
Mahler, H. R., 1, 139, 344
Mair, G. A., 10
Makower, B., 119
Malhotra, O. P., 380–1
Malmberg, C. G., 245
Mandel, M., 162
Mann, T., 71
Marbley, J. L., 297
Marcus, F., 141
Mares-Guia, M., 28, 33, 331, 340–1
Margoliash, E., 10, 18
Marini, M. A., 318–19
Markert, C. L., 343
Markham, R., 295
Markland, F. S., 11
Markovitch, D. S., 347
Markus, G., 306, 373–6
Marler, E., 32
Marr, J., 141
Marsland. D. A., 228, 447
Martin, C. J., 318, 319, 418–19, 449
Martin, M. A., 319
Martin, R. B., 111, 313
Martinek, K., 326–7
Mason, H. S., 67
Mason, S. G., 383, 385
Massey, V., 83–4
Mathias, A. P., 34, 161, 296, 299, 303
Matsubara, H., 261
Matthews, B. W., 10, 20, 33
Matthews, J. E., 227, 446
May, S. C., 26, 33
Meadows, D. H., 12, 297, 299, 300
Mecke, H., 50
Meedom, B., 262
Mehl, J., 339
Meighen, E. A., 374
Meir, G. A., 288
Meister, A., 71
Melander, L., 233, 237
Mella, K., 34, 350

Meloche, I., 256
Menger, F., 84, 86
Menger, F. M., 321
Menten, M. L., 70, 72–3, 111, 164, 166, 167(fig.), 168, 182, 198, 211, 252, 280, 311, 349, 352, 374, 377, 393
Mercouroff, J., 310
Meriweather, B. P., 380
Michaelis, L., 70, 72–3, 79, 84–5, 91, 97–8, 103, 105, 111, 115, 121, 143–6, 152, 164, 166, 167(fig.), 168, 182, 184, 188, 198, 211, 229, 252, 279–80, 287, 295, 311–12, 331, 338, 349, 352, 368, 374, 377, 389, 393–4, 396, 398, 403–405, 407, 409, 418, 420–21
Michel, H. O., 25
Mihalyi, E., 453
Miles, J. L., 153
Millar, D. B. S., 32, 342
Miller, C. G., 67, 312
Milovanovic, A., 327
Milstein, C., 28
Mitsuda, H., 338
Modebe, M. O., 371
Moelwyn-Hughes, E. A., 180–81, 278
Moffet, F. J., 141
Moffitt, W., 13, 313, 315
Mohr, S. C., 375
Moller, K., 343
Money, 401, 403
Monod, J., 363–4, 365(fig.), 368, 370, 375, 377–8, 380
Moon, A.-Y., 315
Moore, D. H., 445
Moore, S., 6, 33, 297
Moore, W. J., 256
Morales, M. F., 71, 73, 119, 176, 182, 199(fig.), 209, 215, 216, 256, 430
Moran, J., 246
Mori, T., 402
Morihara, K., 325
Morimoto, H., 67
Morrison, J. F., 139, 141
Mourad, N., 141
Murihead, H., 10
Murachi, T., 338
Murray, D. R. P., 85–6
Mutakis, W. A., 316
Mycek, M. J., 268
Mygaard, A. P., 139
Myrbäck, K., 146, 147(fig.), 341

Nakamoto, T., 278
Nakamura, K., 320
Nakamura, T., 67
Nakata, H., 86
Nakayama, H., 331
Nakayama. S., 419
Nance, L. S., 350
Nanci, A., 310

Nathan. W. S., 256
Neet, K. E., 33, 254, 285, 310, 361
Neil, B., 139
Neil, G. L., 261
Nelson, J. M., 171
Nemethy, G., 368
Neurath, H., 6, 21, 33, 207–208, 215, 218–19, 256, 316, 330, 338, 413, 422, 440–41
Neveu, M. C., 246, 320
Nielands, J. B., 344
Nielson, G. H., 153
Niemann, C., 86, 111, 113, 166, 171–72, 172–3(fig.), 261, 288–9, 313, 316, 325–8
Nishimura, S., 262
Noller, H., 206, 310, 312
Nord, F. F., 418, 422, 442–3
Nordio, P. L., 12
North, A. C. T., 10, 288
Northrop, J. H., 180–81, 418, 420–2
Novoa, W. B., 344, 346
Nutting, M. D. F., 25–6, 31

Ochoa, S., 269
O'Connel, E. L., 134
Oesper, P., 268
Ofner, P., 258, 279, 345
Ogawa, S., 12
Ogston, A. G., 258
Ogura, Y., 67
Oka, T., 325
Okunuki, K., 262
O'Leary, M. H., 252, 320
O'Neill, S. P., 409
Ong, B., 310
Oppenheimer, C., 177
Oppenheimer, H. L., 33, 315
Orsi, B. A., 141
O'Sullivan, C., 68, 176
O'Sullivan, W. J., 139
Ottesen, M., 13
Ottolenghi, P., 149
Ouellet, L., 176, 181, 185, 199(fig.), 206, 209, 215, 256, 336, 338, 419, 430
Owaga, S., 12
Owen, B. B., 438

Pace, J., 180, 418–19, 422, 430
Packer, L., 387
Page, M. I., 275, 291
Pana, C., 210
Pardee, A. B., 372, 372(fig.)
Park, G. S., 210, 256
Park, J. H., 380
Parker, H., 315–16
Parker, L., 321
Parks, R. E., 141
Parks, P. C., 309, 316–17, 318(fig.), 321
Parsons. S. M., 294

Patchornik, A., 6, 321
Patil, J. R., 141
Pauling, L., 14, 17
Pearson, R. G., 35, 47
Pecht, M.. 402
Peller, L., 117, 134, 138, 149, 183
Pelletier, G. E., 419, 430
Perham, R. N., 380
Perrin, M. W., 223–224
Perutz, M. F., 8
Peterson, R. L., 326
Petitclerc, C., 31
Petra, P. H., 331
Pfleiderer, G., 34, 350
Phillips. D. C., 10. 288. 293–4
Piette, L. H., 67
Pigiet, V., 374
Pivan, R. B., 26
Platt, A., 316
Plocke, D. J., 34
Pocker, Y., 205
Pogell, B. M., 362
Polanyi, M., 224
Polgar, L., 33, 310, 323, 324(fig.), 328
Polissar, M. J., 445–6
Pollock, E. J., 246, 320
Pollok, H., 338
Pontremoli, S., 27
Ponzi, D., 338
Popjak, G., 271
Port, G. N. J., 291
Porter, J. W., 141
Porter, R. W., 371, 374–5
Powell, R., 83
Price, N. C., 380
Price, W. C., 418. 422. 448
Pricer, W. E., 269
Prins, J. A., 50
Purdie, J. E., 326
Purlich, D. L., 139
Putnam, F. W., 268, 413, 422, 440

Quatrecasas, P., 384
Quiocho, F. A., 10, 262

Rabin, B. R., 34, 161, 296, 299, 303, 304, 304(fig.), 305, 361
Rabinowitch, E., 50
Rabold, G. P., 67
Radda, G. R., 380
Raftery, M. A., 294
Ramachandran, G. N., 17, 18 (fig.)
Ramakrishnan, C., 17
Ramel, H., 301
Ramsden, E. N., 34, 284, 285(fig.), 296, 298–9, 304
Rand-Meir, T., 294
Rao. G. J. S., 33, 330
Raval, D. N., 140
Ray, W. J., 29, 30, 139, 310, 311
Redfield, R. R., 6
Redfield, R. R., 6, 297
Reeke. G. N., 10. 262

Reid, T. W., 28, 203, 206
Remick, A. E., 280
Reuben, M. M., 375
Rey, A., 65
Reyben, J., 67
Ricci, C., 27
Richards, D. T., 252
Richards, F. M., 19, 294, 296, 298, 301, 302
Richards, J. H., 253
Richards, W. G., 291
Richardson, D. C., 10
Richardson, D. I., 302
Richardson, J. S., 10
Rideal, E. K., 14, 384
Riesel, E., 401
Riordan, J. F., 27, 34
Rittenberg, D., 268
Roberts, D., 269
Roberts, D. V., 310, 334
Roberts, G. C. K., 66, 297, 299, 300, 303
Roberts, J. D., 10, 316, 323
Robinson, G., 419
Robinson, G. W., 32, 34
Robinson, R. A., 438
Roche, T. E., 141
Roscelli, G., 139
Rose, I. A., 134, 257, 269
Ross. C. A., 299, 303
Rossi, G. L., 8
Rossini, F. D., 245
Rossman, M. G., 10, 341, 349–50
Rothstein, M., 145–6, 418, 420–21
Roughton, F. J. W., 38, 184
Rowe, P. N., 409
Rowley, P. T., 27
Royer, G., 64
Rubin, M. M., 368
Rudolph, F. B., 136, 139–40
Ruegamer, W. R., 450
Rupley, J. A., 316
Rupley, R. W., 294
Russell, T. R., 381
Ruterjans, H., 12, 297
Rutter, A. C., 5

Sack, R. A., 310
Sager, W. F., 321
Sanborn, B. M., 338–9
Sanchez, G. R., 139
Sanger, F., 6, 28, 33
San Pietro, A., 279
Sanwal, B. D., 140
Sarma, R. H., 349
Sarna, V. R., 10, 288
Saroff, H. A., 306
Sasaki, R., 32, 34
Saunders, D., 306
Sawada, F., 299
Scatchard, G., 53, 54
Schachman, H. K., 368, 372–5
Schaffer, N. K., 25–6, 33
Schaltiel, S., 32
Schechosky, S., 139

Schechter, E., 13
Schechter, I., 261, 261(fig.), 325
Schellman, C., 11, 12
Schellman, J. A., 11, 12
Scheraga, H. A., 12, 63, 250, 297, 426
Schevitz, R. W., 10
Schimmel, P. R., 306
Schmeller, M., 169
Schmidt, C. L. A., 162, 166
Schmidt, D. E., 161
Schmidt, P. G., 67, 375
Schneider, F. W., 381
Schoellmann, G., 28, 33, 310
Schomaker, V., 419, 426, 457
Schonbaum, G. R., 27, 204, 205, 206, 310, 312
Schramm, H. J., 311, 328–9
Schreck, G., 347
Schroeder, D. D., 330
Schubert, M. P., 279
Schulman, J. H., 383–4, 402
Schultz, R. M., 327, 328
Schutz, E., 163, 180
Schutz, J., 180–1
Schwartz, G. W., 71
Schwartz, J. H., 28, 34
Schwartz, P., 260
Schwertz, G. W., 32, 121, 140, 200. 206–208. 215. 219. 256. 334, 338, 341, 344–6, 418, 422
Scott, R. A., 426
Scott, R. M., 83
Scrimger, S. T., 332
Seely, J. H., 340
Seers, V. L., 32
Sela, M., 298
Seltzer, S., 253
Seshagiri, N., 80
Seydoux, F., 330, 334–5, 338, 340
Sharon, N., 293
Sharp, A. K., 407, 411–12
Shatiel, S., 32
Shaw. D. C., 33
Shaw, E., 28, 33, 310, 330–31, 340–41
Shen. C., 232
Shen, W.-C., 31, 34
Sherwood, L., 32
Shewitz, R. W., 10
Shibata, K., 336
Shields, G. S., 29
Shill, J. P., 361
Shiner, V. J., 139, 237
Shotton, D. M., 10, 322
Shukuya, R., 332
Shulman, R. G., 67
Shuster, I., 377, 378(fig.)
Sieber, L. C., 10
Siegel, L., 34
Sigler, P. B., 10, 20, 33
Sigman, D. S., 329
Silman, I. H., 383, 401–402
Silverstein, E., 134, 139–40, 343
Simpson, R. B., 419, 422, 425, 451–4, 456, 457, 457(fig.), 458

Simpson, R. T., 34
Singer, S. J., 27
Singhal, R. D., 21
Sireix, R., 67
Slater, E. C., 380
Sluyterman, L. A. A. E., 34, 205, 206
Smiley, I. E., 10
Smillie, L. B., 336
Smith, E. C., 362
Smith, E. I., 215
Smith, E. L., 18, 29, 84, 84(fig.), 256
Smith, J. D., 295
Smith, J. J., 334
Smith, K. E., 372
Smith, R., 331
Smith, S. L., 346
Smyth, D. G., 297
Smythe, C. V., 279
Snoke, J. E., 208, 215, 218, 256, 268
Socquet, I. M., 121, 165(fig.)
Sorn, F., 336
Spackman, D. H., 6
Spancle, T. F., 6
Speck, J. C., 27
Spector, L. B., 71
Spencer, T., 288
Spotorno, G., 332
Springham, S., 33
Springhorn, S. S., 269, 331
Sprinson, D. B., 268
Stachow, C. S., 140
Stark, G. R., 67, 371, 375, 382
Stearn, A. E., 440
Steblay, R., 448
Stein, S. S., 264–6, 268
Stein, W. D., 33
Stein, W. H., 6, 33–4, 297, 304
Steinberg, I. Z., 386
Steinhardt, J., 418–21, 431, 439–40, 449–50
Steitz, T. A., 10, 272, 310, 321
Stephani, R. A., 71
Stern, K. G., 71
Stewart, J. A., 149, 181, 185, 206, 331, 336, 338, 340
Stolzenbach, F. E., 32, 34, 350
Stone, T. J., 12
Storm, D. R., 205, 290–91
Strandberg, B., 10
Strandberg, R., 10
Strumeyer, D. H., 310
Stryer, L., 8
Sturtevant, J. M., 67, 206, 220, 288, 312, 315, 332, 334–6, 338
Sugimolo, E., 32, 34
Sulebeles, G. A., 140
Summerson, W. H., 25–6, 33
Sumner, J. B., 1, 2
Sundaram, P. V., 382, 383, 387, 401–403
Svennsky, G., 141
Swain, C. G., 246, 282–3
Sweeny, J. R., 363

Swen, H. M., 10
Switzer, R. L., 139
Sykes, B. D., 375
Szabó, Z. G., 47

Taft, R. W., 328
Takenaka, Y., 32, 71
Taketu, K., 362
Talalay, P., 279
Tammann, G., 197, 415, 418, 420, 421
Tanford, C., 13, 413, 425, 454
Tanizawa, K., 331
Tashjian, Z., 33
Tatemoto, K., 32
Tchen, T. T., 279
Teipel, J., 32, 377
Tekman, S., 418, 422
Teller, D. C., 32
Teterman, H., 385
Tewes, A., 33, 330
Theorell, H., 32, 114, 123, 139, 209, 210, 256, 345
Thompson, W. J., 381
Thomson, J. F., 139–40, 344, 350
Tilander, B., 10
Timasheff, S. N., 10
Timm, E. W., 256
Tipper, C. H. F., 47
Todd, A. R., 295
Tomalin, G., 244
Tompson, F. W., 68, 176
Tosa, T., 402
Travers, F., 67
Trayser, K. A., 287
Treit, B., 141
Trentham, D. R., 203, 381
Trowbridge, C. G., 331, 339, 341
Trunit, H. J., 384
Tsernoglou, D., 298
Tsuda, Y., 71
Tsuzuki, H., 325
Tweedale, A., 387

Udenfriend, S., 64
Uhr, M. L., 141
Ukita, T., 299
Updike, S. J., 385
Usher, D. A., 302

Valenzuela, P., 64, 316, 319
Vallee, B. L., 27, 34
van Dam, W., 180
Vandendriessche, L., 295
van Eys, J., 32
van Lear, G. E., 7
Van Slyke, D. D., 73
van't Hoff, J. H., 37, 60, 163, 221
Varnum, J. C., 10
Vaslow, F., 26, 31, 81. 268
Velick, S. F., 26, 32, 71, 268, 342, 380
Venkataramanan, B., 67
Vennesland, B., 258, 279, 345
Vestling, C. S., 27
Vishnu, 320

Vithayathil, P. J., 296
Voigt, B., 377
Volkenstein, M. V., 347
Volz, M., 34, 350 .
von Berneck, R. M., 210
von Euler, H., 83, 146, 147, 147 (fig.), 419
Von Hinueben, C., 34

Wacker, W. E. C., 34
Wade-Jardetzky, N. G. W., 348
Wahl. P., 386
Waku, K., 299
Waley, S. G., 146, 147, 147(fig.)
Walker, A. C., 166
Walker, F. M., 26
Walker, P. G., 203. 206
Wall, M. C., 215, 256
Wallenfels, K., 34
Walsh, K. A., 6
Walter, C., 73, 182
Walton, G. M., 363
Walton. J. H., 210
Walz, F. G., 305
Wang, J. H., 283, 303, 304, 305 (fig.), 321, 322(fig.), 323
Wang, S.-S., 338
Warner, R. C., 418, 445
Warringa, M. G. P. J., 29. 182
Wasmund, W., 419
Wasserman, P. M., 34
Watanabe, K., 332
Watasi, H., 67
Waters. W. A., 278
Watson, H. C., 10, 322
Watson, M. T., 451
Watz, F. G., 299
Webb, E. C., 1
Weber, G., 346
Weber, H. H., 174–5
Weber, K., 374
Wedler, F. C., 312, 316
Weil, L., 33
Weiland, P., 32
Weiner, H., 311
Weinstein, S. Y., 326–7
Weinzierl, J. E., 10
Weiss, J., 278
Weissberger, A., 38
Wenckhaus, C., 34
Werbin. H., 230. 232
Wesley, J., 278
Westheimer, F. H., 161, 253, 258–9, 279, 301, 345
Westley, J., 359
Wetlaufier, D. B., 10
Whalley, E., 226
Wharton, C. W., 405
Wharton, W. C., 206
Wheatley, M., 419
Whelan, D. J., 316
Whitaker, J. R., 34, 84, 86, 206
White, J. S., 372
Wieland, T., 385
Wigler, P., 296
Wildnauer, R.. 64, 326

Wiles, R. A., 246
Wiley, D. C., 374
Wilkinson, G. N., 76
Williams, A., 206, 320–21
Williams, D. C., 206
Williams, E. J., 11
Williams, J., 210
Williams, J. O., 141
Williams, R. K., 232
Willstätter, R., 174–5, 278
Wilson, I. B., 28, 84, 200, 201, 201(fig.), 202, 203, 205, 206
Wilson, R. J. H., 385
Winer, A. D., 32, 140, 341, 344, 346
Winstead, J. A., 64, 300
Wiren, G., 10
Wishnia, A., 288
Witkop, B., 6
Witzel, H., 284, 298, 302, 303(fig.)

Wold, F., 64, 300
Wolfe, R. G., 140
Wolfsberg, M., 233
Wolochow, H., 268
Wolthers, B. G., 10
Wonacott, A. J., 10
Wong, J. F. T., 73, 117–18, 134, 182, 367
Wong, L. T., 367
Wood, W. C., 50
Woolf, B., 168, 168(fig.)
Worcel, A., 362
Wratten, C. C., 139–40
Wright, C. S., 10, 11, 322
Wright, G. G., 419, 426, 457
Wright, H. T., 10, 316, 323
Wu, C.-W., 374–5
Wu, H., 413
Wurtz, A., 69, 71
Wyckoff, H. W., 294, 298–9, 302

Wyman, J., 364, 365(fig.), 368, 370
Wynne-Jones, W. F. K., 45, 438

Xuong, Ng. H., 10, 316, 323

Yamazaki, I., 67
You, J., 330, 332, 334–5, 338, 340
Yonath, A., 10
York, S. S., 321
Youatt, G., 26, 31

Zaiser, E. M., 419, 450
Zatman, L. J., 268–9
Zavosdsky, P., 347
Zerner, B., 27, 204, 205, 206, 310, 312
Zernickel, F., 50
Zewe, V., 139–41, 345–6
Zilva, S. S., 419, 449

Subject Index

Abortive (dead-end) complex, 126, 132, 135
absolute specificity, 23
abstraction reactions, 43
acid-base catalysis, 60–62, 279–83
activated complex, 42, 44
activated complex theory, 45–6
activation, 110
  of chymotrypsinogen, 307
  of trypsinogen, 329–30
activation energy, 42–3, 204–9
active centres of enzymes, 24–34
  of chymotrypsin, 309–11, 324–9
  of trypsin, 330–32, 338–41
active site mapping, 261
active site titration, 27
activity coefficients, 51
acylation of enzymes, 77–9
alcohol dehydrogenase, 259
allostery, 352–81
alkaline phosphatase, 203
amino-acid sequence, 6–8
analysis of kinetic data, 35–7
anti-competitive inhibition, 91–110
apoenzyme, 21
Arrhenius law, 42–3, 198–9
aspartate transcarbamylase, 371–6
atoms and radicals in enzyme reactions, 278–9

Beta-pleated sheet, 17
bifunctional catalysis, 283
bovine plasma ablumin denaturation, 445
Brønsted's relationships, 61–2

Calorimetry, 66
carboxypeptidase, 262–3
catalysis, 22–3, 59–62
  acid–base, 60–2, 279–83
chain reactions, 278–9
chemical modification (non-specific labelling), 28–30
  of chymotrypsin, 319
  of ribonuclease, 296
  of trypsin, 330–31
chymotrypsin, 306–29
  active centre, 309–11, 324–9
  chemical modification, 319
  conformational changes, 316–21
  intermediate, evidence for, 311–14
  ion pair, evidence for, 315–19
  specificity, 324–9
  stereospecificity, 328
  X-ray studies, 307–8
chymotrypsinogen, 251–2
  activation of, 307
coenzyme I (nicotinamide adenine dinucleotide, NAD), 259–342
coenzymes, 21
competing substrates, 111–13, 173–5
competitive inhibition, 91–110
complex reactions, 46–9
conformation, 6

conformational changes,
  in chymotrypsin, 316–21
  in lactate dehydrogenase, 348–50
  in ribonuclease, 306
continuous feed stirred-tank (CFST) reactors, 411–12
continuous flow, 38
coulombic forces, 19–20

Deactivation: see Denaturation
dead-end complex, 126, 132, 135
  in lactate dehydrogenase mechanism, 345–6
dehydrogenase,
  alcohol, 259
  glyceraldehyde-3-phosphate, 376–81
  lactate, 259–60, 341–51
denaturation of proteins, 413–14
  agents causing, 413–14
  electrolyte effects, 449–50
  order of reaction, 415–29
  pH effects, 429–45
  pressure effects, 445–9
  temperature effects, 429–45
  by urea, 451–60
dielectric constant, 53–9
differential method, 35
di-isopropylfluorophosphate (DFP), 25–6
diphosphoyridine nucleotide (DPN, NAD), 259, 342
displacement mechanisms, 135–6, 266–72
distortion (strain), 287–8
Dixon plots, 157
double-displacement mechanism, 135–6, 266–72

Eadie plot, 76–7
effectors, 370
egg albumin, denaturation of, 442–5
elementary reactions, 40–42
energy of activation, 42–3, 204–9
enthalpy of activation, 46
entropy of activation, 46, 55, 204–9, 216–20
enzymes,
  active centres, 24–34
  acylation of, 77–9
  binding to, 272–5
  as catalysts, 22–3
  general nature, 1–5
  inactivation, 175–80, 413–66
  specificity, 23–4, 255–62
  structure of, 6–14
enzyme–substrate interactions, 70–1, 160–61, 209–15, 216–17
equilibrium constants, isotope effects on, 233–7
equilibrium dialysis, 26–7
exchange, isotopic, 133–5, 266–71

Facilitate proton transfer, 283–4
feedback inhibition, 371
flow methods, 38, 407–12
free energy of activation, 54
free radicals in enzyme reactions, evidence against, 278–9

frequency factors, 46
    comparison of values, 256

General acid–base catalysis, 61
glyceraldehyde-3-phosphate dehydrogenase, 376–81
    co-operativity, 379, 381
    relaxation studies, 377–8
group specificity, 23

Haemoglobin, denaturation of, 440–42
α-helix, 15–17
Henri equation, 166
heterotropic effects, 372
homotropic effects, 371–2
homeomorphs, 118
hydrogen bonding, 20
hydrogen exchange, 13, 14, 133–6
hydrogen ion concentration,
    and rates, 142–62
    and denaturation, 429–45
hydrophobic bonds, 14, 20, 21, 274
hydrostatic pressure, 220–32, 445–9
    and denaturation, 445–9

Immobilized enzymes, 382–410
insoluble enzymes, 382–410
inactivation (*see also* denaturation), 413–60
    during course of reaction, 175–80
induced fit, 284–7
induction period, 184
infrared spectroscopy, 11–12
inhibition, 89–110, 127–33
    anticompetitive, 91–110
    classical mechanisms, 97–101
    competitive, 91–110
    non-competitive, 91–110
    by products, 170–73
    with second intermediate, 101–10
    by substrate, 84–9, 169–70
    of two-substrate reactions, 127–33
inter-helix bonds, 14
ion-dipole reactions, 56–8
ion pairs, 19, 317
ionic reactions, 50–6
ionic strength effects, 56–9
    on denaturation, 449–50
ionizing groups, 5
    essential and inessential, 152
isoelectric point, 5
isomorphous replacement, 9
isotope effects, 233–53
    on enzyme systems, 249–53
    on equilibrium constants, 233–7
isotope exchange reactions, 133–5, 266–71
isotopic tracer studies, 263–6

Kinetic constants, comparison of values, 256
kinetic isotope effects, 233–53
    on enzyme systems, 249–53
King–Altman method, 79–81, 122

Labelling by substrate analogues and inhibitors, 28
lactate dehydrogenase, 259–60, 341–51
    abortive (dead-end) complexes, 345–6

lactate dehydrogenase—*continued*
    compulsory order, 343–5
    conformation change, 348–50
    ternary complexes, 345
    X-ray studies, 350
Lineweaver–Burk plot, 75–6
lock and key hypothesis, 255, 257
lysozyme, 292–4

Magnetic measurements, 66
Michaelis constant, 73
Michaelis–Menten equation, 72–3
moderators, 90
Moffett parameters, 13
molecular kinetics, 196–232

NAD, 259, 342
neighbouring-group effects, 284
nicotinamide adenine dinucleotide (NAD), 259, 342
non-competitive inhibition, 91–110
non-productive binding, 288–90
non-specific labelling of active centres, 28–30
nuclear magnetic resonance spectroscopy, 12
nucleophile studies, 334

One-substrate reactions, 72–113, 164–7
operational normality, 27
opposing reactions, 81–4, 167–9
optical rotation, 67
optical rotatory dispersion, 13
optimum pH, 142
orbital steering, 290
ordered ternary-complex mechanism, 114, 121–3, 140
    inhibition, 129–32
order with respect to concentration, 37
order with respect to time, 36
order of reaction, 35–7
orientation effects, 290–91

Packed bed, 408–11
pepsin, inactivation, 439–40
pH effects, 142–62
    on denaturation, 429–45
    optimum pH, 142
    in supported enzyme systems, 398–400, 406
pH-stat methods, 65–6
ping-pong mechanisms, 115, 123–5, 141
    inhibition of, 132–3
potential-energy surfaces, 43–4
pressure and enzyme reactions, 220–32
    and denaturation, 445–9
pre-steady state, 181–95
primary bonds, 14
prosthetic groups, 21–2
protein conformation, 8–14
protein denaturation; *see* denaturation of proteins
protein structure, principles of, 14–21
proximity effects, 275–6
pseudorotation, 301–2
pure non-competitive inhibition, 91, 110
push–pull mechanism, 281

Quantum-mechanical tunnelling, 242–5

Radicals in enzyme reactions, 278–9
random ternary-complex mechanism, 114, 119–21, 139
  inhibition, 127–9
rate measurements for enzyme reactions, 63–7
reaction specificity, 24
reactions between ions, 50–56
reactions in solution, 49–59
regulatory enzymes, 370
relaxation methods, 38–40
  glyceraldehyde-3-phosphate dehydrogenase, 377–8
  ribonuclease, 305–6
reversible one-substrate reactions, 81–4, 167–9
ribonuclease, 294–306
  chemical modification, 296
  conformational change, 306
  nuclear magnetic resonance, 297
ricin denaturation, 438–9

Schütz's law, 180–81
secondary bonds, 14, 19–21
secondary isotope effects, 247–8
separation of rate constants k_2 and k_3, 199–204
sequential reactions, 277–8
sensitization by substrate, 180
sigmoid kinetics, 352–81
solvent effects, 53–9, 161–2, 216–20
single-displacement mechanism, 135, 266–72
spectrophotometry, 64–5
specificity, 23–4, 255–62
  absolute, 23
  of chymotrypsin, 324–9
  group, 23
  reaction, 24
  of trypsin, 338–41
steady-state method, 47, 73
steady-state in supported-enzyme systems, 388
stereochemical investigations, 271–2
stereochemical specificity, 24, 257–61
stopped-flow, 38

strain or distortion, 287–8
substrate labelling of active centres, 27–8
substrates, competing, 111–13, 173–5
  supported, 384
supported enzymes, 382–412
  electrostatic effects, 397–8, 404–407
  flow systems for, 407–12
supported substrates, 384

Temperature-jump relaxation, 39–40, 305–6, 377–8
temperature, influence on enzymes, 196–9
  and denaturation, 429–445
  optimum, 196–7
tertiary bonds, 14, 19–21
tetrahedral intermediates, 273
time course of enzymes reactions, 163–95
tobacco mosaic virus, denaturation, 427, 457
transient phase, 181–95
transition state, 42, 44
transition state (activated-complex) theory, 45–6
trypsin, 329–41
  activation, 329
  active centres, 330–32, 338–41
  chemical modification, 330–31
  evidence for acyl intermediate, 333–5
  specificity, 338–41
trypsinogen, 329–30
turnover number, 23, 254
two-intermediates, reactions involving, 77–81
two-substrate reactions, 114–41

Ultracentrifugation, 26–7
ultra-violet spectroscopy, 10
urea denaturation, 451–60

Volume of activation, 222–32
  for denaturation, 445–9

X-Ray crystallography, 8–10, 262–3, 272
  of chymotrypsinogen, 307–308
  of ribonuclease, 298, 300